SEEING THE LIGHT

OPTICS IN NATURE, PHOTOGRAPHY, COLOR, VISION, AND HOLOGRAPHY

DAVID S. FALK

University of Maryland

DIETER R. BRILL

University of Maryland

DAVID G. STORK

Clark University

1817

HARPER & ROW, PUBLISHERS, New York

Cambridge, Philadelphia, San Francisco, London, Mexico City, São Paulo, Singapore, Sydney

Sponsoring Editor: Lisa S. Berger
Project Editor: Steven Pisano
Text Design: Gayle Jaeger
Cover Design: Mark Berghash
Cover Photos: David Falk
Text Art: Reproduction Drawings Ltd.
Production: Debra Forrest Bochner
Compositor: The Clarinda Company
Printer and Binder: The Murray
  Printing Company

Cover: Photographs of television images of a human eye showing the matrix of colored phosphors. The top is a full-color image that is made by a partitive (additive) mixture of blue, green, and red images. These are shown separately in the bottom three pictures. A technique related to the halftone process was used in printing this cover because the colors of the television phosphors are the additive primaries, but the colors of the inks used in the printing are the subtractive primaries (see Chapter 9).

Library of Congress Cataloging in Publication Data

Falk, David S.
    Seeing the light.

    Includes index.
    1. Optics. 2. Optics—Experiments. I. Brill, Dieter R. II. Stork, David G. III. Title.
QC358.F36 1985   535   85-5556
ISBN 0-06-041991-1

85 86 87 88   9 8 7 6 5 4 3 2 1

CONTENTS

CONTENTS

vi

Preface

"My Purpose is, Indeed, a Horse of That Colour," or Why We Wrote a Text for Light and Color Courses

Some years ago we were looking for a course that would be both valuable and appealing to the nonscientist—the liberal arts major, the business major, the social scientist, or any other of the myriad of students who, while perhaps not mathematically sophisticated, have the curiosity and intelligence of all college students. The field of optics seemed to offer an ideal chance to expose these students to science. A huge wealth of phenomena from "the real world," the world of nature and technology that the students already had experience with, could be easily pointed out to them, and they could be made to see, discover, and test the logical relationships that exist between these phenomena. When this course was offered at the University of Maryland (and later at Swarthmore College), we found students from all over campus flocking to it (over forty different major fields in a typical semester). Not only did these students come with a general curiosity, many came with a definite interest—art, vision, photography, holography, illusions, the visible world around them—a large variety of specific interests was already there before we said the first word.

At first we had to keep reminding ourselves that we were not training these students to become scientists (although more than one student who took this course changed his or her major to physics as a result) but rather to appreciate science, just as one does not normally train students to compose in a music appreciation course. As Henry Adams wrote, "He too serves a certain purpose who only stands and cheers." Certainly, as the range of topics in this book illustrates, there is much in the study of light to cheer about. And the more the student understands the relationships between the phenomena and appreciates the methods by which these relationships are discovered, the more fervent his cheers are likely to be. Optics is particularly appropriate here because he can, literally and often easily, *see* what is being discussed. What better way to engage the attention of the many interested students who find the language of mathematics intimidating! To see the light you don't need elaborate equipment or a knowledge of mathematics, you need only to open your eyes. This book, like our course, is meant to be an eye opener for readers with no background in science or mathematics. We pay particular attention to phenomena that illustrate or apply the ideas being discussed, and give frequent examples that occur in everyday life, or that are interesting precisely because of their rarity. These phenomena are essential to our approach, and they also make nice demonstrations that can easily be performed in class.

This book was forced on us by the overwhelming success of the course and the lack of any suitable text. While many excellent books dealt with one or another of the topics of interest, no textbook covered the wide range of related fields, and certainly none emphasized the relationships. We wanted a text that continually reminded the students of the connections between the abstract ideas of physics and everyday life, or phenomena they are familiar with. To illustrate these connections we had to collect from many sources the large number of pictures and figures that are now included in this book.

Moreover, most existing optics books didn't treat these ideas and phenomena in a nonmathematical way. Even when equations were avoided, such books tended either to slip into technical subtleties or to water down the fundamental material. Yet optics is particularly appealing because it contains a wealth of ideas and phenomena that can be discussed without mathematics. In this book we try to develop an appreciation of the beauty and the logic behind the behavior of light by encouraging and building on the reader's curiosity in readily accessible language.

We hope that the students will learn to look, to observe, and to ask, in the words of Artemus Ward, "Why is this thus? What is the reason of this thusness?"

"The Road Through the Forest," or How To Use This Book

Physics should not be studied passively. We believe that, like love and other sports, physics is more fun to participate in than to read about, or even to watch. For this reason almost every Chapter has a few **TRY IT's,** experiments or demonstrations that can usually be performed with a minimum of equipment or materials found around the house. Each is meant to provide an opportunity for the reader to play the game and test out the ideas being discussed. Try it!

Of course, all understanding requires thought. To encourage this, we occasionally interrupt our discussion with questions whose contemplation should materially assist the reader at that point. The reader should pause to think about these **PONDER's**—in the words of Aristophanes, "Ponder and examine closely, gather your thoughts together, let your mind turn to every side of things; if you meet with a difficulty, spring quickly to some other idea; above all, keep your eyes from gentle sleep."

For those who bring at least a rudimentary knowledge of introductory mathematics, we've provided **Mathematical Appendixes** that may give a taste of the quantitative flavor of physics without disrupting the qualitative development. Unlike this optional mathematics, many of the specialized terms used in physics crop up in so many circumstances that they are an indispensable part of contemporary culture, as well as being a convenient shorthand. We've introduced these terms with **definitions** where they first occur, as well as **etymologies*** to help break down difficult words into more memorable roots.

In our study of optics there is no yellow brick road to follow to a unique goal. Rather, there are numerous possible paths that crisscross and branch off into all directions. *Each* of these paths leads to *some* City of Emeralds, and different readers will find different paths more enticing and/or rewarding. We have organized the book to allow for a variety of directions and emphases, according to the interests of students and teachers.

---

*Greek, *etumon*, original form of a word.

The overall organization may be seen in the accompanying **Flow Chart,** which shows the minimum background necessary to study any given chapter. The paths should not be viewed as either rigid or dead ends: for example, one could easily follow Color Perception Mechanisms with Color Photography; or insert Optical Instruments, or The Human Eye and Vision—II, on the way to Holography. The best path to follow will depend on interest, time, and equipment available. Use of this Flow Chart allows one to construct a course of any desired length—a few weeks, a quarter, a semester, or a year.

The amount of time spent on a given topic is also flexible. Almost every chapter has **Optional Sections,** which offer greater depth or interesting sidelights but may be omitted without interfering with the study of subsequent chapters. These sections are marked with asterisks preceding the headings.

The **FOCUS ON's** discuss topics of general interest that tie together ideas and phenomena from more than one chapter. Applications in a single field may even utilize approaches from different branches of the Flow Chart, and thus serve to illustrate the interconnectedness within science and technology.

Finally, each chapter concludes with a number of **Problems**—those labeled PH are somewhat harder than those labeled P, while those labeled PM are designed for students who make use of the Mathematical Appendices. (An **Instructor's Manual** with solutions and other helpful hints is available from the publisher upon request.) All the Problems are designed to exercise the student's brains, for, as the Scarecrow in *The Wizard of Oz* says, "When I get used to my brains I shall know everything."

"For This Relief Much Thanks," or *Acknowledgments*

This book would not have been possible without the cooperation and support of the Physics Department of the University of Maryland. The staff of its Lecture-Demonstration Facility, particularly Richard E. Berg, Bill Brandwein, and Bill Norwood, was a source of continual and invaluable assistance. Joan Wright of the department's drafting facility helped on numerous occasions. Many of our colleagues offered useful ideas, especially John Layman, Victor Korenman, and Arnold Glick, who taught from preliminary versions of this book. Many, many students provided comments, figures, and projects that were essential to the development of both the course and this book.

Additional assistance was provided by the Humboldt-Foundation, the Swarthmore College Physics Department, where one of us taught the course, the University of Maryland Office of the Vice Chancellor for Academic Affairs, the Swarthmore College Office of the Provost, and the Maryland State Traffic Police, who repeatedly provided us with surprise demonstrations of Doppler-shifted radar.

We had particular help with the photographs from Jordan Goodman of the University of Maryland Physics Department; Alistair Fraser of the Pennsylvania State University Meteorology Department; Tom Beck of the University of Maryland Baltimore County Kuhn Library and Gallery; Csaba L. Martonyi of Opthalmic Photography at the University of Michigan Hospital; Marvin Gross of the Villanova University Physics Department; Allen Bronstein of AB Associates; and G. Frederick Stork, who also provided seminal help at the conception of this book.

We were significantly aided by the useful comments and suggestions of many people who read parts or all of the text at various stages. (Needless to say, we take full credit for any remaining errors.) These particularly helpful people were: Fred M. Goldberg, West Virginia University; Van E. Neie, Purdue University; E. C. Parke, Humboldt State University; Richard E. Pontinen, Hamline University; Clarence S. Rainwater, San Francisco State University; Michael J. Ruiz, University of North Carolina at Asheville; Suzanne St. Cyr, New York; James Schneider, University of Dayton; Stephen A. Benton, Polaroid Corporation; Robert M. Boynton, University of California San Diego; Eugene Hecht, Adelphi University; Hollis N. Todd, Rochester Institute of Technology; Vivian K. Walworth, Polaroid Corporation; Patrick F. Kenealy, Wayne State University; Edgar B. Singleton, Bowling Green University; John Z. Levinson, Charles E. Sternheim, and John F. Tangney, Psychology Department, University of Maryland; and Malvina Wasserman, who served as editor for the first three years of gestation of this book. In Shakespeare's words, "I thank you for your voices: Thank you, your most sweet voices."

Finally, for their patience, tolerance, encouragement, and for a number of other reasons that don't concern the reader, we thank Nancy Falk, Birgit Brill, and Nancy Porter.

David S. Falk
Dieter R. Brill
David G. Stork

CHAPTER

1. Fundamental Properties of Light

2. Principles of Geometrical Optics

3. Mirrors and Lenses
   [*Virtual Images]

4. The Camera and Photography
   [†Focusing, Rangefinder,
   Perspective, Film]

5. The Eye and Human Vision—I:
   Producing the Image
   [‡Focusing]

6. Optical Instruments

7. The Eye and Human Vision—II:
   Processing the Image

8. Binocular Vision and the
   Perception of Depth

9. Color

10. Color Perception Mechanisms

11. Color Photography

12. Wave Optics

13. Scattering and Polarization

14. Holography

15. Light in Modern Physics

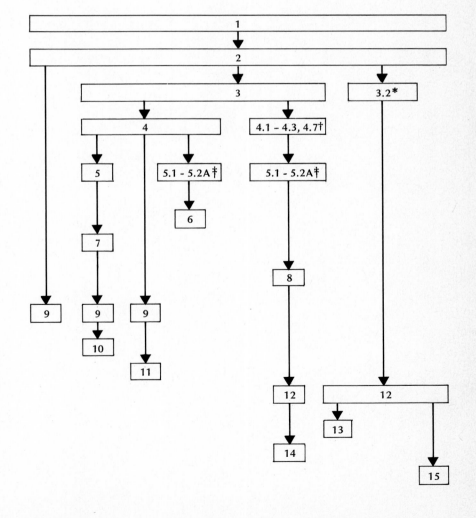

Overview

At college I had to take a required course in physics and chemistry. I had already taken a course in botany and done very well. . . . Botany was fine, because I loved cutting up leaves and putting them under the microscope and drawing diagrams of bread mold and the odd, heart-shaped leaf in the sex cycle of the fern; it seemed so real to me.

The day I went into physics class it was death.

A short dark man with a high lisping voice, named Mr. Manzi, stood in front of the class in a tight blue suit holding a little wooden ball. He put the ball on a steep grooved slide and let it run down to the bottom. Then he started talking about let a equal acceleration and let t equal time and suddenly he was scribbling letters and numbers and equals signs all over the black-board and my mind went dead.

I took the physics book back to my dormitory. It was a huge book on porous mimeographed paper—four hundred pages long with no drawings or photographs, only diagrams and formulas—between brick-red cardboard covers. . . .

Well, I studied those formulas, I went to class and watched balls roll down slides and listened to bells ring and by the end of the semester most of the other girls had failed and I had a straight A. . . .

I may have made straight A in physics, but I was panicstruck. Physics made me sick the whole time I learned it. What I couldn't stand was this shrinking everything into letters and numbers. Instead of leaf shapes and enlarged diagrams of the holes the leaves breathe through and fascinating words like carotene and xanthophyll on the blackboard, there were these hideous, cramped, scorpion-lettered formulas in Mr. Manzi's special red chalk.

> Sylvia Plath, *The Bell Jar*

The intention of this book is to present physics, or at least one aspect of physics, without the "hideous . . . formulas." Instead, we will lean rather heavily on diagrams, drawings, photographs, and the evidence of your own eyes. However, we should admit at once that this is not the entire story. We will discuss physics without using the language of the physicist: mathematics.

Reading physics without mathematics is like reading literature in translation—we lose a great deal by not being able to understand the language (Fig. 0.1). Nevertheless, we still gain for the reading of Goethe and Flaubert, even if we may lose by the limitations of translation. As we try to translate physics into everyday language, you may miss some of the subtleties; some of the inevitability of much of our world; some of the conviction; for, however plausible or beautiful an explanation may be, the test of it is

FIGURE 0.1

1

whether it can be used to predict new results *quantitatively* and precisely.

There is much to be gained by reading physics, even in translation. Like all art, science is an essential part of our culture and presents a description of the world around us, both natural and man-made. It offers us insight into our universe and, uniquely among the arts, it also offers the possibility of harnessing aspects of this universe into a technology that is not only useful, but often has a beauty of its own.

There is an additional difficulty with reading literature in translation. When Lady Murasaki wrote the *Tale of Genji* she assumed, very reasonably, that her readers were familiar with tenth-century Japan. Reading it now, we may find that we lack a certain cultural background, and that it is helpful, periodically, to have someone fill us in. So, too, in reading physics. From time to time we'll have to stray somewhat from our path in order to fill in the background necessary for a full appreciation of our subject.

There is much, we shall see, to appreciate. Perhaps the essence can be summed up in Amanda's view, in Tennessee Williams' *The Glass Menagerie:* "We live in such a mysterious universe, don't we? Some people say that science clears up all the mysteries for us. In my opinion it only creates more!" We'll show you some of the mysteries, and try to clear them up, not to dispel them, but to appreciate them. In doing so, we'll meet new mysteries. This is the way science works: from phenomenon to explanation to new phenomenon. The phenomena and the explanations, the mysteries and the understanding, should lead you to see the world with new eyes, all the more literally because our subject is, in fact, light itself.

The physics of light will very naturally lead to a variety of subjects: art, psychology, philosophy, literature, physiology, and more. We treat these with the same attitude as the heroine of Lawrence Durrell's novel *Pope Joan,* who has come to Rome and begun to give lectures on theology:

The technique of these discourses strongly resembled that of those famous Hambourg bordels where one could find food for every palate, perfumes for every taste—and women speaking all languages and satisfying all appetites. Many a time our heroine began with "The Judgment of God" and ended with the art of cooking. At that time, you see, the processes of the human mind had not been listed and arranged for minor talents to absorb. They had not been classified like reptiles in the bottles of a museum. Theology was literally the only science and it had, like Briareus, a hundred hands with which to draw the elements of ordinary life towards it. Everything of interest came within its scope. And our heroine had by now a comprehensive knowledge of every illegitimate branch of theology.

We will talk about "illegitimate branches" of physics in this book, and we hope to find something for every palate and taste.

Fundamental Properties of Light

CHAPTER 1

1.1
WHAT IS LIGHT?

Light surrounds us during most of our conscious life, but many of us observe it only casually. We take delight in watching a sunset, in seeing multicolored city light reflect in the rippled surface of a lake, in observing sunlight filter through the foliage of a forest and cast bright dancing spots on the ground below. Some of the physical processes involved here (such as reflection and straight-line propagation) are a matter of common sense; but most of us still concur with Samuel Johnson's remark, "We all *know* what light is; but it is not easy to *tell* what it is."

The nature of light has been a topic of great concern and interest throughout history. So important was light that in the third verse of the Bible we find God creating it as the first act of creation, and toward the end of the Bible we find the statement "God is light." It was one of the topics of which the ancient Greeks already had some knowledge, and the study of light really came to flower in medieval times under the Arabs and later under the European scientists. Light was a major ingredient in the development of modern science beginning with Galileo. Yet as late as the eighteenth century Benjamin Franklin could say, "About light I am in the dark," and even today the study of light continues to have a major influence on current physics. However, we shall not follow an historical path to develop the main ideas about light but instead shall use the well-tested method of scientists to discover what light *does*, and thereby understand what it *is*.

A. Which way does light go?

When we use a phrase like "Cast a glance at this picture," we pretend that something goes from our eyes to the picture. People who believe in x-ray vision or the "evil eye" may think this is really how light moves (Fig. 1.1). The Greeks of Plato's time thought light and mind were both made of fire, and that perception was the meeting of the "inner fire" (mind) emitted by the eyes, with the "outer fire" (light)—as Richard Wilbur has written:

The Greeks were wrong who said our eyes have rays;
Not from these sockets or these sparkling poles
Comes the illumination of our days.
It was the sun that bored these two blue holes.

FIGURE 1.1

Superman's x-ray vision comes *from* his eyes, but the rest of us can only see if light comes *toward* our eyes.

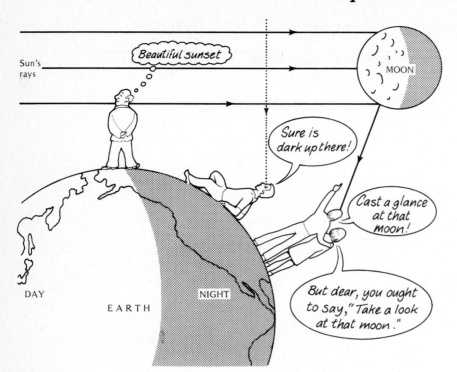

FIGURE 1.2

The void of space looks dark at night even though lightbeams are crossing it.

a beam of light that isn't directed at you, even though it may be passing "right in front of your nose." When you look at the stars at night, you are looking "through" the sun's rays (Fig. 1.2), but you don't see these rays—unless their direction is changed by an object such as the moon, so that they come at you and fall in your eye.

But you *do* see lightbeams like those in Figure 1.3! However, this is only because there are small particles (dust, mist, etc.) in the atmosphere that redirect the light so it

FIGURE 1.3

Scattering in the atmosphere makes shafts of sunlight visible.

Nowadays we know that, in order to see something, light has to *enter* our eyes. But to prove this, would we have to observe light "on its way," as one can observe a football flying in the air? No! To show that light comes *from* the sun, through your window, falls on your book, and then *enters* your eyes, you only need to observe what happens when you draw the curtains: the room becomes dark, but it is still bright outside. If light went the other way, it should get darker outside the window, and stay bright inside, when the curtains are closed.

So we agree that light goes *from* the **source** (sun, light blub, etc.), to the **object** (book or whatever), bounces off or travels through the object (as in the case of glass), and goes *to* the **detector** (eye). Sometimes the source is the same as the object. Light goes directly to the eye from these sources, or **self-luminous** objects. This is how we see lightning bolts, fireflies, candle flames, neon signs, and television.

Why make a special point about this apparently trivial behavior of light? Consider this consequence: if no light gets to your eyes, you don't see anything, no matter how much light there is around. You can't see

FIGURE 1.4

"Irish lights," laser art by Rockne Krebs, shows the straightness and beauty of simple light beams.

can fall into your eyes. The clearer and more free of particles the air, the less of a beam can be seen. In the vacuum of space, a beam that is not directed at the observer cannot be seen at all.

When light falls on most ordinary objects, some of it is redirected **(scattered)** into any direction you please. Thus, no matter where your eye is, you can see the object. You can seen an object *only* if it scatters light. Things like mirrors and windowpanes do not scatter much and are, therefore, difficult to see, as anyone will confirm who has ever tried to walk through a very clean but closed glass door.

The misty or dusty air in which we can see "rays of light" shows us another property of light: it travels in straight lines (Fig. 1.4). The searchlight beam is straight for many miles, and does not droop down due to the earth's gravity, as a material rod or a projectile would (but see Sec. 15.5C). So we agree that light goes in straight lines—unless it hits some object that changes its direction.

B. The speed of light

If light travels from one place to another, how fast does it go?

Clearly it must go at a pretty good clip, because there is no noticeable delay between, say, turning on a flashlight and seeing its beam hit a distant object. To get a significant delay, we must let light travel very great distances.

"Our eyes register the light of dead stars." So André Schwarz-Bart begins his novel, *The Last of the Just*. The physics of the statement is that it takes years for the light of a star to reach us, and during that time the star may burn out. The

light that made the photograph of distant galaxies in Figure 1.5 left the galaxies over a billion years before it reached the film.

FIGURE 1.5

Each star-like object in this photograph of the Corona Borealis is actually a galaxy, over a billion light years distant. In Thornton Wilder's *Our Town*, one of the characters expresses his feelings about looking at such galaxies whose light had left so long ago: "And my Joel—who knew the stars—he used to say it took millions of years for that little speck o' light to git down to earth. Don't seem like a body could believe it, but that's what he used to say—millions of years."

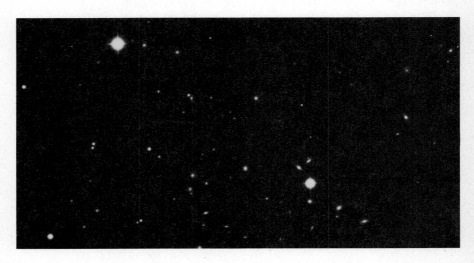

That is, light does take time to get from one place to another. This much seems to have been believed by the Greek philosopher Empedocles, but it was only proved two millennia later. Although more delicate, the proof was analogous to the everyday demonstration that sound takes time to get from one point to another. You can measure the speed of sound by an **echo** technique: yell, and notice how long it takes for the sound to reflect from a distant wall and return as an echo. You'll find that it takes sound about five seconds to travel a mile in air.

Light travels *much* faster than sound. In the thirteenth century, Roger Bacon pointed out that if we see someone at a distance bang a hammer, we *see* the hammer blow before we *hear* the sound. In *Huckleberry Finn*, Huck observes this on the Mississippi River:

Next you'd see a raft sliding by, away off yonder, and maybe a galoot on it chopping . . . you'd see the ax flash and come down—you don't hear nothing; you see the ax go up again, and by the time it's above the man's head then you hear the k'chunk!—it had took all that time to come over the water.

So first you see the ax hit, then you hear it: the light travels faster than the sound.

We use these ideas when timing the delay between lightning and thunder to find out how far away the storm is from us. Five seconds' delay makes a mile because that's how long it takes the sound to get to us, and we neglect the tiny time the light takes to reach us.

Galileo tried to measure the speed of light by the echo technique. He stationed two men on hill tops a mile apart. Each had a covered lantern. The first man uncovered his lantern, revealing the light to the second man. On seeing the light, the second man in turn uncovered his, signaling back to the first man. But light travels much too fast to measure its speed that way. The time delay was almost entirely due to the second man's reaction time in uncovering the lantern.

The first true observation of a time delay due to light's travel was made at the end of the seventeenth century by Ole Roemer when he was studying the moons of Jupiter. He measured the time it took a moon to go around Jupiter (the moon's **period**) by recording the times it went behind Jupiter and disappeared from view. Once he knew these periods, he thought he could then predict when a moon should disappear behind Jupiter. The trouble was, as the earth moved around its orbit away from Jupiter, the moons of Jupiter began to disappear a little later than he calculated, and as the earth moved closer to Jupiter, the moons disappeared a little earlier than calculated. In fact, when the earth was closest to Jupiter, he found the moons to be about 11 minutes ahead of schedule, and when the earth was farthest from Jupiter, about 11 minutes behind schedule. Roemer figured out the reason for the discrepancies: as the earth moved away from Jupiter, the light from Jupiter had to travel farther to reach the earth, and it took time for light to travel this extra distance. This meant that it took about 22 minutes for light to cross from the side of the earth's orbit closest to Jupiter to the opposite side. It is not known whether he actually tried to calculate the speed, but he would have obtained a fairly accurate answer. (A more precise measurement of the time light takes to cross the earth's orbit gives about 17 minutes, or 1000 seconds. Combined with the diameter of the earth's orbit, 186 million miles, this gives an accurate result—186,000 miles per second.)

In the nineteenth century several people got very good values for the speed of light by using improvements on Galileo's technique. In the version due to Albert A. Michelson, Galileo's second man was replaced by a fixed plane mirror, and the first man by an octagonal mirror (Fig. 1.6). The light from a source fell into the apparatus and was reflected by the (originally stationary) octagonal mirror, traveled 22 miles to the plane mirror and 22 miles back to the octagonal mirror. It then bounced off a different face of that mirror into a telescope, where Michelson could see the light source. He then set the mirror into rapid rotation. In general, if the mirror was rotating, then during the time the light traveled the 44-mile path from the octagonal mirror to the plane mirror and back, the octagonal mirror had moved so that the light beam was not reflected into the telescope. But there was a speed of rotation that was just right for the following to happen: by the time the light traveled the 44-mile path, the octagonal mirror made one-eighth of a revolution, so that it was again in the right position to reflect the light into the telescope. (At this proper speed, the light path is the same as when the mirror was not rotating, but face *B* had moved to the position where face *C* was originally.)

At the proper mirror rotation speed, Michelson looked through the telescope and saw the light. This speed was about 530 revolutions per second. During the time the light traveled the 44-mile path, the mirror made about $\frac{1}{8}$ of a revolution. Since light took $\frac{1}{8}$ revolution, or

$$\frac{1}{8} \times \frac{1}{530} \text{ second} = \frac{1}{4240} \text{ sec}$$

to go 44 miles, its speed was (roughly)

$$\frac{44 \text{ miles}}{\frac{1}{4240} \text{ second}} = 186{,}000 \text{ miles/sec}$$

$$\text{or } 300{,}000 \text{ km/sec}$$

Through this and subsequent measurements, the speed of light has become one of the most accurately known constants of nature.

PONDER

What would have happened if Michelson had rotated his mirror at 1060 revolutions per second?

So, although light travels very fast (a distance of $7\frac{1}{2}$ times around the earth in one second), its speed is not infinite. However, it is the fastest thing going. Further, its speed stays the same over billions of years, all over the universe, and no matter how bright the light. The symbol, *c*, is always used to denote

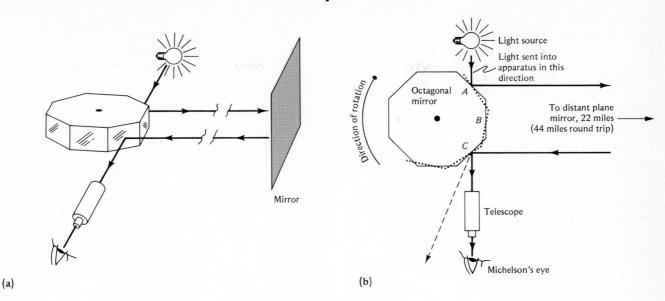

Light source
Light sent into apparatus in this direction

Octagonal mirror

Direction of rotation

A
B
C

To distant plane mirror, 22 miles (44 miles round trip)

Telescope

Michelson's eye

Mirror

(a)

(b)

FIGURE 1.6

Michelson's measurement of c.
(a) Perspective view. **(b)** Top view. Light from the source strikes face *A* of the rotating octagonal mirror. When *A* is in the position shown, it reflects the light toward the distant plane mirror, which then reflects it back. If, while the light is making this trip, the octagonal mirror makes ⅛ of a revolution, then face *B* will reflect the returning light into Michelson's telescope. However, if the mirror rotates a different amount (dotted figure), the returning light will be reflected in a different direction. Knowing the distance the light traveled and the proper speed of rotation, Michelson could determine the speed of light.

this fundamental constant, the velocity of light in *vacuum.** Thus:†

c ≈ 300,000 km/sec

If you compare the speed of different kinds of light in vacuum, you find that the speed is the same for red light, blue light, and white light; in fact, it is also the same for radio waves, x-rays, microwaves, and many other types of rays. From this we may suspect that all these are, in some sense, the same thing as light.

---

*The symbol *c* comes from *celer,* Latin for fast.

†We shall use the symbol ≈ to mean "is approximately equal to."

C. What carries the light?

What is it that carries light from one place to another so rapidly? Let's again compare light to sound. Anyone who has seen the movie "Alien" knows that "In space no one can hear you scream." This is because there is no air to carry the sound. But they can still *see* you! So, sound travels through air and glass and all sorts of things, including materials that are opaque to light—but it does need *some* medium to carry it, it does not travel through a vacuum. Light, on the other hand, does *not* need a medium to carry it; it can travel through a vacuum. Of course it can also travel through some media like glass, air, etc.—but not through others such as wood.

D. What is it that travels?

So far we know that light is something that travels very fast, even through a vacuum. What is it that is being carried so rapidly through the void?

One thing that is surely being carried is **energy.** We know this because we get warm in the sunlight—heat is a form of energy. The energy we get from our food and most of our fuels (e.g., oil, gas, coal) is energy from the sun, brought by light, and stored in the fuel.

Light also carries **momentum** or "push." The momentum of light is extremely small, too small for us to perceive directly. One way to see the effects of the momentum of light may be through the behavior of a comet. A comet is made of dust and ice, particles of which it gives off as it approaches the hot sun. These particles form the beautiful wispy tail. The tail tends to point away from the sun. This has been cited as being due to the pressure of the sun's light. (Currently, however, it is believed that the shape and direction of the tail are due primarily to *particles* emitted from the sun.)

A more direct and controlled proof of light's momentum is similar to the push of a jet of air that makes colorful beach balls hover in the air in some department stores (usually in the vacuum cleaner department). The same sort of thing can be done with light. The push from a powerful laser light beam pointed upward can support a tiny glass ball.

So we add energy and momentum to light's properties. Does that finally tell us what it is? To describe something we usually compare it to something we already know. Historically there were two phenomena that seemed good candidates: **particles** and **waves.** Both can carry energy and momentum. Particles can travel through vacuum, but with any speed (less than *c*), not just one fixed value. Waves have definite speeds (e.g., 335 m/sec for sound traveling in air), but usually they do not travel through vacuum. Since neither candidate fits perfectly, a lengthy dispute ensued that was not settled until this century. We shall say for now that light is a wave (but see Sec. 15.2). Admittedly, light does not look like a wave, so to understand the wave nature of light will require a considerable stretch of your imagination. You will have to imagine wiggles or oscillations of something that you cannot see; that can only be detected, normally, by objects that are too small to be seen; and that can travel through empty space. As a first step, let's make sure we know what a wave is.

1.2
WAVES AND THEIR PROPERTIES

We are familiar with many different kinds of waves, such as water waves, waves rippling along a flag, or waves flowing across a field of grain on a windy day. We also speak of "waves of insects" and of a "heat wave." What do these various waves have in common? Certainly not the material that is waving around, for that is water, cloth, wheat, the density of insects, and temperature in the examples above. A wave is not some special kind of material, but a particular kind of motion by which something nonmaterial is propagated. In general, a ***wave*** is a *propagating disturbance* of some equilibrium, quiescent state. The thing that is being disturbed is usually a "continuous medium" (such as a fluid), but it can also be something that is not itself a material, such as the temperature in our heat wave example. Further, the medium does not itself have to move very far—what is important is that the disturbance should move along. Think of the wave in a field of grain: each stalk moves only a little bit, but the wave moves across the whole field.

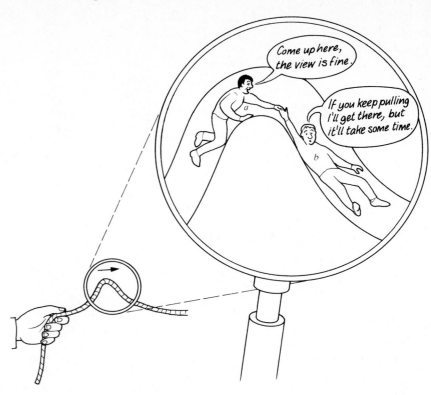

FIGURE 1.7

The hand on the left has just moved up and down. A wave ("hump") propagates along the rope because different pieces of the rope pull on each other—as *a* pulls *b* up, *b* pulls *a* down.

One of the easiest types of wave to demonstrate is a wave on a rope. Move one end of a stretched rope up and down, and a "hump" will travel along the rope, away from your hand (Fig. 1.7). This example also allows you to understand, at least roughly, why the wave moves along: the parts of the rope on top of the hump pull other parts of the rope upward; but since each piece of rope is made of matter and cannot react instantaneously, it takes some time for the parts in front to move up. Thus, it takes time for the wave to travel to the next part of the string.

If light is a wave, what is the medium that is analogous to the water or the rope? When light travels through a vacuum, what is it that vibrates? What is disturbed?

A. Electromagnetic waves

It took a long time for scientists to accept the notion that no material was needed for propagation of some types of waves. In particular, light waves are disturbances in the "electromagnetic field," a nonmaterial physical entity whose quiescent state is the vacuum itself. But when we want to visualize this field we often think of something similar to the stretched rope along which rope waves propagate. Now, what in the world do strings or ropes have to do with electricity?

We know that there are two kinds of electric charge, positive and negative, and that opposite charges attract, likes repel. A negative charge attracts a positive one, that is, the positive charge feels a force even though the two are not touching. So there is something pulling on the positive charge, even if there is nothing material immediately near it. We call this the ***electric field*** (of the negative charge, in this case). Even if no positive charge is actually nearby, the negative charge still has an electric field around it, which is ready and waiting to pull or push on any charge that ventures into its region—like a spider web that is ready and waiting to exert forces on a fly. This ready-and-waitingness can be described by a bunch of **electric field lines,** which point in the direction of the pull or push that a positive charge would feel at each point in space. These field lines exist even in a vacuum, and *they* behave very much like stretched ropes or strings (Fig. 1.8)—we can set up waves in *them.* First, we must visualize these non-material field lines, which represent the force that another charge would feel, and then further visualize waves in these field lines.

How do you grab hold of the field lines in order to shake them? Each field line has a charged body at its end (as it must, since the lines are just a description of the forces exerted by the charge!), which you can grab and move. When you do this, you wiggle the ends of all the field lines that are attached to this charge, and a disturbance will propagate along the lines just as if they were stretched strings. In particular, if you wiggle the charge up and down, you move the ends of all the field lines attached to it up and down. However, the hump that

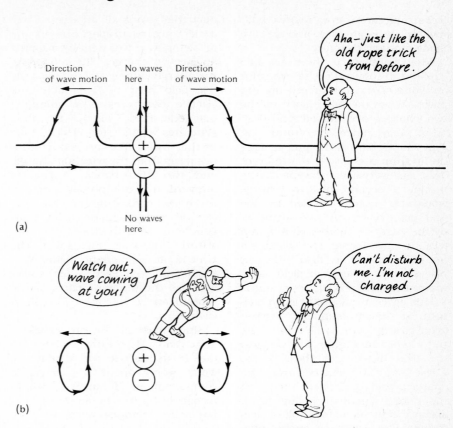

(a)

(b)

propagates along the *horizontal* lines will be largest, whereas a vertical field line will only slide along the direction of its length and, therefore, suffer no propagating displacement (Fig. 1.9a). Hence, no wave is propagated in the direction along which the charge wiggles, and the strongest waves are propagated in directions that are perpendicular to the direction of wiggle. (Thus, in Fig. 1.9 the positive charge moved up and down, the wave propagated sideways.)

FIGURE 1.8

(a) A negative charge with its field lines. The field lines show the direction of force a positive charge would feel if it were at a point on one of the lines. If there happens to be no such positive charge, the field lines still are there. **(b)** A spider with its web. The web is there even if there is no fly caught in it.

FIGURE 1.9

(a) The positive charge at the center has just moved up and down. The hump in the horizontal field lines is largest. The vertical field lines are not affected by this charge motion. **(b)** When the field lines due to both charges are added together, we get loops of field lines. These loops propagate primarily in a horizontal direction, perpendicular to the direction of charge motion.

Most matter is electrically neutral, that is, there are as many positive charges in it as negative charges. Each charge has field lines attached to it, but the lines of a positive charge point *away* from it (since positive charges *repel* other positive charges), whereas the lines of a negative charge point *toward* it (negative charges *attract* positive charges). This directionality makes the field lines different from ordinary strings: if two lines with opposite direction meet, they can cancel so that there is nothing left. That is, a positive charge would feel no net pull. Thus, there is no electric field there. This is why we don't

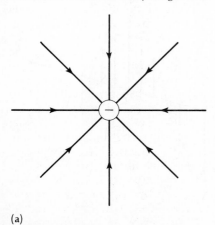

(a)

(b)

feel strong forces from neutral bodies, even though they, and we, are made of many charges.

What happens if there are two equal and opposite charges, but only *one* of them is moved up and down? When the field lines from the two charges are "added up" (Fig. 1.9b), they cancel everywhere except where there is the propagating hump from the charge that moved. That is, everywhere except at the hump, a second positive charge would be equally repelled by the first positive charge and attracted by the negative charge, so it would feel no net force. The result is *no* electromagnetic field in most places, and a *loop* of field lines where the wave propagates.

This gives us a pretty good picture of electromagnetic waves: propagating loops of field lines. As the charge in Figure 1.9 wiggles, the field lines form these loops. These loops are detached from the charges that generated them, so they forget where they came from, and it makes sense to speak of electromagnetic waves as such, independent of their sources. Since the fields and field lines can exist in vacuum, this explains why the waves need no medium through which to travel.

To show what the electric field in such a wave might look like, we have represented the electric field by arrows (with length proportional to the strength of the field) at the bottom right-hand corners of pages 11 to 81. You can watch the wave propagate by flipping the pages.

The only feature of these waves we have so far neglected is that charges in motion mean there is a current flowing, and a current generates a magnetic field. (An electric current creates the magnetic field that pulls that clapper on the bell in your telephone and that turns an electric motor.) The moving charge of Figure 1.9, then, also sets up a propagating magnetic field disturbance. The ring of electric field lines of an electromagnetic wave is therefore accompanied by a ring of magnetic field lines (oriented at right angles both to the electric lines and to the direction of propagation). We may then think of an electromagnetic wave as propagating in the following way: we begin by wiggling an electric charge. The motion of the charge causes a changing electric field near it. This changing electric field creates a changing magnetic field, which, in turn, generates a changing electric field further away. The process continues away from the wiggling charge, and the disturbance propagates outward. It is this propagating disturbance, consisting of changing electric and magnetic fields (perpendicular to each other), that constitutes the wave (Fig. 1.10). The electric and magnetic fields mutually "pull" on each other, somewhat as the different parts of the rope (*a* and *b* in Fig. 1.7) pull on each other.

Since most of the properties of light that will concern us here are due to its electric component, we won't worry about the magnetic component any further, except to recognize its role by always speaking of electro*magnetic* waves.

FIGURE 1.10

Snapshot of the electric fields (black) and magnetic fields (gray) along a single ray of light. The source for this wave is very far to the lower left. The charges of this source have oscillated up and down many times to generate this wave. There are electric and magnetic fields also at other places, but these are not drawn here. If you followed the direction of the electric field beyond the tip of the arrow at *a*, you would move along a field line in a very elongated loop (as in Fig. 1.9b), which comes back through the arrow at *b* and eventually returns to *a*.

How do we detect such electromagnetic waves? If the wave comes to a place where there is another charge, the electric field of the wave pulls the charge up and down, in a way very similar to the way the wave's source charge was moved up and down. So to detect an electromagnetic wave, we just place some charges in its path and find out if they wiggle, much as you might detect water waves by feeling your boat wiggle. Usually the "signal" of the wiggling charges has to be amplified to be detectable, as in your radio. The electromagnetic radio waves wiggle charges in your radio's antenna. The rest of the radio only serves to amplify this current of moving charges, to pick out the interesting part of the wiggles, and to convert it to sound waves. Similarly, your eyes contain miniature "antennas" (certain chemicals) that enable you to perceive light when the charges in your eye are wiggled by electromagnetic light waves. You can try out the complete process of generation and reception of electromagnetic waves by making electric sparks (say, by combing your hair on a dry day) and listening to the "static" caused by them on an AM radio. Sparks occur when electric charges suddenly jump from one body to another. This sudden, fast motion of charges generates an electromagnetic wave, as described above, that is strong enough to be picked up by a sensitive radio. The first man-made radio waves of Heinrich Hertz, which confirmed the idea that electromagnetism can travel in waves, were made by just

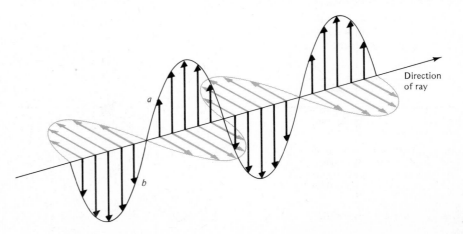

Direction of ray

such a simple spark*—though somewhat bigger than what goes on in your hair. Even bigger sparks, lightning, can be detected over quite large distances as static on your radio.

B. Resonance

If all electromagnetic waves make charges move about, why do we not see radio waves? And why does a radio not receive all stations simultaneously? The reason is that the reception is selective: a radio receiver can be "tuned" to the signal sent by a particular station. The emission and reception of electromagnetic waves is not equally efficient for all waves, but depends on the nature of the emitter and receptor. One phenomenon, called **resonance,**† is responsible for many of these cases of selectivity, from the tuning of a radio to the creation and detection of most colors in nature. Let's first study this phenomenon by itself.

What is the characteristic of a station's signal that the radio selects when you tune it? It is what is marked on your dial: frequency. Frequency describes how often the charges in the antenna wiggle. In general, the **frequency**‡ of any oscillating system is the *number of oscillations* that it performs *per second.* (In the next section we shall explore this concept in more detail.) Knowing this, it is not difficult to think of examples from everyday life that are "tuned" to a particular frequency.

Have you ever sat next to a small child and let your legs dangle? If so, you may have noticed that the shorter the person's legs, the faster they swing back and forth—that is,

they make more swings in a second. (If you don't have a child handy, see the TRY IT.) Is that because younger people are naturally more fidgety, or does it have something to do with the length of the legs? As scientists we look for a general principle, so let's consider other dangling objects—**pendulums.*** It is easy to give examples from everyday life that show that the longer the pendulum, the slower its motion; from large grandfather clocks versus little cuckoo clocks, to a fishing line with fish attached, which swings back and forth faster and faster as the line is wound up at the rod. A pendulum is but one example of the even more general class of

---

*Latin *pendulum*, dangling object. The correct Latin plural is *pendula*.

---

oscillators, which include weights on springs (such as your car), strings in a musical instrument, tuning circuits in your TV, charges in the chemicals in your eyes, and so on. They all have this in common: they like to oscillate at one particular frequency. Anybody who has ever rocked his car out of a snowbank, or swung on a swing, knows that to put energy into such a system (to increase its motion), it pays to do it at the system's preferred **resonance frequency** (Fig. 1.11).

FIGURE 1.11

One resonant system: the swinging towers at Ahmedabad (India). By pushing rhythmically at the resonant frequency, a person at the top of one of these towers can make them both shake noticeably.

---

*Germans still use the same word, *funken*, to mean both spark and broadcast.

†Latin *re* plus *sonare*, to sound repetitively. (A violin resonates at a certain pitch, "repeating the sound" after the bow is removed from the string.)

‡Latin *frequens*, crowded.

This oscillator behavior is shown graphically in Figure 1.12 and is typical of any resonant system. You can try this out yourself using a weight (for instance, some keys) dangling on a string as a pendulum. Hold the string at the top and move your hand from side to side. You will find that there are three regions of your hand's frequency, corresponding to different characteristic pendulum responses. (1) When your hand moves back and forth very slowly (low frequency), the pendulum simply follows it along, with about the same amplitude. We say the two motions are together, or **in phase.** (2) At some higher *resonance* frequency, the response of the pendulum is very large, and it keeps growing. Each swing of your hand puts more energy into the pendulum. Soon the swings become so large that the bob (the keys) wants to "loop the loop" and ceases to be a simple pendulum. (3) Finally, when your hand oscillates very rapidly, the pendulum stays almost still, and any motion it has is opposite to that of your hand. We call this opposed motion **out of phase:** your hand is moving too fast for the pendulum to keep up with it. By the time the pendulum can follow your hand to the right, your hand has already reversed direction.

At resonance you'll find, if you observe carefully, that the bob lags behind your hand's motion, but not so much as to move completely opposite; this is called **lagging in phase by 90°.** What is important is not so much this name, but the effect; because the bob lags behind your motion, you are always dragging it along, hence you put more energy into the pendulum than when it is in phase or out of phase. Actual resonant systems usually have a way of getting rid of this energy, and their resonant response curve (Fig. 1.12) is less peaked and broader the more easily the energy is dissipated (by friction or what have you). If the energy cannot be dissipated quickly enough, the oscillations will grow and the system will eventually break apart, sometimes with spectacular results. Looping the loop is a mild example. A more dramatic one is the wind-driven collapse of the Tacoma Narrows Bridge, which oscillated for hours, building up its energy, until it finally broke (a modern version of the Walls of Jericho). In a smaller way, a favorite substance to break by resonance is glass. Wine glasses have resonances in the audible frequency range, and can be broken if you have (a) a sufficiently intense sound, (b) a glass that doesn't easily dissipate its energy (that is, it rings nicely when you tap it), and (c) a constant frequency sound, so that each vibration of the sound hits the glass just right to push it in the direction it is already moving. Examples abound in literature, from Professor Higgins' "large Wagnerian mother, with a voice that shatters glass," and who must have been able to carry a tune (constant frequency!) whatever her other shortcomings, to Oskar in Günter Grass' *Tin Drum.*

On a smaller scale, an analogous effect can make molecules change their structure or break apart due to the action of light waves; the bleaching of dyes in sunlight is caused by this, as is the first stage of light perception in your eye and in photographic film. Since this happens only at or near the resonance frequency, we do not perceive *all* electromagnetic waves, but only those in the frequency range that we call visible light. Whether we can perceive a particular light depends on the amplitude of the wave (how much of a wave there is), as well as on the frequency. Even a wave of the proper frequency will not break the wine glass, or bleach the dyes, if the wave is not intense enough.

FIGURE 1.12

Resonance curves. The response (amplitude of motion) of a simple pendulum versus the motion of a hand that holds it. The more easily the pendulum can get rid of its energy, the weaker and broader will be the response near resonance. The three curves describe systems of identical resonance frequency, but different amounts of energy dissipation.

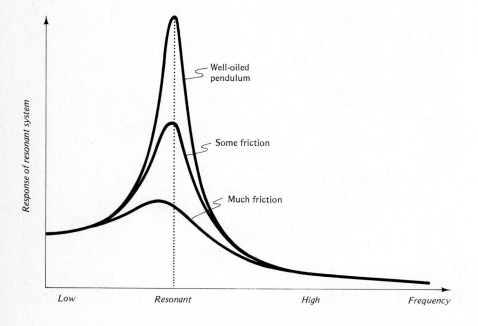

TRY IT

FOR SECTION 1.2B
Resonance in a simple pendulum

Use a piece of string about one meter long, and tie a few keys to one end to form a pendulum. To determine the resonant frequency of this pendulum, hold the string at one end, letting the keys dangle, and tap the keys once to make them swing. Notice how fast they swing back and forth. To be more quantitative, time how long it takes for ten complete oscillations. Now shorten the string by holding it at a point ¼ m away from the keys. Tap the keys and repeat your measurement. Does the shorter pendulum swing at a higher frequency?

Now that you know the resonant frequency for the long pendulum, you can produce a curve something like that of Figure 1.12. While holding the pendulum at the top of the string, move your hand continuously back and forth about a few centimeters. First move your hand very slowly, much slower than the resonant frequency. Do the same thing while you move your hand very rapidly, much faster than the resonant frequency. Now try it at the resonant frequency, the same frequency that the keys swung at when you hit them—the frequency at which the pendulum "wants" to swing (so it will seem natural to swing it at that frequency). Notice the size of the swing in each case. Also, try to observe the relation between the keys' motion and that of your hand. Do the keys move in the same or opposite direction as your hand at low frequency? at high frequency? Can you observe the "lagging in phase by 90°" at resonance?

1.3
NUMBERS ASSOCIATED WITH PERIODIC WAVES

Let's now return to waves in general and see how we characterize their frequency, amplitude, and other properties, and what role these properties of light waves play in our everyday experience.

A. Wavelength, frequency, and velocity

Since resonant systems go through the same motion over and over again—*unlike* the hand that made the single hump traveling down the rope in Figure 1.7—the waves they emit also repeat time after time. If we stand at one point near a rope that carries such a repeating wave, we will see hump after identical hump pass us by in repetitive fashion. A snapshot of the simplest of waves shows that it also repeats in space. That is, if we find a hump at some point in space, then at a given distance (the **wavelength,** for which we use the Greek letter

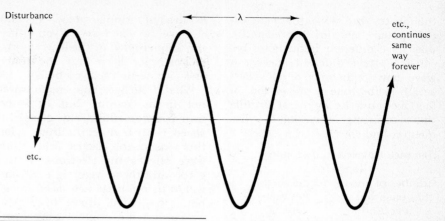

FIGURE 1.13

Snapshot of a sine wave, showing wavelength λ.

λ—"lambda") from that hump in our snapshot, we will find an identically shaped hump, and again at two wavelengths, three wavelengths, etc., on down the wave forever, in either direction. A **sine*** wave is the simplest wave and has the familiar shape shown in Figure 1.13. Although this is the wave most often drawn in textbooks, it is by no means the only kind of wave (Fig. 1.14). The waves in Figures 1.13 and 1.14a repeat, and are therefore called periodic, whereas that of Figure 1.14b is meant to be random and nonrepeating.

The **period**† of a periodic wave is the *time between repetitions* that we find if we stand at one place. To describe the time behavior of a periodic wave we often use not the period, but its inverse, called the **frequency.** The frequency of a wave is the number of times each second that an oscillation occurs at any fixed point in space. So the frequency measures directly how rapidly the wave wiggles. Frequency is usually denoted by the Greek letter, ν ("nu"), and its units are 1/seconds or "(cycles) per second" or "hertz."‡

*From Latin *sinus,* curve or fold of a toga. The word was used in verbally delicate times as a translation of the Arabic *jayb* (bosom).

†Greek *peri* plus *hodos,* path around.

‡Named after the fellow mentioned in Section 1.2A.

A sine wave thus can be described by its frequency ν (time behavior) and its wavelength λ (space behavior). But there is a connection between the behavior in space and the behavior in time. That connection is the wave's speed, v. It tells you how far a crest (or valley) travels in a second. Imagine a coat hanger wire bent into the shape of a sine wave. If you move the wire quickly past your nose (large v), you will experience a high frequency of bumps against your nose (many bumps per second). If you move the wire more slowly (small v), you will experience a low frequency of bumps. Even

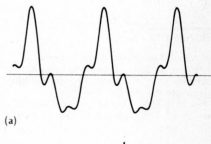
(a)

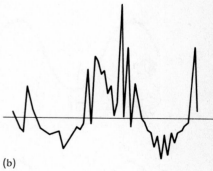
(b)

FIGURE 1.14

Waves other than a sine wave: **(a)** a periodic wave, **(b)** a random wave.

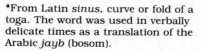

though the two waves have identical spatial behavior (wavelength), they have different frequencies because of their different speeds. A wave moves a distance of one wavelength in the time of one period. In fact, we can write a general formula that connects wavelength, frequency, and speed:

distance traveled in a second = (speed)

$$\begin{matrix}\text{number of crests}\\\text{that pass you in}\\\text{a second}\\\text{(frequency)}\end{matrix} \times \begin{matrix}\text{distance}\\\text{between}\\\text{crests}\\\text{(wavelength)}\end{matrix}$$

or:

$$v = \nu\,\lambda$$

(Fig. 1.15 gives a numerical example.) This is one of the very few mathematical formulas we will need. We need it for the following reason: in a vacuum, as we've stated, the speed of light and all other electromagnetic waves is the same, c, no matter what their frequency or wavelength. For electromagnetic waves in vacuum we therefore know:

ν (in hertz) $\times$ λ (in meters) = c
= 300,000,000 m/sec

FIGURE 1.15

Numerical example of the formula $v = \nu \times \lambda$. Two crests passed you in one second, so $\nu = 2$ hertz. The wavelength (measure it!) is 3 cm. Therefore, v should be 2×3 cm/sec = 6 cm/sec. This is correct, since crest A moved 6 cm in one second.

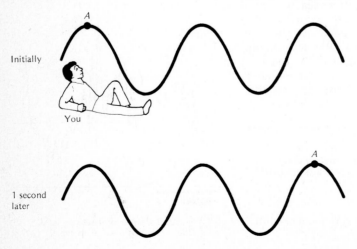

Equivalent formulas are $\nu = c/\lambda$ and $\lambda = c/\nu$. So when we tell you either the frequency, ν, or the wavelength, λ, you can figure out the other one—we needn't tell you both.

What if the light happens to travel not in a vacuum, but in some transparent medium, say, glass? Its speed is then different from c for this reason: the electric field of the wave shakes the electrons in the glass, and these wiggling electrons will in turn radiate new electromagnetic waves. All these little new waves and the original wave combine to make the total electromagnetic wave in the glass. It seems like a miracle that all these waves can add up to a *single* wave, going in a *single* direction. For the moment we shall not explain this miracle but simply say: if it does happen, then the substance is **transparent.** And if a single light wave does exist in such a substance, its speed v is *less* than c, because the wave is somehow burdened by having to wiggle all the charges, so that it takes longer to cover a given distance. Since our formula $v = \nu\lambda$ is still correct for the new medium, at least one of ν and λ also has to change. Which one?

Recall that all the wiggling and reradiating is going on at the same frequency ν, so whatever the result, it should also have the frequency ν. (In Sec. 1.2B, when you wiggled the string with the keys on it, the keys wiggled at the same rate, or frequency, as your hand.) That is, when an electromagnetic wave enters a transparent medium, *its fre-*

quency does not change. Therefore, the wavelength *must* change. Think of the wave crests as cars on a highway; if 100 cars per hour leave Chicago, going to Boston, and 100 cars per hour arrive at Boston, and the flow of traffic is steady, then cars will be passing every place in between at the same rate (frequency) of 100 cars per hour. However, the speed limit at different places along the way may be different: where it is higher, the cars are more spread out. (The first car to reach the region of higher speed limit speeds up, increasing the distance between it and the car behind it—that is, increasing the "wavelength.") At places where the speed limit is lower the cars are more bunched together (shorter wavelength). The frequency of the cars passing any point, however, is the same—unless they junk some cars along the way, or put some new ones on the road (a transparent medium neither junks, nor manufactures, wavecrests and valleys!).

B. More parameters needed to specify a wave

Wavelength, frequency, and velocity are not enough specifications to tell everything about a sinusoidal wave. One additional "spec" is the **amplitude.** It tells you how "big" the wave is, how great is the disturbance from the quiescent state (Fig. 1.16). For electromagnetic waves, it tells how strong the electric field gets. The bigger the wave, the more energy it carries, so the amplitude is a measure of the wave's energy. (We use the cautious words "a measure of" because actually the energy flowing is proportional to the *square* of the amplitude.)

The next specification is the wave's direction. We could specify it numerically as a compass direction, but usually we shall simply use an arrow—a line marking the light's path, with arrows indicating the direction of travel. This line we call a light **ray.**

Now imagine light traveling in every direction away from a small source (Fig. 1.17). There are sur-

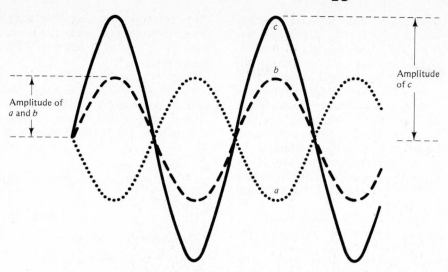

FIGURE 1.16

Three waves of different amplitudes and phases.

faces surrounding the source on which the electric field is constant—say zero, or cresting up, or cresting down. We call these surfaces **wavefronts.** If you drop a pebble in a pond, the circular ripples that you see are the wavefronts moving out. The rays are perpendicular to the wavefronts, and you can deduce one from the other. If we give you the rays, you only know as much about the wave as the wavefronts tell you. For example, you don't know the amplitude or the wavelength. Because electromagnetic waves propagate in vacuum the same way, independently of amplitude or frequency, you do

not need these details for many simple problems. The rays tell you where the light is going, which is the important thing.

Third, we ought to tell you where the wave starts or where it crests if we want to describe it completely. For example, the waves *a* and *b* of Figure 1.16 differ only in that respect. We say that they differ in **phase.** The two important cases we shall be discussing later (Chapter 12) are: (1) two waves *in phase*— their crests coincide as do their valleys, they are "in step" (waves *b* and *c* of Fig. 1.16), and (2) two waves

FIGURE 1.17

A snapshot of a wave traveling away from a small source showing the wavefronts. The rays are also marked. (Real candlelight is unpolarized.)

out of phase—one wave has its crests where the other has its valleys, they are "out of step" (waves *a* and *b* in Fig. 1.16). (Recall the nomenclature of Sec. 1.2B.)

The fourth and final item needed to specify a light wave is the direction in which its electrical field oscillates. This is called the **polarization.** For example, in Figure 1.17 the field oscillates up and down, so its polarization is "vertical." The polarization must always be perpendicular to the ray, because the ray gives the direction of propagation, and the electric field must be perpendicular to that direction. So another possibility for the wave of Figure 1.17, whose ray points to the right, would be "horizontal" polarization. If the electric field direction changes so randomly that no direction of polarization is preferred over any other, we call the wave unpolarized. Most "natural" light, such as sunlight, moonlight, lightbulb light (but not all star light) is unpolarized (see Chapter 13).

C. Seeing the properties of light waves

What role do frequency, amplitude, polarization, etc., play in our everyday experience? Of course, you cannot "see" a light wave wiggle, just as you cannot hear the individual peaks and valleys of pressure that make up a sound wave. Instead, the frequency of sound is heard as its pitch, and for light the frequency is related to its **color.** The light wave itself is not colored. It produces the sensation of color in (most of) our eyes; how it does that is complicated, and we will spend a lot of time discussing it later. Amplitude (or better, intensity, which is proportional to the square of the amplitude) very reasonably corresponds to brightness of the light. When the amplitude is zero, there is no wave at all, we see nothing— hence darkness. A wave with small amplitude corresponds to dim light; a wave with a large amplitude corresponds to bright light. (Actually, how bright the light appears to us depends not only on the light, but

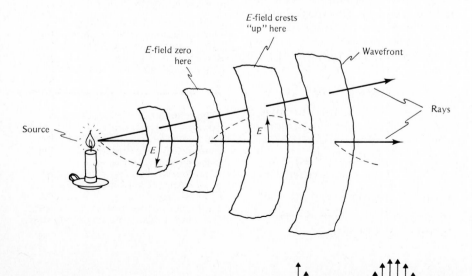

also on our eyes. For now we'll use the word brightness somewhat loosely to correspond to the *amount* of light, whether or not anybody can see it, and return to what can and can't be seen later—Chapters 5 and 7. We'll also use the vernacular word *intensity*, even though *irradiance* is the official word.)

The direction of the ray must, of course, be toward our eye for any ray we see. That seems to exhaust the properties of light as we experience it; there is nothing left for phase and polarization! Indeed our eyes cannot distinguish the phase of light, nor usually its polarization (but see Sec. 13.5). We can experience the phase *difference* between *two* light beams by converting this difference into an amplitude variation (Chapter 12). Similarly, polarizing sunglasses let us experience different polarizations by converting *them* into different amplitudes (Chapter 13).

The sensations of color, brightness, etc., of course apply only if the electromagnetic wave's frequency is in the visible range (in which the detectors of our eyes have resonances). Waves with other frequencies have nothing to do with colors and are called by names that mainly reflect their use. The full range of frequencies that electromagnetic waves may have is called the **electromagnetic spectrum.*** We summarize it next, in order to understand where visible light fits into this larger scheme.

1.4 ELECTROMAGNETIC RADIATION

There are many electromagnetic waves that are familiar: visible light, x-rays, radio waves, ultraviolet light, etc. These differ only in frequency (and wavelength). The reason they all have different names is that they behave rather differently when they interact with matter. This difference is due to

the relationships between their frequencies and the resonance frequencies of the charges in the materials we deal with—our eyes, our skin, our radios, and so forth.

Because the frequencies (and wavelengths) of electromagnetic radiation span such a tremendous range, we use the powers-of-ten notation when discussing them. If you are unfamiliar with this notation, or need some review, see Appendix A before you read the following discussion.

People find it cumbersome to speak of units like "one ten thousandth of a billionth" of an inch, or even "ten to the minus thirteenth" (10^{-13}) inches; so they devise different names. No one gives the distance between cities in inches; we use miles. For people's heights we use feet and inches. But there are 12 inches in a foot, 5280 feet in a mile and so on. That gets messy if we want to convert from miles to inches. It is easier to use the **metric system** where everything goes by powers of 10:

1 meter	$= 1$ m	$\simeq 1$ yard
1 kilometer	$= 1$ km	$= 10^3$m
		$\simeq \frac{5}{8}$ mile
1 centimeter	$= 1$ cm	$= 10^{-2}$m
	$\simeq \frac{2}{5}$ inch (width of a pen)	
1 millimeter	$= 1$ mm	$= 10^{-3}$m
	$\simeq \frac{1}{25}$ inch (width of lead in pencil)	

This much is fine for everyday lengths, but for light we have to use even smaller units of length because its wavelength is so small. Unfortunately, many different names are used. The wavelength of yellow light in vacuum may be called 5750 Å or 575 mμ or 575 nm or 0.575 μm. The units used here are:

$$1 \text{ Å} = 1 \text{ Ångstrom} = 10^{-10} \text{ m}$$

(This unit is often used because a typical atom is a few Ångstroms in size.)

$$1 \text{ nm} = 1 \text{ nanometer} = 10^{-9} \text{ m}$$
(formerly written 1 mμ)

$$1 \text{ μm} = 1 \text{ micrometer} = 10^{-6} \text{ m*}$$

(This is a convenient unit for high-power microscopes, with which one can look at objects as small as a few micrometers.)

There is a method to the prefixes centi-, milli-, etc. They multiply the unit by some power of 10. For example, milli- always means one thousandth. The important prefixes are:

n = nano- (Greek "dwarf")
    $= 10^{-9}$ (thousand millionth)

μ = micro- (Greek "small")
    $= 10^{-6}$ (millionth)

m = milli- (Latin "thousand")
    $= 10^{-3}$ (thousandth)

k = kilo- (Greek "thousand")
    $= 10^3$ (thousand)

M = mega- (Greek "big")
    $= 10^6$ (million)

Thus we can now proudly state that 1 nm $= 10^{-3}$ μm.*

A. The spectrum of electromagnetic radiation

With these preliminaries out of the way, we can now return to the various kinds of electromagnetic radiation that occur in nature. We all know that if we put a prism in a narrow beam of white light (from the sun or from an incandescent light bulb), it spreads the light out into a **spectrum.** But visible light is only a small part of the electromagnetic spectrum. This spectrum extends well beyond the red and violet ends of the spectrum that you can see—so far as to irritate the narrator of Lisa Alther's *Kinflicks*:

I brooded over the vast range of electromagnetic radiations on either side of the tiny band that could be discerned by human sense organs. They made me absolutely furious.

---

*Latin *specere*, to look, see.

*μ is the Greek letter "mu."

*There are other prefixes like deka- = 10, deci- = 10^{-1}, centi- = 10^{-2}, which you could use to sound erudite, for example, by referring to 10 baseball players from St. Louis as a "dekaCards." You could get carried away with this sort of thing. For example, 10^{-2} pedes = 1 centipede, or 1 millipicture = a picture worth only one word, but we wouldn't stoop that low.

Table 1.1 shows the various kinds of electomagnetic radiation that have proved useful so far. Note the very small region of the spectrum around $\nu = 10^{14}$ to 10^{15} cycles/sec* or $\lambda = 10^{-7}$ to 10^{-6} m that is visible light—the frequencies are comparable to the resonance frequencies of the receptors of your eyes. In this region, each frequency (or wavelength) corresponds to a different color in the spectrum of colors from red to violet. This tiny region is, or course, incredibly important to human life. The study of electromagnetic waves in this region is called **optics.***

Much of what we will learn about optics will be applicable to other parts of the spectrum, although the technology is quite different—for instance, devices to generate and detect radio waves are rather different from those for visible light or for x-rays. Thus, for example, not only can you take pictures using visible light, but you can also use other forms of electromagnetic radiation—gamma rays, x-rays, infrared, etc. All you need is a source of the radiation, some method of localizing the rays, and a detector (Fig. 1.18). The picture is a record of how much radiation was received at each point. The main difference between different radiations is that the smaller the wavelength, the

---

*cycles/sec = cycles per second (cps) = hertz. When talking of radio frequencies, one usually uses Mc (megacycle) = 10^6 cycles/sec = MHz = 10^6 hertz, as well as kc (kilocycle) = 10^3 cycles/sec = kHz = 10^3 hertz. Unfortunately, all these forms are used.

*Greek *ops*, eye.

TABLE 1.1 *Electromagnetic radiation*

Detected by			Frequency in hertz	Wavelength in meters	Name	Examples of use
P h o t o n c o u n t e r s	P h o t o e l	P h o t o g r a p h i c e m u l s i o n	10^{23}	10^{-15} (size of nucleus)		
			10^{22}			
			10^{21}		Gamma rays	
			10^{20}			Cancer treatment
			10^{19}			
			10^{18}	$10^{-10} = 1$ Å (size of atom)	X-rays	Materials testing Medical x-rays
			10^{17}			
			10^{16}		Ultraviolet (UV)	
Human eye			10^{15}			Atomic structure Germicidal "Black light," sun tan
			10^{14}	$10^{-6} = 1$ μm (diameter of bacteria)	Visible	OPTICS IR photos, heat lamps
Thermal detectors			10^{13}			"Heat rays," forest fire detection
			10^{12}		Infrared (IR)	Molecular structure, human body radiation
			10^{11}			
	T u n e d c i r c u i t s		10^{10}	$10^{-2} = 1$ cm (size of a mouse)	Microwave	Atomic clocks, Space research
			10^{9}			Radar, microwave ovens (3×10^9 hertz)
			10^{8}	$10^{0} = 1$ m (size of a man)		Radio astronomy TV, UHF: 470–890 MHz, VHF: 54–216 MHz. FM: 88–108 MHz
			10^{7}			International shortwave CB:27 MHz
			10^{6}		Radio frequency (RF)	AM radio broadcast: 550–1600 kHz
			10^{5}	$10^{3} = 1$ km (size of a village)		
			10^{4}			Longwave broadcast
			10^{3}			Long-range navigation
			10^{2}	10^{6} (distance from Washington, D.C. to Chicago)		
			10		Audio frequency	
				10^{8} (distance to moon)		AC power, Brain waves

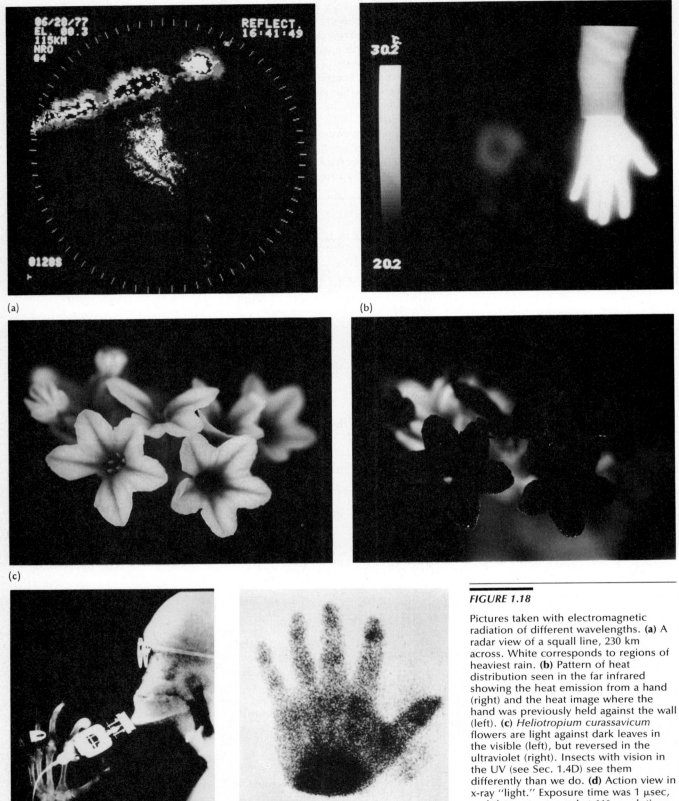

FIGURE 1.18

Pictures taken with electromagnetic radiation of different wavelengths. **(a)** A radar view of a squall line, 230 km across. White corresponds to regions of heaviest rain. **(b)** Pattern of heat distribution seen in the far infrared showing the heat emission from a hand (right) and the heat image where the hand was previously held against the wall (left). **(c)** *Heliotropium curassavicum* flowers are light against dark leaves in the visible (left), but reversed in the ultraviolet (right). Insects with vision in the UV (see Sec. 1.4D) see them differently than we do. **(d)** Action view in x-ray "light." Exposure time was 1 μsec, and the motor turned at 116 revolutions per second. **(e)** Gamma-ray picture of human hand. Source of gamma rays is radioactive tracer inside patient.

smaller the structures with which they interact. For example, the antennas for radio waves (meters to kilometers wavelength) are long wires or high towers, whereas the resonators in your eyes that select light waves (a few hundred nanometers in wavelength) are tiny organic molecules. Therefore, one must be careful in taking pictures with different wavelengths that the wavelength is small enough to detect the objects of interest. If the wavelength is comparable in size to the object, the wave will bend around the object (Chapter 12); if it is much larger, the wave won't be affected by the object. (This is why gynecologists, bats, and whoever else takes pictures with sound use very short wavelengths—high frequencies.)

Radiation of too short a wavelength (i.e., too high a frequency) is also less useful in taking pictures. If the radiation oscillates at too high a frequency compared to the resonance frequencies of the system of interest, the system will not be able to keep up with the oscillations (recall Fig. 1.12). At these high frequencies, then, there is very little interaction between the radiation and the system—the radiation just passes through the system. This is why many objects, such as your skin, that are opaque to visible light, are transparent to the higher-frequency x-rays.

B. How to make electromagnetic radiation

Sources of radiation differ as much as do detectors, particularly because their size usually is comparable to a wavelength. Basically they all are devices to wiggle charges. For very high frequencies we try to get nuclei and atoms to do the wiggling for us, as these very small systems have very high resonant frequencies. At low frequencies, we use electronic circuits instead. One standard way of getting radiation at intermediate frequencies is to make some material very hot. As we heat it, the charges in it oscillate more—heating something means to get the

atoms in it moving more rapidly and randomly. These wiggling charges then radiate—the hotter the material, the faster they wiggle, and the higher the frequency they radiate. The radiation from a heated frying pan usually is not visible, but you can feel it with your hand, which responds to the infrared radiation. If you heat the pan sufficiently, it begins to glow, and becomes "red-hot," hot enough to radiate in the visible. Your eyes are sensitive to radiation of this higher frequency, and you see it as red light.

In fact, a glowing body can give you radiation of any wavelength you like, provided it has the proper temperature. For instance, your radiant face emits radiation primarily with wavelength near 10,000 nm. (Although we cannot see this radiation, the poisonous pit viper can. This snake has deep hollows—pits—on the sides of its head that are sensitive to this frequency range, allowing it to locate its warm-blooded victims.) Heating and cooking goes on at about 1000 nm, and the sun's temperature corresponds, very conveniently, to the visible region at about 500 nm. However, at low temperatures the intensity of the emitted radiation is very low, and at high temperatures there is plenty of intensity but the emitting body tends to burn. The radiation from hot objects is not given off at a single frequency, but as a broad band of frequencies (Fig. 1.19), since the random heat motion has no particular single frequency. We see from Figure 1.19 that the hotter the object, the more light it radiates, and the shorter the wavelength of the predominant radiated light. Thus, the very hot sun radiates primarily in the visible, whereas most of the radiation from a 100-watt bulb is in the infrared, with very little in the visible (making it a rather inefficient light source).

The curves of Figure 1.19 are actually for an ideal object called a **black body,** a body that absorbs all radiation that falls on it (Fig. 1.20). A good approximation to a black body is a cavity with a small en-

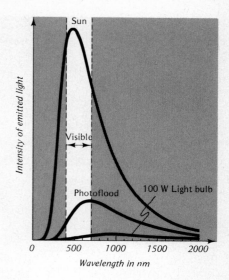

FIGURE 1.19

The spectrum of a glowing black body at temperatures of a light bulb, of a photoflood bulb, and of the sun. The actual spectra of these radiators are only a little different from this idealized black body approximation. (Remember, shorter wavelength means higher frequency.)

trance hole. Light that enters the cavity through this hole bounces around inside but, like a lobster in a lobster trap, rarely finds the hole again, so it does not get out. Your eye is a reasonably good black body because it contains a small opening with a big empty space behind it (Sec. 5.1A).

When a black body is heated it starts to glow, and, therefore, no longer *looks* black. But it continues to be black in the sense that it still absorbs all radiation that falls on it. A small window in a furnace is a good example of a glowing black body. If you try to illuminate this "body" (which is really the window hole) with a flashlight, it will look no different in color or brightness— it absorbs all the light you shine on it. Because the radiation stays inside the furnace for a relatively long time before it manages to escape, it comes to equilibrium with the furnace, and the colors that it emits depend only on the temperature. (Temperatures of the molten metal in foundries, and of ceramic kilns,

FIGURE 1.20

A good approximation to a black body radiator: a heated cavity with a small exit hole (furnace in a steel mill).

are often measured by examining the glowing color.) Most glowing bodies are not truly black bodies. For example, a log glowing in a fireplace is nearly, but not quite, black, as you can check after it has cooled. You can verify its near blackness by shining a light at the glowing coals in your fireplace. (If the coals were really black bodies at uniform temperature, they would all appear equally bright. Inside a well-stoked furnace you see a uniform bright glow and cannot distinguish the individual coals.) Similarly, the sun and light bulbs are not truly black bodies, though the essential features of the light they radiate are not very different from the black-body case.

Finally, we must mention that we cannot fully explain the curves of Figure 1.19 from our understanding of light as a wave. In fact, these curves led to our knowledge that the wave picture is not the entire story (Sec. 15.2A).

C. Light sources

The fact that hot bodies glow, or **in-candesce,*** has played a major role in the history of artificial light sources. From the time Prometheus brought fire down from Olympus until this century, all man-made light sources have been **incandescent** (except for those created by collecting fireflies). Until the advent of adequate generators of electricity in the nineteenth century, these lights were very little more than the fire Prometheus gave us; the only innovations being improvements in fuel from the campfires and torches that formed the earliest light sources.

The next known light source was the oil lamp, used by Paleolithic cavedwellers to allow them to make those magnificent cave paintings. The oil lamp was a dish of stone, shell, or, later, pottery that contained oil and a reed wick. (In the nineteenth century, when kerosene replaced the oil and an air draft was introduced through the wick, the oil lamp began to evolve into the

---

*Latin *in-candescere*, to become white.

lantern we now take on camping trips.)

Several millennia after the first oil lamp, in Egypt or Phoenicia, a candle was made by impregnating fibrous material with wax (from insects or certain trees). The use of tallow, and much later spermaceti (from sperm whales), and improvements of the wick, provided the major changes in this light source.

There were various civic lighting projects. Main streets in towns had previously been lit only by lamps in shops, house entrances, shrines, temples, and tombs. In the larger towns this may have been significant. It is estimated that there was a lamp every one to two meters along the main streets of Pompeii. Around 450 A.D. the first streetlights (in the form of tarred torches) were introduced in Antioch. Lighthouses, primarily to mark the entrances to ports, were another big project. Originally simple hilltop fires, these evolved into towers with bonfires of resinous wood, later coal, and, in the late eighteenth century, candles and whale oil lamps. The first known attempt to concentrate the light from these fires was probably the great pharos of Alexandria (one of the seven wonders of the ancient world, designed by Sostratus of Cnidos in about 280 B.C.), which may have used large mirrors of polished metal and whose light was said to be visible for 35 miles.

Except for a few oddities (the most bizarre being the burning of fatty animals, such as the candle-fish or the stormy petrel), the torch, the oil lamp, and the candle gave the world the bulk of its nighttime light until the nineteenth century. Thus, when the sun went down, it was very dark—". . . the night cometh, when no man can work" (John 9:4). Well into the nineteenth century battles stopped at sundown, which was all to the good, but hospital care ceased also. Witches had to wait for a full moon for a nighttime ramble; the poor went to bed at sunset; only the rich had a nightlife. About a century ago, this began to change with new lighting innovations (a mixed blessing—they brought the 12-hour workday).

The nineteenth century introduced gas lighting. While the Chinese had burned natural gas (piping it from salt mines through bamboo tubes), and coal gas was distilled from coal in 1664, gas was not used very much until it became economically attractive, around 1800. Introducing air, or oxygen, with the gas was found to improve the light. An even brighter light was produced by heating a block of lime to incandescence in an oxy-hydrogen flame, producing the limelight, which was used for "magic lanterns," and soon after mid-century for the theatrical applications that preserve its name today. (The explosive nature of the gas, combined with the flammable scenery, made theatergoing somewhat more of an adventure then.) Introduction in 1885 of the gas mantle, a mesh of inorganic salts heated to incandescence, increased the light by a factor of six over that obtained by just burning the gas, and kept the gas light industry alive into this century. (The improvement is a result of the fact that these salts are not ideal black bodies. Rather they tend to emit somewhat more in the visible, thus making them more efficient in producing useful light. The purpose of the gas, then, is just to heat the mantle, so the gas can be less luminous and less smoky.)

The earliest electric light was the arc lamp in which a spark jumped across two electrodes attached to a large battery. Arc lamps of any practical value had to await the development of big electrical generators in the mid-1800s, and shortly thereafter brilliant arc lights were in wide use. In these lamps, an electric field is set up between two hot electrodes (here, carbon rods) that are separated by a narrow gap (Fig. 1.21). Electrons are emitted from one electrode, pulled across the gap by the electric field, smash into the air molecules in between, knock electrons off them, and create many charged atoms (ions), which in turn are also accelerated by the electric field. All these charges smash into

FIGURE 1.21

An AC carbon arc. Only one of the carbon electrodes would glow if the arc were fed DC, as in the earliest arc lights. Most of the light comes from the very hot electrode tips.

each other and into the electrodes, give off light, and further heat the carbon electrodes. In fact, over 90% of the light comes from the incandescence of these very hot electrode tips. This produces a very concentrated light. While it is much too bright a light for the home, the carbon arc still is found in theatrical spotlights.

The incandescent filament lamp used today had to await the development of the mercury vacuum pump in order to produce the needed quality vacuum. The pump was produced in 1865 and by 1880 Edison had his patent for the **incandescent light** (Fig. 1.22). Enclosed within a glass bulb there is a little coil of thin wire, the **filament,** made of tungsten, because tungsten can become very hot without melting. The filament is about half a meter long if unwound, and its ends are connected to the lamp-

cord. As the current runs through it, the filament becomes hot and radiates. Some radiation is in the visible (Fig. 1.19) but most is in the infrared! You have undoubtedly felt that light bulbs get hot. In fact an incandescent bulb is only about 7% efficient in converting electricity to visible light—the rest of the energy goes into heat. We could get propor-

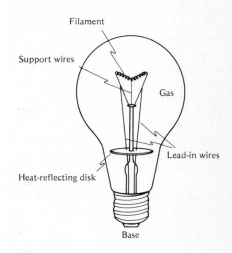

FIGURE 1.22

The main parts of a modern incandescent light bulb.

tionately more visible radiation if we heated the filament more, but the tungsten would then melt or burn. To prevent such burning some of the air is pumped out of the glass bulb. (That's why the bulb goes "pop" when it breaks, and why they needed the vacuum pump.) Actually, there is only a partial vacuum in the bulb. It is filled with a mixture of gases (argon and nitrogen) that do not react appreciably with tungsten. These gases tend to retard evaporation of the tungsten. The evaporated tungsten deposits on the glass (you can see this darkening in an old bulb) and makes the bulb dimmer, and the filament develops mechanically weak, hot spots. When the filament finally breaks, the bulb is "burnt out"; current can no longer flow.

In newer lamps, called "quartz-halogen" or "quartz-iodine" or "tungsten-halogen" lamps, the filament is surrounded by a quartz enclosure containing a little iodine gas (Fig. 1.23). The iodine picks up any tungsten that has been evaporated and redeposits it on the filament. This allows the lamp to operate at a higher temperature, which makes it more efficient since you get more visible light and proportionately less heat. Further, the evaporated filament doesn't blacken the bulb.

The relatively low efficiency of all incandescent sources led, this century, to the development of **fluorescent lamps,** [*] the first departure from the long historical precedent of producing light by heating something. In **fluorescence,** certain substances (called **phosphors**) produce visible light when they absorb ultraviolet ("black") light, and thus make light without using heat. Surprisingly, ultraviolet radiation can be more easily produced efficiently. A glass tube is filled with some gas at low pressure, usually mercury vapor (Fig. 1.24). Electrodes at the

[*]Latin *fluere*, to flow. But not of current: fluorescence was first observed in the mineral fluorite (CaF_2), which was used as a flowing agent to help metals fuse together when melted.

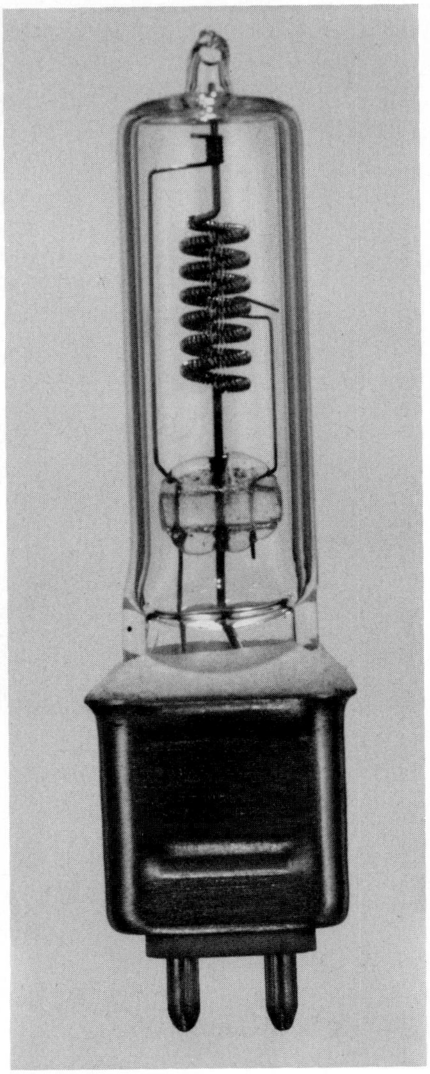

FIGURE 1.23

Tungsten-halogen lamp.

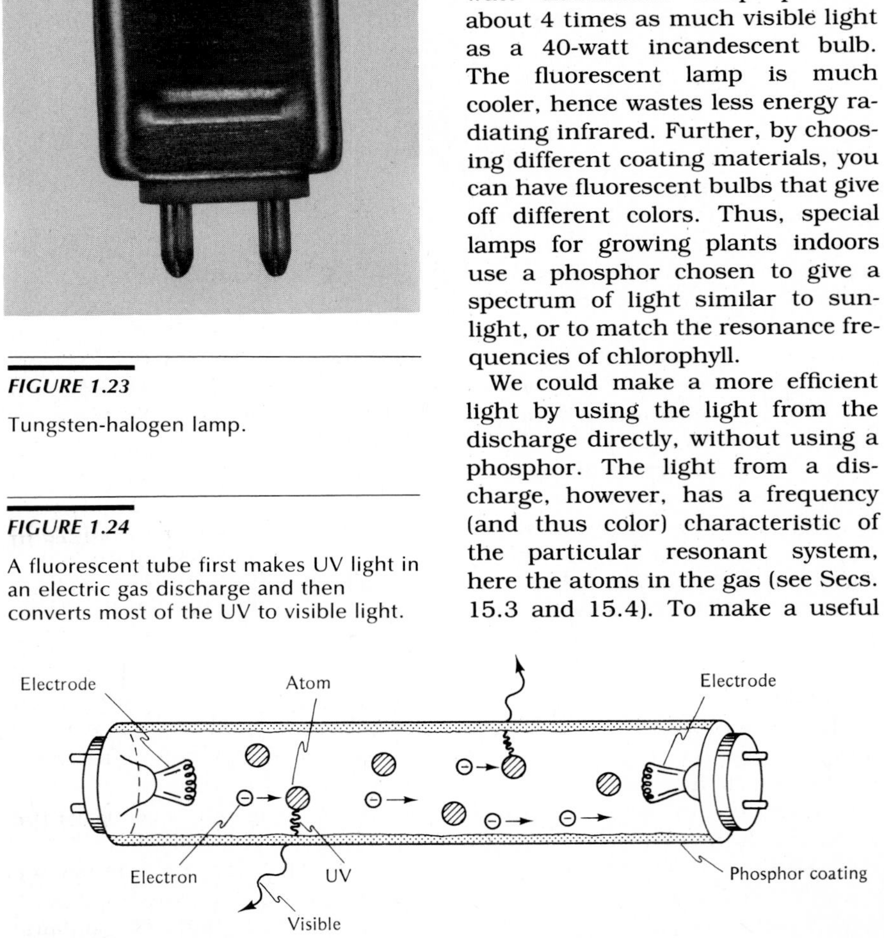

FIGURE 1.24

A fluorescent tube first makes UV light in an electric gas discharge and then converts most of the UV to visible light.

end are connected to an alternating current (AC) source, which drives the charges first one way and then the opposite way. (In the United States, alternating currents go through a complete cycle 60 times a second, so we have 60 hertz AC.) The resulting electric field in the tube pulls some electrons off the electrodes. These electrons collide with atoms, shaking them and the charges on them. The whole process is called a **discharge.** The oscillation of charges on atoms occurs at the atoms' resonance frequency; and this frequency for simple atoms is mainly in the ultraviolet (with some in the visible).

To make a "black light," you coat the tube with material that absorbs the visible but transmits the ultraviolet (UV). On the other hand, to make visible light, the tube is coated instead with material that fluoresces. This way of making visible light is so efficient that a 40-watt fluorescent lamp provides about 4 times as much visible light as a 40-watt incandescent bulb. The fluorescent lamp is much cooler, hence wastes less energy radiating infrared. Further, by choosing different coating materials, you can have fluorescent bulbs that give off different colors. Thus, special lamps for growing plants indoors use a phosphor chosen to give a spectrum of light similar to sunlight, or to match the resonance frequencies of chlorophyll.

We could make a more efficient light by using the light from the discharge directly, without using a phosphor. The light from a discharge, however, has a frequency (and thus color) characteristic of the particular resonant system, here the atoms in the gas (see Secs. 15.3 and 15.4). To make a useful light, we choose atoms that have resonances in the visible, for example, mercury or sodium. The strong coloration of the light emitted by these atoms is avoided in **high-intensity discharge lamps,** where high pressure or impurities broaden the range of frequencies emitted, making a light somewhat more like broadband white.

The production of light due to collisions of accelerated charged particles with atoms is not only a man-made phenomenon. The glowing *aurora borealis* is essentially the same phenomenon. Here charged particles from the sun strike molecules in our atmosphere. The radiation produced depends on the energy to which the charges are accelerated by the earth's magnetic and electric fields, as well as on the molecules they hit.

D. Visible electromagnetic radiation

We'll concentrate now, and for most of the rest of this book, on that tiny part of the electromagnetic spectrum that can make the charges in the receptors of our eyes respond: the visible. Figure 1.25 summarizes some of the important things that go on in and near this region of frequencies. At the top of this figure we indicate the visible region. Short wavelengths (400 nm) look Violet (though we often refer to the short-wavelength end of the spectrum as blue—see Sec. 10.4A). As we in-

---

FIGURE 1.25

Visible light and its interaction with life. The frequencies (and wavelengths) of light are marked along the top.

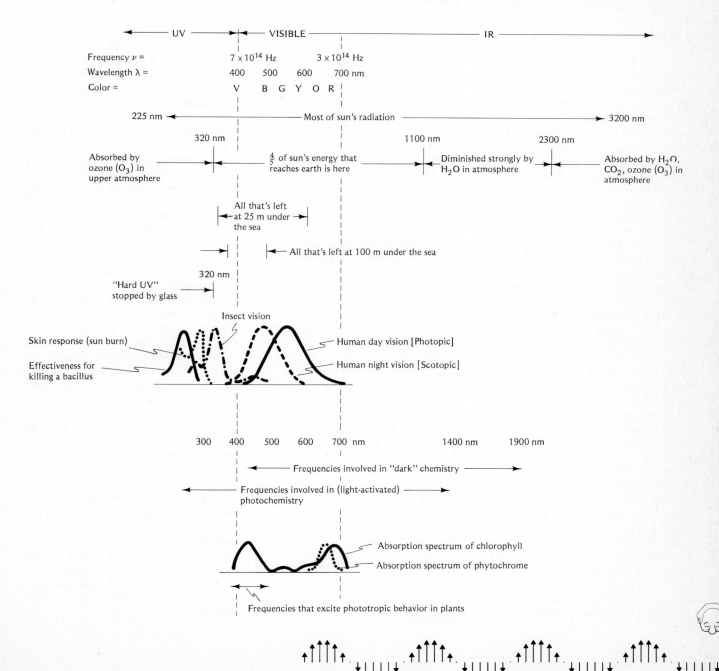

crease the wavelength, the color changes to Blue, then Green, Yellow, Orange, and finally Red when the wavelength gets to the long-wavelength end of the visible (700 nm).

Most of the sun's radiation lies between 225 nm and 3200 nm (see also Fig. 1.19), however, not all of this penetrates the atmosphere and reaches the earth's surface. The short wavelengths (below 320 nm) are absorbed by the resonances of ozone in the atmosphere, the long wavelengths by water (above 1100 nm) and by carbon dioxide and ozone (above 2300 nm). The result is that about half of the sun's radiation reaching the earth lies in the visible.

In the center of Figure 1.25 we have plotted curves to indicate how various systems respond to light of different wavelengths. Thus, human daytime vision is most effective at about 555 nm (yellow-green). Human night vision uses different receptors, which have their maximum sensitivity toward the blue. Actually, under very intense sources, so intense that it feels warm, we can see in the infrared (IR) up to as high as 1100 nm. Further, we could see in the ultraviolet (UV) except that the eye's lens absorbs UV. People who have their lenses removed (for cataracts) are sensitive down to about 300 nm. Insects, however, are most sensitive to ultraviolet light. Since insects have little or no vision in the red and yellow, we can use yellow lights as "bug lights." The yellow light provides useful illumination for our eyes, but the insects don't see it and therefore aren't attracted to it. Conversely, to attract bugs and zap them with high voltage, we use blue or UV bulbs.

Since most vertebrates have vision in the same range as humans, insects can have an interesting kind of protective coloring. In the visible, their coloring can be the same as that of a poisonous insect so that birds will leave them alone, but their UV coloring can differ from that of the poisonous insect so that no mistakes are made when they are looking for a mate.

Notice that your skin is most sensitive to UV of about 300 nm wavelength, but that glass absorbs the UV below 320 nm—we don't sunburn through a window. Notice also that hard UV, below 300 nm, kills bacilli, but that the ozone in the upper atmosphere absorbs this short wavelength UV, thus saving both bacilli and our skin from serious damage.

In the range 250–1400 nm we find not only vision, but all the other light-dependent processes critical to life, because the frequencies of the resonances of chemical bonds occur in this range. Thus, at the bottom of Figure 1.25 we show the absorption spectrum of chlorophyll (actually of several different kinds combined). Since only the green part of the spectrum is not absorbed, it is reflected or transmitted, so chlorophyll looks green. The chlorophyll responsible for the phototropic* behavior of plants absorbs only in the blue. Therefore, if we grow plants in red light only, without blue light, the other chlorophylls will absorb light and the plants will grow, but not toward the light. Also at the bottom of Figure 1.25 is the absorption spectrum of phytochrome—the enzyme that provides the "clock" for plants, determining when they germinate, grow, flower, and fruit, according to the length of the night. It measures the length of day by the amount of light it absorbs in the red. (See the FOCUS ON Light, Life, and the Atmosphere after this chapter.)

Light from sources other than the sun may play a critical role in life. Many organisms, such as the firefly, are **bioluminescent**—emitting their own light as part of their mating rites. (You can elicit a sexual response from a firefly with a flashlight, if you use the correct timing.) Humans, as you know, also find certain types of lighting to be romantic. Probably the most extreme effect of light on human sex life was proposed by Shakespeare. Hamlet,

---

*Greek *tropos*, turn. Therefore phototropic means turning toward the light.

while feigning madness, suggests that it is possible to get pregnant from walking in the sun:

Hamlet:—Have you a daughter?
Polonius: I have my lord.
Hamlet: Let her not walk i' the sun: conception is a blessing; but as your daughter may conceive,—friend, look to 't.

Light certainly seems to play a role in almost everything!

SUMMARY

To see the **light,** it must pass from a **source,** to an **object** (possibly), and then to a **detector.** Light travels at about 300,000 km/sec (in vacuum) and carries **energy** and **momentum.** It is a type of **wave** (*propagating disturbance*), but is unusual in that it can propagate through a vacuum—it is an **electromagnetic wave.** An *oscillating charge* creates a disturbance in the **electric field** (which describes the force on charged particles) as well as in the **magnetic field.** Electromagnetic waves of different frequencies are emitted and absorbed by *resonant systems* (whose response is greatest at the **resonance frequency**).

Periodic waves are characterized by **frequency** (v, number of oscillations per second), **wavelength** (λ, separation between repeating parts), **amplitude** (size or amount of oscillation), **polarization** (direction of oscillation), and the *direction of propagation* (indicated by a **ray**). The attributes of light correspond to our sensations of **color** (frequency, period, wavelength), **brightness** (amplitude), and the direction from which the light appears to come.

Visible light corresponds to a very small range (wavelength 400 to 700 nm) in the **electromagnetic spectrum,** which ranges from the very short wavelength **gamma rays** and **x-rays,** through the **ultraviolet** (UV), the visible, the **infrared** (IR), to the very long wavelength **radio waves.**

Black bodies emit a characteristic **black-body spectrum** in which more energy is emitted from hot objects than from cooler ones. Also, the peak emission occurs at *higher* frequencies for *hot* objects, *lower* frequencies for *cooler* objects. **Incandescent electric lights** consist of hot, glowing **filaments,** while **fluorescent lights** consist of gas glowing in the UV, and encased in a tube with a phosphorescent coating, which converts the UV to visible light.

PROBLEMS

P1 Briefly list some of the properties of light you have learned so far (e.g., how it travels, its speed, etc.).

P2 When you look at the daytime sky, away from the sun, it looks blue. Explain where this light comes from (i.e., what is its source and how does it get to your eyes?).

P3 How do we know that light travels through a vacuum?

P4 Brand new windowpanes and mirrors often come with pieces of tape on them. Why?

P5 (a) Which of the following are *self-luminous* objects (that is, we see them by their own light, rather than light reflected from them)? The sun, the moon, a cat's eye, a television picture, a photograph. (b) In the case of the examples that are *not* self-luminous, what is the source of light that allows you to see them?

P6 Give an example of a resonance. In your example, who (or what) supplies the energy? (Example: A parent pushing a child on a swing. The parent supplies the energy.)

P7 At Cornell, there used to be a narrow suspension footbridge. If you walked across it on a windless day, it scarcely swayed at all. However, if you jogged across it at just the right speed, you could get it swaying wildly. (That's why it's not there any more!) Explain why there should be this difference.

P8 On November 7, 1940 at 10 A.M., a 47-mph wind set the Tacoma Narrows Bridge into (torsional—twisting) vibration. The length of the bridge's main span was about 850 meters. The wavelength of the waves set up in the bridge was also 850 meters. Which of the two pictures in the figure represents more nearly the shape of the bridge that a snapshot taken at the time would show?

P9 Which of the following common everyday *periodic* phenomena are examples of resonance? (a) A cork bobbing up and down in waves in the water. (b) Grandparent rocking in his or her favorite rocking chair. (c) A singer hitting a high note and shattering a glass. (d) The people at a rock concert stomping their feet in rhythm with the music. (e) Floors of the auditorium vibrating and cracking as a result of the audience stomping at the rock concert. (f) A rattle in your car while idling, which stops as the motor increases speed.

P10 Your car is stuck in the snow, and you can't push it hard enough to get it over a hump of ice. It sometimes helps to rock the car back and forth. Apply some of our discussion of vibrating systems to explain why this helps.

P11 The figure below shows a picture of a wave. Use a ruler and measure its wave-length in centimeters.

P12 Redraw the wave of problem P11, and label it "old wave." On top of it, draw a wave of half the wavelength and twice the amplitude, and label it "new wave."

P13 Redraw the wave of Problem P11, and label it "old wave." On top of it, draw a wave of half the frequency and the same amplitude, and label it "new wave."

P14 A light wave traveling in the vacuum comes to a plate of glass. On entering the glass, which, if any, of the following increase, decrease, remain unchanged: the frequency of the wave, its wavelength, its speed?

P15 What is the frequency of your heartbeat while you are resting? Give units with your answer, and tell how you measured it.

P16 Light is an electromagnetic wave. Does that mean that there must be electrons present in the wave for it to propagate? Explain.

P17 Consider a radio wave and a visible light wave. Which has a higher frequency? Longer wavelength? Longer period? Higher speed in vacuum?

P18 (a) The color of light corresponds (in general) to which of the following (there may be more than one answer):frequency, speed, wavelength, intensity, polarization? (b) The brightness of light corresponds (in general) to which of the above list?

P19 (a) As a black body becomes hotter, does it emit more or less radiation? (b) Does it radiate predominantly at a higher or lower frequency?

P20 Identify the type (e.g., UV, visible, IR, etc.) of electromagnetic radiation of each of the following wavelengths in vacuum: 600 nm, 300 nm, 1400 nm, 21 cm, 0.1 nm, 3 km.

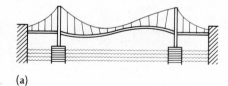

(a)

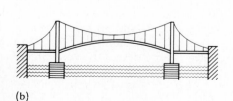

(b)

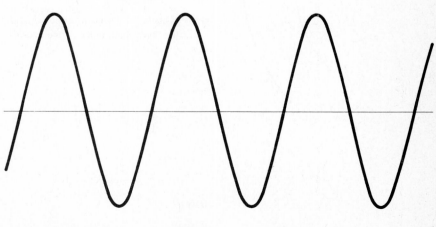

HARDER PROBLEMS

PH1 How could you use a laser beam in a dark room to determine the amount of dust in the air?

PH2 What is the color of the sky on the moon, where there is no atmosphere? Why?

PH3 When soldiers march across a bridge, they are told to break ranks. That is, they are told not to walk in step with each other. Think about the Walls of Jericho and the Tacoma Narrows Bridge, and explain why the soldiers should break rank.

PH4 Cheap loudspeakers often have a resonance in the sound frequency region that we can hear (roughly 50 to 20,000 hertz). You get a much larger output (sound) at the resonant frequency, for a given input, than you do at other frequencies. Better speakers don't have such resonances. Why is it bad to have such a resonance? (Think about what you would hear as someone played a muscial scale.)

PH5 (a) What is an electric field line? (b) What do the arrows on an electric field line convey? (c) How can one physically determine whether an electric field is present at a given position?

PH6 The figure shows some wavefronts of a light wave. Redraw the figure and draw three different light rays on it.

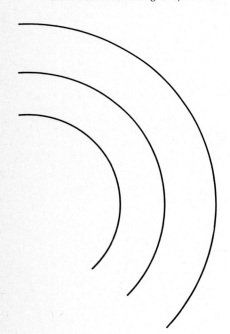

PH7 Describe the physics of a standard incandescent light bulb. (a) What charges wiggle? What causes them to wiggle? (b) Why is there a partial vacuum in the bulb? What gas is present inside? Why? (c) Why don't manufacturers simply make bulbs that last forever? (Sure, they have to stay in business, but what physical constraint mitigates against making a bulb that will last forever?)

PH8 Why can't we see x-rays directly? ("Because our eyes aren't sensitive to x-rays" is correct, but not a suitable answer. Why aren't our eyes sensitive to x-rays?)

PH9 Objects A and B are illuminated by a light source that produces *only* visible radiation. To the eye, A looks bright while B looks dark. Under the same illumination, a photograph is taken of the two objects, using film sensitive only to infrared radiation. In the picture, B shows up brighter than A. Explain how this could be so.

MATHEMATICAL PROBLEMS

PM1 How long does it take light to travel from one of Galileo's hill tops to another, 1.5 km away?

PM2 When a laser beam is sent to the moon, reflected there, and returned to earth, it takes 2.5 seconds for the round trip. Calculate the distance to the moon.

PM3 A rocket probe is sent to pass close to the planet Jupiter. Suppose that at the time the rocket reaches Jupiter, Jupiter is 630,000,000 kilometers from Earth. How long will it then take a radio signal to travel from the rocket to Earth?

PM4 A radio signal takes about 2.5×10^{-3} seconds to travel from Boston to Washington, D.C. Calculate the distance between these two cities.

PM5 (a) What is your height in meters? (b) Express the result of part (a) in millimeters and in nanometers. (c) Which of these three units is more reasonable to use for your height? Explain your choice.

PM6 (a) The note that orchestras tune to, middle A, has a frequency $\nu = 440$ Hz. The speed of sound in air is about $v = 330$ m/s. What is the wavelength of middle A? (b) What is the wavelength of the A one octave lower than middle A? (As you go down an octave, you divide the frequency in half.) (c) Why do you think organs have big pipes and small pipes?

PM7 The figure shows (idealized) the way the Tacoma Narrows Bridge vibrated from 8 to 10 A.M. on November 7, 1940, shortly before it collapsed. The length of the bridge's main span was 850 m. Each up and down oscillation lasted 1⅔ seconds. (a) This means that the frequency of the oscillations was $\nu = 0.6$ Hz. Show how one derives this result from the data given above. (b) What was the wavelength of the waves set up in the bridge? Give the reason for your answer. (c) What was the speed of the waves? (Show your calculation.)

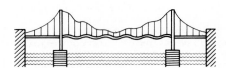

PM8 In Washington, D.C., radio station WRC broadcasts on AM at 980 kHz. Station WETA broadcasts on FM at 90.9 MHz. What are the wavelengths (approximately) of the radio waves used in each case? (Give the units you use: e.g., meters, feet, cubits, versts, etc.)

PM9 The range of wavelengths, in vacuum, of visible light is about 400 nm to 700 nm. What range of frequencies does that correspond to?

PM10 What is the frequency of 575 nm (yellow) light?

PM11 Consider a wave of wavelength 2 cm and frequency 1 Hz. Could this be an electromagnetic wave traveling in vacuum? Why?

PM12 If a black body is heated to a temperature T (in degrees Kelvin, K = °C + 273), the most intense radiation is at wavelength λ (in meters), where

$$\lambda \times T = 2.9 \times 10^{-3}$$

(a) Find λ for room temperature bodies (take $T = 290$ K). (b) What is the frequency of this radiation? (c) What kind of radiation is this (e.g., visible, IR, UV, x-ray, etc.)?

Light, life, and the atmosphere

That our atmosphere is so beneficent as to transmit visible radiation from the sun, but absorb killing ultraviolet, is a consequence of its history. We will outline its development so that we may see the interplay of light, life, and the atmosphere.

In the early days of the earth's development, the atmosphere did not contain many of the key ingredients of life. There was no oxygen (O_2) and consequently, no ozone (O_3). Nor was there carbon dioxide (CO_2). As a result, the ultraviolet (UV) radiation reached the surface of the earth quite easily, much more so than now (Fig. FO.1). This UV provided just the right frequencies (i.e., it excited resonances) for various organic molecules present in the seas to combine into the first living organisms. These primitive organisms had a better chance for survival than they would have now, since there were no other organisms to eat them, and no oxygen to oxidize them. Lacking oxygen, the organisms could only get the energy they needed to live by the process of **fermentation.*** If you've seen beer

ferment, you know that this process produces carbon dioxide (CO_2). These organisms did just that, pumping more CO_2 into the atmosphere than currently is there. The presence of the CO_2 then allowed new organisms to develop that could live by **photosynthesis.*** These new

---

sugar $\rightarrow CO_2$ + alcohol + energy

This process spends its "capital" (sugar) to produce energy. Once the sugar is used up, the process stops. It is a very inefficient process. Only 5% of the chemical binding energy in the sugar is released, and half of that is lost as heat. (Fermentation tends to heat up the surroundings, as anyone who has made yeast bread will have noticed.) The half not lost is stored in the molecule ATP (adenosine triphosphate, the biological "energy currency"), which then, if there is raw material available, offers the energy for new cell production.

*Greek *synthesis,* putting together, hence putting together by light. Photosynthesis is the process:

$CO_2 + H_2O$ + light $\rightarrow$ sugar + O_2

It thus is a new source of sugar from which organisms can derive energy. As such, this is a crucial step in the evolution of life because it allows organisms to take energy from the sun continuously. That is, at this point life developed a (very efficient) solar battery.

---

*Latin *fermentare,* to boil. This process extracts energy by rearranging organic molecules. For example:

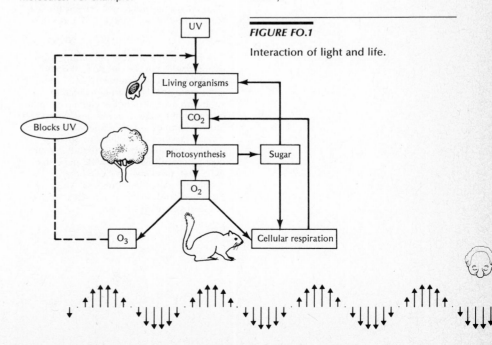

FIGURE FO.1

Interaction of light and life.

organisms removed some of the CO_2, fixing it in organic forms, and at the same time produced molecular oxygen (O_2), which entered the atmosphere, with two critical effects. One was that the sun's radiation converted some of this O_2 into O_3 (ozone). The ozone, as we've seen (Fig. 1.25), blocked the antibiotic UV from the earth's surface, thus permitting living organisms to leave the water for the land (about half a billion years ago). This would not have helped much unless there was a way the organisms could produce energy more efficiently. The process of fermentation produced barely enough energy for survival; none was left for motion. But the second effect of the O_2 in the atmosphere was to allow the much more efficient process of **cellular respiration.**‡ This process uses the O_2 and replenishes the CO_2. Eventually these two balancing processes, photosynthesis and cellular respiration, came into equilibrium, keeping the atmosphere roughly in its present form for ages.

Currently, however, atmospheric CO_2 is increasing, largely because of the destruction of the forests and the burning of fossil fuels, which have raised the CO_2 content in the atmosphere by about 15% since 1850 (over 5% since 1958). This is thought to be responsible for the warming trend in northern latitudes, from 1900 to 1940, by the **greenhouse effect.** The idea is that CO_2, like glass, lets the visible light through. The earth absorbs this, warms up, and radiates in the IR. But CO_2 (again like glass) doesn't let the IR through, so the energy stays inside the CO_2 (or glass) barrier and the earth gets hotter.* But the story is more complicated. Although atmospheric CO_2 continues to increase, a cooling trend seemed to appear in 1950. This may be due to other forms of pollution in the atmosphere that may reflect the incoming visible sunlight (like metal foil, instead of glass) so that less energy reaches the earth in the first place. That is, there are many other important molecules in our atmosphere besides CO_2, such as nitrous oxide, methane, ammonia, sulfur dioxide, and even trace constituents. An important question is whether they *reflect* the sunlight (preventing the energy from reaching the earth's surface), or *transmit* the sunlight (allowing the energy through), or *absorb* the sunlight (thus heating up the atmosphere). The greenhouse effect shows the complication of transmission at one frequency but not at another.

Another critical constituent is the ozone, which protects us from the hard, killing UV. It is unclear how sensitive the ozone concentration is to various man-made effects such as supersonic transports, fluorocarbons (commonly used in aerosol spray cans), atmospheric nuclear testing, or even agriculture (with its effects on the nitrogen cycle). However, a modest change in the ozone content of the atmosphere can significantly change the amount of hard UV getting through, since most of it is currently blocked. Too much of this UV can cause mutants by altering the DNA of which our genes are made, produce skin cancer, and do other ecologically more complicated things. Alternatively, too little UV would prevent our bodies from properly metabolizing calcium.

The effects of human activity on atmospheric constituents, and their effects in turn on climate and ecology through the interplay of light with the atmosphere, constitute one of the more exciting fields of study today. The consequences are, literally, breathtaking.

---

*Latin *respirare,* to breathe. This is the way we get most of our energy. The process is:

$$sugar + O_2 \rightarrow CO_2 + H_2O + energy$$

This is the same process as burning sugar over a flame, but we can do it at body temperature in a more controlled fashion. It is far more efficient than fermentation, capturing over 85% of the chemical binding energy in the sugar. Further, the by-products are not harmful, unlike fermentation, which usually produces something poisonous (alcohol or some acid). With all this extra energy, organisms can now do more than exist, they can develop locomotion and eventually paint the Mona Lisa. Yeast, on the other hand, living by fermentation, do not lead very active lives.

---

*Actually, "greenhouse effect" is a bad name. The major effect of an actual greenhouse is to prevent the rising hot air from escaping.

Principles of Geometrical Optics

CHAPTER 2

2.1
INTRODUCTION

How do we so easily decide where to place a beach umbrella to keep the sun out of our eyes, without worrying about the electric field, the wavefront, the wavelength, and the frequency of the light? The answer is that, for many simple problems it is sufficient to concentrate on the light *rays* (Fig. 1.17), the lines that describe in a simple geometric way the path of light propagation. **Geometrical optics** is the study of those phenomena that can be understood by a consideration of the light rays only. Geometrical optics is useful as long as the objects with which the light interacts are much larger than the wavelength of the light. As our beach umbrella is about a million times larger than the wavelength of visible light, geometrical optics is a very good approximation for this and most other everyday objects. For smaller objects the beam will not propagate in only one direction, but rather spread out in all directions—much as sound waves, with wavelength of about a meter, spread out around obstacles in the street.

In geometrical optics, then, light does not bend around corners. We think of the light as traveling in straight lines as long as it is left alone (Fig. 1.4). This straight-line propagation enables us to locate the beach umbrella so its shadow falls on our eyes.

2.2
SHADOWS

To cast shadows you need light from a fairly concentrated source, such as the sun. The best shadows are cast by light that comes from just one point: a **point source,** which is an idealization, like a ray, that can only be approximated, for example, by a small light bulb or candle. (Even a source as large as the sun or a giant star can approximate a point source if it is far enough away.) You also need a screen, such as a flat, white surface, which redirects incident light into all directions, so that you can see the shadow (the light from the surrounding area must enter your eye).

If an obstacle blocks some of the light rays headed for the screen, the light rays that are *not* blocked still reach the screen and make that part of the screen bright.

Those that *are* blocked don't reach the screen, so the places where they would have hit are dark—a **shadow.** We can figure out the shadow's location by drawing straight lines from the point source to the edge of the obstacle and continuing them to the screen. These lines separate the regions where the rays reach the screen from the region where they are blocked. The resulting shadow resembles the obstacle, but it is of course only a flat (two-dimensional) representation of the object's outline. Nonetheless, we can easily recognize simple shapes, such as a person's profile. Before photography, it was popular to trace people's shadows as silhouette portraits (Fig. 2.1a). Today's x-ray pictures are just shadows in x-ray

---

FIGURE 2.1

(a), (b) Various uses of shadows.
(c) Silhouette of one of the authors. Etienne de Silhouette (1709–1767), France's finance minister for a year, was deposed because of his stinginess over court salaries. He used cheap black paper cutouts in place of conventional decorations in his home and invented a technique for making paper cutout shadow portraits to raise money. When he died, he was penniless and destitute, a shadow of his former self.

(a)

(b)

(c)

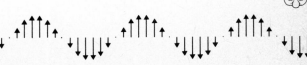

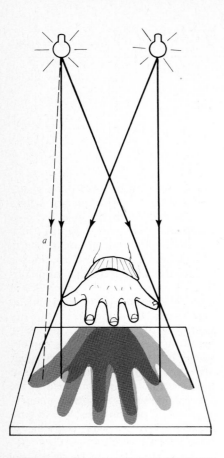

FIGURE 2.2

Two light sources throw two shadows. Their overlap, reached by no ray from either source, is the umbra. The penumbra is illuminated by rays, such as ray *a*, from only one of the sources.

"light," made visible by a fluorescent screen or photographic film.

In these examples, the shadow is seen on an object *different* from the obstacle blocking the light. An object can cast a shadow on *itself* as well. For example, the earth prevents the sun's rays from reaching the other (night) side of the earth. When you watch a sunrise, your back is in such a shadow: your own shadow.

Shadows help artists represent objects more realistically. For example, an unsupported object in mid-air is portrayed detached from its shadow (Fig. 2.1b), long shadows create a twilight (or dawn) mood, shadows of a face on itself give a more three-dimensional appearance, and so on.

What happens if we have *two* point sources? We get two shadows. If the two shadows overlap, the only completely dark region is this overlap. Where they don't overlap, the shadows are only partly dark because they still are struck by rays from the other source. The parts not in any shadow, of course, are struck by rays from both sources (Fig. 2.2).

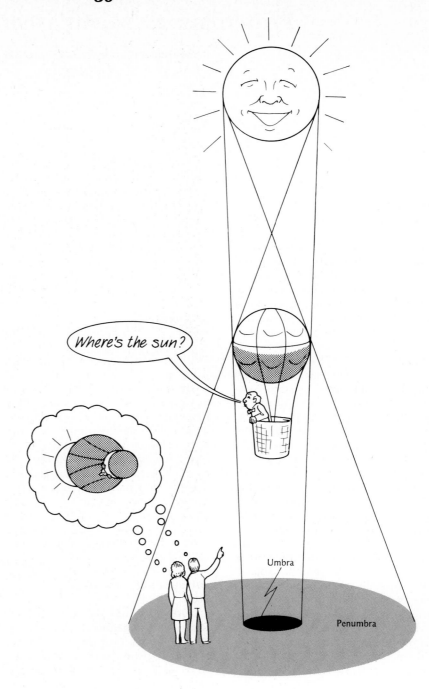

FIGURE 2.3

Looking back at the source from places in the penumbra and umbra.

The story is similar if we have three or more point sources. The only completely dark region is the overlap of all shadows. Other regions on the screen will not be as dark, because fewer shadows from the point sources overlap there. An **extended source** such as a long fluorescent tube can be thought of as a collection of many point sources, each of which casts a shadow. A region where *all* the shadows overlap will be completely dark. It is called the **umbra.*** Regions where the shadows don't all overlap will be partially dark because they still receive light from some parts of the extended source. Such a region is called the **penumbra.†** You can figure out the darkness of the shadow by supposing you are at a point on the screen and looking back toward the source. Where you are, it will be darker in proportion to the amount of the extended source that is blocked from your view. If you stand on the screen in the umbra you cannot see the source at all, but while standing in the penumbra you can see part of the source. Of course you would need a large screen, obstacle, and source in order actually to stand in the umbra or penumbra (Fig. 2.3).

A. Eclipses

We can actually do the experiment just described, by using the sun for the source, the moon for the obstacle, and the earth for the screen, whenever they all lie along a straight line. If the moon's shadow falls on the earth and we stand in the umbra, we cannot see the sun at all, and are enveloped in darkness. This is called a *total* **eclipse‡** of the sun or total **solar eclipse** (Fig. 2.4). In the penumbra we can still see part of the sun, and we are therefore in a *partial eclipse* of the sun.

(This is the first of many optical phenomena associated with the sun that we'll discuss. We must therefore provide you with this important consumer protection:

---

*Latin for shade (cf. umbrella).

†Latin *paene* plus *umbra*, almost a shade.

‡Greek *ekleipsis*, failure to appear.

FIGURE 2.4

A time sequence of photographs of a solar eclipse.

WARNING

DO NOT STARE AT THE SUN.
Failure to heed this warning can cause serious damage to your eyes.

After every solar eclipse, eye clinics have a rash of patients with eclipse blindness (Plate 5.1). However, it is not the eclipse that does the damage, it is the sun. Normally we don't stare at the sun, except fleetingly, nor focus on it particularly carefully, other than at sunset or sunrise when our eyes are protected by the longer path through the atmosphere that the sun's rays then traverse (Fig. 2.60). However, an interest in optical phenomena might tempt you to look more fixedly at the sun, and you would run the risk of permanently damaging your eyes. Constantine VII, the East Roman emperor, apparently went blind looking at an eclipse. Galileo injured his eyes by looking at the sun through his telescope. Should you damage your eyes by staring at

the sun, you would then be in good company, but that would be scant recompense indeed. The best way to examine a solar eclipse is with a pinhole device discussed in Section 2.2B. Because the moon is nowhere near as bright as the sun, you may stare directly at a lunar eclipse to your heart's content. You may also look at other phenomena we'll describe that are *near* the sun.)

We can determine the region of the umbra in an eclipse by drawing two straight lines (Fig. 2.5): one from the top edge of the sun, past the top edge of the moon, and on to the earth *(a)*, and a similar line along the bottom edges *(b)*. Any region behind the moon that lies between these two lines (before they cross) will have the sun completely obscured by the moon, and thus lie in the umbra. The penumbra is bounded by straight lines that start at one edge of the sun and cross to the *opposite* edge of the moon *(c and d)*. That is, lines *a* and *c* give the boundary of the shadow of the moon produced by the source on the sun at *A*. Lines *b* and *d* give the shadow produced by the source at *B*. The shadows produced by all other sources between *A* and *B* lie between these two. Hence, all these

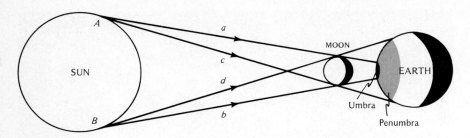

FIGURE 2.5

The umbra and penumbra of the moon projected on the earth, in exaggerated scale—the umbra is only about 200 km wide, which explains why total solar eclipses are rare at any given location.

shadows must overlap between *a* and *b*, and none can overlap outside of *c* and *d*.

An eclipse of the moon (a **lunar eclipse**) is similar, but now the earth is the obstacle and the moon behind it is the screen. We don't stand on the moon, but rather stand on the dark side of the earth and look at the light reflected by the moon.

PONDER

What would a lunar eclipse look like if viewed from the moon?

The Chinese poet Lu T'ung wrote a poem describing an eclipse of the moon that occurred in the year 810:

The glittering silver dish rose from the
*   bottom of the sea,*
. . . . There was something eating its
*   way inside the rim.*
The rim was as though a strong man
*   hacked off pieces with an axe, . . .*
Ring and disc crumbled as I watched
Darkness smeared the whole sky like
*   soot,*
Rubbing out in an instant the last
*   tracks,*
And then it seemed that for thousands
*   of ages the sky would never open.*
*   Who would guess that a thing so*
*   magical*
*   Could be so discomfited?. . . .*
I know how the school of Yin and
*   Yang explains it:*
'When the mid-month sun devours the
*   moon the moonlight is quenched,*
When the new moon covers the sun
*   the sunlight fails.'*

(Of course, it is the earth that "devours the moon," not the sun.)

An understanding of eclipses has occasionally proven useful. During the winter of 1503–1504, Columbus and his men were stranded on Jamaica. The Indians, who had been giving him food, decided to stop. Using his knowledge of an impending lunar eclipse, Columbus informed the Indians that his god wished them to supply food, and, to demonstrate this, would give them a sign from heaven that night. The eclipse began at moonrise and the Indians soon begged Columbus to intercede for them. This he did, when he judged the total phase of the eclipse to be over. The Indians resumed supplying him. (They apparently never asked why his god didn't simply teach Columbus to get his own food, rather than mess around with the moon.) This same theme was used by Mark Twain in *A Connecticut Yankee in King Arthur's Court.*

B. Pinhole camera

Using the principle of looking back from the screen at the partially blocked source, we could construct the umbra and penumbra of any complicated object in the light of an extended source of any shape—but in general the result is complicated and not very enlightening. However, in one particular case the result is again simple, even though the source may be an arbitrary, extended shape. This happens when the obstacle is entirely opaque except for one small hole, a **pinhole.** In Figure 2.6 we have drawn an arrow as the source. (The arrow could represent a standing person, a tree, a building, or whatever. We use an arrow because it is easy to draw and you can tell its direction.) If we follow the rays in the usual way, we see that light from the head *a* of the arrow arrives at only one point on the screen, *A*; all other rays from *a* are blocked by the obstacle. Similarly for the tail *b*, and all points in between. So there will be light on the screen from *A* to *B*, and this light will have the same shape as our arrow. The screen will have an inverted (upside-down) **image*** of the arrow.

Such a device is known as a **pinhole camera.** Though we have much better cameras nowadays, pinhole cameras are still useful occasionally. For example, since only a small amount of light gets through the tiny pinhole, looking at the image of the sun on the screen won't hurt your eyes. Industrial spies have been known to use small cardboard-box pinhole cameras, fitted with film instead of the screen.

---

*Latin, *imago*, imitation, copy.

FIGURE 2.6

A pinhole forms an inverted image.

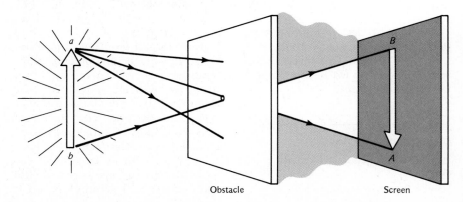

Obstacle   Screen

FIGURE 2.7

FIGURE 2.7

An elaborate camera obscura. To spare the viewer the need for standing on his head, a 45° mirror (Sec. 2.4) is used to bend the light so that the image is projected vertically down on a round table. The viewer can then walk around the table until he sees the image upright. The mirror can be rotated to provide a change of view.

They leave these innocuous looking boxes lying around for a day and then return to pick them up. (Since the pinhole admits so little light, a day's exposure may be just right.) Pinhole cameras are also used where a great depth of field (Chapter 4) is needed, and as cameras on satellites for detecting high frequency x-rays.

A pinhole camera is an example of a ***camera obscura,**** named by its fifteen-year-old inventor in the sixteenth century, Giambattista della Porta (later accused of witchcraft). He had a completely dark room except for a small hole and could then see, on the wall opposite the hole, an inverted image of the outside scene (Fig. 2.7). Much earlier Aristotle noticed round images in rooms that had a small irregular hole in the wall, no matter what the shape of the hole. Aristotle was, of course, looking at the images of the round sun. You can sometimes also see this on the ground under a shade tree, from light coming through the small spaces between the overlapping leaves. Though the small spaces are not usually round, the image is round because the object (the sun) is round (but not always, see Fig. 2.8).

*Latin, dark room. The name also refers to later devices with lenses—cameras without film.

FIGURE 2.8

Images of a partially eclipsed sun. The images are produced by the "pinholes" formed in the small spaces between overlapping fingers, one image for each "pinhole." Notice the crescent shape of each image, due to the partial eclipse. This is a safe way to watch an eclipse.

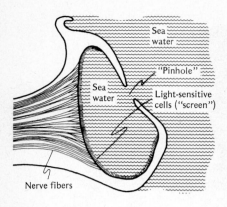

FIGURE 2.9

The pinhole eye of the *Nautilus*.

Even before Aristotle, Nature used the pinhole camera in the design of the eye of the mollusk *Nautilus* (Fig. 2.9). This simplest of eyes is clearly poor for gathering light, so Nature quietly gave up on it: such an eye is found in no other living organism.

An obvious way to cure the problem of the dim image is to increase the size of the pinhole. However if the hole is too big the image becomes blurred—rays from each point on the object pass through different parts of the hole and reach different parts of the screen, thus spreading out (blurring) the image of that point. Too small a hole also blurs the image (in addition to making it dim), because as the size of the hole gets close to the size of the wavelength of light, light ceases

FIGURE 2.10

Pinhole camera photographs of the same object but with different pinhole sizes. Pinhole diameter is largest at **(a)** and decreases, as labeled, for each successive picture. The images in **(a)** and **(b)** are blurred because the pinhole is too large. At **(c)** the smaller pinhole gives a sharper image. At **(d)** and **(e)** the pinhole has about the optimum size. Image **(f)** is blurred because the pinhole is so small that the wave properties of light become important (Sec. 12.5C). Exposure time was increased from **(a)** to **(e)** in order to make each photograph equally bright. The optimum pinhole diameter is about twice the square root of the product of the wavelength and the pinhole-to-screen distance.

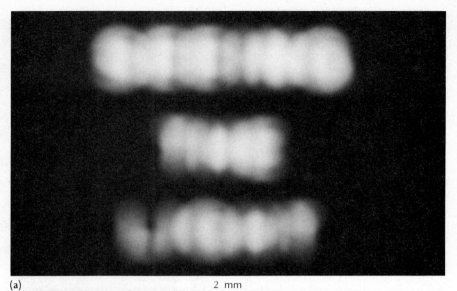

(a)   2 mm

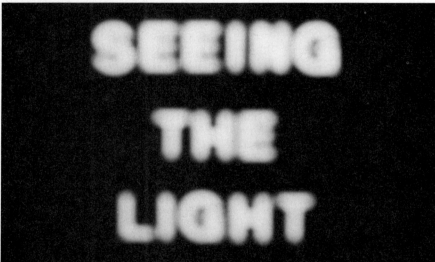

(b)   1 mm

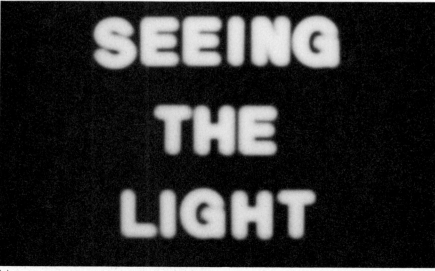

(c)   0.65 mm

(d)   0.33 mm

(e)   0.18 mm

(f)   0.10 mm

to propagate along straight lines (Chapter 12). The optimum size is somewhere in between (Fig. 2.10). For a pinhole camera the size of a shoebox (about $\frac{1}{3}$ meter), the optimum pinhole size is about $\frac{3}{4}$ mm, just about the diameter of a common pin. For a room-sized camera obscura, the best size is 2 or 3 mm. The larger the camera, the sharper and more detailed, but also the dimmer, will be the image of a distant object. To make a pinhole camera and pinhole photographs, see the TRY IT.

TRY IT

FOR SECTION 2.2B
Pinhole cameras, cheap and expensive

You can make a pinhole camera out of any cardboard box, such as a shoebox. Make a small pinhole at the center of one end of the box. To look at the image, cut a large square hole in the opposite end and cover it with tissue paper. The tissue paper serves as a translucent screen so you can see the image on it from the outside. Throwing a black cloth or coat over both the camera back and your head will cut down stray light. First look for the image of some bright object, such as the sun or a lamp bulb. (The image of the sun is easily seen but may be disappointing because of its small size). Using a lamp bulb as the object, note the orientation of the image and its change in size as you move away from the bulb. Then see what happens when you slightly increase the pinhole size. Observing from a shaded area you should be able to see a fairly good image of any brightly lit scene. Next make the hole quite big and watch the image become brighter but more blurred. Then tape a piece of aluminum foil over the large hole, make several pinholes in the foil, and examine the lamp bulb again.

For the image of the sun to be of appreciable size, use a long camera, for example, one made from a mailing tube coated inside with black spray paint. Even more simply, for the sun you can dispense with the box and simply project its pinhole image on a white screen by using a large cardboard with a pinhole in its center (the shadow of the cardboard is sufficiently dark to make the sun's pinhole image visible). This is the recommended way to observe a solar eclipse.

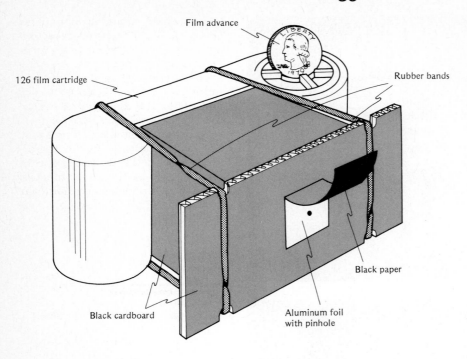

Film advance

126 film cartridge

Rubber bands

Black paper

Black cardboard

Aluminum foil
with pinhole

FIGURE 2.11

Design for a cartridge pinhole camera. The box should have a square cross section, $1\frac{7}{16}$ inches on each side so it just fits into the cartridge, and a length of about 2 inches or more, depending on whether you want a "normal" or "telephoto" pinhole camera (Sec. 4.3A). All edges and corners should be taped with black tape to prevent stray light from entering. The box can be held tightly in the cartridge with rubber bands. Note that you need a black paper flap over the pinhole to serve as a shutter.

If you want to photograph a pinhole image, you can use a 126 film cartridge in place of the viewing screen. Figure 2.11 shows a possible design, and Table 2.1 gives approximate exposure times. Be sure to have the camera on a steady support during the exposure. To advance the film, insert a coin in the round opening on top of the film cartridge, slowly turn counterclockwise, and watch the frame number through the small window in the back of the cartridge. (Stop turning when the third and fourth in each series of frame numbers both appear in the window.)

If you have a camera with interchangeable lenses, you can remove the lens and cover the opening tightly with a piece of aluminum foil. Make one or more pinholes in the foil, and you can take pinhole pictures the same way you usually take pictures with your camera (using a tripod for these longer exposures). Figure 2.12 shows some pictures taken this way.

TABLE 2.1 *Suggested exposure times for cartridge pinhole camera (Assumes pinhole diameter $\simeq \frac{1}{3}$ mm)*

ISO index of film	Film type (example)	Exposure time	
		Bright sun	Cloudy bright
400/27°	Tri-X	$\frac{1}{2}$ sec	2 sec
125/22°	Verichrome	2 sec	9 sec
100/21°	Kodacolor	3 sec	15 sec
64/19°	Kodachrome	4 sec	20 sec

Notes:
1. Also try half and double the suggested time. If the picture improves with one of these changes, continue halving or doubling until the optimum exposure time is found.
2. To measure the proper pinhole camera exposure with a light meter, set the meter at f/11 and find the exposure time it indicates. Multiply that time by 300 to obtain the suggested pinhole camera exposure time.

FIGURE 2.12

Pictures taken with pinhole camera on 35-mm film. **(a)** One pinhole. **(b)** Two pinholes.

(a)

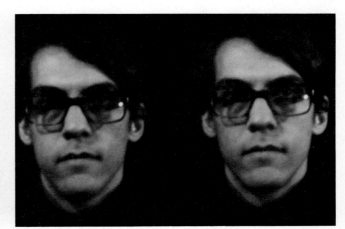

(b)

2.3
REFLECTION

We are familiar with reflected light, just as we are familiar with a ball reflecting (bouncing) off a wall. This common phenomenon, **reflection,** can be treated by geometrical optics. Light is traveling, say in air, in a straight line when it hits a different medium, say a mirror. Its direction then changes suddenly, the old ray stops and a new one starts.

If we want to know *why* a ball is reflected from a wall, we have to look into the structure of the ball—how it is compressed when it first hits the wall, how the resulting elastic forces push it the other way, and so on. Similarly, to understand *why* light is reflected by some substances, we must look at the nature of light. So we must digress a bit from geometrical optics and return to waves.

In Figure 2.13 there are two hands holding a rope. *A* has just started a wave moving to the right. What will happen next? If *B* keeps tension in the rope but doesn't pull up or down on it, the wave *C* will arrive there, shake *B*'s hand up and down, and we simply have a long-distance hand-shaking device. But now suppose *B* decides to pump his hand *down* just at the moment the wave *C* gets there. If *B* pushes his hand down just as hard as the rope is pulling it up, his hand won't move at all. However, because he has exerted a downward force on the rope, he has generated a wave *D*

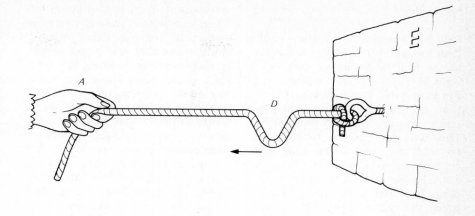

FIGURE 2.14

B, stonewalling the lady's gesture, refused to let his hand be shaken. Consequently, an upside-down wave returns to *A*.

that propagates to the left as a downward wave (Fig. 2.14). But since *B*'s hand never moved, even though it exerted a force, we might as well replace him by a brick wall *E* on which the rope is firmly and immovably attached. So our story from *A* to *E* teaches us that when we send a wave down a rope with a *fixed end*, an *opposite wave* comes back. We say that the wave, *C*, is *reflected* by the wall, because *D* looks just the same as *C* turned upside down (and backward, since the motion is also reversed). In the fancier language of Section 1.2B, we say the wave has undergone a 180° phase change (up to down).

If the rope were not tied to a brick wall, but instead to a much heavier rope that is nearly as difficult to move as the wall, the junction point would then hardly move when *C* arrives, and we would still get a reflected wave. Because the end of the very heavy rope moves, albeit very slightly, a small wave continues on in the heavy rope. We call this the **transmitted** wave. Such reflections, which occur when the medium of propagation (the rope) changes (rather than stops), and which result in a reflected wave that is the negative (180° phase change) of the incident wave, are called *hard reflections.*

The opposite extreme are *soft reflections.* An example is a rope tied to a very thin thread. The thread has hardly any mass and only serves to keep the rope under tension. Under these conditions the junction also reflects waves, but *C* comes back as an *upward* wave, no phase change: because the thread is so light, it is easily pulled up by *C*. It is as if *B* had pulled up on the rope's end, sending an upward wave back toward *A*. Again, part of the wave will be transmitted while the rest is reflected.

Reflections of any kind of wave (including light) occur whenever the medium of propagation changes abruptly. What counts is the change in the wave's speed of propagation. If it changes a lot, a lot will be reflected. If it changes only a little, most of the wave will be transmitted, and the reflected wave will be weak. The reflection is hard if the incident wave is trying to go from a faster medium (say light traveling in air) to a slower medium (say light traveling in glass), and soft otherwise. If the two media have the same wave speed there is no reflection.

FIGURE 2.13

The lady, *A*, keeping her distance, has started a wave propagating toward *B*.

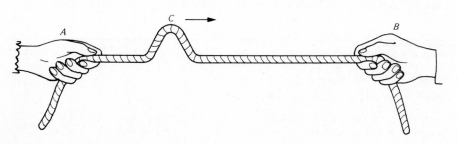

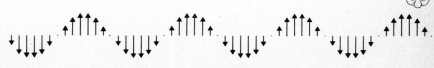

*A. Radar

Reflected waves can be quite useful. If we know how fast the wave travels in the first medium, we need only send out a wave pulse and measure the time it takes to be reflected back to us in order to determine the distance to the reflecting boundary. This echo principle is the basis of **radar.*** The radar antenna is both emitter and receiver. It first sends out a pulse of electromagnetic radiation. If the pulse comes back, then the radar operator knows there is some reflecting object out there. From the echo's time delay, she can determine the distance to that object. Typically radar uses electromagnetic radiation of about a billion hertz.

Bats (as well as the oil birds of Venezuela, Borneo swiftlets, dolphins, and seals) use a similar system, but with ultrasonic *sound* waves (up to 100,000 hertz, well above the maximum frequency of 20,000 hertz that humans can hear). Although folklore refers to them as blind, bats actually "see" very well by means of these short-wavelength sound waves (about ½ mm). This process, called **sonar,†** is also used by humans; for example, to detect underwater submarines, to measure the depth of water, and in certain cameras to determine the distance to the subject (in air).

B. Metals

Let us return to electromagnetic waves and consider reflections more closely. Suppose an electromagnetic wave is traveling in a vacuum and comes to some material, say glass. The electric field in the wave then causes the charges in the glass to oscillate, as we've seen, and these oscillating charges radiate, making a new wave. Some of this new radiation goes backward, becoming the reflected wave. Some goes forward and combines with the inci-

*From *radio detection and ranging.*

†From *sound navigation and ranging.*

dent wave (which is also going forward) to make the transmitted wave (Sec. 1.3). The way these forward-going waves combine, either adding to each other or tending to cancel each other, determines the intensity of the transmitted wave, that is, how transparent the glass is.

But suppose the material is a metal, rather than glass. Metals are good conductors of electricity. That means that they have lots of charges (electrons) that aren't attached to individual atoms, but rather are free to move around in the metal. There are many free electrons in a metal, and, being charged, they set up their own electric field. In response to an electromagnetic wave hitting the metal, the electrons move until the field they set up exactly cancels the electric field of the incident wave. (After all, if there were any electric field left over, more electrons would move until it was completely canceled.) Once the field is canceled, there is no further force to move any more electrons. But if there is no electric field inside the metal, there is no electromagnetic wave there—no transmitted wave. Metals are thus **opaque.** Almost all the energy in the incident wave goes into the reflected wave. That is why metals make very good reflectors.

All of this is only true up to a point. If the *frequency* of the electromagnetic wave is *too high*, then the electrons in the metal cannot move fast enough to keep up with the incident wave and cancel it. The frequency at which this begins to happen is called the **plasma frequency.** For yet higher frequencies, more and more of the wave is transmitted—the metal becomes more transparent and less reflecting.

The plasma frequency depends on the particular metal. In silver, the plasma frequency is a little higher

than the frequency of visible light, so silver is an excellent reflector of all visible light and is consequently used for mirrors (Table 2.2). Gold is similar to silver, but its plasma frequency is a little lower—in the blue region of the spectrum. Thus in gold all visible light except blue is reflected. White light with the blue removed looks yellow (Chapter 9), so gold looks yellow. Copper, also similar, has a still lower plasma frequency, so some of the green, as well as the blue, is not reflected. This causes copper to have its characteristic reddish color.

*C. The ionosphere

The same effect occurs in the **ionosphere**—that layer of our atmosphere in which the ultraviolet light from the sun has stripped electrons off the atoms, producing lots of free charged particles, just like in a metal. Because of the low density of the ionosphere, the plasma frequency is much lower than in a metal—much lower even than the visible (Table 2.3). Thus we can see through the ionosphere; it transmits the visible. However, the ionosphere plasma frequency is well *above* the frequencies of AM radio waves, which are consequently reflected by the ionosphere. That is why AM radio signals carry so far. They can be reflected from the ionosphere, and sent to points on the earth beyond the line of sight (Fig. 2.15). (This doesn't work for FM radio and TV because their higher fre-

TABLE 2.3   *Selected frequencies*

Visible	10^{14} hertz
Ionosphere plasma frequency	10^8 hertz
TV, FM	10^8 hertz
AM radio	10^6 hertz

TABLE 2.2   *Plasma frequencies of noble metals*

Metal	Plasma frequency	Color of metal
Silver	Above the visible	White—good for mirrors
Gold	In the blue	Yellow
Copper	In the blue-green	Reddish

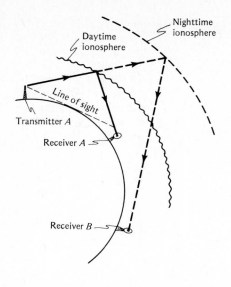

FIGURE 2.15

Reflection of radio signals from the ionosphere makes distant reception possible. The ionosphere rises at night, increasing the range of reception. (Exaggerated for clarity.)

quencies are comparable with the ionosphere plasma frequency.)

Consider now what happens at night. Less solar UV strikes the atmosphere after sunset, so the electrons and positively charged particles in the ionosphere can recombine into neutral atoms. This process occurs primarily in the lower, denser levels of the ionosphere. The result is that the ionosphere rises at night—the charged particles remain free only at the higher altitudes (Fig. 2.15). Thus, at night your AM radio can pick up stations from distant cities, reflected from the ionosphere. As some of these stations come in almost at the same place on the dial as local stations, AM reception may get lousy at night.

D. Mirrors

Returning to geometrical optics, we expect to have reflections whenever light traveling in one medium encounters another medium. Thus we get the familiar reflections from water surfaces or from panes of glass. Glass is, however, not a very good

reflector. When light strikes perpendicularly, only about 4% of the intensity is reflected. For this reason glass is usually used to transmit light, rather than to reflect it.

Silver, we've seen, is an excellent reflector in the visible (close to 100%), 25 times better than glass. But exposed silver tarnishes, and tarnished silver is not a good mirror because the tarnished surface *absorbs* (rather than reflects) the light, and thus looks black. The trick to prevent this is to plate silver on the back of glass (Fig. 2.16). The glass protects the silver surface and provides mechanical strength so that there is no need for a lot of expensive silver. The silver reflects the light. Of course, there is some reflection from the front of the glass (you can sometimes see a weak extra reflection, see Sec. 2.4C), but most of the reflection, by far, comes from the silver. (Aluminum is also used.) High-class mirrors, for use in optical systems, have silver in front in order to eliminate this unwanted reflection. These front-surface mirrors will, of course, eventually tarnish or scratch and therefore must be handled very carefully.

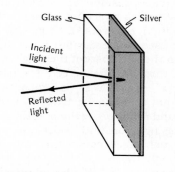

FIGURE 2.16

An ordinary mirror consists of a piece of glass, coated on its back surface with a layer of silver.

E. Half-silvered mirrors

Silver is such a good reflector, we argued, because the electrons inside it move around, and their electric fields then cancel those of the incident light wave. But the electrons need some space in which to do this, hence the light actually

penetrates the silver a little (about $\frac{1}{50}$ the light's wavelength—10^{-5} mm) before it is reflected. For most purposes this penetration can be ignored. However, if the silver is made thin enough, some of the light will penetrate through to the opposite side, and thus some light will be transmitted. That is, you can see through a thin enough layer of silver (or other metal).

One can make mirrors (called *half-silvered*) with glass coated by a layer of silver that is sufficiently thin to reflect half and transmit the other half of the incident light intensity. Such mirrors are useful in a variety of optical devices, such as a rangefinder in a camera (Sec. 4.2D). They are also used in "reflecting" sunglasses, and for "one-way mirrors" (also called "two-way mirrors" or "mirror pane"). These two-way mirrors do not work quite the way they show them in the detective stories on TV—light travels through them in *both* directions. You need the victim, V, to be well lit, whereas the spy, S, sits in the dark (Fig. 2.17). On the light side, V sees plenty of light reflected, because there is lots of light to *be* reflected from the light side. But he sees very little light transmitted from the dark side, because there *is* only a little light on the dark side to be transmitted. So V says, "It's a mirror." Even though the half-silvered mirror does not reflect as much light as a regular mirror would under the same conditions this difference is difficult to perceive (see Sec. 7.3C). On the other side, S sees only a little reflected light because there's not much to be reflected from his side, but lots of transmitted light from V's side, so S says, "It's transparent." If S turns on a light or lights a cigarette, V can see him and the jig is up. If you turn the light on S and off on V, then V sees S, but S sees only himself. In that case, the mirror is "one way" in the other direction.

This last effect is used by magicians to make things appear and disappear, by lighting up one or the other side of such a two-way mirror. Or suppose that in Figure 2.17

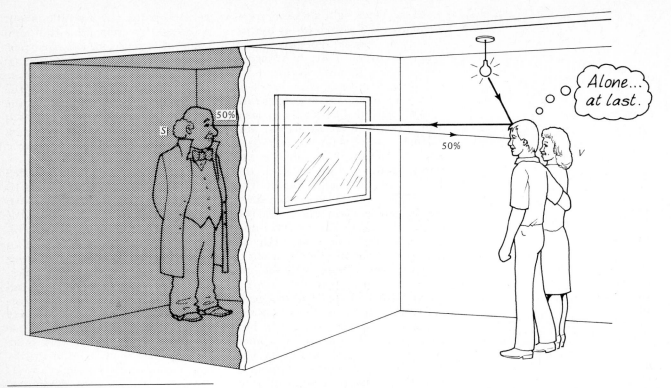

FIGURE 2.17

Illuminated innocents become visible victims of spookish spy behind harmless half-silvered mirror.

there is a weak spotlight on *S*'s face. Then *V* would see a faint, disembodied face next to his own reflection in the mirror. Thus "ghosts" can be made to appear in the mirror—a "reflection" without an object (the opposite of vampires, which are well known not to reflect in mirrors).

Such mirrors are also used in modern architecture. A building may be covered with glass that has a thin layer of metal on it. During the day, most of the light is outside (daylight), so the people in the building can see out, but passersby see a mirror-covered building. At night, the only lights are from the electric lights inside, so the passersby can see in, while the occupants see their own reflections.

PONDER

What do occupants and passersby see at dusk when the light intensity outside is equal to that inside?

The same principle works with semi-transparent gauze curtains often used on windows. Here scattering takes the place of reflection. During the day the bright light is outside, so insiders can see out, but outsiders can't see in. At night the situation is reversed. This idea is also used in the theater. If the light is on the back part of the stage, you see through the screen (called a scrim). If not, it is apparently opaque and can be used as background or for projections. If there is some light on both sides you can get such effects as Siegfried walking through the magic fire. One often sees scrims or half-silvered mirrors (or, more likely, "half-aluminized") on the rear windows of vans driving down the highway.

2.4
REFLECTION AT OBLIQUE INCIDENCE

When light traveling in a given direction hits a smooth surface obliquely,* it is reflected in some

*Latin *obliquus*, not at right angles.

different direction. What determines the direction of the reflected light? We describe this direction by the **angle of reflection,** θ_r (θ is the Greek letter "theta"). It is the angle measured from the perpendicular (**normal***) of the surface to the ray of the reflected light. Similarly we call the **angle of incidence,** θ_i, the angle between the normal and the incident ray. The **law of reflection** states that the two angles are equal,

$$\theta_r = \theta_i$$

(That is, they are equal in magnitude but on opposite sides of the normal, as shown in Fig. 2.18.)

The law of reflection describes how a mirror forms images. We see images the same way we see any object. That is, a ray of red light may be entering your eye in a certain direction, but the conscious information from your eye is not "a red ray is coming at me from 3° left of front." Instead you say "there is a red apple." In recognizing the apple, your brain sorts out the various rays that come into your eye and reaches a conclusion about their or-

*Latin *norma*, rule, carpenter's square.

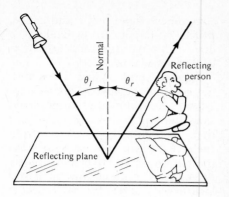

FIGURE 2.18

The law of reflection: $\theta_r = \theta_i$.

igin (the apple) rather than just about the direction from which they entered your eye. Because light usually travels in straight lines, your brain assumes that the light had traveled in the same direction in which it entered the eye (Fig. 2.19). How this processing is done need not concern us here.

Reflections can look extremely realistic (Fig. 2.20), and magicians make good use of this (the TRY IT gives various examples). Similarly in a snapshot of a perfectly smooth lake you can sometimes hardly tell which side is up, that is, which is the real landscape and which is the reflection (Fig. 2.21). However,

FIGURE 2.19 (below)

The reflected ray does not originate where the eye thinks it came from. The light scattered from the observer's nose (the object) really takes a sharply bent path to his eye, since it is reflected by the mirror. But the observer's brain interprets the light as if it had come in a straight line from a part of the image behind the mirror. This is because the reflected ray comes from the same direction *as if* it came from an object behind the mirror: the image, which is as far from the mirror as the object, but on the other side.

FIGURE 2.20

Scene from the Marx Brothers' "Duck Soup." Groucho is unable to tell if he sees his own reflection or Harpo dressed up to look like him.

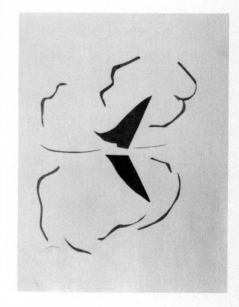

FIGURE 2.21

Henri Matisse, "La Voile." How can you tell it is printed right-side up? (Indeed, the Museum of Modern Art in New York hung this upside down for 47 days, thereby earning mention in the *Guinness Book of World Records*.)

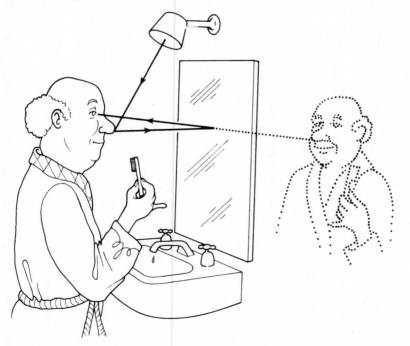

sometimes there is a difference, which is important in making a drawing or painting of a reflection look realistic (Fig. 2.22).

Suppose you look at a tree standing right at the edge of a lake (Fig. 2.23a). As your eye scans down from the top, you see the tree, and then just below the lowest ray from the tree (ray *a*), you see a ray from the bottom of the tree that was reflected by the water (ray *b*). If you look lower you see the ray reflected from the middle, and lower still you see the rays from the top of the tree (ray *c*). Thus, you see both the tree and an inverted (upside-down) image of the tree, symmetrical as most people expect of reflections (the closer your eye to the water, the more symmetrical). But suppose now the tree is up on a hill (Fig. 2.23b) instead of at the water's edge. Now as you look down, you first see the direct rays from the tree (ray *a*), then the direct rays from the hillside (ray *b*). You do not, however, see any reflected rays from the hillside because they are blocked by the bottom of the hill-

---

FIGURE 2.22

Charles C. Hofmann, ''The Montgomery County Almshouse.'' The difference between direct view and reflection is drawn accurately and is a little more obvious than in Figure 2.21.

FIGURE 2.23

(a) A tree at the very edge of the lake has a nearly symmetrical reflection. (b) Here the objects in the scene are at different distances, and the reflection looks different than the direct view.

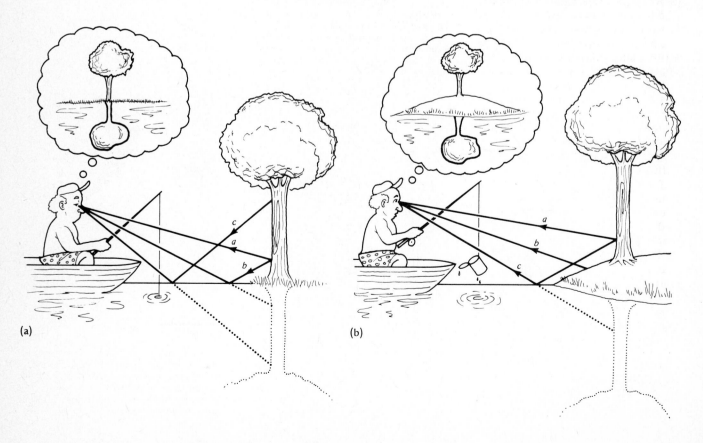

(a)

(b)

side, so the next thing you see is the reflected, inverted, image of the tree (ray *c*). In this case, the reflection does not appear symmetrical.

The law of reflection tells us about the *direction* of the reflected light. The *amount* of reflected light depends on the materials involved and on the angle at which the light hits the interface. If you look straight down into a lake you can often see the bottom—there is little reflection for light incident perpendicular to the surface (only about 2%). But if you look at a point farther away on the lake, you see mainly the reflected light from the sun and sky. So the amount of reflection increases the further the incident ray is from the perpendicular—there is *enhanced reflection* at *grazing incidence*. Manufacturers of wax for cars, floors, and furniture often take advantage of this effect. Advertisement photographs of a newly waxed floor, for example, are usually taken at a large angle, to reflect well on the wax.

The same effect occurs in a puddle. When viewed from straight above, it may appear as an ugly puddle, but when viewed at a distance so as to catch the light coming off at grazing incidence, it is quite reflective. As the artist Ruskin wrote, "It is not the brown, muddy, dull thing we suppose it to be; it has a heart like ourselves, and in the bottom of that there are the boughs of the tall trees and the blades of the shaking grass, and all manner of hues of variable pleasant light out of the sky" (Fig. 2.24).

If the surface of the lake is no longer perfectly smooth but has small ripples, the reflections change and can provide the water with a whole spectrum of colors that vary with the surrounding light. The explanation involves the law of reflection and the increasing reflectivity with increasing angle of incidence. For example, in Figure 2.25 the light from the back of the wave (*a*) is reflected at a glancing angle and

FIGURE 2.24

The nearby water surface is viewed at small angles of incidence and seems transparent, showing a man playing an underwater musical instrument; the more distant water reflects because the light reaching the camera grazes the water.

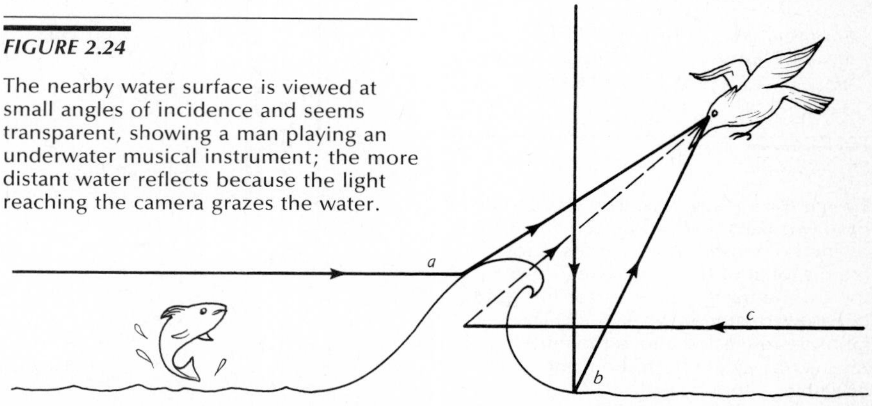

FIGURE 2.25

Reflections of light from an ocean wave.

appears to be the color of the low sky. The light from the front of the wave *(b)* is reflected almost perpendicularly and therefore is weaker. If the overhead sky is not too bright you also see light coming from within the water—either light that penetrates from the back of the wave or that is scattered from the depths of the water *(c)*. The subtle changes in the colors as the wave moves and the lighting changes provide a key part of the wonder of wave-watching.

TRY IT

FOR SECTION 2.4
Magic with mirrors

Although true magicians are sworn to secrecy, we will reveal the principles of some very old mirror tricks so you can try to construct them or variations on them. For a simple device that will allow you to pull a rabbit out of an apparently empty box, you'll need a cardboard box, a mirror, and a rabbit (or whatever you care to make appear). Figure 2.26 gives

the design. Looking in the front flap, the viewer sees the bottom of the box, and its reflection in the mirror. The reflection should look like the back of the box, so the box appears empty. The deception works best if the inside of the box has a pattern, designed so that the edges of the mirror don't stand out. Shadows may be a problem—a broad light source or lots of lights in the room works best. You first open both flaps to show that the box is empty. The top flap must be opened symmetrically with the front flap so that the viewer believes that the front flap's reflection is actually the back flap, and that she can see right through the box. (Place your two hands symmetrically, one on each flap.) Do not allow the viewer to look in the top opening. You then close the front flap, reach in the top flap, and pull the rabbit from the "empty" box.

The nineteenth-century Talking Head (Fig. 2.27) was a variation on this theme. It consisted of a woman whose head protruded through a hole in a table, but whose body was concealed with a 45° mirror that showed the side of the stage, designed to look like the back of the stage. The head could talk, answer questions, and so on, while appearing to sit on a plate, unconnected to a body.

Half-silvered mirrors can be used to turn objects into one another, but, since the amount of reflected light increases as the angle of reflection increases, a 45° piece of window glass works well enough for many tricks. You'll need a large box, a piece of window glass large enough to fit diagonally across it, and two lights, preferably with dimmer switches (Fig.

FIGURE 2.27

The Talking Head illusion.

FIGURE 2.28

Design for a machine to turn a friend's head into a skull. The box has an opening in the front through which the viewer looks. There is another opening directly opposite, in the back behind the glass, through which the viewer's friend inserts his head. To the side a skull is carefully placed, so that its reflection in the glass appears in the same place, to the viewer, as her friend's head.

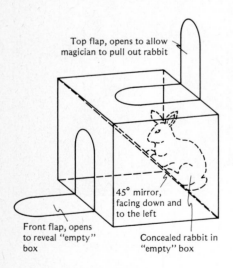

Top flap, opens to allow magician to pull out rabbit

45° mirror, facing down and to the left

Front flap, opens to reveal "empty" box

Concealed rabbit in "empty" box

FIGURE 2.26

Design for a magic box. The box should have two flaps that can be separately opened or closed. A mirror fits across the diagonal of the box, facing down toward the front flap. A mirror tile, sold in hardware stores, works well. The rabbit is concealed above the mirror and thus is not visible from the front opening.

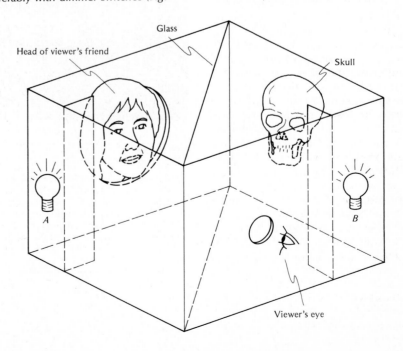

Glass

Head of viewer's friend

Skull

A

B

Viewer's eye

FIGURE 2.29

The Cabaret du Neant.

2.28). If originally light A, beside the friend's head, is turned on and the other light is off, the viewer can only see her friend's head. You then gradually dim light A and simultaneously gradually turn on light B, beside the skull, and the viewer sees her friend's head turn into a skull.

The nineteenth-century Cabaret du Neant (Fig. 2.29) was a glorious elaboration of this trick, featuring coffins and other lugubrious paraphernalia, the proprietor of this "cabaret" providing the appropriate funereal words to create a properly sepulchral atmosphere. Nowadays, the technique is used less morbidly in various fun houses and amusement parks, and even in museum exhibits.

*A. Sub suns and sun pillars

Reflections from ice crystals in the atmosphere cause some infrequent but striking phenomena. When water freezes in the atmosphere it may freeze in a variety of forms, depending upon the conditions—how cold the air is, how suddenly it got cold, etc. A form that concerns us here is a special, very symmetrical kind of crystal, a flat-plate, hexagonal crys-

tal (Fig. 2.30). When these fall through the atmosphere, their large, hexagonal faces tend to become horizontal—the crystals fall flat, just as a dead leaf does when it falls from a tree. If you are lucky enough, there will be a cloud of these essentially horizontal small ice crystals, which can serve as reflecting surfaces. If you then are on a mountain, or in an airplane, so that you are above these crystals, you may be able to see two "suns." One is due to the direct rays from the sun (ray *a*) while the other (the **sub sun**) is due to reflections by the ice crystals below you (ray *b*). The sub sun is less bright and fuzzier than the sun because not all the rays are reflected and because not all the crystals are horizontal (Fig. 2.31). If there are other clouds that block out the direct ray, you may

FIGURE 2.30

A sub sun is formed when sunlight reflects from the horizontal surfaces of ice crystals. The magnifying glass shows a single crystal.

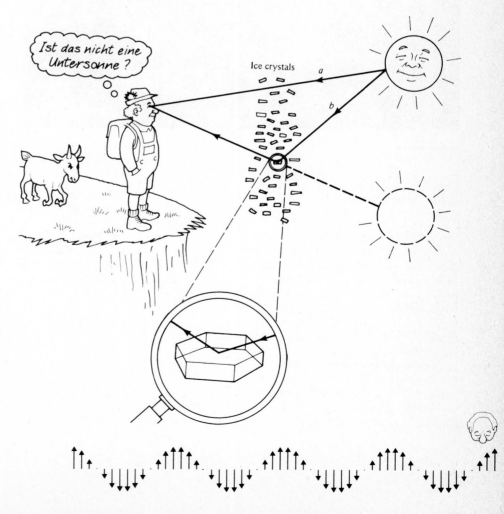

surfaces can be the almost horizontal hexagonal faces of the flat crystals of Figure 2.30. Alternatively, they can be side faces of long, hexagonal, pencil-shaped crystals (as in Fig. 2.74), which tend to fall with their axes close to horizontal. In either case, there are many nearly horizontal reflecting surfaces, and you can get reflections at a variety of angles, depending on the tilt of the reflecting crystal. If there are enough ice crystals, there will be some at various heights that are just at the right angle to reflect the sunlight into your eyes. You then see light coming from all these heights, that is, a pillar of light—the sun pillar (Fig. 2.32). The pillar may be either above or below the sun, but is most often seen when the sun is low and the pillar extends above it (Fig. 2.33).

One often sees a similar effect, without any ice crystals, when the sun or moon is low over a body of water. Here, the role of the nearly horizontal ice crystals is played by the ripples in the surface of the water, and the pillar appears as a swath of light across the water (Fig. 2.34).

FIGURE 2.32

A sun pillar can be seen when the ice crystals are not all exactly horizontal.

FIGURE 2.31

Photograph of a sub sun. (The sun itself is above, as the reflection on the airplane wing indicates.)

see only the sub sun. The sub sun moves with you as your airplane moves around (much as the distant moon appears to move with you as you drive a car—Sec. 8.4) and vanishes suddenly when you pass the ice crystals. It might easily be inferred to be a UFO (unidentified by those who fail to identify it).

The **sun pillar** comes about when ice crystals provide reflecting surfaces that are not exactly horizontal—each crystal is tilted by a slightly different amount. These

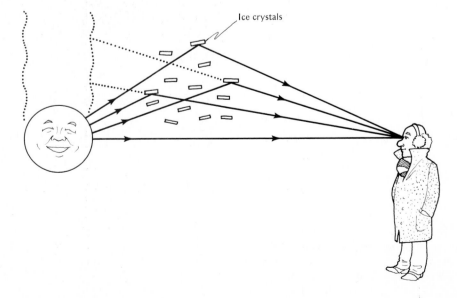

Ice crystals

FIGURE 2.33

The setting sun, with a sun pillar above it. (The horizontal line is a thin cloud layer.)

FIGURE 2.34

Mood photographs of sunsets over a lake usually show a "sun pillar" on the rippled water.

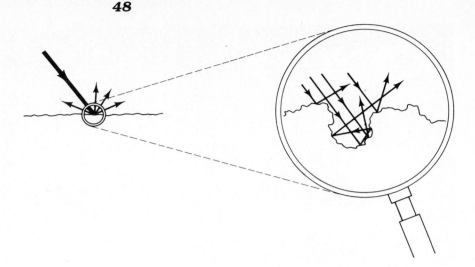

B. Diffuse reflection

We have been talking about reflections from *smooth* surfaces, such as polished metal, glass, or water—so-called **specular*** reflections. However, if the surface is rough, like that of most cloth, we get a smeared-out reflection. Light is reflected in many directions from a rough surface, so it does not produce a "mirror image," but just an overall brightness of the surface. For example, the surface of this paper when greatly magnified is shown in Figure 2.35. The result of the reflections from all the irregularities is a **diffuse** reflection (Fig. 2.36), where the reflected light goes

---

*Latin *speculum*, mirror.

FIGURE 2.35

Scanning electron microscope view of the coated paper used in this book. The actual size of the pictured region is about 50 μm.

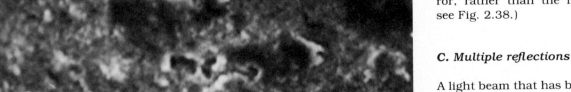

FIGURE 2.36

Diffuse reflection of a beam of light from a rough surface.

in all directions. Most surfaces are rough on a small scale, and that is very useful to us, because otherwise we would have trouble seeing them. For example, the light from your car

headlights hits the diffusely reflecting road surface and is scattered all around, some of it back to your eye, so you see the road (Fig. 2.37a). In the rain, however, the road is covered by a smooth surface of water, which gives a specular reflection, sending light away from your eye (Fig. 2.37b). This is why headlights are less helpful in the rain, and why some paint for road markings contains tiny spheres that reflect light back. Note that the eye treats diffuse reflection (and even specular reflections from very small mirrors) like scattering: it assumes, correctly, that the object is located where the diffuse reflection occurred, rather than farther back along the extended ray, as it does for specular reflections from mirrors. (The driver of the car in Fig. 2.37a "sees" the road at point *a*, whereas the man in Fig. 2.19 tends to "see" the image behind the mirror, rather than the mirror itself; see Fig. 2.38.)

C. Multiple reflections

A light beam that has been reflected from one mirror is just as good as any other light beam, so there is no reason why it cannot be reflected again. And again, and again. . . . You may have seen this effect in a barber shop or a hall of mirrors, where you are between two mirrors that face each other on opposite walls. Looking into one mirror you

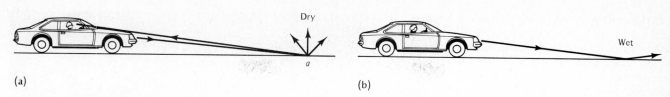

(a) (b)

FIGURE 2.37

(a) Diffuse reflection of headlights by a dry road makes the road easily visible.
(b) More nearly specular reflection by a wet road makes the road hard to see.

FIGURE 2.38

M. C. Escher, "Three Worlds." We see the trees by reflection, the leaves by scattering (or diffuse reflection), and (at a steeper angle) the fish by refraction.

see an "endless" series of images as each mirror produces a reflection of you, and then a reflection of your reflection in the other mirror (the back of your head), and then a reflection of the reflection of your reflection, etc., on and on. An ancient Chinese philosopher used this as an example of the meaning of infinity. Of course, the reflections are not really infinite, for several reasons: (1) often the mirrors are not quite parallel, so each image is a little off from the previous one, and the sequence of images "curves" away so you don't see all of them, (2) the mirrors do not reflect quite all the light, so each successive image gets a little darker, and the infinite end of the series fades out, (3) as John Barth wrote in "Lost in the Funhouse": "In the funhouse mirror-room you can't see yourself go on forever because no matter how you stand, your head gets in the way. Even if you had a glass periscope, the image of your eye would cover up the thing you really wanted to see," and (4) seeing an infinite number of images would mean an infinite number of reflections, hence an infinite path for the light, which it cannot cover in a finite time. In *The Third Policeman*, Flann O'Brien describes the fictional scientist de Selby, who carried this last idea a bit too far:

. . . he constructed the familiar arrangement of parallel mirrors, each reflecting diminishing images of an interposed object indefinitely. The interposed object in this case was de Selby's own face and this he claims to have studied backwards through an infinity of reflections by means of 'a powerful glass'. What he states to have seen through his glass is astonishing. He claims to have noticed a growing youthfulness in the reflections of his face according as they receded, the most distant of them—too tiny to be visible to the naked eye—being the face of a beardless boy of twelve, and, to use his own words, 'a countenance of singular beauty and nobility'.

If the mirrors are not parallel but at some angle to each other, an incident ray will be reflected just a few times. For example, if the mirrors are perpendicular to each other, you get just two reflections (Fig. 2.39). The interesting thing about these perpendicular mirrors is this: no matter what the direction of incidence, the light always is reflected back into that *same* direction (at least for light in the plane of the paper). No matter from where you look into the corner, you always see yourself. Such a corner mirror does not "reverse right and left," so if you pull your left ear, the image pulls his or her left ear (see the TRY IT).

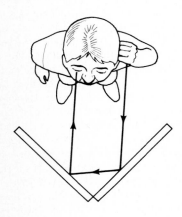

FIGURE 2.39

A corner mirror, made from two plane mirrors.

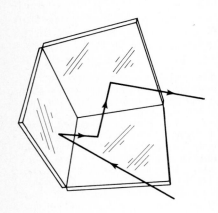

FIGURE 2.40

A corner reflector, made from three plane mirrors.

If we want a device to **retrore-flect*** all rays, whether in the plane of the paper or not, we need a third perpendicular mirror. This arrangement (Fig. 2.40) is called a **corner reflector.** It reflects any beam back in the direction it came from. This can be useful for anything that you want to be very visible in light from automobile headlights (e.g., highway signs, bicycles, joggers). Look closely at a bike or car reflector to see the many small "corners." (These are most easily seen after removing the opaque backing.) Corner reflectors are used on buoys at sea to reflect radar back to the transmitting ship. Corner reflectors were placed on the moon by astronauts so that a laser beam from the earth could be reflected back to the earth (Fig. 2.41). The astronauts could not be expected to align a single mirror precisely enough to reflect the beam to a particular spot on earth, but with a set of corner reflectors there was considerable leeway in the alignment. (The point of bouncing a laser beam back was to measure the transit time and hence monitor very precisely—to a few centimeters!—the earth-moon distance.)

Since most mirrors have a glass surface covering the reflecting silver, you can also get several reflections from one mirror. Most of the light is reflected from the silver, but some is reflected from the front glass surface. At normal incidence

*Latin *retro*, backward.

FIGURE 2.41

(a) Array of corner reflectors placed on the moon. Note that they look black—they reflect the black sky behind the camera; the sun is to the right (note shadows), and the moon has no atmosphere to scatter the sunlight (hence the black sky). **(b)** Telescope at McDonald Observatory, Mt. Locke, Texas, being used "backward" to send a laser beam to the corner reflector on the moon. Note that stars have become short streaks due to the long exposure necessary to make the laser beam visible, even though the beam was made narrower than normal to make it sufficiently bright to be photographed.

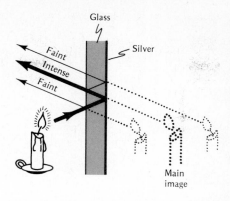

FIGURE 2.42

A single mirror can form several images of a candle. If both the candle and the viewer's eye are held close to the glass surface, the better glass reflectivity at grazing incidence makes the extra images brighter.

the glass reflection is weak (4%), but at grazing incidence ($\theta_i \simeq 90°$) it is comparable to that from the silver. Further, the light reflected from the silver can be partially reflected back to the silver by the front glass surface, so a sequence of images can be obtained somewhat like those from the two parallel barbershop mirrors, but of much more rapidly decreasing brightness. Using a match or a candle as a light source in front of a mirror in an otherwise dark room (Fig. 2.42), you may be able to see three or four images of the candle (Fig. 2.43).

A nice example of multiple reflections is the nineteenth-century toy "X-ray Machine" (Fig. 2.44). You can "see through" any coin with this four-mirror device. A simpler device, using only two mirrors, is the periscope,* which allows one to see over obstacles or around corners. Multiple reflections are commonly used in light devices of modern kinetic art and light shows (Fig. 2.45). Often the reflections are off crumpled aluminum foil, so you don't see images but rather patterns of light that may change as the system rotates.

*Peri, Greek prefix meaning around, about. To make a periscope, see the TRY IT.

FIGURE 2.43

A candle and its multiple reflected images.

FIGURE 2.44

A toy "X-ray Machine." The diagram reveals the secret.

One needn't have specular reflections to have multiple reflections. Diffuse reflections are primarily responsible for the subtle lighting, shading, and coloring captured by artists. Leonardo da Vinci advised painters to notice these reflections carefully, and "show in your portraits how the reflection of the garments' colors tints the neighboring flesh. . . ."

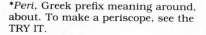

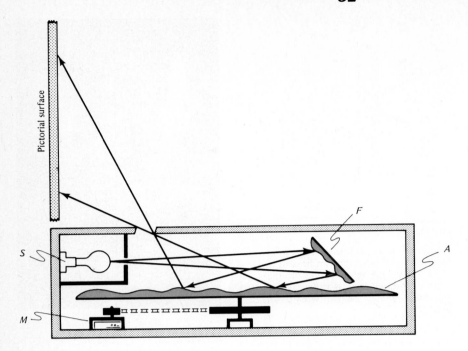

FIGURE 2.45

The "Kinoptic System" by Valerios Cabutsis. Light from source *S* is reflected by fixed mirror *F* onto crumpled aluminum foil *A*. As motor *M* rotates foil, the pattern of light on the screen changes.

TRY IT

FOR SECTION 2.4C
Fun with two small mirrors

Two small pocket mirrors (preferably rectangular, without frames) will enable you to demonstrate several of the ideas discussed in this section. Hold the two mirrors at a right angle, and look at your reflection (Fig. 2.39). (You can be sure that the two mirrors are actually perpendicular to each other by adjusting the angle between the mirrors until your face looks normal.) Move your head a bit, and see if you can avoid looking into your own eyes. Close one eye and see which eye closes in the reflected image. Turn your head slightly to one side. Which way does the reflected image turn? Is this what you are used to seeing in a single mirror? Draw light rays to explain the difference. (If you place an object, such as your thumb, between the two perpendicular mirrors, you'll see more than one reflected thumb. The number of thumbs will increase as you make the angle between the mirrors smaller—see Sec. 3.2B.)

The same two mirrors can be used to construct a **periscope.** These devices, popularized in World War II submarine movies, allow you to see around corners or over obstacles. To make a periscope, you will need a cardboard tube. Make one of square cross section (Fig. 2.46) or use a mailing tube. The size of the tube must be such that the mirrors will fit in as shown in the figure. Slits cut at 45° in

one side of the tube will enable you to get the mirrors in. The mirrors should be held in place with tape (or chewing gum), and the entire tube (including the top and bottom) should be taped shut to eliminate stray light. Two holes, on opposite sides of the tube, one in front of each mirror, should be cut in the tube.

FIGURE 2.46

Design for a periscope.

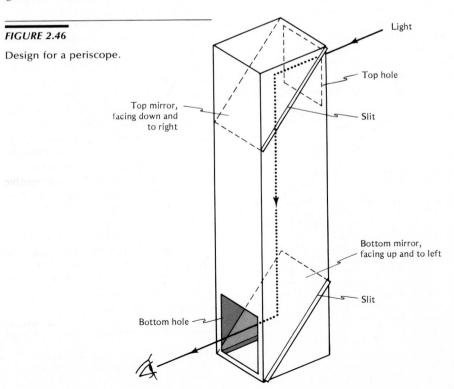

2.5
REFRACTION

We've mentioned that whenever there is a change of medium and an associated change in the speed of light, some of the light is reflected and some of it is transmitted. The transmitted beam usually does not continue to travel in exactly the same direction as the incident beam. This bending of the transmitted beam at the interface between two media is called **refraction*** (Fig. 2.47).

To find out *why* the lightbeam is bent, we must go beyond geometrical optics and again consider waves. The key idea is that light travels more slowly in a denser medium than in a less dense medium (Sec. 1.3A). In glass it travels at about ⅔ the speed it has in air. Now consider a wavefront of a beam (e.g., a crest of the wave) headed toward the glass (Fig. 2.48). A wavefront that starts out at *AA'* will get to *BB'* after some little time, traveling with the speed of light in air. At that moment the left end of the wavefront, near *B*, enters the glass.

---

*Latin *re-fringere*, break away. The light path has a break in its direction.

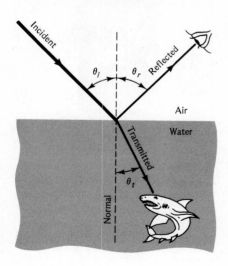

FIGURE 2.47

A beam incident on a transparent medium is split into a reflected and a transmitted (refracted) beam.

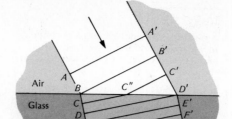

FIGURE 2.48

The wavefronts of a beam entering a slower medium explain why the beam refracts.

During the time the right end (near *B'*) moves the distance $\overline{B'C'}$ in air, the left end, moving more slowly in the glass, moves only the smaller distance $\overline{BC}$. The wavefront then looks like *CC"C'*. As the right end continues to move at the faster (air) speed to *D'*, the left end moves at the slower (glass) speed to *D*. Finally, the entire wavefront is in the glass and moves at the slower speed from *DD'* to *EE'* and beyond.

Thus, the edge of the wavefront that first hits the glass slows down, and the beam pivots about the edge, like a wagon pivoting around the wheel that is stuck in the mud. (Remember that the beam moves perpendicularly to the wavefront.) Hence the angle θ_t of the transmitted beam is *less* than the incident angle θ_i. That is, the beam is bent toward the normal at the surface. If *CC'* and *DD'* are successive crests, you can see that the wavelength (the distance between successive crests) is smaller in glass than in air, as it should be.

Had we sent the beam in the opposite direction, out of the glass, the diagram would be the same except the arrows would be reversed, and we would interchange the names "incident" and "transmitted." The first part of the wavefront to break free of the glass would speed up in the air, and the beam would pivot to the left, *away* from the perpendicular to the surface. So we find that:

Light going from fast medium to slow bends toward the normal, and light going from slow medium to fast bends away from the normal.

This is a qualitative statement of the **law of refraction,** also known as **Snell's law.**

The mathematical form of Snell's law (Appendix B), relating θ_i and θ_t, depends on the ratio of the two speeds of light in the two substances that form the refracting interface (air and glass in our example). Since only this ratio is important, we often specify the speed of light in a medium by comparing it to *c*, the speed of light in vacuum. So, if *v* is the speed of light in some material, we define the medium's **index of refraction** *n* by:

$$n = \frac{c}{v}$$

Thus the larger *v* is, the smaller *n* is. In vacuum *v* equals *c*, so *n* equals 1. For all other media, *v* is always less than *c*, so *n* is greater than 1. The index of refraction is a measure of the density of the medium, as far as the behavior of light is concerned. Using it avoids the unwieldly large numbers associated with the speed of light (Table 2.4).

TABLE 2.4 *Approximate index of refraction for various media*

Medium	Index of refraction
Vacuum	1 (exactly)
Air	1.0003
Water	1.33
Glass	1.5
Diamond	2.4

We can now restate Snell's law qualitatively thus:

Light going from small *n* to large *n* is bent toward the normal. Light going from large *n* to small *n* is bent away from the normal.

(Of course if the light is going *along* the normal, it is not bent at all.) Ta-

ble 2.5 gives an idea how much the beam is bent.

As an example of refraction, suppose we look at a fish in water (Fig. 2.49). The light coming from the fish at A was bent at the surface, away from the normal since n_{water} is greater than n_{air}. But the eye always assumes that the light has traveled in a straight line, so it looks as if the fish is at B. If you stick your spear in at B, you will likely miss the fish. For the same reason, a partially submerged straight stick looks bent—and therefore looks, correctly, as if it missed the fish and went to C instead (Fig. 2.50). This is of course not a refraction of the *stick*, but rather a consequence of refraction of the *light* we use to look at the stick.

FIGURE 2.49

Refraction makes underwater objects appear to be where they aren't.

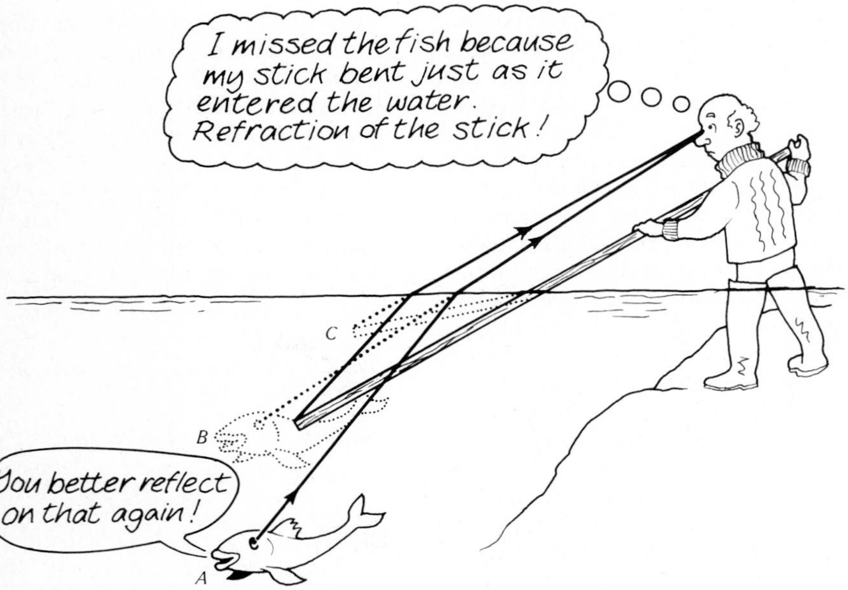

FIGURE 2.50

A pen appears bent when half immersed in water. (Notice the reflection on the surface of the water of the upper part of the pen.)

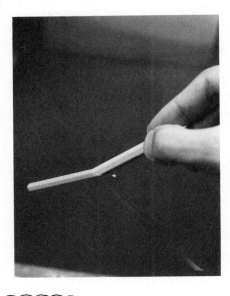

If there are ripples in the water, there will be a different amount of bending at each part of the surface and the fish may appear broken up, or you may even see several fish. A piece of glass made with many facets can, similarly, result in many images when you look through it, as described by John Webster in 1612 in "The White Devil":

I have seen a pair of spectacles fashioned with such perspective art, that lay down but one twelve pence o'th'board, 'twill appear as if there were twenty; now should you wear a pair of these spectacles, and see your wife tying her shoe, you would imagine twenty hands were taking up of your wife's clothes, and this would put you into a horrible causeless fury. (Fig. 2.51.)

As another example, refraction in a thick-walled beer glass makes the walls look very thin, so you think you're getting more beer (Fig. 2.52).

A. Total internal reflection

Notice that some entries are blank in Table 2.5. The reason is that a beam that is trying to enter a faster

FIGURE 2.52

Mug shot shows deception. Not only is there hardly any beer in the foam—which nonetheless reflects a lot of light—but the beer seems to go all the way to the side of the mug, not showing the thick wall. In the photograph, refraction in the lower half of the mug has been reduced by placing the mug in water. (Because *some* refraction occurs at the glass-water surface, the walls still look thinner than they are.)

I missed the fish because my stick bent just as it entered the water. Refraction of the stick!

You better reflect on that again!

TABLE 2.5 *Angle of transmission, θ_t, for various angles of incidence, θ_i, and various media*

From	To	$\theta_i = 15°$	30°	45°	60°	75°	89°
Air	Water	11°	22°	32°	41°	47°	49°
Air	Glass	10°	19°	28°	35°	40°	42°
Air	Diamond	6°	12°	16°	20°	23°	24°
Glass	Water	17°	34°	53°	78°	—	—
Glass	Air	23°	49°	—	—	—	—

See Sec. 2.5A for a discussion of the blank entries.

FIGURE 2.51

(a) Wherein one twelve-pence coin appears as three. Normals to glass-air surfaces are shown as dashed lines. Apparent positions of extra coins are shown in dotted outline, in the directions from which rays *a* and *c* come. The eye also sees a coin straight through, in the direction of the actual coin, via ray *b*. More facets on the glass may produce still more images. **(b)** ''Your wife tying her shoes.''

Eye

a *b* *c*

Air

Glass

Air

Twelve pence

(b)   (a)

medium (smaller n) at a larger angle of incidence would have to be bent away from the normal to more than 90°. But that is impossible; the transmitted beam cannot bend more than 90° from the normal and still stay in the faster medium. Light, therefore, gives up the idea of transmission—nothing is transmitted, and everything is reflected. This is called **total internal reflection.** It only occurs at sufficiently large θ_i at the inside (internally) of a dense medium (larger n) bordered by a less dense medium (smaller n). Of course the law of reflection, $\theta_r = \theta_i$, holds for this as for all reflections. The angle of incidence where total internal reflection first happens is called the **critical angle,** θ_c. From Table 2.5 you can see that the critical angle for glass in air is somewhere between 30° and 45°—actually it is about 42°. So any light hitting a glass-air surface, from the glass side at more than 42° to the normal, is totally reflected (Fig. 2.53).

PONDER

Why isn't there a critical angle for light passing from air into water?

A simple place to see total internal reflection is in a fish tank or swimming pool. The light from the fish in Fig. 2.54 is totally reflected at the top surface of the water, and the swimmer sees the fish's reflection there. Similarly, the fish sees the swimmer's reflection there. If the fish looks up at less than the critical angle, she sees out the top, but everything above the water, down to the horizon, is compressed into the angles between straight up and the critical angle (Fig. 2.55). Below the critical angle the view suddenly shifts to a reflection. It is like seeing the world squeezed together through a hole in a mirror (distorted by ripples).

Total internal reflection is used in some optical instruments to reflect a beam by 90° using a prism cut at 45° (Fig. 2.56). Several of these are used in prism binoculars to increase the path length between the

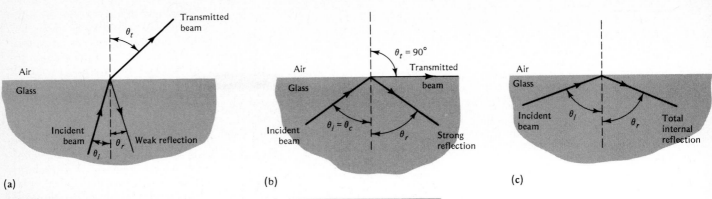

(a) (b) (c)

FIGURE 2.53

Reflected and transmitted beams when incident angle is **(a)** smaller than the critical angle, **(b)** equal to the critical angle, **(c)** larger than the critical angle.

FIGURE 2.54

The view from under water is strange indeed.

FIGURE 2.56

Use of the total internal reflection to change the direction of a light beam, **(a)** by 90° ($\theta_i = \theta_r = 45°$), **(b)** by 180° using a Porro prism. **(c)** Two Porro prisms as used in binoculars (actually, they would be touching).

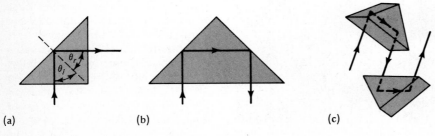

(a) (b) (c)

FIGURE 2.55

An underwater photograph of a doll standing in water, showing both direct and reflected views of her legs and hands.

lenses and also to invert the image (Sec. 6.4B).

PONDER

Would the devices of Fig. 2.56 work under water?

*B. Fiber optics

Another application of total internal reflection is *fiber optics*—the use of thin flexible glass or plastic fibers as **light pipes** (Fig. 2.57a). You send light down the fiber, which has a larger index of refraction than its surroundings. If the light in the fiber hits the surface at an angle of incidence greater than the critical angle, it is totally internally reflected, and there are no losses due to light escaping. The light continues bouncing back and forth down the fiber, even if the fiber bends. As long as the fiber does not bend too sharply (so the incident angle is always greater than the critical angle), the light is gradually bent with the fiber.

You see this effect in fountains where the "fibers" are streams of water, in advertising displays, and in "decorative" lamps (Fig. 2.58). These light pipes are easily adjust-

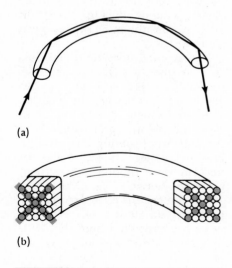

(a)

(b)

FIGURE 2.57

(a) A glass or plastic fiber can be used as a light pipe. (b) Many light pipes packed together can transmit an image.

FIGURE 2.58

A fiber optic lamp.

able light sources for miniature photography, and are used as tiny window lights on spacecraft models in science fiction films. (The TRY IT gives another example.)

A bunch of such fibers, suitably clad so the light does not leak from one fiber to the next, can be used to transmit *images.* At the original object, some fibers are exposed to light, and others are not. At the other end we get an array of fibers lit up in some places, dark in others, corresponding to the original image (Fig. 2.57b).

If the fiber is used to transmit a light signal consisting of a rapid series of pulses, it is important not only that all the light from a given pulse reach the other end of the fiber, but that it all arrive there at *one time.* Otherwise the pulse becomes spread out, and might blur with the next pulse. **Graded fibers** with varying index of refraction can be made so that the time for rays going straight down the center is the same as that for rays bouncing from side to side. Such fibers are replacing metal wire for telephone lines and for data transmission

(Fig. 2.59). In general, the higher the frequency of a wave used for communication, the more information can be transmitted per second. Since the frequency of visible light is much higher than radio frequencies, these fibers can carry many more conversations at the same time. Further, they don't radiate signals, so they can't be bugged without directly connecting to them.

The ability to "bend" light and thus see around corners without the rigidity of a periscope allows doctors to see and photograph the inside of vital organs in living persons. Some of the fibers are used to get light inside the living organ, and the rest of the fiber bundle to get the image out.

Another example, at the opposite end of the scale of technological sophistication, is the "wet tee shirt" phenomenon. Here light pipes of water are formed through the cloth, allowing an image of something in contact with one side of the cloth to be transmitted to the other side. (The water also provides a more gradual change in index of refraction between the air and the cloth, so there is less reflection.) The principle works both ways—you can get a sunburn in a wet shirt.

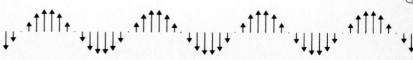

FIGURE 2.59

Photograph of a 12 × 12 array of light pipes, used to transmit laser light coded with telephone conversations. Such a cable can carry 40,000 voice circuits.

Nature also thought of this idea. Light is transmitted down the eyes of insects by total internal reflection through a bundle of light pipes, called *ommatidia*, which take the light incident on the eye and bring it to the light-sensitive cells. A similar process occurs in the cones of our eyes (see Sec. 5.3B). Bundles of fibers are even used in oats and corn, where they transmit light into the seedling plant to assist various photochemical processes.

TRY IT

FOR SECTION 2.5B
Light trapped in glass by total internal reflection

A piece of window glass can be used as a light pipe to demonstrate total internal reflection. Look at a small light source, such as a lightbulb, with the glass tilted between your eye and the bulb so that the light reflects from the glass to your eye. As you tilt the glass more, to make the reflection be almost at grazing incidence, the reflection should become quite good, indistinguishable from a reflection in a mirror. However, if you look at the edge of the glass toward you, you should also be able to see light emerging that has been trapped and reflected back and forth within the glass. This light didn't enter through the face of

the glass, but rather from the opposite edge (the edge closest to the bulb). Check this by covering that edge with your finger, without blocking the surface reflection. If the light bulb is far enough away, you may also be able to see light emerging near the sides of the edge closest to you. This light will have been internally reflected from the side edges of the glass.

*C. Mirages and atmospheric distortion

We have been saying that the index of refraction of air is 1. That is not exactly true—it is about 1.0003, and depends on the air's temperature, density, and other properties. (For example, the index changes from 1.0003 to 1.0002 when the air temperature is increased by 100°C.) These properties of air can change from place to place, so when light propagates through the atmosphere it usually experiences a continually varying index of refraction.

When the index of refraction of a medium varies smoothly (rather than abruptly as it does at the surface of glass or water), parts of each wavefront of light move faster than other parts (as in Fig. 2.48), so the wave bends. Here, however, the smooth variation in the index of refraction causes a smooth change in the wave direction (rather than the abrupt change of Fig. 2.48). In other words, the light ray bends in a curve.

This gradual bending of light occurs frequently. One example is that we see the sun after it has set—when it is beyond the horizon. The atmosphere bends the light rays, but we interpret things as if the light were coming in a straight line, so we "see" the sun above the horizon, even though it is actually below the horizon (Fig. 2.60).

You can see the effects of air temperature on index of refraction more dramatically where the air is shimmering above a hot jet engine or candle flame. Light passing through the rising and wiggling hot air is bent by varying amounts, depending on the nonuniform temperature of the air through which it passes. Frequently, a dark asphalt

FIGURE 2.60

We can see the sun after it has set below the geometrical horizon, even if we're not in love. The atmosphere is denser toward the bottom, less dense toward the top. The gradual change in density produces a gradual change in index of refraction, which bends the light.

FIGURE 2.61

(a) A (rather extreme) mirage created by cool air above hot air. To analyze the mirage in detail, look through the magnifying glass; since ray *a* travels faster through the hotter air than ray *b*, *a* gets ahead of *b*, hence the beam gradually bends upward. The observer sees the tree both directly by means of rays *d*, *e*, *f*, through the cool air, and indirectly by means of the bent rays *a*, *b*, *c*. Thus, he sees a (somewhat compressed) "reflection" of the tree. In this particular example, the parts of the tree below ray *d* are invisible to the observer.
(b) Photograph of a mirage. The (compressed) reflected image prevents us from seeing the bottom of the mountain.

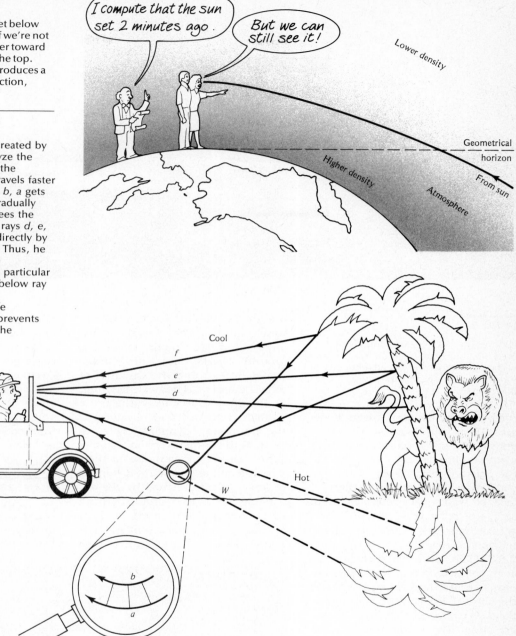

(a)

road will get very hot in the summer sun. When this happens, the air next to it may become much hotter than the higher air. This, in turn, can cause light to be bent so much that it appears reflected (Fig. 2.61)—a **mirage.*** In a rough way, you can think of the effect as a gradual total internal reflection at

*French *se mirer*, to be reflected.

(b)

FIGURE 2.62

Photograph of a mirage as frequently seen on a hot road. Note the snow in the background. You don't need a hot day for a mirage—only a temperature *difference*.

the boundary between hot and cool air. When you look at the road ahead, you see a reflection in it of whatever lies ahead. You also see the object directly. Since you associate such reflections with the surface of water (the most common source of reflections on roads), you get the impression that there is water on the road ahead of you (Fig. 2.62). (The water would seem to be at the place where the bent rays appear to cross the road surface, near the point W in Fig. 2.61.) Of course, the water never materializes as you get close. The closer you get, the more sharply the rays must be bent to reach your eyes. But, as the light is only bent by a certain amount, and not more, it misses your eyes when you get close and the "reflection" vanishes. This is rather disappointing if you happen to be thirsty in the desert.

The appearance of "water" under conditions of hot ground has suggested to Alistair B. Fraser an explanation for the parting of the Red Sea for Moses. The idea is that nobody was in the Red Sea, but rather

out in the desert, and the waters were a mirage. The "waters" apparently receded at night, when the ground cooled and the mirage would retreat, and the Children of Israel "crossed." The next morning, as the ground became warmer, the "waters" returned and "swallowed" the pursuing Egyptians. This, at least, is the Jewish version (Exodus 14:20–28). The Egyptians, for their part, presumably saw the Children of Israel being drowned by the advancing "waters" and, figuring there was no need to chase them anymore, went home. Similarly, the British lost the Turks in a mirage

during the First World War, and had to call off the battle.

There are various types of mirages possible, some of them quite spectacular (Fig. 2.63). The details differ depending on whether the hot air is above or below the cool air, and on how sharp the transition is from cool to warm air. If warm air is below (say, at the surface of a

FIGURE 2.63

A series of photographs of a ferryboat.
(a) The undistorted boat.
(b) to **(g)** Various mirages distort the boat in different ways.

(a)

warm lake on a cool morning), light is bent up, as in Figure 2.61, so that objects may disappear below the apparent horizon even though they are actually quite close. A person walking on a flat island in the lake may appear to be walking on water when the ground under his feet is depressed below the observer's horizon by such a mirage (cf. bottom part of tree in Fig. 2.61). If warm air is above cool air, you can see ships that are located "beyond" the horizon (like the sun in Fig. 2.60), seemingly floating in air, like that of the Flying Dutchman. (These effects are best seen if you look at the horizon with your eye very close to the water.) If the air temperature changes sufficiently gradually, the ground (or, more typically, the surface of the water) can appear to loom up like a wall, greatly magnified and sufficiently distorted so that it may look like mountains, or castles in the sky. This **fata morgana** mirage derives its name from Morgan le Fay, King Arthur's evil sister who could actually make castles in the sky.

More recently, in 1906, Commander Robert E. Perry discovered an area of peaks and valleys west of Greenland, which he named "Crocker Land." In 1913, Donald MacMillan confirmed the discovery, only then to watch Crocker Land disappear as he approached it. The warmer air over the colder air had produced a fata morgana, and the conditions were common enough in the Arctic that two explorers (and many Eskimos) had seen it. Indeed, so common is this effect that it has been suggested that various early discoveries, where explorers set off in just the right direction, such as the Celtic discovery of Iceland and Eric the Red's discovery of Greenland, might have been aided by such Arctic mirages, which would enable these people to see these places beyond the usual horizon.

Mirages in combination with the earth's curvature can distort the apparent shape of the sun; it can look flatter, have wiggles in its outline, or even have horizontal gaps in it (Fig. 2.64). Effects of the curvature of Venus are described, with

FIGURE 2.64

Photograph of setting sun shows both flattening and corrugations of solar limb due to uneven inversion layer.

considerable poetic license, in Tom Robbins' *Even Cowgirls Get the Blues:*

On Venus, the atmosphere is so thick that light rays bend as if made of foam rubber. The bending of light is so extraordinary that it causes the horizon to tilt upward. Thus, if one were standing on Venus one could see the opposite side of the planet by looking directly overhead.

2.6
DISPERSION

So far we have pretended that a substance like glass has one index of refraction for all kinds of light, no matter what the frequency. But the index of refraction, $n = c/v$, is specified by the speed of light in the glass, which depends, in turn, on the way the charges in the glass respond and radiate when they are wiggled. We have seen that the amplitude of the charges' motion depends upon the frequency with

which they are wiggled; the closer one gets to a resonance frequency, the more the charges will oscillate for the same applied force. Thus, the index of refraction will depend on frequency.

For glass and many transparent substances the resonance frequencies are in the ultraviolet (UV). (You don't get suntanned sitting in the sun behind a window because glass is not transparent in the UV—see Fig. 1.25.) As the (visible) frequency with which we wiggle the charges gets closer to the resonance (UV) frequency, the charges oscillate more violently and radiate more. We can then expect a greater effect on the light beam in the glass, which is the combination of the initial beam and the radiated waves. So the speed of light in glass should differ more and more from its vacuum speed. That is, the speed should get smaller and the index of refraction larger as the light changes from red to blue toward the UV:

$$1 < n_{red} < n_{blue} < n_{UV}*$$

*The notation A < B means A is less than B.

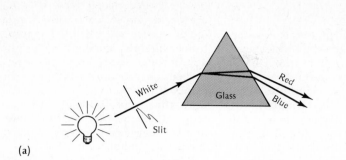

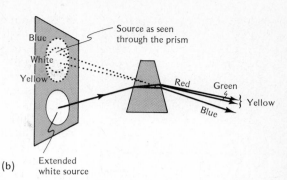

(a)

(b)

The dependence of n on frequency is called **dispersion.**† As expected, n is larger at higher frequencies for glass (as well as for diamond and water—Table 2.6). This difference of the index of refraction of a piece of glass for different frequency (or color) light means that the amount of bending that occurs when light hits a surface depends on the color of light—light is spread into the different colors when incident on a glass prism. Since n_{blue} is greater than n_{red}, the blue light is bent more than the red at each surface of the glass (Fig. 2.65a, Plate 2.1).

Newton showed that white light is made of all the different colors of the spectrum by breaking up white light with a prism and then recombining it into white light with another prism. However, if you look through a prism and expect to see everything in brilliant color, you will be disappointed. You will find that objects viewed through a prism acquire color only at their borders. For example, if you have a white area surrounded by black, and view it through a prism as in Figure 2.65b, you will see blue along the top edge of the white area, and mainly yellow with a little bit of red

†Latin *dispersio*, spreading out.

FIGURE 2.65

(a) Dispersion of colors by a prism—narrow source (also see Plate 2.1).
(b) View of *extended* white source through a prism.

along the bottom edge. The reason is that the border in blue light is shifted away from the border in the other colors, because the glass has its greatest dispersion at the blue end of the visible. Thus, the top edge looks blue. The rest of the colors of the spectrum are usually not well separated; all these other colors together combine to give yellow at the bottom edge (Sec. 9.4B), with the red peeping out just a bit. Any point well within the white area still is seen in all colors, and hence looks white, even though that light comes from different points in the original white object. Similarly, most of the black area receives no light, and hence still looks black. In order to examine the spectrum of a light source, therefore, you should place a narrow slit in front of it, and view the slit through your prism.

***A. Diamonds**

Diamonds have a number of properties that account for their high price (not the least of which is that their supply is controlled by the dia-

mond industry). Diamonds, as gems, have a long history of intrigue, passion, and superstition, and many powers have been ascribed to them. Diamonds could prevent nightmares, loosen your teeth, provide strength in battle, and repel phantoms. But the properties that concern us here are the ones that are responsible for their "brilliance," "fire," and "flash."

Diamond has a large index of refraction, $n \simeq 2.4$, so that the critical angle for total internal reflection is about $24.5°$—much smaller than that for glass. This accounts for the **brilliance** of diamonds. A diamond is cut (Fig. 2.66) so that almost any ray of light that hits it from the front strikes one of the rear surfaces at an angle greater than $24.5°$, is internally reflected to another surface, and eventually is reflected back out the front. Viewed from the front, then, the diamond is bright—brilliant. However, if you look from the back through a cut diamond at a light source, the diamond will appear black—almost no light passes out the back.

The brilliance of diamond can be imitated by glass if the glass is backed with metallic foil or silvered like a mirror, as that will also return the light to the front. However, the glass still has a major visual shortcoming compared to diamond; diamond is highly dispersive—the index of refraction varies considerably with frequency (Table 2.6). The blue is bent *much* more than the red, so white light is spread out into a broader spectrum than in glass. This accounts for the **fire** of diamonds—the beautiful colors.

No matter how you look at a cut

TABLE 2.6   *Index of refraction for various media*

| Frequency (hertz) | Wavelength (nm) | Color | Index of refraction | | | |
			Glass zinc crown	Glass light flint	Diamond	Water
4.57×10^{14}	656	Red	1.514	1.571	2.410	1.331
5.09×10^{14}	589	Yellow	1.517	1.575	2.418	1.333
6.91×10^{14}	434	Deep blue	1.528	1.594	2.450	1.340

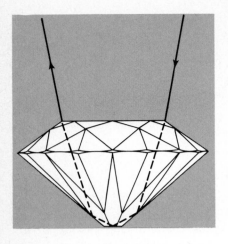

FIGURE 2.66

Most of the light entering a diamond is eventually reflected back out the front.

diamond, the chances are that you will get rays reflected from some source of light in the room, and it will be spread out into its spectral colors. As you move your eye slightly or rock the diamond, the view changes and you see some other rays that reach your eye by some other path (Plate 2.2). This causes the **flash** of diamonds; the motion causes the diamond to sparkle as the light reaches you from different sources. Diamonds are best shown off in rooms with lots of small lights or candles and mirrors, and must have been at their best when borne by the beauties in the ballroom of Versailles.

It is important to note that the bending of light and total internal reflection in diamond is due to the ratio of the index of refraction of diamond to that of *air*—the greater the ratio, the greater the refraction at a boundary. If the diamond is badly mounted it can get a film of oil ($n_{oil} \simeq 1.4$) on its back side, which lowers the ratio. Therefore, the critical angle for total internal reflection in an oily diamond is much higher than 24.5°. Since the diamonds are cut assuming $\theta_c = 24.5°$, the increased θ_c means a loss of light that should have been totally internally reflected—a loss of brilliance. The moral is: keep the back, as well as the front, of your diamond clean.

B. Rainbows

Because of the dispersion of water (Table 2.6), droplets of water can break up the sun's light into a spectrum, much as a prism does. This accounts for the **rainbow** (Fig. 2.67, Plate 2.3). Its formation involves not only dispersion of light on entering and emerging from the drop, but internal reflection as well. The result is that the drop reflects the sun's light, but the different colors emerge at different angles since, for example, blue light is bent more than red. If you are to see both red and blue light, these lights must come from different raindrops; those reflecting red light to your eye are at the higher angle, because red emerges more downward than blue. Thus, the rainbow is red at the top (outside) and blue at the bottom (inside), with the other colors in between.

An entire arc of water drops will look red to your eye: all those drops lying on a cone from your eye of about 42° around the direction of the sun's rays (Fig. 2.68). Similarly, drops along arcs at smaller angles reflect other colors to your eye. The entire rainbow thus appears as an arc ranging between 40° and 42°. Since this angle is measured from your eye, it is your own private rainbow. The light from these par-

FIGURE 2.67

A light beam is dispersed twice and reflected once by a raindrop, letting the eye see a rainbow. (The dispersion is exaggerated here.)

ticular water drops reaches *your* eye, and misses everyone else's. Others must see the light coming from different drops, those at the appropriate angles from *their* eyes. Indeed, as you move your eye, the rainbow moves with you to maintain the same angular relationship—it follows you as you move, and you never can get any closer to the pot of gold at its "end."

Notice in Plate 2.3 that outside the **primary rainbow** there is a **secondary rainbow.** This is caused by rays that undergo *two* reflections in the water drop (Fig. 2.69). Here, because of the extra reflection, the blue emerges more downward than the red. Hence in the secondary rainbow, the colors are reversed; the blue (at 54°) is outside the red (at 50°).

Look closely at Plate 2.3 and notice that the region between the two rainbows is darker than the region inside the primary bow. To explain this darker region (called Alexander's Dark Band), we must discuss some subtleties omitted in Figure 2.67. In that figure only one incident ray was drawn to each drop. Of course, there are many such rays hitting all over each drop (Fig. 2.70a). Each of these rays is refracted, dispersed, and reflected, as was the one drawn in Figure 2.67, though the one drawn there was a special ray. It was the ray that comes out at the steepest angle (e.g., 42° for red). All other rays of the same wavelength come out at shallower angles (Fig. 2.70a), but most of them come out *near* the steepest ray. Hence, the rays of a

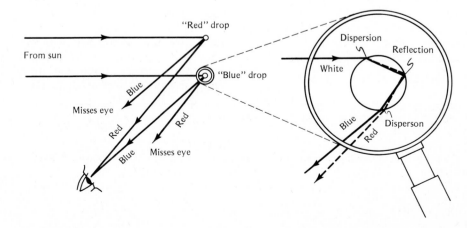

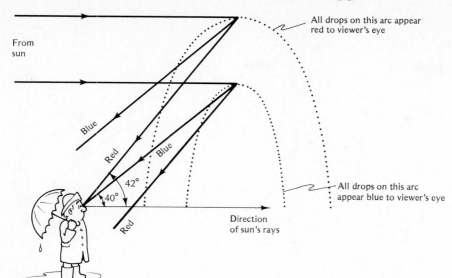

From sun

Blue

Red

Blue

42°

40°

Red

Red

Direction of sun's rays

All drops on this arc appear red to viewer's eye

All drops on this arc appear blue to viewer's eye

FIGURE 2.68

Only the drops along the rainbow's arc send light to a particular eye. All the drops on the upper arc appear red—the blue light they send out misses the eye. Similarly, all the drops on the lower arc appear blue.

wavelengths lie between these two. Inside the primary rainbow, then, all the wavelength disks overlap giving white light, so it looks brighter there. (Because the secondary rainbow is reversed, the outside is its brighter side, but that is a weak effect.)

Finally, since the sun must be behind you and the rain in front, the presence of a primary rainbow tells you something about the weather. As old English folklore has it: "A rainbow at night, fair weather in sight; a rainbow at morn, fair weather all gorn." At night (evening) the sun is in the west, so the rain must be in the east. As the weather in the northern hemisphere travels from west to east, the rain has passed, the good weather is coming. A rainbow in the morning has rain to the west, heading toward you.

FIGURE 2.70

(a) Light from the sun falls on the raindrop at all levels. Most but not all of the red rays from this light are returned by the raindrop at about 42°. In the magnified view, the arrow indicates the special ray that is returned *at* 42°. **(b)** The eye sees a red disk, brightest at the rim. **(c)** Superimposed on the red disk is a smaller blue disk (as well as disks of intermediate colors).

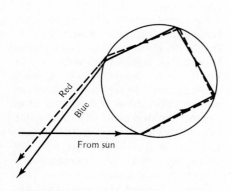

Red

Blue

From sun

FIGURE 2.69

The secondary rainbow is caused by two reflections and two refractions in each raindrop. The extra reflection reverses the order of the colors.

given wavelength form more than an arc; they form a disk that is brightest at the rim and fades toward the middle (Fig. 2.70b). The disks from the different wavelengths overlap, with the red (being the widest angle disk) sticking out the most. For example, the blue disk overlaps the inner part of the red disk (Fig. 2.70c). All the other

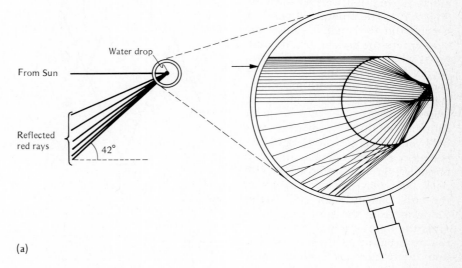

Water drop

From Sun

Reflected red rays

42°

(a)

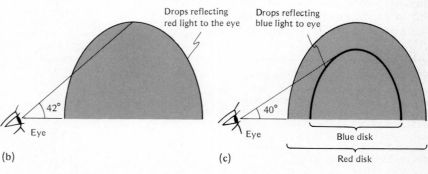

Drops reflecting red light to the eye

Drops reflecting blue light to eye

42°

Eye

(b)

40°

Eye

Blue disk

Red disk

(c)

FIGURE 2.71

A private rainbow for each person.

(a)

*C. Sun dogs, 22° halos, and more

Hexagonal ice crystals in the atmosphere refract light more nearly like a prism than raindrops do. The same flat hexagonal plates floating horizontally that cause sub suns or sun pillars by *reflection* off their horizontal surfaces can also *refract* light entering their vertical surfaces (Fig. 2.72). The amount of bending depends on the angle at which the light enters the crystal, but no matter at what angle it enters the hexagonal ice crystal, the light will be bent by at least 22°, with most light bent by about 22°. (22° is about the angle between the tips of your thumb and little finger when you hold your hand at arm's length and spread out your fingers as much as possible.) So in addition to the sun itself, you may see two images of the sun, one on each side, about 22° away from the sun. Dispersion gives the images colors, but usually not as brilliant as those of a rainbow (more rays are bent by more than the minimum amount, so the different colors mix together and tend to wash out). Either of these two images is called a **sun dog***

---

*Probably from "to dog"—to follow. The sun dog is always near the sun. Sun dogs are also referred to as mock suns or parhelia. Presumably, the sun with two sun dogs (as in Fig. 2.73) is what Shakespeare refers to in *Henry VI, Part III* when Edward says, "Dazzle mine eyes, or do I see three suns?"

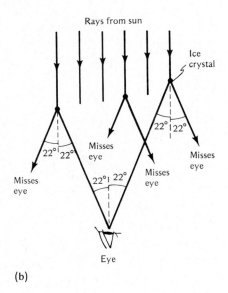

(b)

FIGURE 2.72

A sun dog is caused by refraction in hexagonal ice crystals. **(a)** As seen by an observer. **(b)** The path of a light ray through a crystal, as seen from above. Only those crystals that are at 22° from the eye send light to it. Light from all the other crystals misses this particular eye.

(Fig. 2.73). Sun dogs are visible only when the sun is close to the horizon, so that the rays can go through the horizontal, rather thin, plates of ice.

Ice crystals also occur in *long* hexagonal columns, which tend to float with their long axes horizontal (Fig. 2.74). Again the sun's light is bent by at least 22° if it goes through the hexagonal crystal as shown. This time you can look up toward the overhead sun, and since the crystal axes may point in any horizontal direction, you will get some light from all directions 22° away from the sun—you see a ring of light around the sun: the **22° halo** (Fig. 2.75). Since some light is bent by more than 22°, you see some light outside the circle, but since no light is bent less than 22°, it is dark just inside the circle—you see a halo that fades off outside the circle. The halo may be slightly reddish on the inside because of dispersion.

If the sun (or moon) is not exactly overhead, you may still see the 22° halo if the ice crystals are more randomly oriented. This occurs for somewhat shorter ice crystals, which tend to tumble as they fall. Other times you may see only a part of the halo. You may also observe

FIGURE 2.73

Photograph of the sun flanked by sun dogs on each side.

FIGURE 2.74

The 22° halo, caused by tiny pencil-shaped ice crystals. As in Figure 2.72b, only those crystals that are at 22° from the eye send light to it.

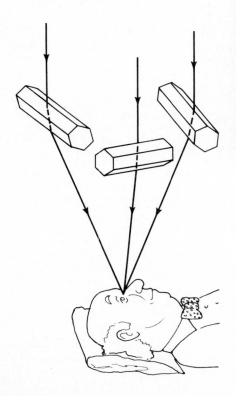

FIGURE 2.75

Photograph of a 22° halo.

strange arms leading off the 22° halo above and below the sun (or moon). They are due to other, less common, reflections and refractions off these crystals. For example, reflection off the *ends* of the long crystals of Figure 2.74, when the sun is low, produces a low arc in the sky called a parhelic circle.

These phenomena are relatively rare, but may be seen from time to time (more easily seen around the moon, against the dark night sky). To photograph them, try to get an obstacle (such as a tree branch) in front of the moon or sun, which would otherwise be far too bright and completely overexpose your film during the time it takes to properly expose the halo or arc. (You may also need a wide-angle lens—Sec. 4.3A.)

SUMMARY

As long as light encounters only obstacles much larger than its wavelength, **geometrical optics** may be used to describe the behavior of light **rays**. An obstacle in front of a **point source** casts a **shadow,** whose location can be determined by tracing the rays that just pass the edge of the obstacle. An **extended source** may give a region where all shadows overlap **(umbra)** and a region of partial shadow **(penumbra),** which receives light from some points on the source. **Eclipses** of the sun and moon are examples. From the penumbra we see a **partial eclipse.** By blocking all the light except that passing through a small hole, we make a **pinhole camera.**

In addition to traveling in straight lines, light can be **reflected** by objects. The reflections are used in **radar** (or **sonar** with sound waves). **Metals** are excellent reflectors of light at frequencies below their plasma **frequencies.** (The **iono-sphere,** with a low plasma frequency, reflects AM radio waves.) **Mirrors** are made by coating glass with silver metal. If the silver is thin enough, we have a **half-silvered** mirror, which appears transparent to someone on the dark side of the mirror, but appears reflecting to someone on the bright side. For light striking a reflector at an angle, the *angle of incidence equals the angle of reflection*. The reflection gets *stronger* at *grazing incidence.* **Sub suns** and **sun pillars** are reflections of the sun from flat plate, or long pencil-shaped, hexagonal ice crystals, falling almost *horizontally*. Light can be reflected more than once, as in a **corner reflector,** which **retroreflects** the rays. Reflections can be **specular** (obeying the law of reflection) or **diffuse** (reflections that go in all directions from a rough surface).

Light can also be **refracted—** bent as it enters a new medium with a different **index of refraction** ($n = c/v$). **Snell's law** states that *light going from small* n *to large* n *is bent toward the normal* to the surface; *light going from large* n *to small* n *is bent away from the normal.* If the **angle of incidence** is greater than the **critical angle,** when light is traveling *in* the *slower medium,* the light will be **totally internally reflected.** This principle is used in **fiber optics** to guide light beams. The bending of light as the index of refraction changes is responsible for a variety of **mirages,** depending on how the air temperature changes. The amount of refraction depends on the *wavelength* of the light **(dispersion).** A diamond's **brilliance, fire,** and **flash** are due to its *large index of refraction (large critical angle),* high *dispersion,* and the *cut* of the diamond, chosen to take advantage of these. **Rainbows** result from *dispersion* of the light entering and leaving water droplets, and one (for the **primary rainbow**) or two (for the **secondary rainbow**) *reflections* inside the droplets. *Refraction and dispersion* in hexagonal ice crystals produce **sun dogs,** the **22° halo,** and other meteorological optical phenomena.

PROBLEMS

P1 Stand a pencil point up on a piece of paper and illuminate it, from one side only, with a broad light source (a window, a fluorescent tube, etc.). With another pencil, trace the shadow formed, indicating which regions are umbra and which penumbra.

P2 Suppose the earth, in Figure 2.5, were farther from the moon, so that lines a and b crossed before reaching the earth. What would the eclipse look like when viewed from a point on the earth between those two (crossed) lines? Would that point be in a region of umbra or penumbra? (It may help to make a sketch and draw straight lines back from the viewing point to the sun to see which parts of the sun are visible, if any.)

P3 In "The Rime of the Ancient Mariner" Coleridge writes: "And on the bay the moon light lay,/And the shadow of the moon." (a) Under what conditions does the shadow of the moon touch the earth? (b) Is there any moonlight visible under these conditions? Why?

P4 Why did Etienne de Silhouette use a candle (point source) when making his silhouette tracings rather than a larger (and perhaps brighter) source?

P5 A camera using a lens has a much larger hole than the pinhole camera has. (a) Without knowing anything about lenses, why would you want to have a large hole rather than the tiny pinhole? (b) We've seen what happens to the image in the pinhole camera when the pinhole is made larger. What do you think the purpose of the lens is in a regular camera?

P6 Draw a 10-cm line across the top center of a piece of standard 8½ × 11 inch paper. Toward the left end of the line, print the word "Pinhole," and toward the right end, print the word "Camera." Now, at the bottom of the page, draw a square, 15 cm on each edge, with a 5-mm hole in the center of the top side of the square. This will represent your pinhole camera. The bottom edge of the square will be the film. The words at the top of the page are your objects. (a) Using a good straight edge, carefully find the location of the image of the word "Pinhole" on the film. Repeat for the word "Camera." (b) Notice that the pinhole is fairly large (5 mm). This would give a blurred image—the image of a point would be a spread out blur patch. Construct the blur patch that is the image of the dot on the i in "Pinhole." (c) To decrease the blur, one can make the pinhole smaller. What would be the *major* drawback of a pinhole camera whose hole was small enough so that the blur were acceptably small?

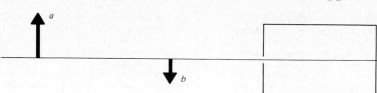

P7 The figure shows a pinhole camera photographing two arrows. (a) Which arrow will have the largest image? (b) Which arrow's image will be pointing up?

P8 The light of the pinhole image of the sun comes from the sun, reflects off the white screen, and gets to your eye. The eye, thus, still gets the sun's light. Why is the pinhole image less likely to hurt your eyes?

P9 If a cloud drifts over the sun from right to left, which way will it seem to drift over the sun's image in a pinhole camera?

P10 The figure shows a pinhole camera with the image of an arrow. The actual arrow is located somewhere outside the camera, is 2-cm long, and is vertical. Use a ruler to redraw the figure and construct suitable rays to (a) draw the position of the actual arrow that would result in the image drawn, and (b) draw two arrows, each of length 3 cm, that would result in the *same* image.

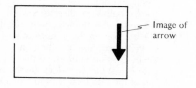

Image of arrow

P11 Write down at least five words that still make sense when reflected by a plane mirror. The reflected word need not be the same as the original. The words may be written or printed, in small or capital letters.

P12 Three different setups of eye, mirror, and object are shown in the figure. In which of the setups can the eye see a reflection of the object? Redraw the setups with appropriate light rays to support your answer.

P13 The man in the figure is looking at himself in a triple mirror found in a clothing store. Redraw the figure and show two *different* rays of light that go from his right ear to his left eye. One ray should hit only one mirror, the other ray should hit the other two mirrors. At each reflection, draw the normal to the mirror, and label θ_i and θ_r.

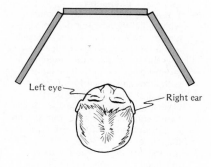

Left eye — — Right ear

P14 Briefly explain how an oceanographer might use sonar to measure the depth of the ocean.

P15 In some places, housepersons like to "spy" on what is going on in the street without leaning out the window. They attach a mirror to the window frame on a bracket so it sticks a little way into the street. (Look for these the next time you're in Europe.) (a) Draw a diagram showing how houseperson H, without getting up from his or her chair, can see the visitor V at the neighbor's home. Draw how the mirror must be placed, and show how light gets from V to H's eye. (b) Suppose H has no mirror, but does have a large 45°-45°-90° prism. H mounts the long face of the prism in the same position as you have drawn the mirror, and obtains a good reflection of V. What phenomenon is H using to get this

reflection? (c) Why can H hear V ringing the neighbor's bell when H cannot see V directly?

P16 At nighttime, we can often pick up radio stations from cities very far away. (a) Why? (b) Why does this occur for AM radio, but not FM radio?

P17 As you look at an ordinary bathroom mirror at large (grazing) angle, you can often see three, four, or more images of an object. Why?

P18 A student, who obviously didn't read this book, had a van whose window he had covered with half-silvered mirrors (actually aluminum on plastic). When he checked it out during the day, he assured himself that, while he could see out, no one could see in. That night he parked his van in a dark parking lot, climbed inside with his lantern and a friend, and enjoyed what he thought was the privacy of his own home. A passing police officer saw him through the half-silvered mirrors, and promptly busted him. Why could the police officer see in?

P19 When peering into a dark house from outside, you often get right up to the window and cup your hands around the side of your head. Why?

P20 In Figure 2.37, we have shown how light from the headlights of a car gets to the driver's eye. Redraw the figure for a very foggy day. (Fog consists of many small droplets of water. Draw some of these droplets.) Draw how the light now gets to the driver's eye, and explain why it is harder for her to see the road ahead.

P21 Blue light bends more than red light when entering glass from air because: (a) red light travels faster than blue light in glass, (b) blue light travels faster than red light in glass, (c) red light travels faster than blue light in air, or (d) blue light travels faster than red light in air. (Choose one.)

P22 (a) Relate the fire, flash, and brilliance of a diamond to physical properities of the diamond. (b) Why does a diamond look black when you attempt to look through it from the back at the only light in an otherwise dark room?

P23 (a) What is the critical angle? (b) Which pair of materials has the greater critical angle, air-water or air-diamond?

P24 Why does the sun appear above the horizon when it is actually below the geometric horizon during a sunset?

P25 James Morris, in *Heaven's Command*, describes a view from Grosse Isle in the Gulf of St. Lawrence as follows: "in the early morning sun the islands are inverted in mirage, and seem to hang there suspended between sky and water." In the early morning, the air gets warm faster than the water, so there is warmer air above the cooler air that is lying just above the

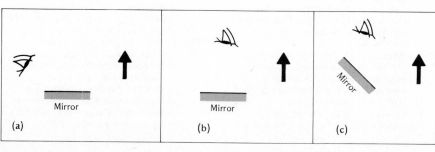

(a) Mirror

(b) Mirror

(c) Mirror

water. Sketch the air, water, and islands and show how the light rays bend, resulting in the mirage Morris describes. In particular, why do the islands seem "suspended?"

P26 In what direction can you see a rainbow early on a rainy morning (east, west, north, or south)? Why?

P27 In a riddle by the fifth-century Anglo-Saxon poet Aldhelm, the rainbow speaks of itself: "I am born red by the sun in the vicinity of a watery cloud. I shine here and there in the northern skies, but I do not climb through the southern skies." Explain the last sentence.

P28 Draw a sketch showing where you might see a rainbow in your lawn sprinkler in the late afternoon. Show the position of the sun, the sprinkler, and you.

P29 (a) Draw a suitable sketch showing where the ice crystals would have to be if you were to see a "sun" pillar from a bright, distant streetlight. (b) Draw another sketch to show where the ice crystals would have to be if you were to see a "sun" dog from the same streetlight.

HARDER PROBLEMS

PH1 (a) Draw a diagram to explain a solar eclipse, showing regions of umbra and penumbra. (b) What do you see when you look up while standing in the umbra? The penumbra? (c) What would you see if you were standing on the moon, looking at the Earth? (d) During a solar eclipse, some regions see a total eclipse, some a partial eclipse, and some no eclipse at all, yet everyone sees a lunar eclipse the same way. Why?

PH2 (a) Use a diagram to explain a lunar eclipse. Label the umbra and penumbra. (b) What would you see looking up at the moon? (c) What would you see if you were standing on the moon looking up at the earth?

PH3 The figure shows an extended light source and a screen. You have an object 1 cm in height, which you are allowed to place in various positions. When the object is very close to the screen (as shown), it will throw an umbra about 1 cm in height, and a very small penumbra. Redraw the figure, and show where to place the object so that: (a) the penumbra will be larger than the umbra, and

(b) there will be no umbra, only penumbra. Show the rays you use to construct the umbra and penumbra in the two cases, using different colors to keep the two cases apart.

PH4 If there were a radio station on Mars, would we be more likely to receive the Martian broadcast if it were at AM or FM frequencies? Why?

PH5 An eye *E* wants to see a bug *B* that has crawled down a long, narrow, dark tube. *E* can't use a light *L* to illuminate *B* directly because *E* and *L* would get in each others' way. Fortunately, a half-silvered mirror *M* is available. Draw where *M* should be placed to illuminate *B* allowing *E* to see it. Draw a ray from *L* to *M* to *E*, verifying that *E* can see *B*. (Photographers sometimes use this "coaxial illumination.")

PH6 The figure shows a pinhole camera with a slightly different design; it has a mirror mounted at 45° in the back, and a ground-glass screen for viewing is at the back top of the camera. (A ground-glass screen is a piece of glass with a rough surface.) An observer looks at the image of the usual object—an arrow. (a) Redraw the figure and trace the rays from the tip and from the tail of the arrow through the camera to the screen. (Be sure to use the law of reflection at the mirror.) Hence draw the image of the arrow on the screen. (b) Does the observer see the image upside down or rightside up? (c) Suppose the camera, pinhole, and glass are removed, but the object, mirror, and observer remain in place. In the same drawing, draw the image of the arrow in the mirror, where the observer now sees it, and label it "Image."

(b) If you rub some oil into the crack, it becomes much less visible. Explain. (The index of refraction of oil is close to that of glass.)

PH9 People walk into sliding glass doors because they see right through them. Birds fly into the same doors because they see the sky's reflection in them. Explain.

PH10 In *Tess of the D'Urbervilles*, Thomas Hardy writes: "As the looking glass was only large enough to reflect a very small portion of Tess's person at one time, Mrs. Durbeyfield hung a black cloak outside the [window] casement, and so made a large reflection of the panes, as it is the wont of bedecking cottages to do." How does Mrs. Durbeyfield's homemade mirror work?

PH11 Some sunglasses have half-silvered mirrors on them. (a) Why can the wearer see through them, but no one else can? (b) If you take them off, the reflection looks brighter and sharper from the front than from the back. Why? (Remember, glass absorbs some light.)

PH12 (a) A vengeful underwater fish is planning to poke a stick at a fisherman standing on the shore. Should the fish push the stick above, below, or directly toward where he sees the fisherman? Explain your answer, using a diagram. (b) The fish now uses a powerful laser as a weapon. Should he aim the laser above, below, or directly toward where he sees the fisherman? Explain.

PH13 Place a coin in the bottom of an empty teacup that is sitting on a table. Position yourself so that the coin is just barely hidden by the near rim of the cup (that is, you cannot

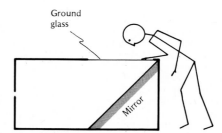
Ground glass

PH7 (a) Do you think a pedestrian would be more visible in heavy fog wearing dark or light clothing? Why? (See Problem P20.) (b) Should the motorist use the high- or low-beam headlights?

PH8 When a crack develops in glass, and the two edges of the crack separate by a small amount (as almost always happens), the crack becomes quite visible when viewed obliquely through the glass. (a) Explain why this happens—why the crack looks shiny.

see the coin). Hold your head in the same place while you fill the cup with water. Now you can see the coin. Explain, using a diagram. What physical principle is responsible?

PH14 Even *before* you've drunk your whiskey, a swizzle stick placed in it may appear bent or broken when viewed from the side. Draw a diagram of the glass, as seen from above, to explain this phenomenon. Draw the glass as a circle, and the (vertical) stick as a heavy black dot

Object

Light source

Screen

somewhere inside (but not at the center of) the circle. Draw a ray from the bottom half of the stick to your eye outside the glass (a ray that passes through the whiskey) and one from the top half of the stick, which is not in the whiskey. Show where the bottom and top halves of the stick would appear to be.

PH15 Draw a glass rectangle with a ray of light incident, at an angle, on one of the long sides. Continue the ray through the glass and let it enter an eye somewhere beyond the glass. How does the apparent position of the source of the ray compare with its actual position?

PH16 (a) Describe the appearance of a simple mirage. What atmospheric conditions are necessary for one? (b) What conditions, on the other hand, might allow Erik the Red to see Greenland hundreds of miles from Iceland, where he was staying?

PH17 You lie on your back in a bathtub, without any water in it, and look at a cup resting on a soap dish. Now somebody fills up the tub with water, so the water level is over your head, exactly at the top of the soap dish and the bottom of the cup. Sketch what your view of the cup and soap dish is after the water is added and before you drown. To help you do

this, you should draw a side view, showing how the rays from the cup and soap dish reach your eye.

PH18 Explain, using a diagram, how reflection off the ends of the long crystals of Figure 2.74, when the sun is low, produces a horizontal line in the sky—a parhelic circle.

PH19 Use a top-view diagram to explain Figure 2.52. Draw a ray from a point P on the outside edge of the beer (inside the glass) to a viewer's eye. Of course, the ray must bend when it gets to the outer surface of the glass. Draw the normal to this surface at the appropriate place, and make sure the ray bends in the correct direction. Show the apparent position, P', of the point P, as seen by the eye.

PH20 A glass beaker has a small round hole in its side. A laser beam shines horizontally through the beaker, out the hole, and strikes a screen. The hole is now plugged up and the beaker is filled with water. When the plug is removed, the water streams out the hole. The laser beam still shines through the hole, but no longer strikes the screen. Rather, there is a region of light on the floor where the streaming water is landing. Explain what has happened to the light, drawing a diagram and showing a light ray's path.

MATHEMATICAL PROBLEMS

PM1 Suppose you are 2 m tall and are standing halfway between a candle on the floor and a white vertical wall. (a) How high is your shadow on the wall? (b) If you shake a rope up and down at a frequency of 2 Hz and amplitude $\frac{1}{2}$ m, with what frequency and amplitude does your shadow shake the shadow rope?

PM2 (a) The speed of light in a certain glass is 200,000 km/sec. What is the index of refraction of this glass? (b) The index of refraction of a certain type of diamond is 2.5. What is the speed of light in this diamond?

PM3 Explain how you could use Snell's law to measure the speed of light in diamond.

PM4 A beam of white light in air is incident on a diamond, with an angle of incidence of 45°. (a) Using Table 2.6, calculate the angle at which red light is transmitted in the diamond. (b) Do the same for deep blue light. (c) What is the angle, from red to deep blue, into which the white is spread?

PM5 Use Table 2.4 to calculate the critical angle for light going from: (a) water toward air, (b) glass toward air, (c) diamond toward air, and (d) glass toward water.

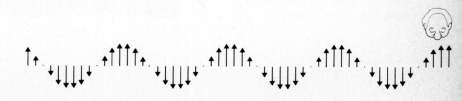

Mirrors and Lenses

CHAPTER 3

3.1
INTRODUCTION

In this chapter we want to apply the principles of reflection and refraction to situations of man's own creation. Here a central problem is to put together optical systems that deliver clear, undistorted *images*. The mirror and lenses used in such systems are most economically and simply manufactured if their surfaces have *spherical* shape, that is, the same shape as a part of an appropriately sized sphere. Therefore, we will mainly discuss such spherical mirrors and lenses.

To find the image in a particular mirror or lens of a point on an object, we must find out what happens to the light rays from this object point that strike the mirror or lens. This procedure is complicated because we have to apply the law of reflection or refraction to each of these rays. Fortunately, there are a few special rays whose reflections or refractions obey simple rules. The process of **ray tracing** uses these rays to find the image. Once we have located the image by this process it is an easy matter to find the path of any other ray. We'll illustrate this approach with the simple and familiar case of the flat **plane mirror,** and then go on to the trickier (and more interesting) case of curved surfaces.

3.2
VIRTUAL IMAGES

When a light ray from an object is reflected by a plane mirror and reaches your eye, your eye and brain make no allowance for this reflection; rather your brain traces rays back in straight, unbroken lines to the point they *seem* to come from. Thus, when you look into a plane mirror, the light seems to be coming from an image *behind* the mirror (Fig. 2.19). (A young kitten, when exposed to a mirror for the first time, will often try to go around behind the mirror in order the get at the "other" kitten.) That is, in the region in front of the mirror, the light behaves exactly as if it were coming from an actual object located at the position of the image (Fig. 3.1). How can we locate the

---

FIGURE 3.1

Where does the object stop and the virtual image begin? Or, ray tracing done with wooden beams. "Untitled," Robert Morris.

---

image, using the law of reflection? Let's trace a few rays.

A. Locating the image

Consider an object (the arrow PQ in Fig. 3.2) and an observer's eye, E_1, both in front of a mirror. Light shining on the arrow tip, Q, will be scattered in all directions from it. We can, therefore, consider point Q as a source of light. Consider one ray, QA, leaving the tip. By applying the law of reflection at point A, making the angle of reflection equal to the angle of incidence, we can determine the direction in which the ray travels after it is reflected. The reflected ray happens to go to the eye E_1. (Had we chosen a ray going to a different point on the mirror, say D, the reflected ray would go someplace else.) Where does the eye E_1 see the image of Q? As the eye

72

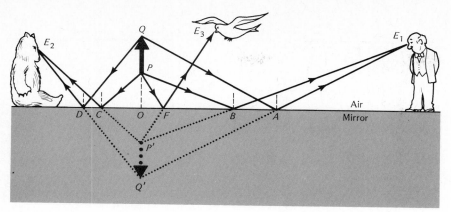

(a)

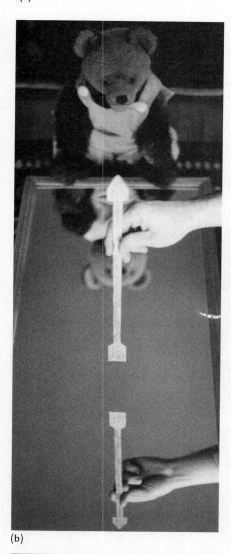

(b)

FIGURE 3.2

(a) Rays from an object that reach three different eyes can be used to locate the image. (b) Photograph taken from the position of eye E_3.

assumes that the light reaching it has always traveled in a straight line, it concludes that the source of the ray, AE_1, must be somewhere along the straight line, AE_1 extended. So far we cannot tell where along that line the image lies. If we repeat the construction, however, for a second eye (or the same eye moved), say E_2, we can construct ray QDE_2. Where do the *two* eyes see the image of the arrow's tip? E_1 sees it as lying somewhere along the straight line, AE_1 extended. E_2 sees it as lying somewhere along the straight line, DE_2 extended. The only point that lies on both these straight lines, and on which both eyes can agree, is their intersection Q'. The image of the arrow tip, Q, then must lie at the point Q'. It is, in fact, as far behind the mirror as Q is in front of it. (The distance from Q' to O is the same as from Q to O.)

In the same manner, we see that the image of the tail of the arrow, P, lies at P'. (Note that we didn't *have* to find rays that, after reflection, went to E_1 and E_2. Any pair of rays would do.) We have now located the image of the entire arrow. It lies between P' and Q'. A third eye, E_3, will also see the image of the tail at P' (try the construction). In fact, an eye at *any* position will agree that the image is at $P'Q'$—we have a true image. Knowing this, we need not bother with the law of reflection in order to determine how a ray gets to E_3. We simply draw a straight line from the image point, P', to E_3 (intersecting the mirror at F). Light

then actually travels in a straight line from P to F, and again in a straight line from F to E_3. The law of reflection is then automatically obeyed. The light rays always remain in the air, but they *appear* to come from the image $P'Q'$ behind the mirror. Such an image is called a **virtual image,** because the light doesn't really come from it, it only appears to do so. A virtual image is found by extending actual light rays back to it. There is no light at the point Q'—it might be embedded in a brick wall on which the mirror hangs. Nevertheless, it looks to all eyes as if Q' were the source of light.

You can convince yourself that your image is behind your bathroom mirror (rather than *on* it) by letting the mirror become steamed up and then tracing the outline of your head on the mirror with your finger. You will notice that it is only half the actual size of your head.

PONDER

Why does this mean that the image is behind the mirror, twice as far as the mirror is from you? (Notice that if you stand farther from the mirror, your image will just fit into the same traced outline, even though it now looks smaller.)

If you remain skeptical, you can measure the distance to your image if you have a camera with a range finder or focusing device. (Another way is described in the TRY IT.)

TRY IT

FOR SECTION 3.2A
Locating the virtual image

This way to locate the position of the virtual image produced by a plane mirror requires a small mirror and two identical pencils. Stand the mirror vertically on a table top. Stand one of the pencils upright, a few inches in front of the mirror. The pencil should stick up higher than the mirror, so its image appears cut off. Now position the second pencil behind the mirror so that it appears to be

an extension of the image of the first pencil. *Move your eye from side to side to make sure that the second pencil and the image of the first pencil look like one continuous pencil, no matter where your eye is.* The second pencil is now at the position of the virtual image of the first pencil, and you can measure its distance from the mirror. Compare that to the distance from the first pencil to the mirror. *(If you use a thick glass mirror, remember that the reflecting surface is the* back *surface of the glass.)*

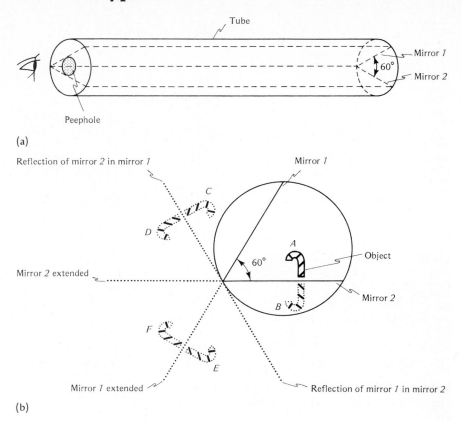

(a)

(b)

*B. Kaleidoscopes

As an example of a plethora of virtual images, let's consider the **ka-leidoscope*** (Fig. 3.3a). This is a long tube containing two mirrors running the length of the tube and set at an angle to each other. You look through a peephole between the mirrors at objects at the other end. If the angle between the mir-

---

*Greek *kalos*, beautiful, plus *eidos*, form, plus *skopeo*, see. Invented by David Brewster.

(c1)

(c2)

(c3)

(c4)

FIGURE 3.3

(a) Construction of a kaleidoscope. (b) A kaleidoscope view. An object, *A*, is placed between the two mirrors. You see the object directly. The object is also reflected in mirror *2*, so you also see a virtual image at *B*. As this is an image, the light appears to come from *B* just as if there were an object there. Hence both the object, *A*, and the image, *B*, are reflected in mirror *1* and their virtual images are located at *C* and *D*. (*D* is an image of an image.) Thus, looking in the peephole between the mirrors, you see what appear to be objects at *C* and *D* (outside the actual tube). Again, these are reflected in mirror *2*, and the new virtual images occur at *E* and *F*. The final pattern observed is then the object, *A*, and the five virtual images, *B* to *F*, arrayed in the threefold symmetric pattern shown. (c) Photograph of object reflected in two mirrors: (1) mirrors at 60°, (2) mirrors at 45°, (3) mirrors at 36°, and (4) mirrors at 50°, a nonintegral fraction of 360°.

rors is 60°, 45°, or 30° (or any other angle that goes into 360° an integral number of times) you see a symmetric array of these virtual images. For 60° you see a hexagonal field of view with a pattern of threefold symmetry, consisting of the object and five images (Fig. 3.3b). Figure 3.3c shows how some of these images come about.

If the mirrors are reasonably reflective, the pattern can be quite good, making it difficult to distinguish the original object from the images. When glass mirrors are used, they are usually front-surface mirrors to avoid the extra reflection one normally gets from ordinary glass mirrors, particularly at grazing angles. This is important in a long kaleidoscope, where most of the reflections are at grazing angles.

These toys are supposed to have been used by the weavers of Lancashire to inspire new design patterns for their cloths. They are often used in light shows to project interesting patterns. Walk-in kaleidoscopes (Fig. 3.4) are a source of amusement in fun houses. They are, truly, devices to "multiply variety in a wilderness of mirrors" (T. S. Eliot, "Gerontion").

FIGURE 3.4

A walk-in kaleidoscope made of mirrors on a wooden frame: Lucas Samaras, "Mirrored Room."

3.3
SPHERICAL MIRRORS

It was easy to locate the image in a plane mirror by following several rays from an object to the mirror. We can also apply this technique to **spherical mirrors**—pieces of reflecting material in the shape of part of the surface of a sphere. (We draw such surfaces on paper as parts of a circle. You will have to imagine them rotated about their central axes to visualize the spherical surface.) Such mirrors have quite different properties and different uses from plane mirrors, as we'll see.

Let's first understand the properties of spherical mirrors by examining a few special rays for which the law of reflection results in rather simple rules. The rules that we will write down are actually idealizations. But spherical mirrors are *not* ideal optical systems: the images they form are, in general, not perfect but slightly blurred. However, as long as both the position and angles of the rays that reach the mirror are sufficiently close to an **axis** (a line passing

through the center of the sphere), our approximate rules will be quite good and the images formed quite acceptable—considering only these so-called **paraxial** (or near axis) rays gives us a pretty good picture of what actually happens. In Section 3.4B, we'll discuss the blurring that results from the nonparaxial rays.

FIGURE 3.5

(a) Light rays illustrating the three rules, incident on a convex spherical mirror. Note direction of the rays—they are *not* all incident parallel to the axis! In this two-dimensional diagram and throughout the rest of the book, the three-dimensional spherical mirror appears as a part of a circle. **(b)** A shiny copper bowl makes a fine convex spherical mirror.

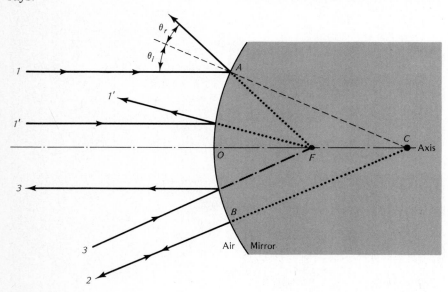

(a)

(b)

A. Convex mirrors

Figure 3.5 shows a spherical **convex** mirror (a mirror that bulges *toward* the source of light), with the center of the sphere marked C, and an axis that intersects the mirror at a point marked O. We have also drawn a number of paraxial rays. (Strictly speaking, these rays are not paraxial; they are too far from the axis, and some are at too steep angles. We draw them that way so that you can see them clearly.)

First consider ray 1, which comes to the mirror parallel to the axis. To apply the law of reflection, we need the normal to the surface at the point A, where the ray hits the surface. This is easily found because any radius of a sphere (a straight line from its center to its surface) is perpendicular to its surface. Thus we simply draw a straight line from the center, C, through the point A, and we have the normal at A (shown by a dashed line). The reflected ray is then constructed so that $\theta_r = \theta_i$, according to the law of reflection. When we now extend the reflected ray backward, we see that this extension crosses the axis at some point, labeled F. Were we to repeat this construction for another ray coming parallel to the axis, say ray 1', we would find that the reflected ray 1', when extended backward, intersects the axis at the same point F (at least within the paraxial-ray approximation). The point F is called the **focal point** of the mirror. The distance from O to F is called the **focal length,** f, of the mirror. For spherical mirrors, the focal length is just one half of the radius of the sphere (see Appendix C for a proof):

$$f = \overline{OF} = \tfrac{1}{2}\,\overline{OC}$$

so if we know the radius of the sphere, we know the focal length. We now have a general rule:

RAY 1 RULE
All rays incident parallel to the axis are reflected so that they appear to be coming from the focal point, F.

Within our approximation, this allows us to draw such a reflected ray quickly. We simply locate F and then draw the reflected ray as if it came from F. We need only a straightedge to do this—we don't need to measure angles once we've located F.

Now consider ray 2, which is originally aimed at the point C. As such, it arrives perpendicularly to the mirror at B, and therefore is reflected directly back on itself. (That is, $\theta_i = \theta_r = 0$.) This gives us, then, another general rule:

RAY 2 RULE
All rays that (when extended) pass through C are reflected back on themselves.

Finally, consider ray 3, which is aimed at F. It is just like ray 1 or ray 1′, except that it is traveling in the opposite direction. The angles θ_i and θ_r, being equal, don't depend on the direction the ray is traveling. So, if you reverse the arrows on a ray of type 1, you get a ray of type 3, which must therefore be reflected

parallel to the axis. Hence, our third general rule:

RAY 3 RULE
All rays that (when extended) pass through F are reflected back parallel to the axis.

With these three general rules, we are now able to use the techniques of ray tracing to locate the image of any object placed in front of a convex spherical mirror.

B. Locating the image by ray tracing

Using the three convex mirror rules, we can now construct the image of an object, PQ. As before, we treat the object point Q as a source of light. In Figure 3.6a, we've drawn ray 1, parallel to the axis from Q, reflected as if coming from F. In Figure 3.6b, we've drawn ray 2, aimed at C from Q, reflected on itself. Both these rays appear to come from the intersection of their extensions, the

point Q′. Thus, Q′ is the image of the point Q. In Figure 3.6c, we check this result by drawing ray 3, which leaves Q headed toward F, and is reflected back parallel to the axis. The extension of this reflected ray also passes through Q′, confirming our result. Now we can confidently mass produce other reflected rays that originated at Q simply by drawing them as if they came from Q′ (Fig. 3.6d). An eye, E, looking at the mirror sees the image of Q at Q′, as a result of the ray drawn.

What about the rest of the image? We could repeat this construction for every point on the object between Q and P (except the point P itself), but as the object PQ was per-

---

FIGURE 3.6

Construction of the image in a convex mirror: **(a)**, **(b)**, and **(c)** show the use of ray 1, ray 2, and ray 3 rules respectively; **(d)** shows how you can mass produce other rays—for example, one going to eye E—once you have found the image.

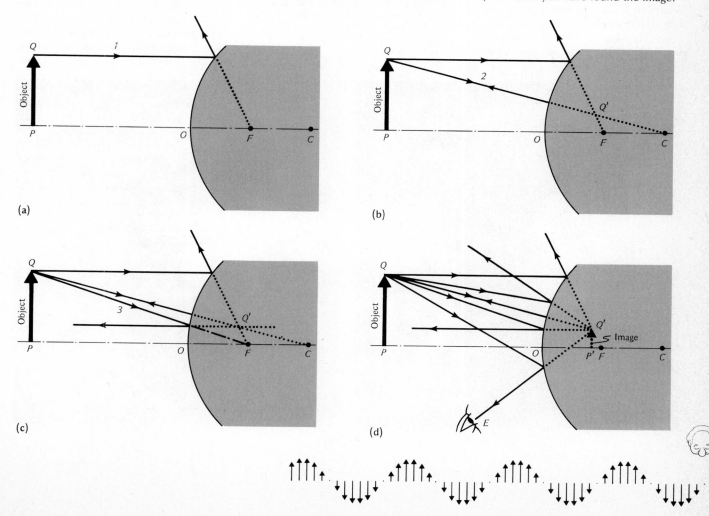

pendicular to the axis it is simpler to drop a perpendicular from Q' to find the image point P'. The image then lies between P' and Q' as shown in Figure 3.6d. (What happens to the three rays if the object point lies on the axis, e.g., point P? As our rules don't help us find the image point in such a case, we must always use an object that is off the axis.)

Consider now the image $P'Q'$. As in the case of the plane mirror, it is a *virtual image:* no light actually comes from it. It is also *erect.* However, unlike the plane mirror case, the image is *closer* to the mirror than the object. Further, the image is *smaller* than the object. The eye sees a smaller image than it would if the convex mirror were replaced by a plane mirror.

Because the image is smaller than the object, we see more of it in the convex mirror than in a plane mirror of the same size. A convex mirror is thus a **wide-angle** mirror, giving you a view of a wide angle of the world around the mirror. It is commonly used in stores to protect the merchandise from shoplifters. Some trucks and vans use such wide-angle mirrors as rear view

mirrors that give a broader view (Fig. 3.7). The point is succinctly made by the Japanese poet Issa: "Far-off mountains/mirrored in its eye—/dragonfly." He contrasts the tiny convex mirror of the dragonfly's eye with the immense object (the mountains) visible in this "wide-angle mirror."

If the object is very distant (such as a star), the rays from it that hit the mirror must be traveling in essentially the same direction, toward the mirror—they must be parallel to one another. In Figure 3.8 we have drawn a bundle of such parallel rays at an angle to the axis of the mirror. As none of these rays is parallel to the axis, there can be no ray *1.* However, it is possible to select rays

---

FIGURE 3.7

Photograph of a convex mirror used to give a wide-angle view. The plane mirror on which it is mounted shows the normal view. **(a)** The camera is focused on the image in the convex mirror, just behind the mirror (see Fig. 3.6). **(b)** The camera is focused on the image in the plane mirror, as far behind the mirror as the object is in front. From a distance, your eye can focus on both images simultaneously.

2 and 3 out of the bundle of parallel rays; one aimed at C and one at F. These two rays are sufficient to locate the image at Q'. Note that Q' lies on the plane through F perpendicular to the axis. This plane, called the **focal plane,** shows up only as a line in our figure, but actually extends in front of, and behind, the paper on which the figure is drawn. This given us another useful rule:

PARALLEL RAYS RULE
Rays parallel to each other intersect on the focal plane.

The virtual image of the distant star is then at Q'. Any other reflected ray may then easily be constructed by drawing a straight line from Q' to the point on the mirror where the incident ray hits. We have drawn three such rays (other than rays 2 and 3), omitting the lines between Q' and the mirror so as not to clutter the diagram. Stars in other directions would be imaged at other points on the focal plane.

The method of ray tracing we have illustrated, if done carefully, can give you the correct size, position, and orientation of the image

(a)

(b)

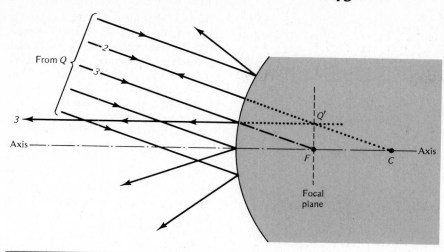

FIGURE 3.8

An object Q, too distant to be shown, sends parallel rays to a convex mirror. Its image, Q', lies in the mirror's focal plane.

of any object. Using a ruler, you can measure the results on your diagram to find these image properties reasonably accurately. You need only know the focal length, f, of the mirror to perform the construction (see also Appendix D).

*C. Deformations in convex mirrors and anamorphic art

If the object doesn't lie in a plane perpendicular to the axis, or if the mirror is not spherical, the image in a convex mirror will be deformed. The deformations so produced have artistic interest. In Victorian times, a convex mirror was often hung on the wall to give an exotic image of the room around it and the people nearby. Parmigianino, Escher, and other artists have drawn self-portraits in such mirrors (Fig. 3.9).

Later it became fashionable to reverse the process. Instead of using a convex mirror to deform things, one painted a deformed picture that, when viewed with the help of a particular convex mirror, appeared undeformed. Such **anamorphic art*** developed in the six-

*Greek *anamorphoo*, transform.

teenth century, was popular for several hundred years, and is currently undergoing a revival. As Shakespeare described it

For sorrow's eye, glazed with blinding tears,
Divides one thing entire to many objects;
Like perspectives, which, rightly gazed upon,
Show nothing but confusion, eyed awry,
Distinguish form

FIGURE 3.9

M. C. Escher, "Hand with Reflecting Globe."

picture becomes the inner part of the virtual image (Fig. 3.11). These, too, may be made using an appropriately transformed grid, but they may also be made photographically (see the second TRY IT).

Another type of anamorphic art is made to be viewed without a mirror. Greatly elongated pictures, such as the streak across Holbein's painting "The Ambassadors," when viewed at a glancing angle from the side appear undistorted (Fig. 3.12). This is called a slant anamorphic picture. (This is the type of anamorphic art Shakespeare refers to in the passage cited above.)

FIGURE 3.11

A conical anamorphic photograph, made by the method of the TRY IT. The conical mirror in the center reconstructs the undistorted image from the anamorph surrounding it, so you see the cat's head in the central circle surrounded by the deformed image.

FIGURE 3.10

Cylindrical anamorph, with mirror that reconstructs the image.

Anamorphic art falls into different classes, depending on the way the image is "decoded." One common type requires the use of a cylindrical mirror (a shiny cylinder, such as a tin can). After placing the mirror in the center of the distorted picture, you look into the mirror and see a virtual image of the picture, now undistorted (Fig. 3.10). Concentric circles around the mirror become straight lines in the image, the outer circular boundary of the picture becoming the top of the image. Anamorphic pictures are drawn by means of a special grid, as shown in the first TRY IT.

Conical mirrors can also be used. These are placed, point up, in the center of the conical anamorphic art, and are viewed (with one eye only) from directly above the cone's point. These pictures are particularly perplexing because they are "inside-out"; the outer part of the

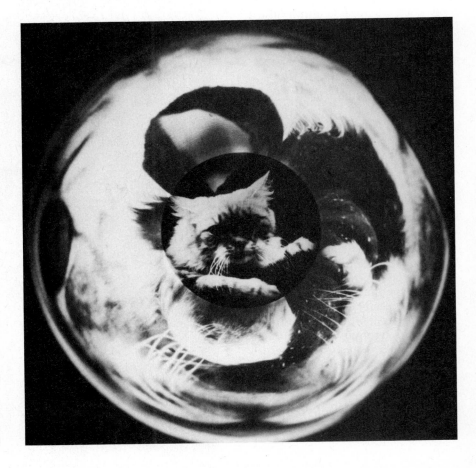

These pictures can carry messages that are not readily apparent to a censor, but are easily accessible to their true audiences. They have been used for this purpose ever since their inception. One of the authors' introduction to anamorphic art came in the form of an apparently abstract design in a freshman handbook. When viewed at a slant, however, the design became a very clear (and earthy) student comment on the university administration (an obviously slanted view).

FIGURE 3.12

Hans Holbein's "The Ambassadors." View the streak across the foreground of the picture from the upper right. (Reproduced by courtesy of the Trustees, The National Gallery, London.)

First TRY IT

FOR SECTION 3.3C
Cylindrical anamorphic drawing

To make your own cylindrical anamorphic drawing, you'll need about a square-foot sheet of aluminized mylar (from a graphic arts or plastics store). Figure 3.13 demonstrates the construction technique. An undistorted drawing is made on a rectangular grid. Then each line in the undistorted drawing is translated into a corresponding line on the curved grid.

Make your undistorted drawing on a separate piece of graph paper 12 × 14 squares in size. Use a photocopy or tracing of Figure 3.14 as your curved grid. After completing your transformed drawing, curl and tape your mylar sheet into a cylinder that just fits on the circle inside your curved grid. Stand this cylindrical mirror on the central circle and view the reflected image. Redraw any lines in your drawing that don't appear correct. This is a bit tricky; you must draw while looking at the reflection. When you're satisfied with the anamorphic drawing, erase the grid lines, or trace the drawing onto another sheet. If you color the drawing, be careful to avoid any ridges due to brush strokes.

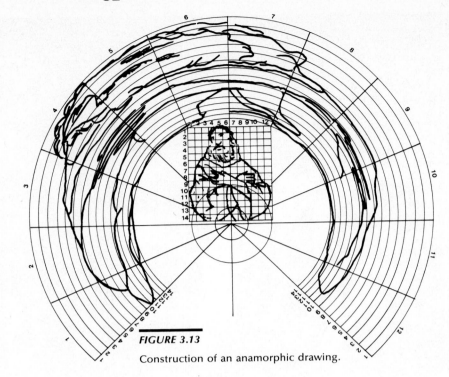

FIGURE 3.13

Construction of an anamorphic drawing.

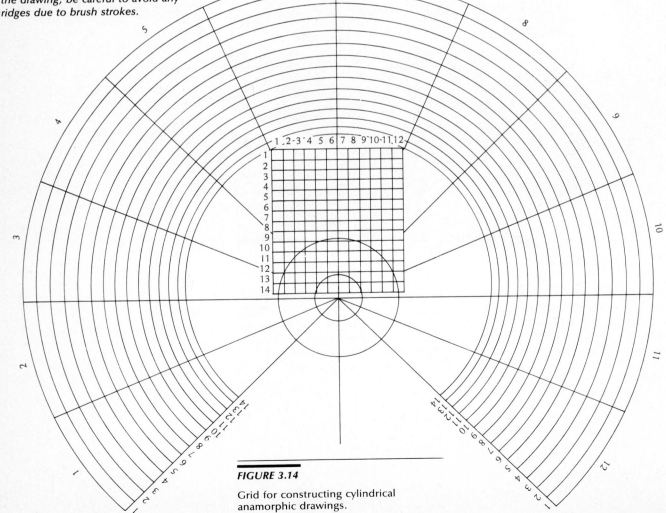

FIGURE 3.14

Grid for constructing cylindrical anamorphic drawings.

Second TRY IT

FOR SECTION 3.3C
Conical anamorphic photograph

If you can make black and white photographic enlargements (8 × 10 inches or larger), you can make a conical anamorphic photograph. Besides a standard enlarger and photographic chemistry, you'll need a ringstand and clamp or other supporting device; a variable aperture or a set of cardboards, each with a different size hole; and a shiny cone of apex angle about 30°. (A cone made from aluminized mylar, taped, is acceptable.)

Choose a negative that has a razor sharp image, in which the subject is fairly small and located at the center of the frame. It is best to have a fairly dark, uniform region at the exact center. (For such a negative, unwanted distortions due to an imperfect cone apex will not be visible on the final print.) Cut a piece of cardboard to fit snugly in your print-paper holder. From the center of the cardboard, remove a circular hole the size of the base of the cone, to form a placement mask. Put your negative in the enlarger head, turn on the projection lamp, and center the print-paper holder directly below the enlarger lens. Lay the placement mask in the holder, put the cone base down in the hole, then remove the mask. Next, use the ring stand and clamp to hold the variable aperture above the cone. Adjust the aperture opening so that light from the enlarger strikes the cone, but not the print-paper holder directly. Focus the distorted image. In Figure 3.15 the placement mask, hinged to the print-paper holder, is shown tipped away, as for focusing and for exposing the paper.

For your exposure of the print, use the placement mask to center the cone on the print paper. Expose for the "usual exposure period," the period you would use were the cone not present. Then, slightly reduce the size of the hole in the aperture and expose for the "usual time" again. Repeat this procedure about ten to fifteen times until the aperture hole is as small as possible. If your negative is dark at the very center, you can skip the last few exposure periods. This unusual dodging technique guarantees that each area of the paper receives enough light for a proper exposure.

PONDER

If light from the enlarger were allowed to strike the entire cone for the duration of the exposure, why would areas of the paper near the cone receive much more light than areas farther away (assuming a negative of uniform density)?

To view your print after development, place the cone on the center of the print and look down on the cone, using only one eye (Fig. 3.11).

FIGURE 3.15

Set-up for making an anamorphic print. The shiny cone is in place, and the placement mask is tipped back, for exposure.

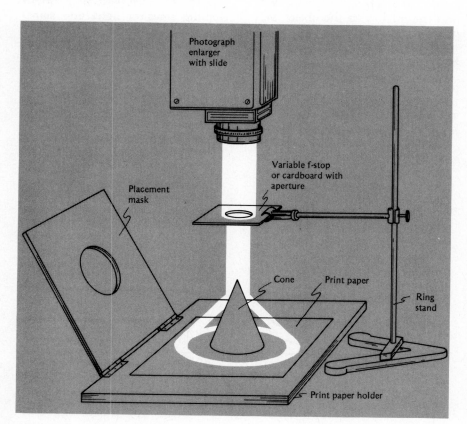

D. Concave mirrors

The reflecting surface of a **concave** mirror bows *away* from the light source, like a cavity. This means that the center, C, of the sphere is in front of the mirror. Following the same steps as we did for the convex mirror, we apply the laws of reflection to ray 1 of Figure 3.16 and discover that the focal point, F, also lies in front of the mirror. (Again $\overline{OF} = \frac{1}{2}\overline{OC}$ for a spherical mirror.) In order to distinguish this case from the convex case, we say that the focal length, f, is the *negative* of the distance $\overline{OF}$. Thus f is *negative* for *concave* mirrors, *positive* for *convex* mirrors.

Except for the fact that the points C and F are in front of the mirror, the general rules for convex mirrors can be taken over directly for the concave mirror. In Figure 3.16, we illustrate each of the three types of rays.

Having F and C in front of the mirror allows for new possibilities not available with convex mirrors. For example, suppose we put a point source of light at F. All rays from this source are then rays of type 3, as they all come from F. Consequently, all such rays will be reflected back parallel to the axis of the mirror. This gives us a way of constructing a parallel beam of light: simply put a small light source at the focal point of a concave mirror. For example, the con-

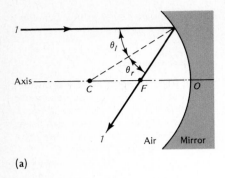

(a)

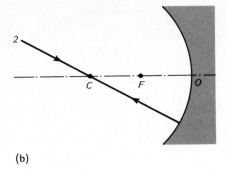

(b)

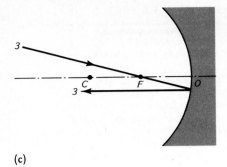

(c)

FIGURE 3.16

Three rays obeying their respective rules, for a concave mirror.

cave mirrors in flashlights and headlights make such parallel light beams.

Conversely, as all incident rays parallel to the axis are reflected to pass through F (ray 1), quite a bit of light energy from a distant source, such as the sun, can be directed to one point. In the third century B.C., Archimedes may have used a concave mirror, constructed of the shiny shields of many soldiers, to burn the attacking Roman fleet. The idea was to focus all the sun's energy striking the large area of the mirror onto the small area of the sun's image on one of the boats. This concentrated energy would set the ship on fire. Whether or not Archimedes did this, such devices are now often used as concentrators in solar heating devices—you can buy small ones to light cigarettes or campfires. (See the FOCUS ON Solar Power.)

Having F and C in front of the mirror has some other rather amusing consequences. We can get rather different types of images depending on the location of the object, but all can be found by the rules of Section B. In our example (Fig. 3.17), we have placed the object between points C and F. Rays 1 and 3 are then easily drawn follow-

ing the ray tracing rules. Notice now that, to the eye, E, ray 1 not only appears to be *coming from F* after reflection, but it actually *passes through F*. Similarly, ray 3 is not only aimed at F from Q, it actually *goes through F* on its way to the mirror. The rule for ray 2, however, requires some comment. Starting at Q, no ray can head toward the mirror and pass through C. However, if we draw a ray from Q headed toward the mirror *as if* it had come from C, then this ray is still perpendicular to the mirror and hence is reflected back on itself, as ray 2 normally is. We can thus treat an object lying between C and the mirror. (If the object were closer to the mirror than F, similar comments would apply to the Ray 3 Rule.)

The three reflected rays of Figure 3.17 actually intersect (without having to be extended backward). Therefore, the image lies at Q', in *front* of the mirror. We locate the

entire image $P'Q'$, as usual, by dropping a perpendicular to the axis from Q'. It is not a virtual image because the rays from Q really *do* pass through Q'. Such an image, through which the light actually passes, is called a **real image.** To the eye, E, it looks just as if there were an object at Q' blocking the view of the mirror behind. However, if the eye moves to E' and looks toward Q', the "object" disappears because there is then no mirror behind Q' to reflect light to E'. As light rays actually cross at a real image, you can place a screen there that will catch a projected image for you to see. The pattern of bright rippling lines of light one sometimes sees reflected from water is due to this effect (Fig. 3.18). The ripples in the water form little concavities that reflect the sunlight. Because there are many ripples, there are always some of them that image the sun on any nearby surface. Since each concavity tends to

---

FIGURE 3.17

Construction of the image in a concave mirror by ray tracing.

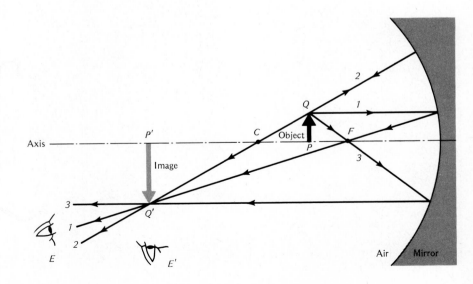

FIGURE 3.18

Rippling lines of light reflected and focused by the uneven surface of the water. Lines can be seen near the surface of the water as well as on the wall, in the shadow of the fence.

be trough-shaped—long, rather than spherical—the image of the sun is elongated into a line. The many ripples produce many such images, which intersect, shimmer, and change as the ripples flutter across the surface of the water.

In Figure 3.17 the image formed by the concave mirror is a real image. It is located in front of the mirror—closer to E than the object is! It is inverted, pointing in the opposite direction from the object. It is also **magnified;** it is larger than the object. These properties of the image will be changed if the location of the object is changed. For example, if the object were at $P'Q'$, the image would be at PQ. (Why?) In that case the image would still be real and inverted, but smaller than the object and closer to the mirror.

If the object were closer to the mirror than the focal point F, the image would be virtual, behind the mirror, erect, and magnified. Such mirrors are used as shaving or makeup mirrors, allowing you to

see a magnified image of your face, provided you are close enough. If you have this type of mirror, examine your image as you move your face closer or farther away from it. Alternatively, you can use the inside of the bowl of a large spoon. In that case it's easier to use your finger tip as the object. Turn the spoon over and look at the back of the bowl as an example of a convex mirror, and compare the two cases.

An unusual example of a concave mirror in nature is the eye of the plankton *Gigantocypris* (Fig. 3.19).

This deep sea crustacean has no lens in its eye, but rather a concave nonmetallic mirror (see Sec. 12.3C). While mirror eyes are relatively rare in the animal kingdom, they do occur and we'll see another example in Section 6.4D.

FIGURE 3.19

Drawing of *Gigantocypris*. The dark curves are the reflecting eyes, with the retina in front of the reflector. They look through transparent windows in the carapace.

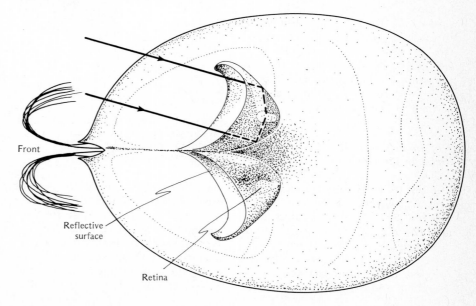

Front

Reflective surface

Retina

3.4
SPHERICAL LENSES

We can also create virtual or real images by *refraction*—bending of the light as it enters a new medium. Let's consider two media, air and glass, with a boundary between them shaped like part of a sphere. Suppose light traveling parallel to the axis strikes this boundary from the air side (Figs. 3.20a and c). We have drawn the normals—they are just radial, straight lines from the center of the sphere, C. According to Snell's law, the light bends toward the normal so that the transmitted angle, θ_t, is less than the incident angle, θ_i. The two rays in Figure 3.20a, traveling in air parallel to the axis, become converging rays in the glass. This surface is a **converging** surface. In Figure 3.20c, the rays become diverging in the glass. This surface is a **diverging** surface. Notice also that turning the surface around, so that the light originates from the glass side, does not change the nature of the surface: reversing 3.20a gives 3.20b, still a converging surface. Reversing 3.20c gives 3.20d, still a diverging surface.

A surface like that in Figure 3.20a remains converging when the glass is replaced by water. For this reason, one often sees a pattern of bright rippling lines of light on the bottom of pools, what T. S. Eliot described as "the light that fractures through unquiet water." This pattern is produced by refraction of sunlight at the water surface, much the same as the pattern of lines produced by reflection that was described in Section 3.3D.

A. Converging and diverging lenses

Combining the two surfaces of Figures 3.20a and b, we get the **lens** of Figure 3.21a. We see, by successive applications of Snell's law, that light originally traveling parallel to the axis is made to converge by refraction at the first surface (points A and B), and made even more con-

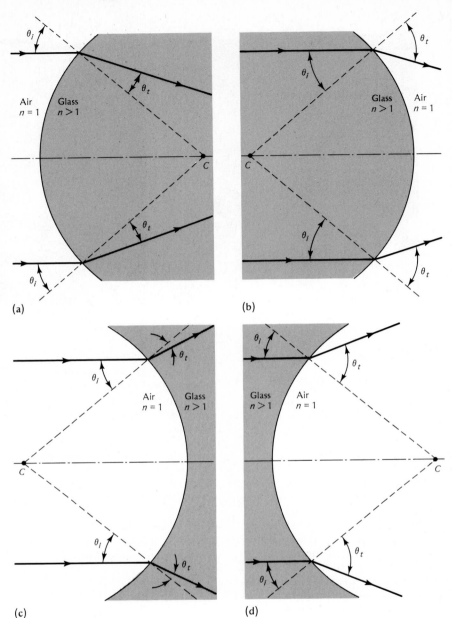

(a)

(b)

(c)

(d)

FIGURE 3.20

Effect of a spherical glass surface on light rays incident parallel to the axis:
(a) and **(b)** converging surfaces,
(c) and **(d)** diverging surfaces.

vergent by refraction at the second surface (points C and D). Emerging from the glass lens, these rays intersect at the point labeled F'. In fact, all paraxial incident rays parallel to the axis, including a ray along the axis itself, will intersect at F' (see the first TRY IT). The point

F' is called the **second focal point** of this lens. (The location of this point depends on the curvature of both lens surfaces and on the index of refraction of the lens and of the surrounding medium.) Such a lens is called a **converging** or **focusing lens** (see the second TRY IT). It can be used as a burning glass with which to start fires. The parallel rays of the sun will be brought to a focus at F' by this lens, forming a small image of the sun there. All the energy carried by these light rays is

(a)

Axis

Air
n = 1

Glass
n > 1

Air
n = 1

Lens

F'

A

C

B

D

(b)

F

Lens

FIGURE 3.21

A converging lens, consisting of two converging surfaces. **(a)** Rays parallel to the axis are focused at *F'*. **(b)** Rays originating at *F* are also made to converge and emerge parallel to the axis.

then concentrated in the sun's small image, and will heat that spot (Fig. 3.22). Focusing lenses are quite old. The early ones were glass spheres filled with water. The Greek comic playwrite Aristophanes suggested that burning glasses could be used to annul promissory notes by melting the letters off the wax notes.

A converging lens also has a **first focal point,** *F*, in *front* of the lens (Fig. 3.21b). All rays originating at *F* and passing through the lens will emerge parallel to the axis. To see that this must be true, notice that Figure 3.21b is simply Figure 3.21a flipped around, with the arrows on the rays reversed.

We can make a **diverging lens** by combining the two diverging surfaces of Figures 3.20c and d. In Figure 3.23a, we see that the incident rays parallel to the axis are made to diverge. To an eye on the right-hand side of the lens, these divergent rays will *appear* to be coming from their point of intersection (when extended backward). We again label this point *F'* and call it the second focal point. For the diverging lens, then, *F'* lies in front of the lens (the reverse of a converging lens, where parallel incident rays actually go *through F'*).

Flipping Figure 3.23a over and reversing the arrows (Fig. 3.23b) shows that the first focal point, *F*, of a diverging lens lies behind the lens. All rays aimed at *F* that hit the lens emerge parallel to the axis.

FIGURE 3.22

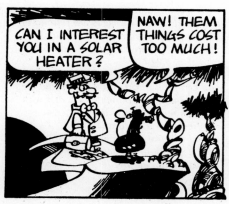

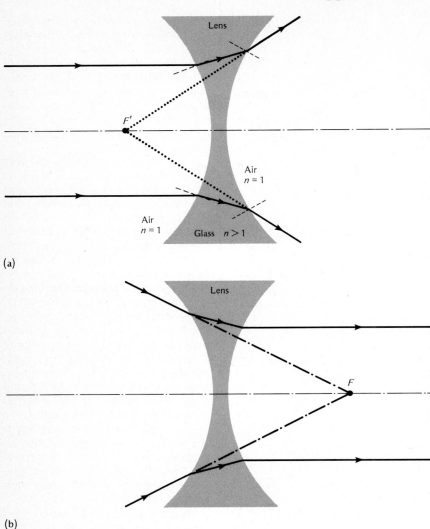

(a)

(b)

FIGURE 3.23

A diverging lens, consisting of two diverging surfaces. **(a)** Rays parallel to the axis seem to come from F' after they pass through the lens. **(b)** Rays converging toward F are also made to diverge and emerge parallel to the axis.

First TRY IT

FOR SECTION 3.4A
Focusing of parallel rays

You can verify that a converging lens, such as a magnifying glass, will focus parallel rays. Using the sun as a source, hold a comb perpendicular to a piece of paper so that the shadows of its teeth form a set of long parallel lines on the paper. Now hold the lens so that the light passes through the comb, then through the lens, and hits the paper. Notice that the previously parallel shadows become converging lines that meet at the focal point of the lens. Use a ruler to measure the focal length of the lens.

You can also try this with a cylindrical lens (one shaped liked a cylinder rather than a sphere). A glass of water makes a nice, converging, cylindrical lens.

Second TRY IT

FOR SECTION 3.4A
The focusing ability of a lens

With the aid of the lens in your eye, you can verify that a focusing lens will take rays going in different directions and focus them. In a piece of aluminum foil, make three small pinholes in a triangle, separated from each other by about 2 mm. These will select the three different rays from a point source that are going in three different directions.

First, replace the pinhole in your pinhole camera by these three pinholes (see the TRY IT for Sec. 2.2B). How many images of a distant, small, bright object do you see in this case, where there is no lens between the object and the screen?

Now replace the pinhole camera by your eye, which does have a lens. Hold the three pinholes next to your eyeball and look through them at the distant object. How many images are formed on the "screen" in your eye (the retina)? That is, how many images do you see now (Fig. 3.24)?

If you have a single-lens reflex camera, you can look through the viewfinder to do the same experiment. Look first with the lens removed and the lens opening covered with the aluminum foil containing the three pinholes. Now remove the foil, replace the lens, put the foil right over the lens, and look again.

FIGURE 3.24

Optics of an eye viewing through multiple pinholes. The lens of the eye creates one image from rays that passed through different pinholes.

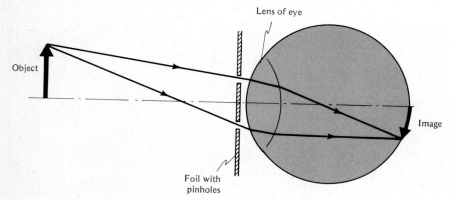

B. Dew heiligenschein and another type of retroreflector

A drop of water makes a good converging lens. Its second focal point, F', is just a short distance beyond the drop. If a dew drop is held at this distance in front of a blade of grass by the fine hairs on the grass, sunlight is focused on the blade of grass (see Figs. 3.25a and b, and the TRY IT). The blade of grass reflects this light back toward the sun, reversing the path of the incident light—the dew drop on the grass forms a retroreflector.

With the sun behind you, the only retroreflected light you will see will be from the grass straight in front of you, around the shadow of your head. Hence, there will be more light coming from the region around your head's shadow than from any other region of the grass, a phenomenon called **heiligenschein.** * The sixteenth-century artist Benvenuto Cellini concluded that this was evidence of his genius. Actually, a nearby friend would see the glow only around her own head's shadow, and conclude that she is the genius.

As glass is somewhat more refracting than water, a glass sphere has its focal point just at the edge of the sphere. Therefore, glass spheres with white paint behind them make excellent retroreflectors. They are used in highway signs, license plates, and the paint used for the white lines on highways, because they send so much of the incident light from your headlights back to your eyes (Fig. 3.25c).

*German, light of the holy ones.

FIGURE 3.25

(a) Water droplet and blade of grass acting as a retroreflector. Only *one* incident ray is shown, and only a few of the many, diffusely reflected rays due to this one incident ray. (b) Dew heiligenschein around the shadow of the photographer's head. (c) Glass beads used as retroreflectors make this jogger's vest visible in car headlights.

TRY IT

FOR SECTION 3.4B
Focal point of a water drop

You can easily determine the focal point of a water droplet. Make a droplet at the end of an eyedropper or a drinking straw. Bring a piece of paper behind the droplet to catch the sun's rays that have passed through the droplet. Bring the paper closer until you have the sharpest image. (The paper will then almost touch the water, so view the sun's bright image from the side so you don't block the sun's rays with your head.) How far is the focal point from the water droplet?

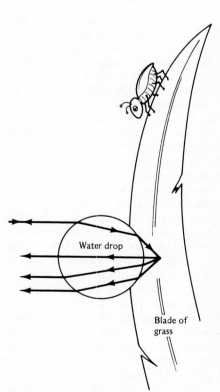

(a)

Water drop

Blade of grass

(b)

(c)

C. Ray tracing for thin lenses

Once we know the location of the focal points F and F', we can construct images by ray tracing, much as we did for mirrors. The situation is simplest if the thickness of the lens is much less than the distance from the lens to either focal point. We will confine our attention to such **thin lenses** and, as before, to paraxial rays.

For thin lenses, the distance from the lens to F is equal to that from the lens to F' and is called the **focal length** of the lens, f. To distinguish between converging and diverging lenses, f is taken as *positive* for *converging* lenses and *negative* for *diverging* lenses. Another simplification of thin lenses is that any ray passing through the *center* of the lens is undeviated—it continues straight through the lens without being bent.

With this information, we can set up ray tracing rules as we did for mirrors. Again we choose three special rays with particularly simple properties to help us locate the image. We first illustrate the technique for a converging lens, but we'll phrase the rules so as to cover both types of lenses. In Figure 3.26a, we have drawn a lens and the two focal points F and F', each at the distance f from the lens. We've also drawn our standard object PQ. For our first ray from Q, we refer back to Figure 3.21a (and Fig. 3.23a) and write:

FIGURE 3.26

Three stages of construction by ray tracing of the image formed by a converging lens. The lens is shown relatively thick so that you can see it but treated as if it were just the vertical plane through its center.

RAY 1 RULE
A ray parallel to the axis is deflected through F' (or as if it came from F').

We've drawn such a ray from Q, labeled ray 1 in Figure 3.26a.

Ray 2 passes through the center of the lens:

RAY 2 RULE
A ray through the center of the lens continues undeviated.

This rule is illustrated in Figure 3.26b. We'll call ray 2 the **central ray.**

Finally, we obtain the third rule by examining Figure 3.21b (and Fig. 3.23b):

RAY 3 RULE
A ray to the lens that (when extended, if necessary) passes through F is deflected parallel to the axis.

Ray 3 is illustrated in Figure 3.26c.

Any two of these rays are sufficient to locate the image point Q'. The third ray serves as a check. Again, we find P' by dropping a perpendicular to the axis from Q'. The resultant image in Figure 3.26 is a *real* image (the light actually converges to this image). The image is inverted. Further, in this *particular* case (we took the object distance, from the object to the lens, to be twice the focal length), the image is the same size as the object.

A screen in the plane of $P'Q'$ would show an illuminated inverted image of the object. This is the principle of a slide projector (and of a photographic enlarger). The object, in that case, is the transparency illuminated by the light source in the projector. The lens then projects the inverted image on the screen (thus, the slide must be put in the projector upside down to get an upright image).

Of course, for a slide projector we want the image much larger than the object. This is achieved by mov-

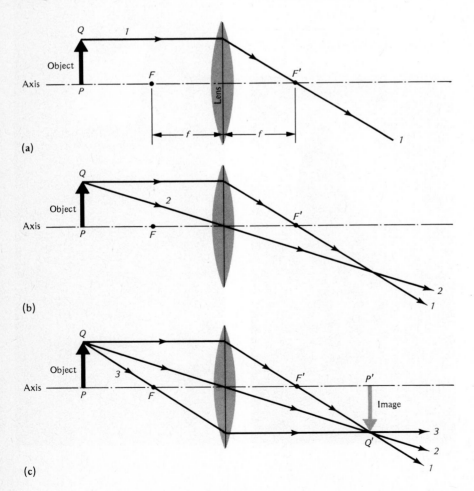

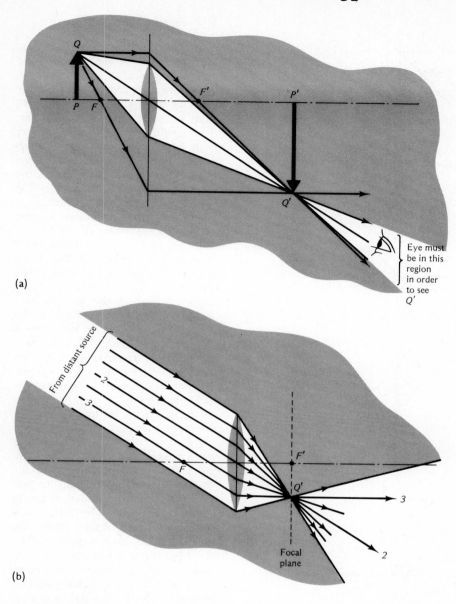

(a)

(b)

Eye must
be in this
region
in order
to see
Q'

From distant source

Focal
plane

**Rays parallel to each other
are imaged on the focal
plane.**

Because the focal plane passes
through F', if you want to take a
picture of a distant object, you
should put your film at the focal
plane.

In Figure 3.28 we apply the ray
tracing rules to a diverging lens (f
negative, F' in front of the lens, F
behind it). Compare rays 1 and 3 to
the rays in Figures 3.23a and b, respectively. The image in this case is
virtual. Any eye to the right of the
lens will see rays apparently coming
from Q', although most rays do not
actually pass through that point
(only ray 2 does). The image here is
erect and smaller than the object.

Thus, knowing only the focal
length, f, of a lens, we can construct
the image of any object formed by
that lens. The focal length of a lens
is given in one of two ways. The
lens on your camera may have
printed on it, "f = 50 mm." As f is
positive, this is a converging lens.
However, the prescription for your
eyeglasses may read, "−2 D." The D
stands for *diopters,* a different
measure of focal length. The number of diopters (called the *power of
the lens*) is related to the focal
length as follows:

Power (in diopters)

$$= \frac{1}{\text{focal length (in meters)}}$$

For example, a −2 D lens is one
with a focal length of $\frac{1}{2}$ m or
−50 cm (the negative sign means
that it is a diverging lens). The
more the lens bends the rays, the
higher its power. So power measures the everyday notion of the
"strength" of a lens. Once we know
the focal length (in either form) we
can easily locate F and F' and then
trace rays to find the image of any
given object (see Appendix E).

FIGURE 3.27

(a) Another example of ray tracing, with
some special twists. Here the lens is too
small to transmit all three tracing rays.
You draw them anyway, pretending the
lens is larger. Once the image is found,
it is easy to draw the cone of rays
(unshaded) that actually make it through
the lens. This cone converges on the
image, crosses there, and then diverges
from it. In order to see the image, the
eye must be within the region of the
cone of light on the right of the image.
(In a projector, there is a screen at $P'Q'$,
so the light does not continue to the
right beyond Q' but instead is scattered
by the screen so that all eyes to the *left*
of $P'Q'$ can see the projected image.)
(b) Parallel rays incident on a converging
lens are focused on the focal plane.

ing the object closer to F, as in Figure 3.27a. The closer the object is
to F, the larger (and farther) the image will be. This figure also illustrates another point: what to do if
the actual lens (or mirror) is too
small to catch the three rays used
in ray tracing.

Figure 3.27b shows how to trace
a bundle of rays parallel to one another. Here, as in Figure 3.8 you do
not have any incident rays parallel
to the axis, so you cannot construct
ray 1. Rays 2 and 3, however, are
sufficient to locate the image and
give us the same rule we had for
mirrors:

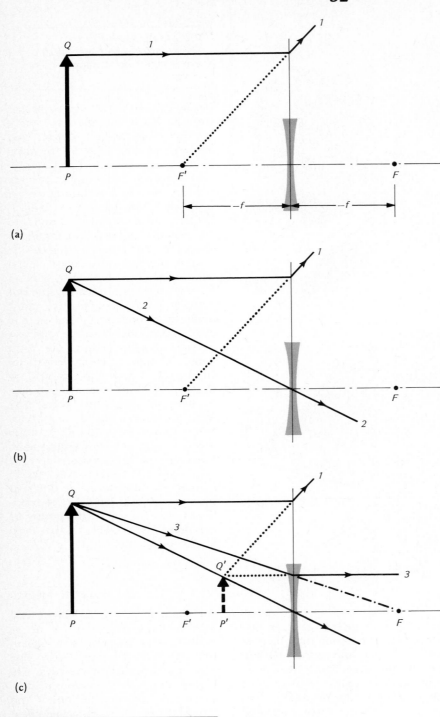

(a)

(b)

(c)

FIGURE 3.28

Three stages of construction, by ray tracing, of the image formed by a diverging lens.

*D. Fresnel lenses

In lighthouses, for stage lighting, and whenever a single lens must intercept much of a source's light, one wants a lens with both a large diameter and a short focal length (that is, very curved). Together, these requirements imply a very thick lens. But large thick lenses create problems. They are heavy, take up space, absorb more of the light sent through them, develop large stresses, and often crack.

To avoid these difficulties, Augustin Fresnel conceived of the idea of removing most of the interior glass from a thick lens (Fig. 3.29a). After all, the refraction occurs only at the surfaces of the lens. Therefore, he imagined cutting the lens in sections. By removing most of the glass in each section, he would have a lens with a spherical front surface and a set of circular steps on its back (Fig. 3.29b). Fresnel then collapsed the rings to a common flat back surface (Fig. 3.29c), producing a thin lens with the power and diameter of a large, thick lens.

It is true that this **Fresnel lens** has some spurious refraction at the risers. However, the applications for which it is used do not require precision lenses, so these drawbacks are not important. In fact, such lenses are usually made cheaply by pouring glass into a mold or by stamping them out of plastic.

Glass Fresnel lenses are commonly found in lighthouses and in spotlights for theater lighting (Fig. 3.30). (Spotlight lenses are designated by their diameter and focal length. Thus, a 6 × 9-inch lens has a six-inch diameter and a nine-inch focal length.) Stamped plastic Fresnel lenses are cheap and fit where ordinary lenses of short focal length would not. Novelty stores carry them as curiosities for your window, but they also find use in overhead projectors, camera viewfinders, traffic lights, and numerous other applications. You can recognize them by noticing the pattern of finely spaced rings. Using the same principle, stamped plastic **Fresnel mirrors** are flat but behave like concave or convex mirrors.

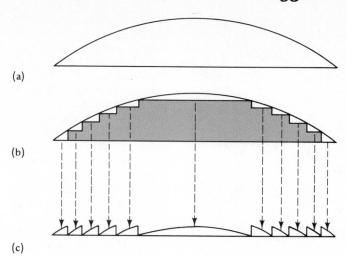

(a)

(b)

(c)

FIGURE 3.29

(a) A thick, converging lens (with one flat side). **(b)** Parts of the glass of this lens that have no essential effect on the bending of light (shown shaded). Remember that these sections are really rings oriented perpendicular to the paper. **(c)** After removing the nonessential glass and rearranging the lens, you get a Fresnel lens.

FIGURE 3.30

(a) Photograph of a Fresnel lens designed for use where Fresnel first intended: in a lighthouse. **(b)** Photograph of a Fresnel spotlight.

(a)

(b)

The traffic-light application is of interest in its own right (Fig. 3.31). A lamp, backed by a concave reflector, illuminates a ground glass screen. A Fresnel lens is located about a focal length in front of this screen. The entire apparatus is surrounded by a housing, the back of which (including the lamp and reflector) can be removed. The task is to adjust this light so that it can be seen in some lanes of traffic (say the left-turn lane), but not in other, adjacent lanes. You set the traffic light in place, remove the back, and look in from the back. You see an inverted image of the road on the screen. (Why?) With opaque mask-

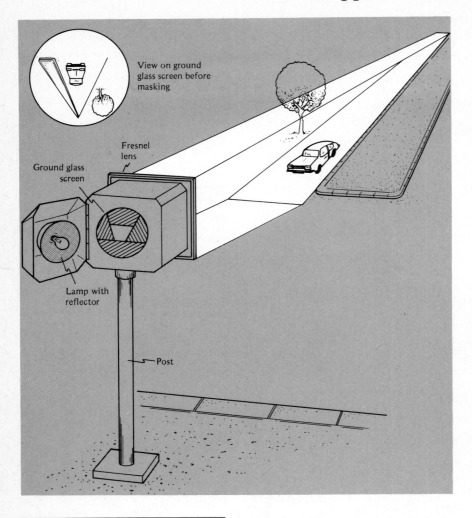

View on ground glass screen before masking

Fresnel lens

Ground glass screen

Lamp with reflector

Post

FIGURE 3.31

An "optically programmed" traffic light.

ing tape, you mask off those parts of the image that correspond to the lanes in which you don't want the traffic signal to be seen. When the lamp is replaced, the direction of light travel is reversed; the masked ground glass becomes the object (illuminated by the lamp), and it is projected on the traffic lanes, like a slide in a slide projector. When you stand in a lane where the traffic signal is not meant to be seen, the masked part of the ground glass is projected on you, and the traffic light looks dark. In the proper lane, however, the light shines to your eyes through an unmasked part of the ground glass, and you see the signal.

E. Compound lenses

We often have occasion to use combinations of several lenses. Such **compound lenses** are common in photography, almost all optical instruments, and the human eye. To apply the techniques of ray tracing to multiple lens systems, we simply take one lens at a time and apply the rules over and over for as many lenses as we have. Let's illustrate this with the two-lens system of Figure 3.32.

We begin by ignoring the second lens and finding the image produced by the first lens alone. This **intermediate image,** $P'Q'$, is located in Figure 3.32a by drawing the three rays labeled 1, 2, and 3. (We've drawn all the rays to the right of the second lens with dashed lines to indicate that they

actually will be modified by the second lens.)

Now that we have located the point Q', we can draw other rays that begin at the object point Q, as they must all go to Q'. In particular, we can draw those rays that will be useful for ray tracing through the second lens (Fig. 3.32b). Ray 1', aimed at Q', arrives at the second lens traveling parallel to the axis. (It coincides with ray 3, which left the first lens traveling parallel to the axis.) Ray 2' is aimed at Q' through the center of the second lens. Finally, ray 3', also aimed at Q', travels between the lenses as if it came from F_2, the first focal point of the second lens. (We have traced rays 2' and 3' back through the first lens to the object point, Q, so you can see their actual paths. This is not necessary, however, to construct the final image.)

Having located the three rays incident on the second lens, we now apply the ray tracing rules to each of them in Figure 3.32c. Thus, ray 1' is deflected through F_2', ray 2' is undeflected, and ray 3' is deflected parallel to the axis. They intersect at Q'', allowing us to locate the **final image,** $P''Q''$.

In sum, we first located the intermediate image, $P'Q'$, using only the first lens. We used this image to construct the three rays needed for ray tracing through the second lens. Having found the three rays, we then ignored the first lens and applied the ray tracing rules at the second lens to find the final image. If there were still more lenses, we would treat $P''Q''$ as an intermediate image and use it to locate the three rays for the next lens. (In general, the next lens may or may not deflect the light before it reaches this intermediate image. The techniques used here still apply.)

The situation becomes simple, however, in one special case: if *two thin lenses* are so close as to *touch* each other, they form a combination that behaves just like another thin lens; one whose *power* is the *sum of the powers* of the individual lenses. (See Appendix F for a proof.) Thus, suppose we have a 1-diopter

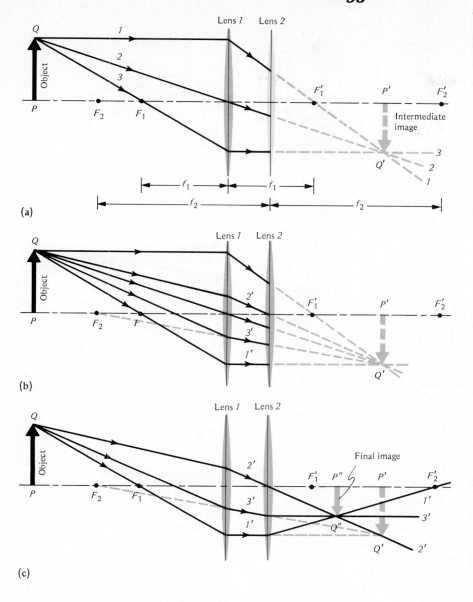

(a)

(b)

(c)

FIGURE 3.32

Construction by ray tracing of the final image of a compound lens consisting of two converging lenses. We have indicated all the focal points and the focal length of each lens. **(a)** Construct the intermediate image, due to the first lens alone. **(b)** Find the three rays incident on lens 2 that are needed for constructing the image *it* forms. **(c)** Use ray tracing rules on lens 2 to find the final image.

lens (focal length is 1 m) and a 4-diopter lens (focal length is $\frac{1}{4}$ m = 25 cm). Placing them together produces an equivalent single lens of 5 diopters (focal length is $\frac{1}{5}$ m = 20 cm). This rule works for negative as well as positive powers, and is the reason that power is a useful way of expressing focal length. In the TRY IT we show how you can measure an eyeglass prescription with the aid of this rule.

TRY IT

FOR SECTION 3.4E
Measuring your eyeglass prescription

Your eyeglass prescription is simply the power (in diopters) or your lenses. If you do not suffer from astigmatism (see Sec. 6.2B, not Sec. 3.5D), your eyeglass lenses are spherical lenses and you need only measure the focal length of each lens to determine its power. We suggest one method, but if you can think of others, try them to see if they all give the same result.

PONDER

How can you tell by sight whether your lens is converging or diverging? How can you tell by touch?

Converging Lenses: *Since objects are imaged on the focal plane, use your lens to form the sharpest image of the sun on a piece of paper, as you would with a burning glass. Measure the distance from the lens to the paper to obtain the focal length. Suppose the distance is 62.5 cm = 0.625 m. The power of that lens, and, therefore, its prescription, is then $\frac{1}{0.625}$ = 1.6 diopters.*

Diverging Lenses: *Find a magnifying glass and measure its focal length by the technique described above. Place it next to your eyeglass lens and measure the focal length, and hence the power, of the combination. (If it isn't converging, find a stronger magnifying glass.) The power of your lens is the power of the combination minus that of the magnifying glass alone (Sec. 3.4E). Suppose the combination has a power of 1.0 diopters (f = 1 m) and the magnifying glass has a power of 4.0 diopters (f = $\frac{1}{4}$ m). Your lens power is then −3.0 diopters.*

**3.5
ABERRATIONS**

Up to now we have assumed that all rays hitting our mirrors or lenses are paraxial, and we have ignored the fact that the glass in the lenses is dispersive. With these assumptions, spherical mirrors and lenses form sharp, perfect images. In actuality, however, the images formed are slightly blurred and distorted, and may have variously colored edges. That is, spherical lenses and mirrors, even if ground and polished perfectly, do not produce perfect images. The images' deviations from perfection are called **aberrations.***** By combining various lenses, lens designers attempt to minimize those aberrations that are most important for the particular application in mind, often at the expense of others. For instance, the lens design for a telescope, used primarily for looking at objects very close to its axis, is quite different from that for a wide-angle camera lens, which accepts light from objects very far off-axis.

Clearly, with a bigger diameter lens (gathering more light), more of the accepted rays deviate from paraxial, and, therefore, the aberrations become more important. Cheap cameras have small diameter lenses for which most aberrations are insignificant. Expensive cameras have larger diameter lenses, which may require more than a dozen lens elements to correct for the aberrations. You pay, in large measure, to get rid of the aberrations.

A. Chromatic aberrations

Glass is a dispersive medium, bending blue light more than red light (Table 2.6). Figure 3.33a shows parallel white light being dispersed by a glass lens. The blue and red light have different focal points, F_B' and F_R'. The other colors are focused between. If we place a screen at F_B', we find a sharply focused blue dot surrounded by a halo of the other colors. The largest halo is red (Plate 3.1). Moving the screen to F_R' produces a sharply focused red dot surrounded by a halo of the other colors, with blue spread out the most. This colored blurring is called **chromatic aberration.**

Such a lens in your camera would produce colored outlines on your

*Latin, *ab-errare* to stray away. Some of the light rays stray from the ideal focus.

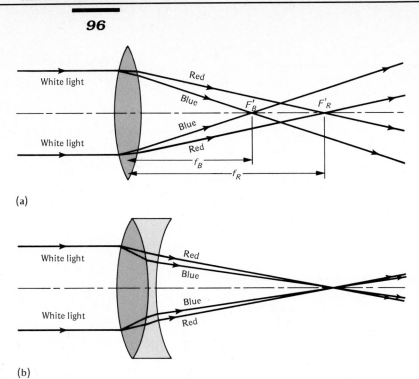

(a)

(b)

FIGURE 3.33

(a) Chromatic aberration of a converging lens. **(b)** Elimination of this aberration by an achromatic doublet.

photographs. Even if you used black and white film, the incident white light would be spread out into a blurred image.

To correct for this aberration, one uses an **achromatic doublet:** two lenses, cemented together, and made of different kinds of glass that have different dispersions (Fig. 3.33b). The converging lens, usually made of crown glass, converges the blue more than the red. The diverging lens, usually of flint glass, diverges the blue more than the red. If the converging lens has the greater power, the combination will be converging and, because of the differences in the glasses used, the red and blue can be focused at the same point. The other colors may not be focused exactly at that point, but will be closer together than in the uncorrected lens of Figure 3.33a. Even the cheapest cameras usually use an achromatic doublet to eliminate this aberration.

Mirrors do not exhibit chromatic aberration because the law of reflec-

tion is the same for light of any color. Telescope designers often avoid chromatic aberration by using mirrors instead of lenses.

B. Spherical aberrations of lenses

When we relax the paraxial condition and accept rays far from the axis of a spherical lens, we find that they do not all pass through the same focus. Figure 3.34 shows an exaggerated example of this. No matter where you put your film or screen, the image is not a sharp point, but is spread out forming a circular blur, called the **circle of confusion.** The smallest circle of confusion is at the plane marked A. This is where you would place the screen to obtain the smallest blur. This blurring is called **spherical aberration.** The TRY IT (after Sec. 3.5C) tells you how to demonstrate this phenomenon.

For a given *object distance*, you can correct for spherical aberration by using a lens with surfaces that are not exactly spherical. To minimize this aberration for a *range* of object distances you need a compound lens.

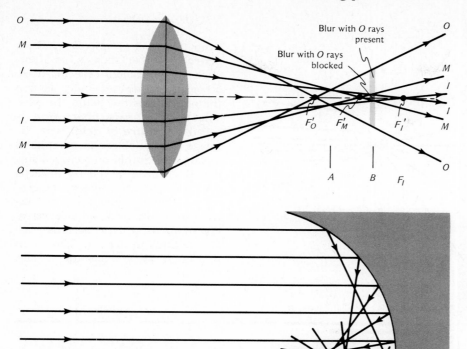

FIGURE 3.34

Spherical aberration in a converging lens. The inner rays, *I*, closest to the axis, have the farthest focal point, F_I'. The outer rays, *O*, are bent most, so they have the nearest focal point, F_O'. The rays in the middle, *M*, are focused between these extremes, at F_M'.

A simpler and cheaper way of diminishing this problem is to use a single, smaller diameter lens. If you block the outer rays, *O* in Figure 3.34, you can move the screen to the plane marked *B* and have a smaller blur than you did at *A*. (With the *O* rays present, the blur at *B* would be huge.) Thus, cheap cameras avoid the expense of multielement lenses to diminish spherical aberration by using a small diameter lens.

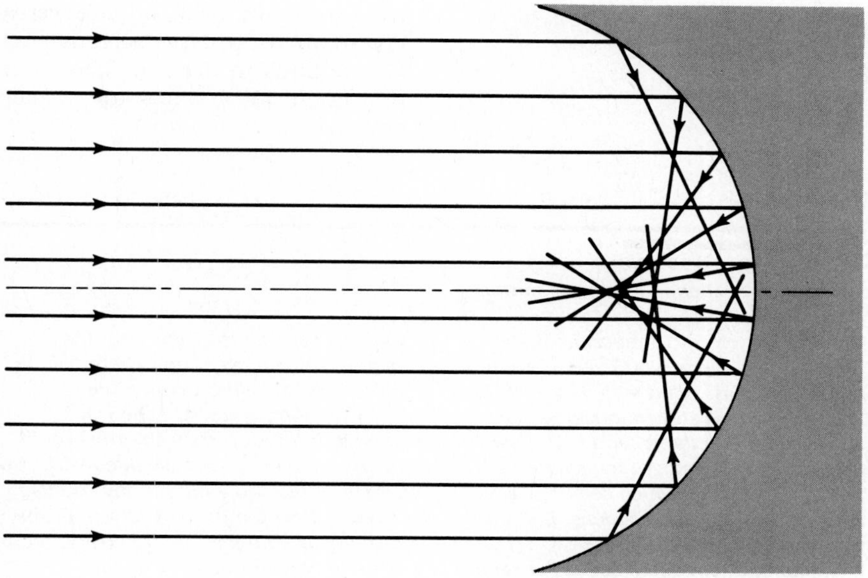

C. Spherical aberrations of mirrors

Spherical aberration also occurs for spherical mirrors. Nonparaxial parallel rays incident on such a mirror are not reflected to a sharp focus, but rather spread out in a pattern like the one illustrated in Figure 3.35. (This pattern can easily be seen, as described in the TRY IT.) Note that the more central rays are almost brought to a good focus. When the peripheral rays are needed for brightness, mirrors that are supposed to focus incident parallel rays, as in a telescope, are made flatter at larger distances from the axis, in the shape of a paraboloid (Fig. 3.36). Such ***parabolic reflectors*** are quite common. They are used in headlights and searchlights to take light from a point source and make a parallel beam. Microwave and radar antennas, which may be sending or receiving, are also parabolic. Parabolic microphones, used to pick up sound waves from distant sources, have a small microphone located at

FIGURE 3.35

Spherical aberration in a concave mirror.

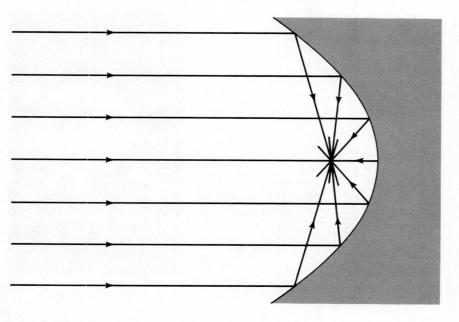

FIGURE 3.36

A parabolic reflector has no aberrations if the object is on the axis and very far away.

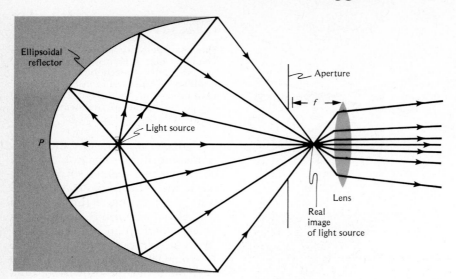

FIGURE 3.37

An ellipsoidal spotlight.

the focus of the (sound) reflecting paraboloid.

To bring most of the light from a nearby source to a focus without aberrations, you need an **ellipsoidal reflector.** It has two points called foci, and all light from one focus is reflected to form a real image at the other focus, without aberration. If you replace a part of the ellipsoid by a lens placed in front of that real image, you can make a beam for use as a theater spotlight (Fig. 3.37). There are two advantages of this scheme. First, you can use a short focal length lens without worrying about the lens hitting the light bulb. Also, you can put an aperture behind the real image, and the lens will focus that aperture on the stage. You then have a spotlight beam that, when it is incident on the stage, has any shape you like—the image of the aperture shape.

FIGURE 3.38

(a) Spherical aberration in a water glass lens (side and top views). (b) Photograph of spherical aberration pattern in a teacup reflector.

*D. Off-axis aberrations

Even if a lens or mirror makes a perfect image on its axis, as a parabolic or ellipsoidal reflector does, other aberrations generally occur if the object does not lie on the axis. One such **off-axis aberration** is called **curvature of field.** Here the image of a flat object does not lie in a plane, but rather on a curved surface (Fig. 3.39). If you try to focus the image on a flat screen, the center will be in focus while the extremes are blurred, or vice versa. For this reason, large screens such as one finds in drive-in theaters or projection TV systems are usually curved.

TRY IT

FOR SECTIONS 3.5B AND C
Spherical aberration in lenses and mirrors

To see a pattern of spherical aberration produced by a lens, you need a candle and a clear glass filled with water. Arrange them as shown in Figure 3.38a, turn off all other lights, and look down on the table at the light transmitted by the glass of water, which serves as a lens. (This is a cylindrical lens, not a spherical lens, but the spherical aberration pattern you'll see is the same.) You may have to move the candle around a little before you get a good spherical aberration pattern. Dust on the table will scatter the light from the pattern to your eye. The pattern will be most visible against a dark table.

To see the pattern of spherical aberration produced by a mirror, you'll need a china cup partially filled with dark coffee or tea and a nice sunny day (Fig. 3.38b). Allow the parallel light from the distant sun to reflect off the inside rim of the cup, which will serve as the ''spherical'' mirror. The pattern should be visible on the surface of the dark liquid in the cup, and should look like Figure 3.35.

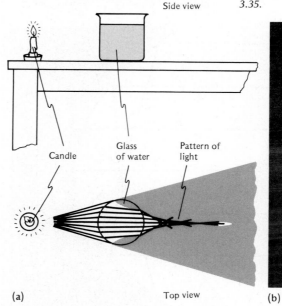

(a)

Top view

(b)

Object

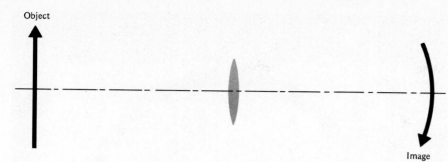

Image

FIGURE 3.39

Curvature of field of a converging lens.

FIGURE 3.40

A small circular spot of light *on the axis* is projected by a lens to form the faithful image on the left (no coma). An identical *off-axis* spot of light produces the image with coma, on the right.

Coma, another off-axis aberration, is a modification of spherical aberration for off-axis objects. For a point source, the blur due to spherical aberration that surrounds an on-axis image shifts to one side when the image is off-axis. The result is that the image of a point source has a flaired tail, something like a comet* (for which the aberration is named) (Fig. 3.40).

Astigmatism is a difference in focal length for rays coming in different planes from an off-axis object. For example, the rays that we have drawn in Figure 3.26 all lie in the plane of the paper. If there is astigmatism, rays from Q that *leave* the

*Greek *kometes,* long-haired.

plane of the paper and hit the lens are brought to a slightly different focus than Q'. The sharpest image of the point Q is then a short line, perpendicular to the paper at Q', or a short line in the plane of the paper at the other focus. Somewhere between these two points, the image of Q is smeared into a small circle (Fig. 3.41). (It is unfortunate that a lens defect of the eye, caused

FIGURE 3.41

The astigmatic image of a point object in three planes, perpendicular to the axis. When the screen is at *A*, the image is a vertical line, and it is a horizontal line when the screen is at *C*. A screen at *B* gives the smallest circular image. (The photograph was taken by triple exposure, changing the position of a single screen between exposures.)

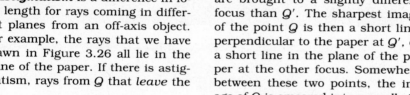

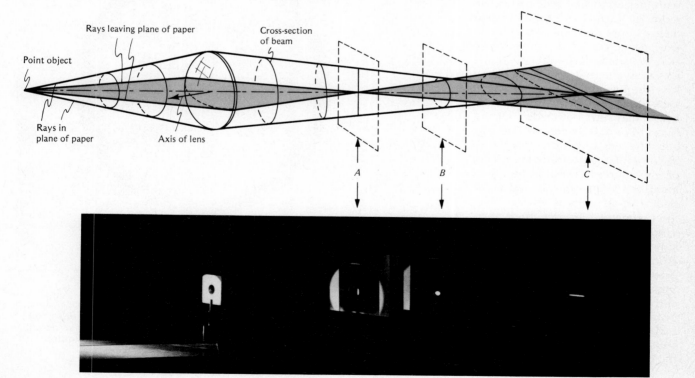

Rays leaving plane of paper

Cross-section of beam

Point object

Rays in plane of paper

Axis of lens

A   B   C

by nonspherical curvature of the lens, is also called astigmatism—see Sec. 6.2B.)

The final off-axis aberration is **distortion.** The image of an object that extends off-axis is pulled out of shape, as if first drawn on rubber and then stretched. Figure 3.42 shows a rectangular grid under two types of distortion.

The amount and nature of aberrations depend on where the **stops** in the optical system are located, that is, how *the dimensions of the beam are limited*—for example, by an aperture or by the outer edge of a lens. The location of the stop determines which parts of the object can send rays to the periphery of the lens, where they will be bent too much. In particular, distortion changes from *barrel distortion* to *pincushion distortion* as the stop is moved from in *front* to *behind* the lens. The diaphragm in a camera, as well as your eye's iris, which serves as a stop, is placed *between* the lenses. The pincushion distortion of the front lens is then balanced by the barrel distortion of the rear lens, giving a relatively undistorted image.

These aberrations often are minor effects, but for precise work, as required in modern optical instruments, they become significant. Lens designers try to reduce these effects, usually by adding additional lens elements, but also by choosing lens *shapes* carefully. For example, a focusing lens with a given focal length may have both sides bulging outward (a **double-convex lens**), or only one side bulging outward and the other flat (a **plano-convex lens**) or even curving inward slightly (a **meniscus lens**)—just so long as the middle of the lens is thicker than the perimeter. However, their aberrations will be different. Indeed, simply turning a plano-convex lens around changes the aberrations it produces. The size of these aberrations also depends on the location of the object, so the design must depend on the use of the lens.

Many of these aberrations may be seen with a magnifying glass and the help of the TRY IT.

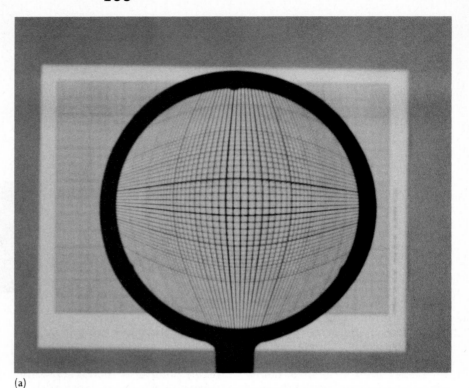

(a)

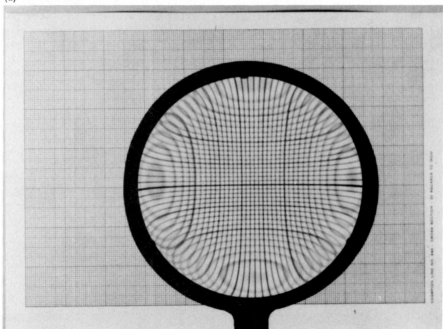

(b)

PONDER

Convince yourself, using a similar construction to that of Figure 3.21a, that a plano-convex lens, facing either way, is a focusing lens.

FIGURE 3.42

Graph paper with square rulings as seen through a lens that exhibits distortion. (Photographs taken by method described in the TRY IT.) **(a) Barrel distortion. (b) Pincushion distortion.**

TRY IT

FOR SECTION 3.5D
Aberrations of a magnifying glass

You will need a focusing lens, such as a magnifying glass, preferably of large diameter. Image the grid of a window with sunlight shining through onto a piece of paper. Can you get every point of the window grid in focus at once, or does one position of the paper provide a better focus for some parts, and other positions for others? What aberration is this? Do straight lines of the grid image as straight lines? If not, what type of distortion do you have? Does twisting the lens slightly affect the aberrations? Why should it?

Make a point source by covering a light source with aluminum foil that has a pinhole in it. Image this source on a piece of paper. How sharp an image can you make? Is there spherical aberration? Can you see any colors around the image? What happens to the image when you twist the lens? What kind of aberration can you see?

Look through the lens at the lines on a piece of ruled paper. The lens should be just far enough from the paper (somewhat beyond a focal length) so that the image is inverted. Your eye must be far enough from the lens so that the image is in focus for you. What kind of distortion do you see? Now move the lens closer to the paper (a little less than a focal length away) and examine the distortion again (also see Fig. 3.42).

How can there be distortion without any stop evident? The answer is that the pupil of your eye is usually the stop when you are looking through the lens. Because of this stop, a lens sometimes gives a better image when you look through it than when you use it to project an image. You can see this effect best with a large lens. Try it!

SUMMARY

Lenses and **mirrors** form **images** depending on the relative location of the object. The images are located using the *rules of* **ray tracing.** Light actually passes through **real images,** but only seems to come from **virtual images.** A plane mirror produces virtual images behind its surface. *Multiple reflections* may produce many images, for example, in a **kaleidoscope.**

Both *spherical* lenses and mirrors are characterized by their **focal lengths** (f). A spherical *mirror* has only one **focal point** (F); it is midway between the mirror and its *center of curvature* (C). Lenses have two *focal points* (F,F'). For a *thin lens* they are on opposite sides and equidistant from the lens. *Positive focal length lenses* are **converging** lenses, while *negative* focal length lenses are **diverging** lenses. Converging lenses can be used to produce a *retroreflector* **(heiligenschein).** **Fresnel lenses** are made thin by eliminating much of the glass (or plastic). The **power** (in **diopters** = 1/f, f in meters) of a compound lens, made of two or more thin lenses in *contact,* is equal to the *sum of the powers* of its component lenses.

Aberrations are imperfections of the image produced by a lens or mirror. **Chromatic aberration** in lenses results from dispersion and can be corrected using an **achromatic doublet. Spherical aberration** can be corrected by using a lens or a mirror of nonspherical shape, or by allowing, through the use of **stops,** only **paraxial** rays (near the axis and nearly parallel to it) to be focused by the lens or mirror. A *parabolic mirror* has no aberrations for distant on-axis sources. **Off-axis aberrations** in lenses (**curvature of field, coma, distortion** such as **pincushion** or **barrel distortion)** can be reduced by proper placement of stops. They depend on the lens shape (**double-convex, plano-convex, meniscus,** etc.).

PROBLEMS

P1 Copy the figure, which shows a top view of a (flat) face looking in a mirror. (a) By drawing rays, show where the image of the right ear is. To locate the image the proper distance behind the mirror, you will

Mirror

Eyes
Ears

need two rays from the right ear that obey the law of reflection.
(b) Complete the image and use it to draw the ray from each ear to the right eye. (c) The person with this flat face wants to mark the outline of her face on the steamed-up mirror. She closes her left eye and, using only her right eye, writes an R where she sees her right ear and an L where she sees her left ear. Clearly indicate where she writes those letters on the mirror.

P2 An arrow is viewed through a kaleidoscope that has an angle of 45° between its mirrors. Draw two lines meeting at a 45° angle to represent the mirrors, and, between them, an arrow pointing at one of them. Sketch all the images of the arrow.

P3 (a) Use ray tracing to find the image of a 2-cm arrow in a concave mirror of focal length 2 cm, when the arrow is 6 cm in front of the mirror and perpendicular to the axis. (b) Is the image real or virtual? Upright or inverted? Larger or smaller than the object?

P4 Repeat Problem P3, but with the object a 1-cm arrow, only 1 cm in front of the mirror.

P5 Repeat Problem P3 for a convex mirror of focal length 2 cm, with the object 2 cm in front of the mirror.

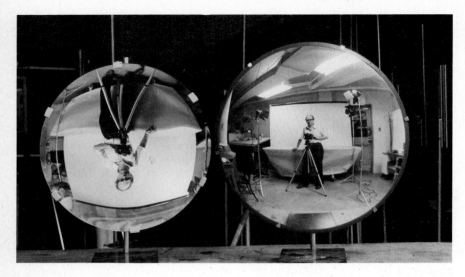

P6 In the photograph, identify which mirror is convex and which is concave.

P7 (a) When you use a lens as a burning glass, it forms a small bright spot. Is the lens then forming an image, and if so, of what? (b) Why does it burn? (c) What would the bright spot look like during a partial solar eclipse?

P8 Here is a problem that occurs almost every day. You are stranded on a sunny desert isle with only a standard flashlight but no batteries. How could you use some part of the flashlight to ignite a small dry leaf, and thus start a fire?

P9 (a) Use ray tracing to find the image of a 2-cm arrow due to a converging lens of focal length 5 cm, when the arrow is 10 cm in front of the lens and perpendicular to the axis. (b) Is the image real or virtual? Upright or inverted? Larger, smaller, or the same size as the object?

P10 Repeat Problem P9, but with the object only 2.5 cm in front of the lens.

P11 Repeat Problem P9, but for a diverging lens of focal length −5 cm, with the object 10 cm in front of the lens.

P12 The word "LIGHT" appears as shown in the figure, when viewed through a lens. What kind of distortion is this? Where was the stop, probably?

LIGHT

P13 Do front surface mirrors have chromatic aberration?

P14 Images in a very thick glass mirror may be slightly colored by dispersion. Draw two vertical lines, separated by about 2 cm, to represent the front (glass) surface and the rear (silver) surface of the mirror. About 2 cm in front of the mirror, draw the object—a white *point* source. (a) Carefully draw red and blue rays to represent the red and blue light. Show that there are separate red and blue images, and indicate their locations. (Hint: To locate each image you'll need two rays. Choose as one the ray perpendicular to the mirror.) (b) Where would the image be if the glass were removed, but the silver and everything else remained in the same positions? Would the image be colored in this case?

P15 A certain thin lens has an extremely large amount of chromatic aberration. Its focal length is 3 cm for blue light and 4 cm for red light. A 2-cm arrow is 8 cm in front of the lens and is illuminated with white light. (a) By ray tracing construct the images for red and blue light. (b) Draw a sketch indicating what you would see on a screen that is placed at the red image.

P16 An object (the usual arrow) is located at the focal plane of a thin lens of focal length *f*. On the other side of the lens, at a distance $\frac{1}{2}f$ from the lens and parallel to it, is a plane mirror. (a) Use ray tracing to find two rays leaving the lens. (b) Use the law of reflection to determine how these rays are reflected. (c) Use ray tracing again to follow the reflected rays back through the lens, and thus find the final image.

P17 In *Tess of the D'Urbervilles*, Thomas Hardy writes:
Then these children of the open air, whom every excess of alcohol could scarce injure permanently, betook themselves to the field-path; and as they went there moved onward with them, around the shadow of each one's head, a circle of opalized light, formed by the moon's rays upon the glistening sheet of dew. Each pedestrian could see no halo but his or her own, which never deserted the head shadow, whatever its vulgar unsteadiness might be; but adhered to it; and persistently beautified it, till the erratic motions seemed an inherent part of the irradiation, and the fumes of their breathing a component of the night's mist; and the spirit of the scene, and of the moonlight, and of Nature, seemed harmoniously to mingle with the spirit of the wine.
Explain the phenomenon Hardy describes.

HARDER PROBLEMS

PH1 (a) What is the maximum number of images of himself that can be seen by the man in Problem P13 of Chapter 2? (b) If the man raises his right hand, which of these images will raise their right hands, and which their left?

PH2 How was this photo taken? Discuss in detail how many mirrors were used, at what angles, the unequal brightness of the images, which image was formed by direct light (without mirror reflection), and the reason for the double image at the bottom.

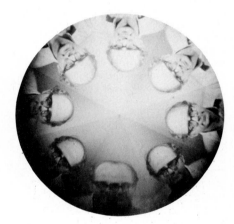

PH3 A shaving or makeup mirror is a large concave mirror. In terms of its focal length *f*, what is the *range* of distances from the mirror at which you can put your face and get an enlarged image in the mirror?

PH4 Will an underwater air pocket in the shape of a plano-convex lens act like a converging or like a diverging lens?

PH5 When you wear goggles underwater, objects usually appear larger than normal. Trace rays from the top and bottom of an object *PQ* to an eye to verify this statement. Draw a vertical line to represent the front of the goggles, with the eye on one side of the line, and water and the object on the other. Use Snell's law to trace the rays at the front of the goggles.

PH6 (a) Without goggles, things appear blurred underwater to you. Why? (b) The goggles form a plano-concave pocket of air in front of your eye—the flat part is formed by the glass (or plastic) plate, while the concave part is formed by your cornea. Use your answer to problem PH4 to explain why diving masks allow you to see clearly underwater.

PH7 The figure shows a periscope used to view a bomb being defused from a safe position behind a thick wall. (a) Find the image of mirror 2 seen by the eye, by locating the images of points A and B in mirror 1. (b) Find the final image of the arrow as seen by the eye. If you do this by ray tracing, you must draw the rays *very* accurately. Preferably do it by first finding the intermediate image of the arrow in mirror 2, and then finding the final image in mirror 1. Before starting to draw, think about where the intermediate image will be, so you leave plenty of room for it. Make sure the distance from the eye to the final image is equal to the distance that the light actually travels from the object to mirror 2 to mirror 1 and then to the eye. (c) Draw the rays that define the field of view of the eye—that is, the rays that just hit the edges A and B of mirror 2. What happens to this field of view if the periscope is made longer to see over a higher wall?

figure out the only place the lens can be placed to produce the image. (b) Now that you have located the lens, use ray 1 to locate the focal point F'. (c) Similarly, use ray 3 to locate the focal point F. (d) Since you know where the image is, you know how any ray from the object behaves. Draw one ray from the point of the object arrow to an eye located on the axis 5 cm to the right of the lens. (e) Does the image look larger or smaller to the eye than the object did without the lens? Is the image real or virtual? Is the lens converging or diverging?

PH9 A spotlight consists of a light source (say an incandescent bulb), a concave mirror, and a converging lens. The bulb is located at the center of the mirror, and the lens is a distance equal to its focal length in front of the bulb. Draw a sketch of such a spotlight. (a) Where is the image of the bulb in the mirror? (b) Where is the final image of the bulb after the light has passed through the lens? (c) Draw a typical ray that goes directly from the bulb through the lens. (d) Draw a typical ray that goes from the bulb via the mirror through the lens. (e) Explain why each of the elements is used in a spotlight.

PH10 A *beam expander,* used to widen a laser beam, consists of two lenses. A laser sends parallel light into the expander. In order to be able to see the effect, draw a beam of parallel

PH11 A 1-cm arrow is located 1 cm in front of a converging lens of focal length 2 cm. On the other side of the lens, a distance 4 cm from it, is a second converging lens of focal length 6 cm. Use ray tracing to locate the final image produced by the two lenses. First find the image produced by the first lens alone, then use that image as a new object for the second lens, as if the first lens were no longer there.

PH12 An object (the arrow, as usual) is located at a distance 2f in front of a converging lens. (a) Use ray tracing to find the image. (b) Now draw a plane mirror at 45° to the axis, crossing the axis at a distance 4f behind the lens, and locate the final reflected image. (c) Repeat parts (a) and (b), but with the mirror at a distance f behind the lens.

MATHEMATICAL PROBLEMS

PM1 By finding two suitable congruent triangles in Figure 3.2a, prove that the distances $\overline{QO}$ and $\overline{Q'O}$ are equal.

PM2 Show that the central ray through a thin lens is undeviated, as follows. Consider the center of the lens as a slab of glass with *parallel* sides. By applying Snell's law (Eq. B5) twice, show that any ray entering one side of such a slab will emerge from the other side parallel to its original direction. The displacement between these parallel beams will be small for a thin lens.

PM3 The focal length of lens 1 is -25 cm, and the focal length of lens 2 is $+75$ cm. What are the powers (in diopters) of the two lenses? When the two lenses are brought together to form one compound lens, what is the resulting power? What is the focal length of the compound lens?

PM4 (a) Repeat Problem P3a without ray tracing, by using the mirror equation. (b) What is the magnification?

PM5 (a) Repeat Problem P4a without ray tracing, by using the mirror equation. (b) What is the magnification?

PM6 (a) Repeat Problem P5a without ray tracing, by using the mirror equation. (b) What is the magnification?

PM7 (a) Repeat Problem P9a without ray tracing, by using the lens equation. (b) What is the magnification?

PM8 (a) Repeat Problem P10a without ray tracing, by using the lens equation. (b) What is the magnification?

PM9 (a) Repeat Problem P11a without ray tracing, by using the lens equation. (b) What is the magnification?

PM10 Verify your result of Problem PH8 by seeing if it satisfies the lens equation.

PM11 Repeat Problem PH11 by using the lens equation twice, for the same two steps that you used when doing the problem by ray tracing.

Mirror 2

B

A

Mirror 1

ARROW BRAND BOMB

PH8 On a page turned sideways draw a horizontal line to represent the axis. Near the middle of this line, draw a 3-cm arrow to represent an *image* created by some lens not yet drawn. At a distance 2 cm to the right of this image, draw a 2-cm arrow to represent the *object* that produced this image. (Both arrows should point up.) Your job is to figure out what kind of lens will produce this image, and where it must be placed. Do this in steps as follows: (a) Use ray 2 to

light 2-cm wide to represent the laser beam. The first lens is a diverging lens of focal length -3 cm. (a) Draw the two rays that represent the edges of the beam after they pass through the first lens. (b) A second lens is located 6 cm after the first lens. It makes the rays parallel again. Should the second lens be a diverging or converging lens? Draw a suitable lens and the final beam. (c) What should be the focal length of the second lens?

FOCUS ON. . . .

Solar power

To produce solar power, you must collect energy from the sun, and extract that energy for a useful purpose, such as heating your home or making electricity. At best, any solar **collector** can absorb all the energy in the sunlight that it intercepts. Although the energy flux from sunlight that arrives at the ground varies greatly with weather conditions, it seldom exceeds a kilowatt per square meter. Thus, to yield an output of, say, 5 kilowatts, the collector surface must be at least 5 m^2. For maximum efficiency, of course, the collector should be oriented perpendicular to the sunlight's direction of incidence.

Some of today's large-scale collectors use a technique not much different from Archimedes' (Sec. 3.3D). In the **solar tower** (Fig. FO.2), a large flat area is covered with mirrors, each reflecting a beam from the sun onto a boiler on top of a tall tower, where water is heated to steam that drives generators of electricity. Instead of Archimedes' soldiers, automatic clock-driven mechanisms, called **heliostats,** are used to make each of the hundreds or thousands of mirrors follow the sun. (The many synchronized heliostats constitute a significant cost of these systems.)

Another solar collector that concentrates in a small area all the energy that falls on a large surface consists of a spherical or parabolic

---

FIGURE FO.2

Photograph of the Central Receiver Test Facility near Albuquerque, NM. Over 200 mirrors (some of which are shown in the foreground) are used to direct up to five million watts of sunlight toward the top of the tower, where different types of collectors can be tested.

FIGURE FO.3

This parabolic reflector, in Bhavnagar, India, is made of many small plane mirrors. The sunlight is concentrated on a Sterling engine (similar to a steam engine, but it uses no water) that pumps water from a well.

reflector (Sec. 3.5C). The boiler, or *absorber,* is at the focus of the reflector. To make a large reflector surface sufficiently cheaply, it is often made of individual pieces of plane mirrors (Fig. FO.3). As in the case of the solar tower, a heliostat is needed to keep the reflector-absorber combination pointed at the sun.

If not as much concentration is needed, one uses *cylindrical* collectors. These may consist either of a cylindrical mirror or a cylindrical Fresnel lens (Sec. 3.4D) to concentrate the light onto a cylindrical absorber. If the cylinder axis is oriented in the east-west direction, these collectors need not be moved to follow the sun. A simpler type of solar collector that does not concentrate the energy at all involves just an absorbing *flat plate.* Such collectors are used on rooftops to heat up water flowing through them.

The reasons for using concentrating, rather than flat-plate, collectors involve questions of heat transfer and energy conversion. If solar energy is to be used for heating, the heat can be developed at low temperature; warm air from a solar greenhouse, or warm water from a flat-plate collector on your roof, is quite adequate for most home requirements. However, suppose that the *output* is to be electrical energy, but that the solar energy is to be *absorbed* as heat (rather than by direct conversion—Sec. 15.2B). This heat must then be developed at high temperature in order to be efficiently convertible to electricity, so concentrators are used. There is a limit, however, to how high a temperature can be developed by concentrating the sun's rays, for the following reason. A solar parabolic concentrator forms an image of the sun, which acts as a black body at the sun's surface temperature, 5500°C, at best. Anything placed at the focus of any concentrator, therefore, can be heated to a temperature no higher than this. In practice, the limit is much lower due to heat losses and imperfection of the image. For example, if the heated surface is large, it is likely to lose heat rapidly. Also, if the temperature of the heated surface is high, it is likely to lose heat much more rapidly than if it is kept relatively cool (e.g., by rapid circulation of the water in a flat-plate collector).

Heat losses are classified into three types: *conduction, convection,* and *radiation.* Conductive loss is due to heat transmission in matter, such as the collector supports, where there is no motion of the matter itself. Convective loss is due to heat carried away by some moving material, usually air. Radiative loss is due to electromagnetic radiation, which carries away some of the energy. Conductive losses are minimized in flat-plate collectors by backing them with insulation material, such as styrofoam. Convective losses are reduced by covering the collector with one or several sheets of glass at a distance of a few centimeters from the collector, which prevents rapid motion of the air—even at a larger distance, as in a greenhouse, a piece of glass or plastic can still aid significantly in reducing convective losses. (However, each glass-air surface reflects at least 4%, and more at oblique incidence. A comparable amount may also be absorbed by the glass.) Radiative losses can also be controlled by the glass cover's greenhouse effect (see the FOCUS ON Light, Life, and the Atmosphere). For more efficiency in combating radiative losses, the collector can be given a *selective surface*—one that does *not* behave like a black body. This surface should absorb in the visible, where the energy of sunlight is greatest. However, it should be a poor radiator in the infrared so it doesn't radiate away its energy as it heats up. Such a surface can be made by coating a metal (which radiates poorly in the visible and infrared) with a layer of semiconductor (which absorbs in the visible, but is both transparent and does not radiate in the infrared).

Clearly, the surfaces of solar collectors are very important. Keeping them clean, free from dust, and protected from oxidation, corrosion, and hailstorms is a major problem in any large-area collection scheme.

The Camera and Photography

CHAPTER 4

4.1

INTRODUCTION

*Photography** resulted from the fortunate and ingenious union of the optics of the camera and the chemistry of light-sensitive materials made into film. For several centuries, people amused themselves with cameras of various kinds, from small portable models used as artists' aids, to large rooms you could enter and see realistic, moving, full-color projections of the scenery and life going on outside (Fig. 2.7). But the key step of recording such images permanently was not taken until about 150 years ago, when the precursors of our modern film were developed. Since then, a steady improvement in both cameras and film has resulted in a wealth of sophisticated photographic devices: from the instant camera that delivers a sharp, colorful photograph moments after a button is pushed, to specialized astronomical cameras for recording distant galaxies, to cameras capable of revealing the different stages of the explosion of a balloon as a rifle bullet pierces it. The veracity of the medium is summed up by the Japanese ideographs for "photography": *sha-shin,* literally "copy truth." The ubiquity of photography and its power to capture not only the ordinary, but the distant, the transient, the colorful, the otherwise unobservable, has greatly expanded our knowledge (not to mention our aesthetics). In 1925, when cameras were primitive by today's standards,

the artist and photographer László Moholy-Nagy wrote: "We have—through a hundred years of photography and two decades of film—been enormously enriched. . . . WE MAY SAY THAT WE SEE THE WORLD WITH ENTIRELY DIFFERENT EYES."

Let's see how a camera manages to turn an eye on the world.

A. The essential parts of a camera

To produce a photograph, a **camera** must project on its film a good image of controlled intensity for a controlled amount of time. A somewhat old-fashioned camera (Fig. 4.1) is simple and large enough to illustrate all the essential parts clearly. The bulk of it is a *light-tight box,* which allows only the desired light to fall on the film. It has a **lens**

FIGURE 4.1

(a) Photograph of a large-format camera. The brass cylinder contains the lens. The back of this camera can carry either a ground glass screen for viewing or a film holder. **(b)** The essential parts of this camera.

(a)

(b)

---

*Greek *photo,* of light, plus *graphein,* recording.

in front to project a real image on the film, and a device to move this lens in order to focus the image on the film. The camera has an adjustable **diaphragm** to control the intensity, and a **shutter** to control the duration, of the light falling on the film. The back of the camera carries the **film** and devices to advance or change it between exposures. This fundamentally simple gadget allows us to do something fairly incredible: to preserve a permanent image of the world around us with almost no effort on our part.

The minimal effort required has always been one of the attractions of photography. In 1838, shortly after he perfected his process of making a permanent photograph, Daguerre commented on the ease of the process: "the little work it entails will greatly please the ladies . . . ." By 1889, George Eastman, the founder of Kodak, used a less sexist version of this statement to popularize his camera: "You push the button, we do the rest."

Many ingenious mechanical components have been invented to make picture-taking easy. For example, the bellows allows the lens to move while keeping the box light tight, cartridges let you put film into a camera without carrying a darkroom tent with you (as the early photographers did), and sprockets advance the film by precise amounts. You can easily figure out the purpose of many of these mechanical parts, so let's consider some of the less obvious optical components of a camera.

4.2
FOCUSING THE IMAGE

In Section 3.4C, we saw how to locate the real image produced by a converging lens. To record this image, we put our film in the plane of the image. If the object were at a different distance from the lens, the image would be located on a different plane and we would have to relocate the film (or the lens)—the process of **focusing.**

A. Depth of focus, depth of field

Suppose you are taking a photograph of a friend and you have focused her image onto the film. No matter how careful you are, her image generally will not lie exactly on the film and thus will be slightly blurred there. Luckily, you needn't have the image perfectly focused for the photograph to come out acceptably sharp. A little bit of blurring is not noticeable to your eyes, and in any event, all film has limitations that prevent it from recording absolutely sharp pictures, and there is always some blurring due to lens aberrations and motion of the object or camera. All of this means that if the image is not exactly on the plane of the film, the photograph may be only negligibly blurred. Thus there is a *range of film locations*, called the **depth of focus,** for which the photo of an *object at a given distance* comes out acceptably sharp (Fig. 4.2a).

FIGURE 4.2

(a) Depth of focus. For the range of film positions shown, the image of the fixed object will be acceptably in focus.
(b) Depth of field. Objects within the range shown are acceptably in focus for the particular lens-film distance chosen.

Suppose now that you wish to photograph *two* friends, each at a *different* distance. If you move your film to make one friend's image in focus, the other friend may be slightly *out* of focus. If your friends are at nearly the same distance, however, both images will be reasonably sharp. The *range of object distances* that result in an acceptable photograph, for a *given film location*, is called the **depth of field** (Fig. 4.2b). If one person is very close but the other is very far away, you cannot focus on both of them simultaneously—they both do not lie within the depth of field of your camera and hence at least one of their images will be significantly blurred. For this reason we needed two photographs for Figure 3.7. (Notice that the scratches on the plane mirror are visible only when the camera is focused on the image that is nearest to them.)

The depth of field depends on how far away you are focusing. For instance, no matter how far separated your friends are, if they are *both* very far away, their images will lie close to the focal plane of the lens and will be acceptably in focus—the depth of field is largest for distant subjects.

Adequate depth of field helps solve many focusing problems. Not only does it allow us to photograph

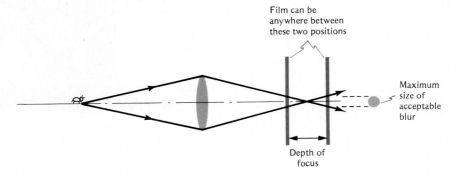

(a)

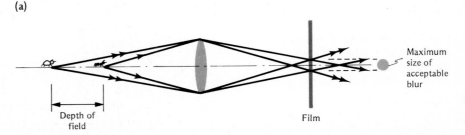

(b)

three-dimensional (rather than only flat, planar) objects, it also lets us sometimes get away without having to focus at all **(fixed-focus camera).** Suppose a camera lens is mounted at some fixed distance from the film, such that the image of an object 3 m away is exactly on the film. Suppose further that the depth of field is from 1.5 m to infinity. Such a camera works fine, without the need to focus, for snapshots of people and landscapes (because the polite photographer does not approach people closer than 1.5 m). Many inexpensive cameras are made this way.

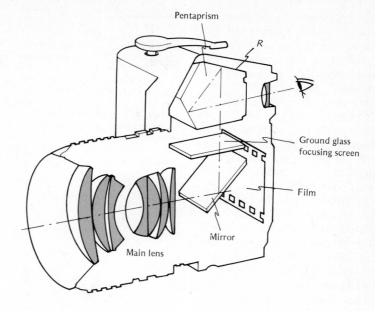

B. The view camera

To enable focusing on close as well as on distant objects, the lens (rather than the film) is usually made movable within the camera body. Moving the lens moves the image, so it can be manipulated to fall on the fixed film position.

One way to know when the camera lens is adjusted to its proper position is to replace the film by a ground glass screen and look at the image to see if it is sharp (Fig. 4.3). After focusing, the ground glass screen is replaced by a film holder, a slide that covers the film is removed, and then the film can be exposed. Although this somewhat lengthy procedure does not suit most people's fast-moving life style, such **view cameras** are still used for exacting work.

C. The single-lens reflex

A modern version of focusing by viewing the image directly is found in **single-lens reflex (SLR) cameras** (Fig. 4.4). Instead of exchanging ground glass and film, the image is first projected on the ground glass by a 45° mirror* that flips out

*Which reflects, hence "reflex."

FIGURE 4.3

(a) The traditional view of a traditional photographer looking into the back of the camera, with a black cloth thrown over his head to exclude stray light so he can see the screen. (b) The actual view that the photographer may have as he looks into the back of his view camera.

FIGURE 4.3

(a) The traditional view of a traditional photographer looking into the back of the camera, with a black cloth thrown over his head to exclude stray light so he can see the screen. (b) The actual view that the photographer may have as he looks into the back of his view camera.

(a)

(b)

FIGURE 4.4

The main parts of a modern single-lens reflex camera. The distance of the light path from the lens to the film is the same as from the lens to the focusing screen by way of the mirror. The pentaprism inverts the image on the focusing screen before the photographer sees it. The roof *R* of the pentaprism consists of two faces that provide extra reflections perpendicular to the plane of the figure in order to reverse the image in that direction as well.

of the way just before the picture is taken, allowing the film to be exposed. To view the horizontal ground glass **focusing screen** directly, the photographer would have to look down into the camera, and he would see a left-right reversed image. To avoid this image reversal, the viewfinder usually is provided with a **pentaprism,** a device that rotates the image for horizontal, nonreversed viewing.

The SLR's viewing system also may incorporate other interesting tricks to make the focusing easier: the ground glass, which scatters light in all directions, is sometimes replaced by a Fresnel **field lens,** which directs most of the light toward the eye, giving a more nearly uniform, bright image, especially around the edges (see Sec. 6.6A).

Another trick relies on the fact that a prism displaces the apparent location of an object by an amount that depends on the distance of the object from the prism, and on the orientation of the prism. If you view this page through a prism, it will look displaced. Reversing the direction of the prism (so the base of the wedge is toward the left instead of the right) reverses the direction of displacement. As the prism gets closer to the paper, the amount of displacement gets smaller. When the prism rests on the paper, there is no displacement (for a thin prism). Therefore, if a pair of prisms that slant in opposite directions (a **biprism**) is mounted on the focusing screen, the images seen through these prisms are displaced in opposite directions, *unless* the images lie exactly *on* the focusing screen. Unless the image is properly focused, then, you see a **split image** (Fig. 4.5). Since your eye can detect a break in a straight line with remarkable precision, this allows you to focus very accurately.

On some cameras, many small biprisms **(microprisms)** are distributed all over the viewfinder's field of view. The image is then broken up everywhere when it is not in focus, with the result that it appears to go out of focus much more quickly when the camera is defo-

FIGURE 4.5

(a) A biprism of the type often mounted in the center of the ground-glass screen of an SLR. **(b)** Effect of one of the prisms: When image *I* is on the focal plane, it is seen undisplaced (left). When it is below (or above) the focal plane, it is seen displaced, as *I'* (right). **(c)** Photograph of the resulting split-image effect, when in focus (left) and out of focus (right). (Note the ring of microprisms around the central biprism.)

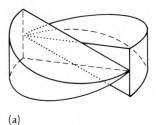

(a)

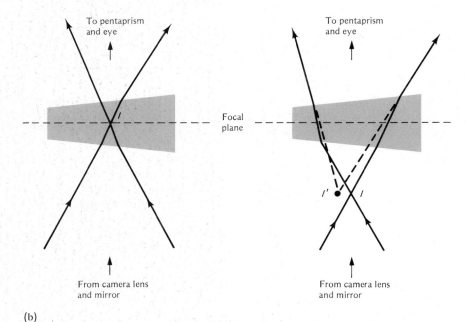

(b)

(c)

cused than without the micro-prisms. This of course also permits precise focusing.

D. The rangefinder

Cameras other than SLR's employ a different means of focusing. It is easy enough once and for all to calibrate the device that moves the lens. Often the lens moves in and out on a screw fitting, and the object distance corresponding to the various settings is marked on the rotating focusing ring (Fig. 4.6a). To focus by this scale we only need to measure the object distance, for example by pacing it off. One cannot pace to all subjects, though (e.g., down a steep cliff, or across an alligator-infested swamp).

Where no great accuracy is necessary, the *apparent size of a standard length*, such as a typical person, can be used: if the person's head just fills the viewfinder, she is quite close; if the frame shows her from head to toe, she is at some intermediate distance; and if she is much less than a frame's size, she must be pretty far away, effectively at infinity. On simple cameras the focusing scale is often marked only according to these three situations (Fig. 4.6b), with satisfactory results.

For more critical work one uses a **rangefinder** based on the principles of surveying *(triangulation),* as shown in Figure 4.7a, where P measures the angle between his line of sight *(PS)* and B's line of sight *(BS)* as well as the (alligator-free) distance from P to B. With the help of a little mathematics, he then deduces the desired distance $\overline{PS}$.

A rangefinder follows the same principle (Fig. 4.7b) but replaces B by a rotatable mirror M_B, P by a half-silvered mirror M_P, and P's mathematics by a built-in scale that converts the angle of M_B into the distance $\overline{PS}$. You fool around with M_B until the image seen reflected in M_P and M_B coincides exactly with that seen directly through M_P. In fact, M_B can be directly coupled to the camera's lens position so that the camera is always focused on the

(a)

(b)

FIGURE 4.6

Photograph of distance scales on photographic lenses. **(a)** Distance scales in meters and feet. Pointers not only indicate the distance of best focus, but also bracket the depth of field. Note how the close distances are spread out, and the far distances are crowded together on this scale, corresponding to the way the lens crowds together the images of distant objects. **(b)** Size of common objects as seen in viewfinder may serve as distance indicator.

place where the two rays seen through the rangefinder intersect.

The rangefinder can do double duty if it is also coupled to the view-finder window (the window through which you frame your picture) to correct for **parallax.** Suppose the viewfinder is 5 cm above the camera lens. Then the field seen by the viewfinder is 5 cm higher than what the camera photographs. These 5 cm make no difference in a picture of a distant landscape, covering many meters or kilometers; but a closer object might miss its top 5 cm in the finished photo. Tilting the viewfinder by an amount determined by the rangefinder setting can eliminate this parallax discrepancy.

(a)

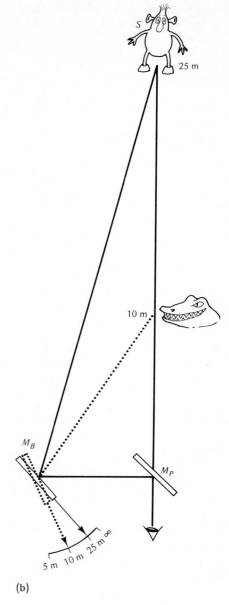

(b)

FIGURE 4.7

(a) The rangefinder problem: Our photographer *P* wants to measure the distance $\overline{PS}$ to a subject *S*. A bystander *B*, a little distance away, is also staring at *S*. **(b)** Solution to the rangefinder problem.

Rangefinders have been automated for cases where speed (or laziness) is of the essence. Some cameras send out an infrared beam and hunt for its reflection from the object by a displaced rotating mirror, similar to M_B of Figure 4.7b. Others use the echo technique discussed in Section 1.1B, not with light but with a slower sound pulse (sonar).

PONDER

Which of these automatic rangefinders would work on the moon (where there is no air)?

4.3
EFFECTS OF FOCAL LENGTH

High-quality cameras can not only move their lenses for focusing, they can remove them entirely and replace them with lenses of different focal lengths; or they have a zoom feature, which smoothly changes the focal length of a lens. The point is to be able to change the *size* of the image without having to change the camera's position.

A. Telephoto and wide-angle lenses

Consider an object sufficiently distant that the image is at the lens' focal plane, to a good approximation. Since the central rays (ray *2*) are undeviated by any lens, the size of the image must be proportional to the lens' focal length (Fig. 4.8). So a lens of larger focal length gives a larger image, as if you were photographing through a telescope. A long focal length lens is a ***telephoto lens,*** and is what the photographer of Figure 4.7a should use (rather than trying to get closer to *S*). A telephoto lens makes a big image of a small portion of the scene, so it

---

TABLE 4.1 *Focal length and angle of view for 35-mm cameras*

Focal length of camera lens	Wide angle			Normal	Telephoto		
	17 mm	28 mm	35 mm	50 mm	85 mm	135 mm	300 mm
Diagonal angle	104°	75°	63°	47°	29°	18°	8.2°
Horizontal angle	93°	65°	54°	40°	24°	15°	6.9°
Vertical angle	70°	46°	38°	27°	16°	10°	4.6°

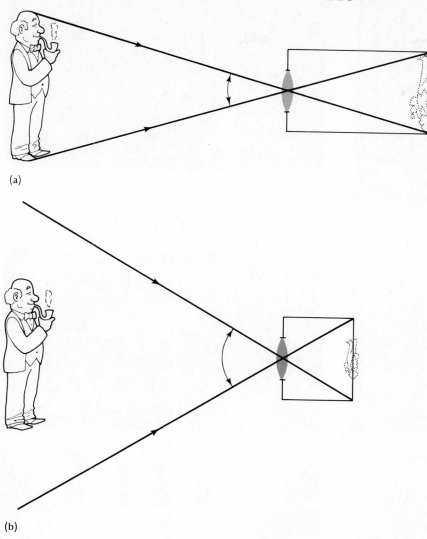

(a)

(b)

FIGURE 4.8

Image size and angle of view depend on lens-film distance. The two lenses have different focal lengths and must be at different distances from the film to produce a sharp image of a given object. **(a)** The long focal-length lens produces a large image of the object and has a small angle of view. **(b)** The short focal-length lens produces a small image of each object and has a large angle of view.

corresponds to a small **angle of view** (Fig. 4.8a). Angle of view is specified in degrees—the angle subtended at the camera by the entire scene recorded, customarily measured along the diagonal. Table 4.1 gives this angle for various lenses on a 35-mm camera.

The opposite effect is achieved by a lens of *short* focal length. Here the image of any object is smaller, hence more objects will appear in the same film frame; the picture will take in a larger angle of view. Such a short focal length lens is called a **wide-angle lens.** It makes an image of a larger portion of a scene (Fig. 4.8b), and, therefore, is useful when you cannot step back to encompass all of a large object in your picture, for example, to snap a large building from across a narrow street. Lenses of an intermediate focal length, neither telephoto nor wide angle, are called **normal** (Fig. 4.9). Experience shows that pictures appear normal when the focal length of the lens roughly equals

the diagonal of the film frame. For example, as 35-mm frame has dimensions 24 × 36 mm, and a diagonal of 43 mm. So 40 to 50 mm is a normal lens for a 35-mm camera. Whereas a $3\frac{1}{2}$ × $4\frac{1}{2}$ inch Polaroid picture has a 12-cm diagonal, which is close to the focal length of the normal lens on a Polaroid camera.

Another type of special lens is called for when you want to photograph small objects. One way to get a large picture is to approach the object closely. However, the closer the object is to the lens, the farther the image is. For close-up work, the lens-film distance needs to be considerably greater than the focal length (e.g., for 1:1 reproduction it should be *2f*, see Fig. 3.26), and the usual focusing range of the lens does not extend that far. Hence, you can either put the lens on a special **extension tube** or **bellows** (to increase the lens-film distance), or you can use a close-up lens (Sec. 4.4D). Another problem associated with close-up work is that you must get uncomfortably close to the subject, for example when trying to photograph the pistil and stamen deep in a flower. You could use a lens with long focal length (as long as that of a telephoto lens) whose aberrations have been corrected for close-up work. Such a **macro lens** demands a large film-lens distance, but allows you to move back comfortably from the subject.

If you want an image that is really large, larger than the object, you can get a reversal ring, which allows you to reverse your telephoto lens, so that the end usually facing the film faces the object. This is done because the lens aberrations are corrected assuming a distant object and a nearby image. The lens is equally well corrected if you run the light rays backward, as long as the "front" is always facing whichever is more distant, object or image. If the object is at about the focal length, the image must be considerably farther from the lens than its normal distance. In this way you can get considerable magnification, but, as we'll see in Section 4.5B, you also sacrifice some light.

f = 8 mm

f = 28 mm

f = 50 mm

f = 75 mm

f = 100 mm

f = 150 mm

f = 200 mm

f = 400 mm

FIGURE 4.9

FIGURE 4.9

Photographs taken by a 35-mm camera with lenses of different focal lengths, as given below the pictures. Notice that the telephoto pictures are the same as smaller regions of the *f* = 50-mm picture.

B. Perspective

It is sometimes said that wide angle and telephoto lenses distort **perspective.** This is not strictly true; a telephoto picture is the same as a small portion of a normal picture taken from the same place (see Fig. 4.9). You could make the two pictures identical by enlarging the central portion of the normal picture. But enlarging the normal picture only increases the sideward dimensions, not the apparent depth. So it is as if in the original scene everything kept its *distance* from you but grew in the *lateral* directions by the enlargement factor. Judged by your normal conditioning of space perception, objects along the line of sight look crowded in such photos. That is, under normal viewing conditions, a *telephoto lens* gives a *larger image* (side to side), as if you were closer, but the *same perspective* (depth relations) that you had from the point at which you took the picture (see the TRY IT and Appendix G).

Conversely, in a *wide-angle picture* it is as if everything in the picture had *shrunk* in the *lateral directions* only, but kept its distance along the line of sight. A person's nose then seems to bulge out far too much toward the camera, and looks too large compared with the size of the head. If you place your eye sufficiently close to a wide-angle picture so that the picture subtends the *same* large angle at your eye as the original scene did at the camera, everything looks normal again (Fig. 4.10).

FIGURE 4.10

Object distance and focal length were changed simultaneously in these photographs so as to keep the size of the main subject constant. The size of objects at other distances varies, changing the perspective: **(a)** telephoto perspective, **(b)** normal perspective, **(c)** wide-angle perspective. If you hold your eye about 5 cm from **(c),** the perspective should look normal again. (You probably cannot focus clearly on such a short distance, unless you use a magnifying glass.)

(a)

(b)

(c)

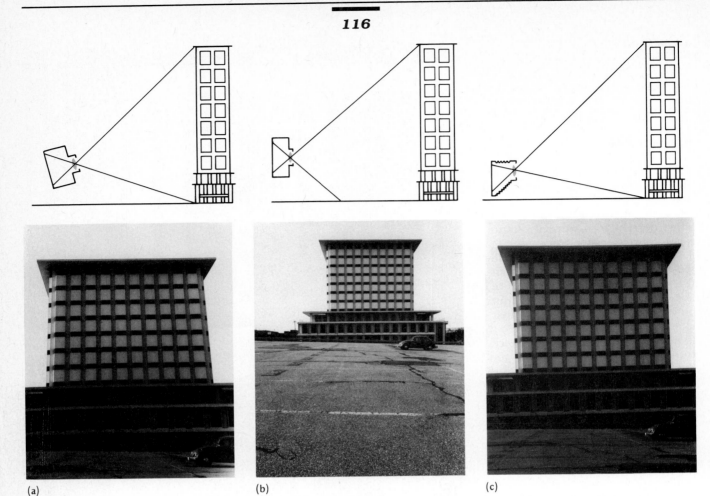

(a)   (b)   (c)

FIGURE 4.11

Converging lines—problem and solution. The upper row shows the camera arrangement, the lower row shows the corresponding photographs. **(a)** Camera tipped up, film not parallel to building, converging lines. **(b)** Camera horizontal, film parallel to building, using wide-angle lens; photograph shows building as desired in upper half, too much foreground in lower half. **(c)** Camera with PC lens, film parallel to building. Photograph is equivalent to an enlarged version of the upper half of photo **(b)**.

These wise remarks don't help much, however, if the only way to photograph that tall building across the street is by using a wide-angle lens, and the picture then shows strongly converging lines, with the base of the building much too large and the top too small. You can't ask your valued architect client to view the picture only 5 cm from his nose. Fortunately, when the trouble is es-sentially in only one plane (the fa-cade of the building) you can do something about it (Fig. 4.11): don't tip the camera up! If you have a really wide-angle lens, half of your picture will then show the building with fine parallel lines, and the other half will show the pavement of the street in great detail (that half you cut off and throw away). Equiv-alently, you could slide the film frame vertically with respect to the lens so it only covers the part of the picture you really want. This is where the flexibility of a view cam-era with bellows (or a **_perspective control (PC) lens_** for an SLR) is handy, allowing you to slide the film frame independently of other parts of the camera. The second TRY IT for Section 4.4B gives an example.

Another way to correct for con-verging lines is in **_enlargement._** Since an enlarger is just a camera in reverse, the image it produces would coincide with the building if the enlarger were placed where the camera was originally. If you want to catch the image on an easel in-stead, you have only to tilt the easel by the same amount with respect to the enlarger as the building was tilted with respect to the camera in order to get rid of the converging lines. You're probably more familiar with the converse process: a projec-tor with a correctly proportioned slide and a screen that is not at right angles to the beam. This gives diverging lines where parallel lines should be in the projected image, and makes the top of the frame larger than the bottom. The effect is called **_keystoning_** and happens very frequently in amateur slide or movie shows (see Fig. 6.24). There the center of the screen is fre-quently higher than the projector, so that the latter must be tipped up. This keystoning appears not to be annoying enough to warrant PC lenses in projectors.

TRY IT

You can get a remarkably accurate idea how the finished picture of a scene will look by viewing the scene through a properly positioned frame. Into a piece of cardboard or heavy paper, cut a window of size 12 × 18 cm. Hold the frame at normal reading distance (25 cm) from one of your eyes, and look through it at the scene, closing the other eye. You will see the perspective of a "normal" lens. Hold it at arm's length (50 cm); the view will be that of a moderate telephoto (100-mm lens on a 35-mm camera; Fig. 4.12). To see the perspective of a wide-angle lens, hold the frame correspondingly closer to the eye (for example, at 17.5 cm for the view of a 35-mm lens on a 35-mm camera). Incidentally, this example also shows you that in order to see a picture taken with a normal lens correctly at normal reading distance, the picture should be blown up to 12 × 18 cm (about 5 × 7") size.

PONDER

The dimensions given above can all be figured out by proportions from the fact the a 35-mm frame is 24 × 36 mm, and the normal lens has f = 50 mm. If you are mathematically inclined, check our numbers!

It will be difficult to see the distorted perspective with the frame, because everything still subtends the same angle as without the frame, but you can see the distortion with the help of a large magazine photograph that shows a lot of depth. Check that it appears normal when you look at it from normal reading distance. Moving your eye closer to the picture will make it appear with a telephoto perspective—the same effect that you would have if it had been enlarged more before it was printed—because when you move closer, the two-dimensional picture looks bigger, but the depth relations within the picture don't change. You can enhance the effect by cutting a small rectangular window in a piece of paper and placing it over the picture. A small window (e.g., 6 × 9 cm—the "wallet" snapshot size), viewed from half the normal distance, will correspond to the same scene photographed by a lens with twice the focal length of that actually used. Note that now there is a flattening effect, typical of telephoto pictures.

FIGURE 4.12

*4.4

CAMERA LENSES

A lot of good physics can be learned from the problems of camera lens designs and their solutions. We'll just touch on some of the interesting features here.

A. Aberrations of camera lenses

Aberrations (Sec. 3.5) constitute the main problem of lens design. One method to correct aberrations is to *stop down the lens*, that is, to cover all but the central part of the lens and make it more like a pinhole, giving a geometrically perfect image. Even though this means that the lens will gather less light, many aberrations are cured this way in simple snapshot cameras. However, not all aberrations respond equally well to this treatment. For example, distortion due to an improperly placed stop remains even if the hole in the stop is very tiny. Fortunately, it is not necessary to correct all aberrations fully—they are not equally objectionable in a photograph. For ex-ample, distortion by itself does not blur the picture, and unless the periphery of the picture contains lines we know should be straight, a little distortion will not be noticed. Thus, in designing a lens it is possible to make compromises, and in fact no photographic lens is perfect in all respects.

To illustrate the possibilities, consider the simplest kind of snapshot camera lens (Fig. 4.13). It is just a single lens and a stop, and would seem not to allow much variation for creative lens design. Yet, having decided on a particular focal length, we are still free to choose the *lens*

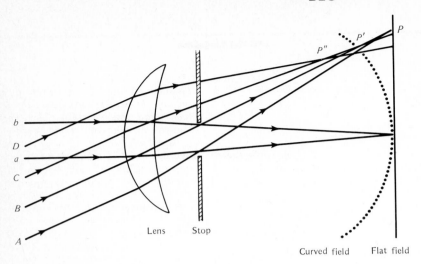

coating of the lens surfaces (Sec. 12.2C) makes them nearly 100% transparent, allowing the use of many lenses. Further, ray tracing calculations for complicated lens systems are now possible on computers. These developments have made possible the excellent reasonably priced lenses of today. **Graded index** and **aspheric** lenses incorporate the next big advances, namely varying index of refraction in lenses with nonspherical surfaces (as in your eye—Sec. 5.2B). These surfaces are better suited than spherical surfaces to avoid spherical aberration.

FIGURE 4.13

A meniscus lens with a stop to flatten the field. Due to aberrations, different parallel incident rays A, B, C, and D do not intersect in one point after emerging from the lens. A and B intersect at P; B and C at P'; and C and D at P". If there were no stop, the circle of least confusion for all four rays would be near P', leading to a curved field. The stop selects rays A and B, intersecting at P, hence flattens the field. (Rays a and b, parallel to the axis, locate the film plane.)

shape (biconvex versus plano-convex versus meniscus), the *index of refraction* of the *glass*, and the *position* of the *stop*. This does not give us enough leeway to get rid of *all* the aberrations, but a lot can be done even with these few elements, as seen by the example shown in the figure.

If you want a lens that has a larger opening to admit more light, then the aberrations need to be further reduced. This can be done only by using several lenses spaced some distance apart. The stop is then located *between the lenses*, allowing some of the aberrations of the front components to cancel those of the back components. Astigmatism is the hardest aberration to correct; a lens that eliminates it (as well as other aberrations) is usually billed as an **anastigmat.** Three or four lens elements, as in the examples shown in Figure 4.14, are about the least you need to eliminate the important aberrations for a normal

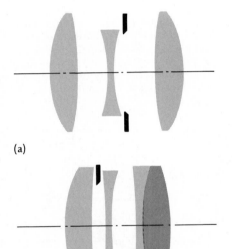

(a)

(b)

FIGURE 4.14

(a) A Cooke Triplet. (b) A Zeiss Tessar.

lens of good light-gathering ability; up to a dozen components and more (Fig. 4.19) are used for very large-aperture or particularly wide-angle lenses.

Modern lens design has many possibilities not previously available. Previously, only a small number of lenses could be used, because the small reflections at each lens surface added up and directed light either back out of the camera or to the wrong place on the film. Today,

B. Compound lenses

You can't measure the focal length of a *thick, compound lens* by the usual method of projecting an image of a distant object, because although it's clear enough where the focal point is, it is not clear from where, in or on the lens, you should measure *f*. However, you can define the focal length of a thick lens by the size of the image it produces (see Sec. 4.3A). That is, a compound lens is said to have a focal length of 50 mm if the image it makes of any object is the same size as a simple lens of $f = 50$ mm would make of the same object. (The first TRY IT explains how you can use this principle to measure the focal length of your camera lens.) For a compound lens it is not necessary that *any* part of the lens actually be at the focal distance from the film. Thus, one can make telephoto lenses that do not stick out as far in front of the camera as would correspond to their focal length (Fig. 4.15), and wide-angle lenses that do not get as near to the film as their short focal length would suggest. Typically, the body of an SLR is about 45-mm deep, and a lens any closer to the film than that distance would interfere with the flip-up mirror. Nevertheless, shorter focal length lenses are commonly available, but they must be thick lenses—the central ray does not pass through the center without bending. Unless they are

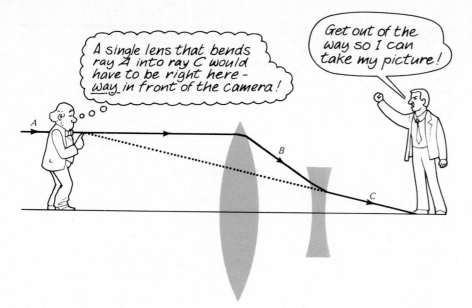

FIGURE 4.16

Frames from the wide-screen movie film "Hello Dolly." When projected through a suitable lens the images would look normal. (You can see how they would look by tilting the book and looking at the figure from the bottom.) The lines on the left are the sound track.

FIGURE 4.15

Principle of a telephoto lens. By extending the final emerging ray backward to the incident ray, we find the position and focal length of an equivalent single lens. Note that the single lens would have to stick out much farther in front of the camera than the telephoto combination.

rectilinearly corrected, such lenses result in barrel distortion. **Fish-eye lenses,** which have extremely wide angles of view, produce extreme examples of this distortion (see the $f = 8$-mm photo in Fig. 4.9).

If lenses with unequal curvature in horizontal and vertical directions are incorporated in the design, the resulting combination can have unequal horizontal and vertical focal lengths, but still have the same focal plane. This type of lens makes a deformed image; for example, if the horizontal radius of curvature, and hence the horizontal focal length, are smaller than the vertical ones, then the picture will be compressed in the horizontal direction. Such **anamorphic lenses** are used to take "wide-screen" movies on standard-size movie film (Fig. 4.16). The projection reverses the process (with the help of an anamorphic projection lens) to give a wide-screen, undistorted image. The second TRY IT shows how to make an anamorphic pinhole camera.

First TRY IT

FOR SECTION 4.4B
Measure the focal length of your camera lens

The focal length, f, of any lens can be obtained from the **magnification**, M. Suppose an object of size s_o, and at a distance x_o from the camera, makes an image of size s_i. Then $M = s_i/s_o$ (a number that is generally much less than 1), and $f = x_o[M/(1 + M)]$. (See Appendix H for a derivation of this formula.) So your task is to measure s_i, s_o, and x_o, and substitute into the formula to find f.

If you have a SLR camera, an easy way to do this is to mark off s_o as a horizontal distance of 3.5 m (= 11′ 6″) at eye level (e.g., on a blackboard), view and focus it through your camera viewfinder, and step back until the marks just fill your field of view horizontally. The 35-mm frame is about 24 × 36 mm. When the image fills the frame, s_i will be about 36 mm, and one of the factors you need in the formula will be about $M/(1 + M) \simeq 0.01$. Now measure the distance x_o from the camera to the blackboard. The focal length is then simply $f = 0.01 x_o$. For example, you may find $x_o = 5$ m, hence $f = 0.05$ m = 50 mm. (Note that f comes out in the same units you used to measure x_o.)

If you want to test lenses with focal lengths far different from 50 mm, the necessary viewing distance, x_o, may turn out rather inconvenient. Table 4.2 gives object sizes suitable for wide-angle and telephoto lenses, and the corresponding factors $M/(1 + M)$ to convert x_o to f. The table also lists suitable object sizes in inches (in case you don't have a metric tape measure) and the factor to convert x_o, measured also in inches, to f, measured in mm.

Your result may disagree slightly with what is marked on the lens. One uncertainty, particularly for thick lenses such as telephotos, is from where in the lens you should measure x_o. A good rule is to measure x_o from a point a distance f from the back of the camera. Since you know (or can find) f approximately, you can find this point with sufficient accuracy.

A more important source of error is that your viewfinder frame may not be the same size as the film frame (most viewfinders show you up to 10% more or less than the actual picture frame). If you know how much more or less the viewfinder shows, increase or decrease the object size by the same percentage. If you don't know, you can still obtain quite accurate results if you can use one lens of known focal length to calibrate. For example, if you know your normal lens has f = 50 mm, view a blackboard from a distance $x_o = 5$ m, and mark on the blackboard the edge of the region you can see. This object size then replaces the standard object of 3.5 m for your camera. It can now be used to measure the focal length of other lenses. This method is probably most useful to find the effective focal length of your normal lens when modified by a supplementary or converter lens, where larger differences from the design value are more likely to occur.

Second TRY IT

FOR SECTION 4.4B
Anamorphic pinhole camera

Many of the special camera lenses we discussed can be imitated by a pinhole camera. You can get wide-angle and telephoto effects simply by choosing the pinhole-film (or -screen) distance appropriately, the pinhole being always "in focus." (Of course, it will not have the "optimum size" discussed in Sec. 2.2B for all film distances, but this is unimportant unless the distance varies by a large factor.) You'll get the perspective of a focal length, f, if you chose the pinhole-film distance equal to f.

You can also get the effect of a PC lens by moving the pinhole to an off-center position in front of the film plane. To "straighten out" the converging lines of a tall building, locate the pinhole above the center line and make sure that the film is in a vertical position. Three-dimensional objects included in such pictures can still appear deformed, however. For instance, if you photograph two balls, one on the camera's axis, the other off, the resulting on-axis image will appear properly circular whereas the off-axis image will appear elliptical. This kind of deformation frequently occurs in the periphery of wide-angle photographs.

For even more striking anamorphic effects, you can make an anamorphic pinhole camera. Follow the design described in the TRY IT for Section 2.2B, but cut a slot in the top and bottom of the box, about halfway between front and back. Into the slot insert another piece of cardboard so that it is parallel to the front of the box. This cardboard should have a large hole in it, which you will cover with aluminum foil, like that in the front of the box (Fig. 4.17). Now experiment with openings of various designs in the two aluminum foils. For a normal anamorphic "pinhole," use a razor blade and ruler to cut a narrow slot (less than $\frac{1}{2}$ mm in width) into each of the aluminum foils. Mount the two foils so that one slot is horizontal and the other vertical. Look through the slots to make sure they are long enough that light gets through both of them to all places on the film or screen, and that the light always seems to come through a point-like (or tiny square) aperture only. View or expose in the normal way (Fig. 4.18). You can also try other arrangements of openings, for example, slots at some angle other than 90° or curved slots.

TABLE 4.2   *Conversion factors to measure focal lengths of 35-mm camera lenses*

Type of lens	Object size in meters	α	Object size in inches	β
Wide angle	3.5	10	142	$\frac{1}{4}$
Normal	1.75	20	70	$\frac{1}{2}$
Telephoto	0.85	40	34	1

Note:
The formula given in the TRY IT (and derived in Appendix H) is:

$$f = x_o[M/(1 + M)]$$

where f is the focal length, and x_o is the object distance, and they must both be in the *same* units. Since you will want f in millimeters, but will measure x_o in some other units (meters or inches), you will have to put in some extra factors to convert the units. To save you that trouble, we have calculated the factors for you. We write the formula for the two cases as:

f in millimeters, x_o in meters: $f = x_o \alpha$, or
f in millimeters, x_o in inches: $f = x_o \beta$

and tabulate the values of α and β.

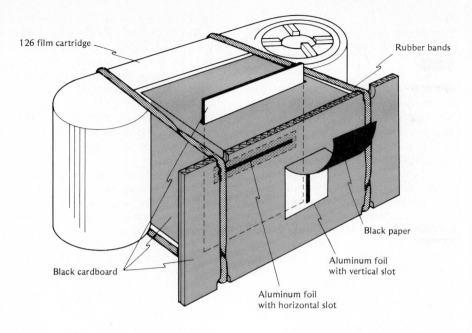

126 film cartridge

Rubber bands

Black paper

Aluminum foil
with vertical slot

Black cardboard

Aluminum foil
with horizontal slot

FIGURE 4.17

Design of an anamorphic pinhole
camera.

FIGURE 4.18

Anamorphic author: photograph taken
using camera like one in Fig. 4.17.

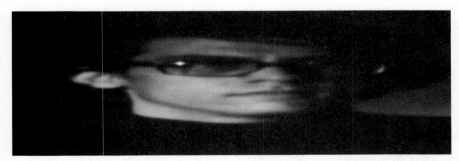

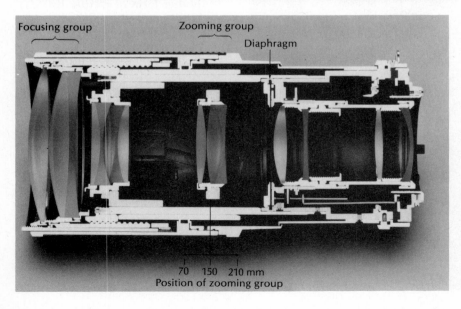

Focusing group

Zooming group

Diaphragm

70   150   210 mm
Position of zooming group

C. Zoom lenses

By using several lenses, one or more
of them movable, you can obtain a
lens of *variable* focal length. Consider a positive and a negative lens.
If they are close together, the negative power of the diverging lens simply subtracts from the power of the
converging lens, and the combination behaves as a thin lens. But if
the diverging lens is some distance
behind the converging lens (as in
the telephoto case of Fig. 4.15), the
full power of the latter makes light
rays converge for some distance;
only when the rays hit the negative
lens does the diverging action start.
Hence, the spaced combination has
a larger power than the close combination—that is, a shorter focal
length. If the two lenses are
mounted so that their separation
can be changed, you have a **variable focal length lens.** If, further,
as the effective focal length
changes, the entire combination
moves so that the *focal plane stays
fixed* on the film plane, you have a
zoom lens (Fig. 4.19). Such a lens
can "zoom" between normal and
wide-angle pictures in a continuous
sequence. If movies are shot while
zooming, the center of the scene
will seem to expand, very much as
if the camera were swooping down
on it. (Actually of course it is only a
magnification effect, without
change of perspective—so it is more
as if the viewer's seat were being
moved closer to the screen.) Zoom
lenses are also convenient in still
photography, enlarging, and slide
projection, to provide a rapid
change of focal length without the
need for refocusing.

D. Close-up and
converter lenses

You can achieve the purposes of
some of the special lenses with sim-

FIGURE 4.19

A zoom lens (Vivitar Series 1 Macro) at
$f = 70$ mm and $f = 210$ mm. The lens
elements move and also change their
spacing in just such a way as to keep the
image on the film plane.

ple additions to your normal lens. **Close-up lenses** are converging lenses, placed in front of the normal lens. Since the powers of the two lenses add (Sec. 3.4E), you have a shorter total focal length, and, therefore, can focus on closer objects. (You can also think of the close-up lens as a simple magnifying glass through which you are taking a picture with a normal lens—Sec. 6.2D). Supplementary telephoto and wide-angle lenses are small telescopes or inverted telescopes through which your normal lens photographs. A **teleconverter,** however, is a negative lens that you mount *between* a normal lens and the camera, so that the combination is a true telephoto lens of the type shown in Figure 4.15. You can even get a **_fisheye converter,_** which consists of a spherical, convex *mirror* that you mount in front of your camera lens. This gives you a wide-angle view of the world behind you, with yourself as the center of attraction (Fig. 4.20).

FIGURE 4.20

Wide-angle photograph taken in a convex mirror. Compare with Figure 3.9 and note that the place of Escher's head is here taken by the camera that took the picture. See also Figure 3.5b. Note the circles of confusion resulting from light points in the (out-of-focus) background.

4.5
DEVICES TO CONTROL LIGHT

Ours is a changing world, as has been observed by Heraclitus. For photography this means that if you expose your film a long time, you will get a blurred picture—hence the **shutter,** a device that controls the exposure time. In the early days, when a lot of light was necessary to take a photograph, it was quite sufficient to use the lens cap as the shutter; it was left off for maybe half an hour. Street scenes looked deserted because anything moving registered only as a faint blur, and portraits had to be taken using elaborate chairs with holders designed to immobilize the subject's head during exposure. Subjects were cautioned not to move their eyes during the exposure, lest they appear in the photograph with clear, white, Orphan Annie eyes.

Nowadays, however, it is possible to **stop action** with short exposure times—that is, to photograph the object without getting streaks in the direction of motion (Fig. 4.21). Some of the earliest developments of fast shutters were a result of Eadweard Muybridge's successful efforts to stop action in order to settle a bet: do all four legs of a run-ning horse simultaneously leave the ground (Fig. 4.22)? Today, exposure times of less than a millisecond are possible thanks to fast shutters, sensitive film, and lenses that can admit more light.

At this point a word about *speed* is appropriate. In photography the term **_fast_** often refers to various *different* features that enable a picture to be taken in a short time. A *shutter* is *fast*, reasonably enough, if it opens and closes in quick succession. A *film* is *fast* if it is sensitive to small amounts of light (such as are transmitted by a fast shutter); and a *lens* is *fast* if it gathers a lot of light (enabling film to be exposed at fast shutter speeds).

A. The shutter

A good shutter should not only block all light from the film when closed, it should also expose all parts of the film equally throughout the exposure. (The TRY IT shows you how to test the shutter speeds of your camera.) One popular design is the **between-the-lens shutter,** consisting of one or several lightweight blades that are opened and closed by springs. Being near the lens (in fact, in it), this shutter does not throw a moving shadow while opening or closing, so the edges of the film get as much light as the center.

Another form of shutter throws a precisely controlled shadow on the film, by moving a *curtain* with a *slot* in it at a constant speed just in front of the film (**focal-plane shutter**—Fig. 4.23). Here the film is exposed strip by strip. The narrower the width of the slot, the less the time it spends in front of any one point on the film, hence the shorter the exposure time. Most cameras with interchangeable lenses use focal-plane shutters to save the expense of one between-the-lens shutter on each of the lenses.

FIGURE 4.21

The action-stopping power of a shutter. The shutter speeds are given below each photo. (Also see stars in Fig. 2.41b.)

$\frac{1}{1000}$ sec

$\frac{1}{500}$ sec

$\frac{1}{250}$ sec

$\frac{1}{125}$ sec

$\frac{1}{60}$ sec

$\frac{1}{30}$ sec

(continued)

$\frac{1}{15}$ sec

$\frac{1}{8}$ sec

$\frac{1}{4}$ sec

$\frac{1}{2}$ sec

FIGURE 4.22

Sequence of photographs by Muybridge. Note that the horse does lift all four legs off the ground simultaneously.

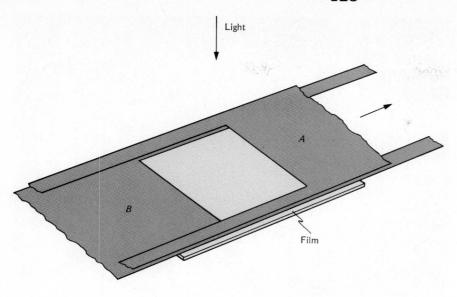

FIGURE 4.23

The principle of the focal-plane shutter. The two parts of the curtain always move across the film at the same speed, but part *B* is delayed by different amounts behind part *A* for different exposure times, creating a moving slot of different widths.

FIGURE 4.24

Photograph illustrating the distortion of a rapidly moving car due to a focal-plane shutter, here moving vertically. Modern focal-plane shutters move much faster, so the distortion is usually negligible, except for extremely rapidly changing scenes such as in Figure 4.26.

Additionally, focal-plane shutters admit light from the entire lens throughout the exposure, whereas between-the-lens shutters cover up a portion of the lens during the opening and closing phases. However, because the focal-plane shutter exposes different parts of the film at slightly different times, you can get deformed images of rapidly moving objects (Fig. 4.24). The focal-plane shutter is a modern variation of older cameras used for wide-angle pictures. In these cameras, the lens slowly turned sideward, gradually exposing the whole width of the stationary film. Such cameras were often used for photographs of large groups, and there was always one joker who, by running from place to place ahead of the camera, would appear two or three times in the picture.

Another way of controlling the exposure time is to limit the amount of time that the subject is illuminated. If the illumination of the object lasts for a shorter time than the opening time of the shutter, then

the illumination time is the exposure time. This happens when you use an *electronic flash*, which lasts only for a few milliseconds. The duration of the flash can be varied by electronic control, usually coupled to an electric eye that senses the light reflected from the scene to make sure that the flash lasts just until the film has "had enough."

Extremely short exposure times, capable of stopping a bullet in flight, can be achieved with the illumination times of 10^{-6} sec, obtainable from modern electronic flash units. In these lights, electrical energy is stored for a while and then discharged all at once through a gas, creating a short, brilliant flash. This process can be repeated at regular intervals. In that case, the resulting device, emitting periodic flashes, is called a **stroboscope.** If the shutter of a camera is left open for several of the flashes, the photograph will show a multiple exposure of the object frozen in successive stages of its motion—a valuable way of analyzing moving objects of all sorts, from bullets to athletes (Fig. 4.25).

FIGURE 4.25

Photograph of a bullet in flight, exposed for 10 μsec.

TRY IT

FOR SECTION 4.5A
Measure your shutter's exposure time

We describe here a way to measure the exposure time that makes use of your TV. Television pictures are "painted" on the screen by scanning (see Section 6.3B) successive horizontal lines. A picture consisting of alternate lines of one complete frame is scanned in $\frac{1}{60}$ second, and the remaining lines of that frame are scanned in the next $\frac{1}{60}$ second. (In Europe this time is $\frac{1}{50}$ second.) Thus, it takes $\frac{1}{30}$ second to scan a complete picture, but the surface of the tube is covered from top to bottom once every $\frac{1}{60}$ second.

To test a fast exposure speed (above $\frac{1}{60}$ second), take a picture of your TV screen so it fills most of the frame. (Include in the picture a small card on which the shutter setting is written, so you will know it after development.) From the developed picture you can judge what fraction of the TV picture is exposed. The shape of the bright region will be different, depending on the type of shutter you have.

If you have a between-the-lens shutter, you will get a horizontal band of properly exposed picture across the TV tube image. You may get a strip of picture in the middle of the tube or part of the picture may be at the top and part at the bottom. (Count only the well-exposed part of the picture—the whole screen may actually show more faintly because it continues to glow even after the scanning beam has moved on.) For example, you may find that your exposure gave a strip of picture $\frac{1}{4}$ as high as the full TV screen. Therefore, the exposure time was $\frac{1}{4} \times \frac{1}{60}$ sec $= \frac{1}{240}$ sec. If this is the result when your shutter was set at $\frac{1}{250}$ sec, it is working pretty well, as shutters go. For more accuracy, count the number of scan lines in your photo, and compute the exposure time from the fact that each line takes

$$\frac{1}{30} \times \frac{1}{525} \text{ sec} = 6.34 \times 10^{-5} \text{ sec}$$

since in North America there are 525 lines in a frame.

If you have a focal-plane shutter, the picture will be rather different, because as the TV picture is scanned from top to bottom, the shutter scans it horizontally. The resulting photograph is a diagonal bright band that is essentially a plot of the shutter's slot position (horizontal) versus time (vertical), so it tells you everything about the shutter motion. If the shutter moves at a constant speed, as it should, the band is straight (rather than curved). The horizontal width of the band measures the width of the slot in the shutter's curtain. The actual exposure time is the time the slot spends in front of any one point on the film. Hence, as for the between-the-lens shutter, the exposure time is measured by the vertical thickness of the band (Fig. 4.26).

To test the slower exposure speeds, try moving a thin object, such as a pencil, across the TV screen during the *exposure. The result should be a sequence of shadows of the object, one for each $\frac{1}{60}$ second the shutter was open. For example, if you see four shadows (Fig. 4.27), the exposure time was:*

$4 \times \frac{1}{60}$ *sec* $= \frac{1}{15}$ *sec*

A somewhat better procedure is to photograph a glow lamp, which flickers with each AC current pulse through it, that is, at $\frac{1}{120}$-second intervals. Such lamps are used, for example, to measure the speed of revolution of hi-fi turntables. Photograph such a lamp while "panning" the camera (e.g., rotating it downward) so successive flashes of the lamp expose different points on the negative. By counting the number of images you get, you can compute the slower exposure times, as above.

B. Stops

We have already met stops as means to control aberrations (Sec. 4.4A). A **stop** is an opaque barrier with a hole—a device to limit the width of a light beam. Clearly a stop affects the light intensity at the film, since only the light passing through the hole can get to the film. (If you double the *area* of the hole, you let in twice the amount of light during a given time.) In good cameras the stop is made from several thin steel blades that can be rotated to make a hole of *variable* size, intercepting more or less light (Fig. 4.28).

Some of the mystique of cameras is connected with the scale used to indicate the size of the variable stop. The numbers on the camera's adjustment ring do not give you the diameter or area of the hole directly; instead they are something called **f-numbers**, which are defined as:

f-number

$$= \frac{\textbf{focal length of lens}}{\textbf{diameter of stop's hole}} = \frac{f}{d}$$

We have written f for the focal length and d for the diameter of the stop's hole, or **aperture.** For example,

if $f = 50$ mm and $d = 12.5$ mm, then $f/d = 50/12.5 = 4$

so the lens is set at an f-number of 4, which is written as f/4. Whereas,

if d is only 4.5 mm, then $f/d = 50/4.5 \simeq 11$

FIGURE 4.26

Photograph of a TV screen, taken using a focal plane shutter at $\frac{1}{250}$ sec.

FIGURE 4.27

Photograph of a moving hand in front of a TV screen taken at $\frac{1}{15}$ sec.

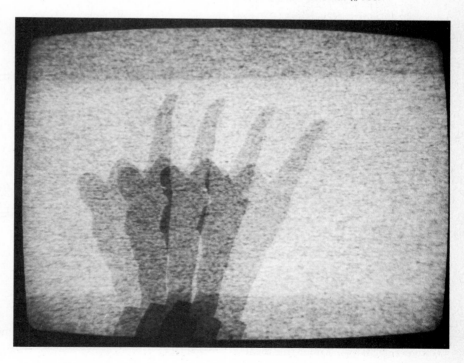

so the lens is set at f/11. The *smaller* the *diameter*, the *larger* the *f-number!* When the lens is set at its smallest f-number, its stop is "wide open," and it gathers the most light. This smallest f-number (or the only f-number of a nonadjustable lens) is called the **lens speed.** As the f-number goes down, the speed (and price) goes up.

is really a formula for the diameter, into which you substitute the numerical value of the focal length, *f*. (The slash / is then to be taken in its usual meaning of a *fraction* symbol—hence the f-number is sometimes called the f-ratio.) For example, our f = 50-mm lens set at f/11 has the diameter

$$50 \text{ mm}/11 = \tfrac{50}{11} \text{ mm} \approx 4.5 \text{ mm}$$

in agreement with the above.

What does the f-number have to do with the *intensity* of the light on the film? If you increase the f-number by some factor, you cut down the light by the *square* of that factor. For example, if you double your f-number, you get $\tfrac{1}{4}$ the light on the film. The reason is that it is the *area*, rather than the diameter, of the stop that counts in gathering light. It may seem annoying that the f-number gets bigger as the aperture gets smaller, and that the light on the film depends on the square of the f-number. Nonetheless, the f-number is a very useful way of stating the relative light intensity at the film plane because it does not matter what the focal length of the lens is. Lenses having

(a)

(b)

FIGURE 4.28

Photographs of a variable stop.

Conversely, if you know the focal length and the f-number of a lens (these are marked on most camera lenses), you can then figure out the stop diameter (should you ever need it). You pretend that the f-number

FIGURE 4.29

f-numbers and relative light intensity on the film. **(a)** A long focal length lens has an aperture set to f/2. The meter shows the relative light intensity on the film plane. **(b)** The lens of **(a)** is stopped down to f/4—the diameter of the aperture is one half that shown in **(a)**. Because one fourth the amount of light can now get through the lens (and the image remains the same size), the light intensity on the film plane is one fourth that in part **(a)**, as shown by the meter. **(c)** A short focal length lens produces a small image. The aperture here is the same as in **(b)** and thus the same light energy as in **(b)** strikes the film. This light, however, is concentrated over an image one fourth the area of that in **(b)**, and thus the light intensity on the film plane is four times that in **(b)**. The f-number for this short focal length lens is the same as in part **(a)**, and hence the light intensity on the film plane is the same as in **(a)**.

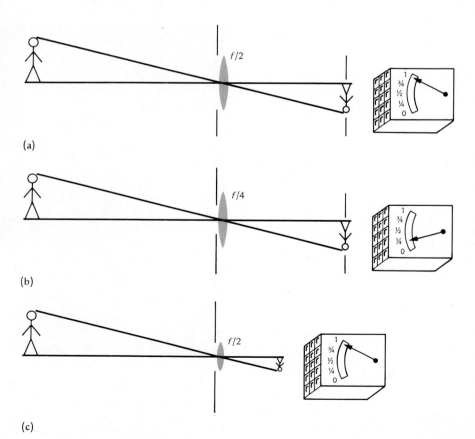

(a)

(b)

(c)

different focal lengths focused on the same subject will have the *same* average light *intensity* at their respective film planes if the lenses are set at the *same f-number*. To see this, suppose instead we make the *apertures* of these two lenses the same. Then, the light intensity on the film plane of the long-focal-length lens is *less* than on the film plane of the short-focal-length lens. This is because the long-focal-length lens produces a *larger* image on its focal plane, spreading out the same amount of light energy as that entering the short-focal-length lens. By reducing the aperture of the short-focal-length lens so as to have the *same f-number* as the long-focal-length lens, the amount of light admitted is decreased so that the average light intensity on the film plane is the same for both lenses—the effects of the smaller hole and of the smaller film distance exactly compensate (Fig. 4.29).

Occasionally (for example, when an extension tube is used for *close-up work*) the film is considerably farther from the lens than the focal length, so the convenient compensation no longer applies. Then the light intensity at the film is no longer given directly by the f-number setting of the lens. (When the film is farther from the lens, the image is larger, so the light is spread out over a larger area. This means that a given region of the film gets less light.) Usually a correction is made by using an *effective* f-number for computing the exposure time.

Finally, we mention a subtlety we neglected so far in our discussion of the f-number: to calculate it via *f/d*, you should use for *d* not really the diameter of the hole in the stop (which is normally between lens elements and hence hard to measure), but the *apparent size* of the hole diameter as seen through the *front* of the lens (the **entrance pupil**); after all, that is what the incoming light "sees" as the opening through which it has to pass. The TRY IT shows you how to measure the entrance pupil and calculate the f-number for your camera.

TRY IT

FOR SECTION 4.5B
Measure the f-number of your camera lens

If you know the focal length, f, you only need d to compute f/d, the f-number. You don't need to get at the actual stop between the lenses to measure d, because d is the diameter of the entrance pupil—the virtual image of the stop as seen when looking into the lens from the front. Attach a millimeter ruler across the front of the lens, and open the shutter in the "B" position. Make sure you can see the stop (e.g., by opening the back of the camera) and view at arm's length from the front to minimize parallax (Sec. 8.4). Be careful not to move or rotate the lens as you read the ruler at the two ends of the stop's diameter. (If the stop is not quite circular in diameter, estimate the average diameter.) Compute f/d. If the f-number is marked on the lens, agreement should typically be better than 10%. On SLR cameras you will be able to see the diameter of the lens without opening the shutter, because it is illuminated by light entering the viewfinder; however, to get other than the largest opening you must use the depth of field button. Usually, taking the lens off the camera also leaves it at its widest opening, allowing you only to measure the speed of the lens, that is, the lowest f-number setting.

C. The f-number sequence

The f-numbers help us get the right amount of light on the film. Fortunately, the film does not care whether we get the exposure *exactly* right, but for good results we should get it correct within a factor of two. Therefore, the scale of stops on adjustable cameras is marked in a standard sequence of f-numbers, designed so that a change from one f-number to the next in the sequence corresponds to a *factor of 2* change in light *intensity*. The f-numbers themselves change by a

factor of about 1.4, because the area of the stop then changes by a factor $1.4 \times 1.4 = 1.96 \approx 2$. This standard sequence is given in Table 4.3. If we want to know how much more light we get, say, at f/2 than at f/22, we merely need to count the number of steps between these settings in the standard sequence (7 steps) and go up a factor of 2 for each step; that is $2 \times 2 \times 2 \times 2 \times 2 \times 2 \times 2 = 128$. So we get 128 times as much light in this case. Alternatively we could do it from first principle:

Intensity ratio = ratio of area
    = ratio of squares of diameters

Now recall that the diameter is the f-number regarded as a fraction, hence:

$$\text{Intensity ratio} = (f/2)^2/(f/22)^2$$
$$= (22/2)^2 = 11^2 = 121$$

(This is close enough to the other answer, 128, since the numbers in the sequence are rounded off.)

D. Effect of f-number on picture quality

All lenses have some residual aberrations, which can be reduced by stopping down, but this improvement by itself is hardly enough to motivate a variable stop. More important is that stopping down improves the depth of field. The geometry of Figure 4.30 makes it clear why this is so; any point on the film is illuminated by a cone of light converging to it from the lens. If an object point is not quite in focus, the apex of the corresponding cone is not quite on the film plane, so the film plane cuts the cone in a little circle (the **circle of confusion**), which represents the blurring of the image. (See the out-of-focus spots of light in the background of Fig. 4.20.) The base of the cone of

---

TABLE 4.3   *Standard sequence of f-numbers*

. . . 0.7, 1, 1.4, 2, 2.8, 4, 5.6, 8, 11, 16, 22, 32, . . .

←—— Larger aperture   Smaller aperture ——→
       More light   Less light

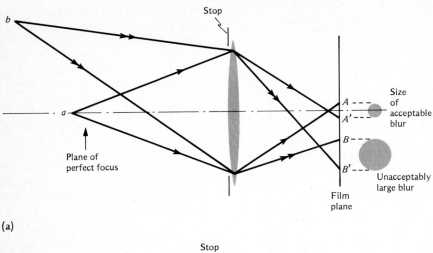

(a)

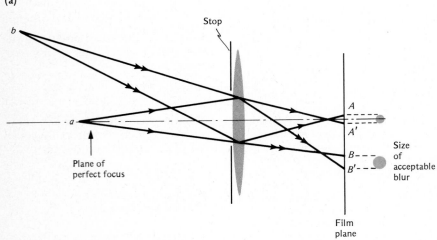

(b)

(Fig. 4.31). For this reason, for each setting of the stop, your camera shows the corresponding depth of field (Fig. 4.6).

The depth of field depends not only on the f-number, but also on the focal length of the lens; longer focal length lenses have less depth of field at the same f-number setting than shorter focal length lenses. Thus, cheap pocket cameras (small camera size, hence small f) with slow lenses (large f-number) have large depth of field and can afford to have fixed focus. On the other hand, sometimes a shallow depth of field is desirable, for example to focus just on the picture's main subject and to soften the background. This is often used in TV to draw the viewer's attention to a particular actor. Similarly, a small depth of field is useful whenever an undesirable foreground is unavoidable, such as the bars on an animal cage in a zoo, or blemishes on a pane of glass through which you have to photograph. With the foreground close, the lens wide open, and perhaps a fairly long focal length, you can sometimes blur the foreground sufficiently so that it is not noticeable in the photograph (Fig. 4.32).

FIGURE 4.30

The effect of stop size on the depth of field. **(a)** The circle of confusion of object a has diameter AA', smaller than the acceptable blur size; a is in focus. Object b, however, has an unacceptably large blur size (BB'); b is out of focus. **(b)** The smaller stop reduces the size of each circle of confusion. Now both a and b are acceptably in focus.

light is at the stop. If you decrease the hole in the stop, you make the cone more sharply pointed. The little circle at the film plane is reduced in proportion to the size of the base. So if you can tolerate a certain blur, that is, a certain size circle of confusion, then the narrower and more pointed the cone of light, the farther your object can be from the place of perfect focus. In other words, the more the lens is stopped down, the greater the depth of field

FIGURE 4.32

Here a shallow depth of field is used to blur an undesirable foreground. The out-of-focus wires of the cage are almost invisible, except in the upper left corner.

(a)

(b)

(c)

(d)

(e)

(f)

(g)

FIGURE 4.31

35-mm photographs with increasing stop sizes, decreasing depth of field: **(a)** f/16 at $\frac{1}{15}$ sec, **(b)** f/11 at $\frac{1}{30}$ sec, **(c)** f/8 at $\frac{1}{60}$ sec, **(d)** f/5.6 at $\frac{1}{125}$ sec, **(e)** f/4 at $\frac{1}{250}$ sec, **(f)** f/2.8 at $\frac{1}{500}$ sec, **(g)** f/2 at $\frac{1}{1000}$ sec.

4.6
EXPOSURE

We are now ready to put together the various factors affecting light and contemplate taking a picture. In a snapshot camera, the f-number and exposure time are fixed, and the picture either comes out properly exposed or not. The only things you can vary are the speed of the film and the light conditions (which may include flash). In what follows we'll assume that the film and light conditions are fixed, but that you have a camera in which you can adjust *exposure time* and *f-number* separately.

To find a correct exposure for a given type of film, you need to know the light intensity in the scene you are photographing. One generally uses an **exposure meter,** which samples some light from the scene and converts the light's energy into electrical energy (Sec. 15.2A). The electrical energy can then be used to control a pointer on a meter, from which you can read off the intensity; or it can be directly coupled into the camera to set the proper exposure. The light sampling can be done by letting the exposure meter "look" at the scene directly, but preferably it should "look" at the image formed by the lens, so that the meter measures the light that actually gets through the lens. In some cameras the exposure meter even measures the light reflected from the film during exposure and adjusts the exposure time accordingly.

The correct exposure, however, does not give a unique camera setting; there are many combinations of f-number and exposure time that will give a correct exposure. This is due to the film's **reciprocity;** the film cares primarily about the total light energy given to it, and that is the *product* of intensity (energy per second) and exposure time (seconds). This product is called the **exposure.** The f-number determines the intensity, the shutter setting determines the exposure time, and their product needs to be constant for equivalent exposures. So, you can get a sequence of equivalent exposures by stepping *up* the intensity by factors of 2 while simultaneously stepping *down* the exposure time by the same factors of 2.

For example, on a bright day a proper camera setting may be $\frac{1}{60}$ second at f/11 (the snapshot camera setting). Equivalently, you could expose at half the time, $\frac{1}{120}$ second, and twice the intensity, which means the next smaller f-number in the standard sequence, f/8. Repeating this you find many equivalent settings (Table 4.4). If you were to shift the entire lower row to a different position under the upper row of numbers, you would get another set of equivalent settings, useful for a different film or for different light conditions. Some cameras have such a shifting arrangement built into their scale adjustment rings, so that you need turn only one ring to change between equivalent settings.

If you have one correct camera setting, why would you want other, equivalent settings? There are different advantages offered by different settings, and which ones are important depend on what you are photographing. For example, if the object is in rapid motion, you may want a short exposure time in order to stop (or "freeze") the motion (see Fig. 4.21). In that case the lens must be opened wider, which decreases the depth of field. In other cases the depth of field may be more important, so you would shoot the picture at a large f-number and a long exposure time (see Fig. 4.31). Of course this assumes that you have some choice of settings. Sometimes, for example in "available light" photography, you must get as much light as possible on the film and sacrifice both depth of field and action-stopping power. If the light level is really low, long exposure times of seconds or minutes may be needed. When shooting movie film you don't have that option. In order to shoot scenes in color under candlelight, the movie "Barry Lyndon" used specially designed, large-aperture lenses, having f/0.95! (For long exposure times, however, the reciprocity of the film may fail—you may need *more* than the intensity predicted by the method of Table 4.4.)

It is interesting to contemplate all the operations that automatically take place in a modern SLR camera when "you press the button." To prepare the camera for picture taking, the mirror flips out of the way, and the lens stops down to the preselected value (the lens is usually kept wide open for viewing, no matter what f-number has been set). Then the shutter opens, and electrical contacts are made if you are taking a flash picture. In some cameras, the exposure meter starts to collect light and, when it has enough, causes the shutter to be closed. Finally, the mirror flips down, letting you again see in the viewfinder the scene you have just photographed. All this happens in the blink of an eye—so fast that you hardly notice the viewfinder picture blacking out.

4.7
FILM

The whole point of the camera is to record the image by preserving it on film. In this chapter we confine our discussion to black and white film, turning to color film in Chapter 11.

The process occurring in the film between exposure and final image belong mainly to chemistry, but it behooves us to understand at least in outline form what is going on. Actually, some of the earliest photographs relied on physical rather than chemical development, and it is interesting to explore this and other features of the history of pho-

TABLE 4.4 Equivalent settings

Exposure time (seconds)	$\frac{1}{1000}$	$\frac{1}{500}$	$\frac{1}{250}$	$\frac{1}{120}$	$\frac{1}{60}$	$\frac{1}{30}$	$\frac{1}{15}$
f-number	f/2.8	f/4	f/5.6	f/8	f/11	f/16	f/22

tography in order to appreciate some of the ingenuity that was necessary to bring us the films of today.

A. Principles

Many substances react to light; examples are cheap paper that yellows, dyes that fade, and skin that is tanned by the action of sunlight. But these reactions are all extremely slow, so the first steps toward successful photography were finding substances that are more sensitive to light and ways to treat them that make extremely faint images more visible.

A few, unique substances were found to be suitably light sensitive, and remain unique even today: silver compounds of chlorine, bromine, or iodine, collectively called **silver halides.** The discovery that such silver compounds are appropriate for photography was made independently during the 1830s by W. H. Fox Talbot, J. N. Niepce,* and J. M. Daguerre.

In most photographic recording, the light strikes the silver halide, breaking a chemical bond and yielding metallic silver in the form of microscopic clumps, called **nuclei.** Where a lot of light falls on the film, many such nuclei are formed; at places that receive less light, fewer are formed. At this point, however, no image can be seen on the film as the nuclei are too small. We have a **latent image**—the information of the image is recorded, but it is not yet in a visible form. The problem is now to make the image visible—the film must be **developed.** If the film is developed so that the final image is white (or clear) where the original subject was bright, we have a **positive** image. Conversely, if the final image is dark where the original subject was bright, we have a **negative** image.

There are two major techniques of development. The first, **physical development,** was used by Daguerre and Talbot, and survives to-

*Pronounced "nyeps."

day in photocopiers (Sec. 15.2B). The second, **chemical development,** is responsible for the myriad photographs that surround us today. After the film is developed, thus making the image visible, the film must be made insensitive to additional light to prevent further, unwanted changes. It was found that this could be accomplished with sodium thiosulfate **(hypo),** which dissolves and washes away the remaining, unexposed silver halide, so that the image becomes permanent. This process, used with either kind of development, is called **fixing** the image.

*B. Daguerreotypes

Together with Niepce, Daguerre had been experimenting with photography on silver plated sheets of copper that had been polished and then treated with iodine to make a layer of silver iodide on their surfaces. Daguerre found by accident that mercury vapors could bring out a previously invisible image on an exposed iodized silver plate. He had invented **physical development,** a kind of amplification similar to the growth of water droplets in the trail of a jet airplane, or growth of sugar crystals in a glass of honey. This type of growth proceeds relatively rapidly once it has found a place to start—a nucleus.

When the plate with the latent image is bathed in mercury vapor, the mercury preferentially forms droplets around the silver nuclei, because silver and mercury can amalgamate, that is, stick easily to each other. The droplets continue to grow, amplifying the image by this physical process of condensation, and rendering it visible. After being fixed, the plate contains mercury where the light hit it, and silver everywhere else.

Mercury does not look very different from silver; the only difference is that the silver plate is polished and shiny, whereas the mercury droplets give a rough, diffusely reflecting surface. Therefore, **Daguerreotypes,** as these early photographs were called, were mounted

under glass in lockets of black velvet. The glass protects the easily smeared mercury image, and the locket when opened gives a black surface to reflect·in the polished silver. Where there was no light during exposure, and, therefore, no mercury on the daguerreotype, you see the reflection of the black velvet in the shiny silver—a dark region. Where there was light during exposure, the mercury diffusely reflects any incident light, and hence looks white (Fig. 4.33). So daguerreotypes give positive images (although when improperly viewed they could also look like negatives), but you get only a single one that cannot be easily copied.

Daguerreotypes with their sharp and detailed images were an immediate success. As the inventor and artist Samuel F. B. Morse described them: "the exquisite minuteness of the delineation cannot be conceived. No painting or engraving ever approached it." Millions of daguerreotype portraits were taken, as well as landscapes and other scenes. Among the most famous are Mathew B. Brady's documentary photographs of nineteenth-century America. However, the limitations we mentioned eventually made daguerreotypes obsolete.

C. Modern film

Today's black and white film derives from the negative-positive process invented by Talbot. Talbot's negative film was paper coated with silver chloride (Fig. 4.34). Although many positive prints could be made from a negative, they showed some of the unavoidable fiber structure of the paper negative and were not sharp. The problem remained of finding a more transparent medium than paper to carry the silver halide—a medium that would also allow liquid developers to move around and reach the silver, while keeping the latter in place for a sharp image.

The answer was to put the silver halide in an **emulsion** that adheres well to glass plates. Early emulsions had to be made on the spot and be

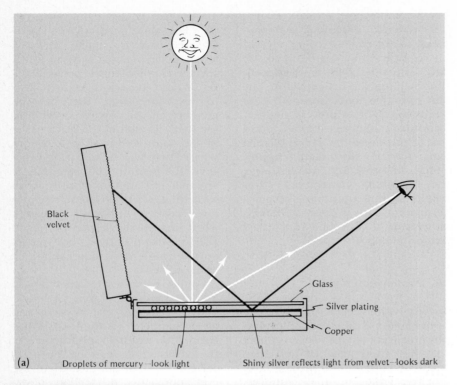

(a)

Black velvet

Glass

Silver plating

Copper

Droplets of mercury—look light   Shiny silver reflects light from velvet—looks dark

(b)

(c)

FIGURE 4.33

(a) Proper method of viewing a daguerreotype to see a positive image. (b) Example of an early daguerreotype, properly illuminated. (c) Improperly illuminated, it shows a negative image.

kept wet so that the developers could then diffuse to the silver halides. An important advance for increased convenience was the use of gelatin as the emulsion material, because it can be exposed while dry and will later absorb the developer solution, much like blotting paper takes up ink.

Toward the end of last century, photography was ready to change from small shop operations to become a large industry. The first large film manufacturing, developing, and printing service was provided by George Eastman's Kodak Company. It produced not only the traditional emulsions on glass plates, but also on a flexible base that could be rolled up. Roll film for many exposures could then be stored in a "Kodak" (as his cameras were called), rather than introducing one plate at a time in cumbersome plate holders. (The early flexible base was highly flammable nitrocellulose. Today's **safety film** uses plastic instead.)

Another important turn-of-the-century development was the creation of photographic emulsions sensitive to a wider range of wavelengths. That is, silver halides by themselves form silver nuclei only under the influence of blue or UV light. Hermann W. Vogel found that the emulsion could be made sensitive to other regions of the spectrum, as well as more sensitive overall, by adding certain chemicals to it—**sensitizers.** This allowed film makers to introduce orthochromatic film (sensitive to all colors but red) and later panchromatic film (sensitive to all colors).

Modern black and white film consists of three layers (Fig. 4.35): a clear plastic base coated on one side by an emulsion and the other by an **antihalation backing.** The purpose of the latter is to absorb any light that isn't completely absorbed by the emulsion. This light would otherwise be reflected by the back of the base, return to the emulsion layer, and reexpose it, resulting in a halo* or fog around the picture of

---

*Hence the name, antihalation.

(a)   (b)

FIGURE 4.34

Detail of one of Talbot's photographs: **(a)** a negative made in 1844, **(b)** a positive print made the following year.

any concentrated bright light. This backing is removed during development. Let's now consider this process of development.

FIGURE 4.35

Schematic cross section of a modern black and white film.

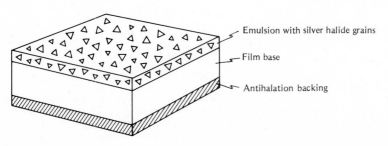

Emulsion with silver halide grains

Film base

Antihalation backing

D. Chemical development

In modern film, small silver halide crystals of various sizes are embedded in the gelatin emulsion. Light striking such a crystal generates one or several silver nuclei. In chemical development (Fig. 4.36), the trick is to find a chemical that tends to convert *(reduce)* the entire halide crystal into silver metal. This reducing agent will work faster where there are already silver nuclei. Thus, the reducing agent will convert the silver halide into metallic silver first in those crystals that have been exposed to light. If the reducing agent isn't too strong, the film can be removed from it before the other, unexposed, silver halide crystals are reduced (i.e., before the film is "overdeveloped"). Such reducing agents do not work well in an acid environment. Therefore, they are mixed with other chemicals to provide a suitable environment and are then called *developers.* After development, the reduced silver is usually formed as tangled masses of filaments, called **grains,** which absorb light in nooks and crannies and hence look dark (Fig. 4.37).

For fine, detailed pictures, film with smaller silver halide crystals is used. This film typically requires a longer exposure time (it is slower film) because it needs an exposed

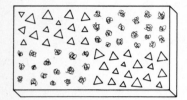

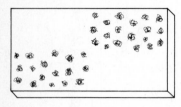

(a)

(b)

(c)

FIGURE 4.36

The microscopic changes taking place in silver halide crystals during chemical development. **(a)** After exposure: The dot represents the nucleus formed in the exposed crystals—the latent image. **(b)** After development: The exposed crystals have been converted to dark silver. **(c)** After fixing: The unexposed crystals have been removed.

FIGURE 4.37

Microphotograph of the initial stages of chemical development of a silver halide crystal.

nucleus in *each* crystal in the exposed area.

In order to stop the development before the unexposed crystals are reduced to metallic silver, the developer is washed away in the **stop bath** (usually an acid, which makes the developer inactive). At the end of the development a negative is formed, since the places that originally had light are converted to dark regions of metallic silver grains.

Of course, where there was no exposure, silver halide is still present after the developer and stop baths. It appears light, but if it is exposed to light for a long time, it will eventually turn into dark silver even without the benefit of a developer (see the TRY IT). To prevent this, the film is fixed with hypo, which is then washed away with water. If the negative thus formed is on a clear film base we have a negative **transparency,** and if it is on white paper backing **(print paper),** we have a negative **print.**

To get a *positive* print we expose a photographic print paper, by contact or projection, to the negative image of a photographic negative transparency. The result of the printing exposure, after development in the same way as the film, is a positive picture (a negative of a negative), white where the original film was exposed to light, dark where it wasn't (Fig. 4.38). In this way we can make any number of copies from one negative.

There is also a way to get a positive transparency directly from the film that was exposed in the camera, by the process of **reversal development** (Fig. 4.39). Here the film is first developed as usual, but not

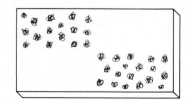

FIGURE 4.38

Diagram of the microscopic structure of the positive print made from the negative of Figure 4.36.

fixed. Instead, the silver image is removed by a process called **bleaching,** leaving only the unexposed silver halide crystals. The latter are then exposed by letting light fall on the entire film (or by chemical means) and then turned into silver grains in a second development. So now the unexposed parts of the emulsion are dark grains, whereas the exposed parts, having been bleached away, are clear and transparent—just what is needed for a positive transparency.

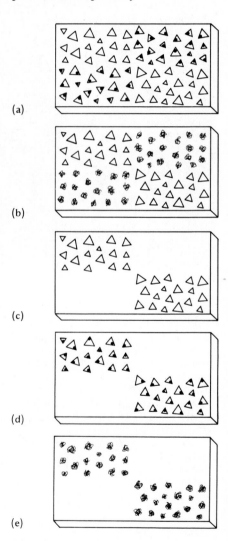

(a)

(b)

(c)

(d)

(e)

FIGURE 4.39

Changes in silver halide crystal during reversal development: **(a)** after exposure, **(b)** after first development, **(c)** after bleaching, which removes the silver grains, **(d)** after second exposure, **(e)** after second development.

TRY IT

Photography without development

The latent image on modern film is invisible; moreover, if you tried to look at it, you would have to illuminate it and hence spoil it. If you have some print paper you can convince yourself that light by itself does have an effect on the paper and can make an image. The idea is to use a sufficiently high exposure to form so many silver nuclei that eventually whole, visible crystals will be reduced to silver even without developer (but you won't get good blacks because the silver is produced in a different form than when chemically developed).

Find a small, flat object of recognizable shape (e.g., a key or a high-contrast negative) and tape it to a sunny window. Tape the print paper over the object, with the shiny emulsion side facing the sun. Leave it for several hours or days, occasionally lifting a corner to see whether an image is appearing. Note the color of the image when it first appears (after about an hour in direct sunlight) and any later changes. The color that first appears will be explained when we discuss scattering in Section 13.2.

This same "photography without development" occurs in eyeglasses made with **photochromic** glass. In these, the lenses are clear indoors, but darken in bright sunlight. The glass contains microcrystals of silver halide (10 nm across—small enough not to scatter visible light). When exposed to bright light, the silver halide breaks into silver atoms and halide atoms. The silver, powered by the light (mainly the UV), drifts together into large, light-absorbing clumps, causing the glass to darken. That is, in bright sunlight, the eyeglasses become sunglasses. When the eyeglasses are removed from the energizing sunlight, the silver clumps break apart and the silver recombines with halide, reforming the silver halide microcrystals and leaving the glass again transparent. In this way, the glasses are as dark as the ambient light requires. If you have such glasses, try taping a key over one lens before you take them outside.

*E. Other development techniques

Instant photography can today give us a finished picture in much less time than it took in the early days for just the exposure. Here the exposure is made on an emulsion that is similar to ordinary black and white emulsions; but ahead of each frame the film carries a pod of viscous chemicals that perform the function of both developer and fixer. After exposure, the film is pulled through rollers that squeeze it against a support paper and break the pod, spreading its contents between film and support paper (Fig. 4.40). The fixing action of the chemicals is designed to be slow, so that the film is first developed normally. A moment later the fixer starts to convert the remaining, unexposed silver halide crystals into *soluble* silver *ions* (i.e., positively charged silver atoms). These diffuse out of the emulsion, toward the support paper, which contains microscopic development *nuclei.* Chemicals from the pod make these development nuclei negatively charged, and, therefore, attractive to the diffusing silver ions. Thus, this dissolved silver "plates out," as the usual dark silver grains, onto

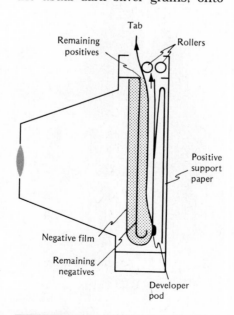

FIGURE 4.40

Diagram of Polaroid film pack in camera.

the support paper (a kind of physical development). The originally unexposed regions become dark on the support paper. The crystals in the exposed regions were turned into metallic silver and were not dissolved by the fixer. Therefore, no ions could diffuse to the paper, which remains white in those regions. The result is an "instant" positive, for the whole process takes but a few seconds.

Other applications use the fact that, while we have to use silver for *light sensitivity* we can replace it by something else for *image formation.* For example, colored dyes can be attached to the developed silver **(toning),** or to the regions *not* occupied by the silver **(tinting).** Alternatively, there are developers that harden the gelatin near the developed silver grains. If the film is then washed in warm water, the gelatin is washed away except at places where it has been hardened. Thus a "relief image" of gelatin is formed, which can be used in various ways, for example to etch a metal base preferentially at the places where there is no hardened gelatin. This is the principle used to make the printing plates for photographs such as in this book, or for the pages in books reproduced by photo-offset.

Another unusual development technique employs the **Sabattier effect,** which allows you to produce photographs whose values of light and dark are not simply related to those of the original subject. To achieve this effect, you first expose a normal negative film to a scene. Then, in the darkroom, you halt the development process prior to completion. At this point, the film contains a faint negative image of the scene. Now, you expose the entire film to light, for instance by turning on the room lights for a moment. Exposed silver grains of the faint negative image act as a mask, blocking the room light from exposing silver halide crystals beneath them. But places on the film that do not have exposed silver crystals have no such mask, and the bright room light exposes all the silver halide there. This second, overall ex-

FIGURE 4.41

(a) Print from a normal negative. (b) The Sabattier effect: A print from a negative that has had an overall exposure during the development stage.

posure, therefore, leads to a *positive* image of the original scene (a negative of a negative). Then, you complete the development process. Thanks to the Sabattier effect, you have superimposed a positive and a negative image on the same photograph (Fig. 4.41). (This effect is often mistakenly called **solarization,** which also leads to superposition of positive and negative images. True solarization, however, results from extreme overexposure during the exposure stage and is nearly impossible to achieve with modern films.)

F. Film sensitivity: H & D curve

When light exposes a silver halide crystal sufficiently to make it developable, the whole crystal is always converted into a silver grain. It would therefore appear that we can only record white or black; how does gray come about? The trick is to use crystals of various sizes. Strong light affects all crystals, but weaker light will primarily hit the larger crystals. A gray area results when only the larger crystals are developed to silver grains, but not the smaller ones. In other words, the grains are equally black in either case, but the weaker light causes fewer grains to appear. On a large scale this looks less black, that is, gray.

The amount of light that passes through a region of a final developed negative, therefore, depends on the exposure of that region. We can plot this relative amount of light passed versus the exposure as an **H & D curve***. Because the range of both exposure and the relative amount of light passed is

*Named for Ferdinand Hurter and Vero Charles Driffield, who did the initial research in this field in 1890.

(a)

(b)

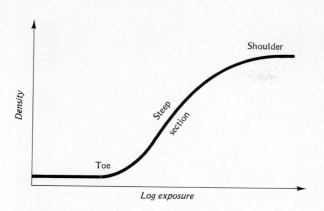

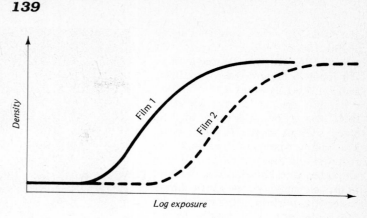

FIGURE 4.42

A typical H & D curve. Low densities (few developed silver grains—negative clear) are plotted near the bottom; high densities (many developed silver grains—negative black) are plotted near the top.

great, we use *logarithmic** scales, in which an equal step on the axes corresponds to an increase or decrease of the plotted quantity by an equal *factor*. The logarithm of the ratio of the incident to the transmitted light intensities is called the (optical) **density** of the developed negative. The logarithm of the exposure has no special name. (The density is closely related to the number of developed silver grains, though scattering and other effects complicate this relationship.)

Figure 4.42 shows the H & D curve for one film, using some standard developing technique. (For a different way of developing, e.g., at a different temperature, the curve would be different.) The unavoidable background of grain density even at very low exposures is called **fog.** For very small exposures the density is unaffected; the few extra grains due to the low exposure are swallowed up by the fog. At some higher exposure we come to the first noticeable rise in density, the toe of the H & D curve. A little beyond the toe the curve rises more steeply. Finally, at very high exposures, we reach the shoulder of the plot, beyond which all the emul-

sion's crystals have been developed so no further increase in density is possible, no matter how large the exposure—the developed negative has reached its maximum blackness.

G. Film sensitivity: speed, contrast, and latitude

For choosing an appropriate film and determining the proper exposure, most photographers use only certain aspects of the information contained in an H & D curve: the speed, contrast, and latitude of the film. We've seen how the location of the toe, steep section, and shoulder of the H & D curve each tells us something about the characteristics of the film. For example, the toe tells us the minimum exposure necessary for any kind of an image, so it is one rough measure of the speed of the film. Of course, for an acceptable image, the grain density should not only barely exceed the fog density, but should be at least several times as large. Thus, the **speed** of the film measures the exposure necessary to be well above the toe; qualitatively, the lower the exposure at the toe (or the more to the left the steep section lies), the faster the film (Fig. 4.43). There is no one property of the H & D curve that best measures film speed under all circumstances, and many different ways of measuring it, called **exposure indexes,** were used in the past. But today the industry has settled on one standard by international agreement, and only the scales of the indexes differ. The

FIGURE 4.43

H & D curves for two films of different speed but equal contrast.

three indexes that you usually find marked on cameras, flash units, and film are the ASA, DIN, and the ISO indexes.

The ASA* index is a "linear" index of speed. That is, an ASA 100 film is twice as fast as an ASA 50 film, which is, in turn, twice as fast as an ASA 25 film. The DIN† index on the other hand, is a "logarithmic" index; each factor of two increase in speed corresponds to an *additive* increase by three of the DIN index. Thus, a DIN 21 film is twice as fast as a DIN 18 film, which is, in turn, twice as fast as a DIN 15 film. The relation between the two indexes is such that, for example, ASA 100 is the same as DIN 21. For a rough idea of the sensitivities in absolute terms, we may say that outdoors on a sunny day one correct setting is 1/(ASA exposure index) seconds at f/11.

The ISO‡ index is determined by methods differing somewhat from the other two, but the numbers that result are the same as the ASA and DIN numbers for all but a very

---

*Greek *logos*, reckoning, plus *arithmos*, number. Appendix I explains more about logarithmic scales for those who have not encountered them before.

---

*For American Standards Association, now the American National Standards Institute, which sets standards for many industrial products.

†For *Deutsche Industrie Norm*, the German national standards institute.

‡For International Standards Organization, a federation of all national standards bodies of the world.

few films. Thus for example, a speed of ISO 100/21° is equivalent to ASA 100 or DIN 21. (The degree symbol ° is used to denote the logarithmic scale of the DIN index.)

If you are exposing correctly, then, you'll be operating on the steep section of the H & D curve. The brighter parts of the image correspond to points closer to the shoulder, and the dimmer parts to points closer to the toe of the curve. Since this portion of the curve is almost straight, you might expect that if object *2* is twice as bright as object *1*, the negative at the image of object *2* would pass half as much light as the image of object *1*. This would give you a faithful (though reversed or negative) representation of the light values of the scene. But this is only so if the steep section rises at a 45° angle (or, as mathematicians say, this section has a **slope** of unity). If the curve is steeper, each factor 2 increase in intensity in the original corresponds to *more* than a factor 2 increase in "darkness" of the negative; and if the curve is shallower, the increase in "darkness" would be *less* than a factor of 2. The slope of the curve is called by the Greek letter gamma: γ. The gamma is a measure of the **contrast** of the image (Fig. 4.44): large gamma, steep curve means that the variation of light in the photographic image will be larger than that of the original (darks darker and lights lighter), that is, higher contrast. Small gamma, shallow curve means less variation of light in the image than in the original, that is, lower contrast.

What is the use of nonfaithful emulsions with gamma different from unity? In some situations we want to increase the contrast of an original. For example, when reproducing line drawings we want only complete black or complete white in the image, so the steep region should be *very* steep, and we would choose a high-contrast, high-gamma film. Conversely, a very high-contrast original might reproduce better on a film of low contrast (low gamma). The contrast of a neg-

(a)

(b)

FIGURE 4.44

(a) Low-contrast and **(b)** high-contrast photographs.

ative can be corrected in printing; a high-contrast negative should be printed on low-contrast paper and vice versa. However, even in the finished print, the aim is usually not an absolutely faithful representation of the brightness relations of the original scene. Usually, it is more important to make use of the full range of brightness values available, so a good rule of thumb is to select the contrast of the paper so that there is one completely white spot and one completely dark spot on the print. In other words, we do not want pictures with too low a contrast, which have an overall grayish cast, nor with too high a contrast, where details in the shadows and highlights cannot be recognized. Also, as we'll see in Section 7.4B, contrast affects perceived lightness, so for brilliant pictures one usually prints with higher than "natural" contrast. For example, black and white (positive) movies are typically printed with $\gamma = -1.4$. (It may go against the "grain" that the gamma of positives is negative and vice versa, but that's the way photographers like it!) Masters of photography, such as Ansel Adams, may take as long as a day to experiment and print one photograph—choosing print paper, and exposure and development procedures to ensure the best contrast, thereby allowing remarkable subtleties in lightness to be visible.

Speed and contrast measure the position and slope of the H & D curve's steep section. In addition it is important to know how long the steep section is. This is called the **latitude** and is measured by the range of exposures between toe and shoulder of the H & D curve. The larger this range is, the greater the range of light intensities that can be photographed in one picture, or the greater the range of camera settings that will give an acceptable negative of a scene of moderate intensity range. Therefore, in black

and white negative films, a large latitude is desirable (and this usually implies a lower contrast, since the density can have only a finite range). For print papers, the contrast is the more important characteristic.

These properties of the film depend primarily on the size of the silver halide crystals. For a high latitude you want a mixture of widely different crystal sizes, which will respond to widely different intensities. For high speed you want a large number of crystals, with many of them of large size. This is done for example in x-ray film, where it is important to keep the x-ray exposure of the patient low. (Loss of sharpness is not critical for these shadow images.) It may seem easy to give any emulsion high speed and high latitude, but this can conflict with other characteristics of the film that do not show up in the H & D curve. For example, large crystals lead to large developed grains of silver, but it is important to keep the grain size small **(fine grain)** if the negative is to be considerably enlarged. Also, the emulsion should not be too thick, because it would then exceed the region of best focus, and light may scatter within the emulsion, again leading to a decrease in sharpness of the image.

SUMMARY

A basic **camera** consists of a **lens** to form an image on the light-sensitive **film,** a **diaphragm** to limit the intensity of light admitted, and a **(between-the-lens** or **focal plane) shutter** to limit the duration of the light admitted. The camera-subject distance, needed for **fo-**cusing, is determined by a **rangefinder,** which uses optical triangulation; **split prisms,** which reveal when the image is on or off the film; **sonar,** which uses a sonic echo technique; or informed guesswork. **Telephoto** lenses have a longer **focal length,** smaller **angle of view** (the angle subtended by the scene recorded), greater **magnification,** smaller **depth of field** (near-far range of subjects acceptably in focus) than do **wide-angle** lenses. Stopping down increases depth of field and reduces aberrations. The **f-number** (focal length/effective diameter of aperture) indicates the relative light intensity at the film possible under given conditions. High f-numbers require a longer **exposure time** to give the same overall **exposure** (intensity × time). The lowest f-number obtainable with a given lens is called the **speed** of the lens.

Black and white film consists of three layers: a clear plastic **base,** an **antihalation backing,** and an **emulsion** that contains tiny light-sensitive **silver halide crystals.** During exposure these crystals absorb light, form a **nucleus** (several silver atoms), and thus produce a **latent image.** (The antihalation backing prevents unabsorbed light from reflecting back to the emulsion and degrading the image.) In the chemical **developer,** black filaments of pure silver form around such nuclei. The chemical **stop** arrests this process before unexposed crystals become black. The process of **fixing** washes away unexposed crystals, resulting in a **negative.** **Reversal development** achieves a **positive.** Here, before fixing, the image is **bleached,** exposed overall, and developed again. High-**speed** (ASA, DIN, ISO) films have more **grain** than do slow films. A high-**contrast** film generally has smaller **latitude** (the useful, steep region on an **H & D** curve) than does a low-contrast film.

PROBLEMS

P1 To focus a camera, you move the lens back (closer to the film) or forward (farther from the film). When the lens is closest to the film, are you focused on a near or a far object?

P2 List the components of a simple camera and describe their functions.

P3 To get a wider angle picture, you can change the lens on your camera to one with a different focal length. (a) Should you use a lens with a longer or shorter focal length? (b) Prove your answer with a ray tracing diagram.

P4 What is depth of field and how is it controlled in a camera?

P5 You have a 35-mm camera and the following interchangeable lenses: 28 mm, 50 mm, and 150 mm. Which one of these lenses is best suited to each of the following situations? (a) You're standing near a building and want to photograph as much of it as possible. (b) You're in the bleachers at a baseball game and want a picture showing the pitcher as big as possible. (c) You want to photograph a fisherperson holding the fish toward the camera, so the fish looks large compared to the person.

P6 An exposure meter indicates that, if a correct setting for a photograph is f/5.6 at $\frac{1}{100}$ second, then f/11 at $\frac{1}{25}$ second is also correct for the same conditions. (a) Explain. (b) Give two other correct settings.

P7 The following f/stops and shutter speeds all give the same exposure: (verify!) (1) f/2 at $\frac{1}{200}$ sec, (2) f/2.8 at $\frac{1}{100}$ sec, (3) f/4 at $\frac{1}{50}$ sec, and (4) f/5.6 at $\frac{1}{25}$ sec. Which exposure setting is best suited if you want to photograph the following? (a) A dog running by. (b) Flowers in a garden, some close, others relatively far away. (c) A sleeping gibbon in the zoo, without the wire bars showing in the photograph.

P8 (a) Show that the photographic image of a three-dimensional object can appear deformed if that object is far off the cameras axis. Do this by drawing three parallel lines. One of them represents the film plane of your camera, the next should have a small hole in it to act as a pinhole, and the centers of the three-dimensional objects will lie on the third line. Draw a circle to represent a sphere in the middle of your object plane. Construct the image of this sphere on your film by drawing the relevant rays. Repeat this for a same size sphere well *off* the axis of your camera. Compare these two images. (b) Describe in words what a pinhole photograph of two basketballs would look like if one ball were on the camera axis and the other well off the axis.

P9 What is a sensitizer and what does it do?

P10 In physical development, additional material is deposited on silver nuclei to make them visible. (a) Give an example of physical development. (b) How are silver nuclei made visible in chemical development?

P11 Which of the two films whose H & D curves are given in Figure 4.43 is faster?

P12 Film *1* is faster and has higher contrast than film *2*. (a) Draw an H & D diagram to show these properties. Label curves for your two films. (b) State two possible applications appropriate for each of your films.

P13 The H & D response curves for three kinds of film are shown in the figure. Which film would be best for each of the following situations? In each case, tell which film characteristic (speed, contrast, or latitude) is the important one. (a) Photographing in a situation where there is very little light. (b) Photographing in a situation where there is a large variation in light level. (c) Photographing in a situation where you want to enhance small differences in light level.

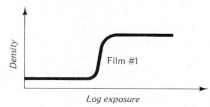

Film #1

Log exposure

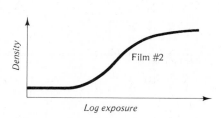

Film #2

Log exposure

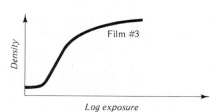

Film #3

Log exposure

P14 Why might eyeglasses made of photochromic glass (see TRY IT for Sec. 4.7D) not work as well as sunglasses when you are driving? That is, why might they not turn dark inside your car?

P15 (a) Describe what a "solarized" photograph (i.e., one made employing the Sabbatier effect) looks like. (b) What does a photographer do to make such a solarized photograph?

HARDER PROBLEMS

PH1 When taking photographs under water (using a camera and goggle system), you should use a wide-angle lens to make the picture appear normal. Why? (See Chapter 3, Problem PH4.)

PH2 This problem is designed to simulate the "telephoto perspective." You will need two pieces of standard $8\frac{1}{2} \times 11$-inch paper. Draw an axis in the center of each piece of paper, running the long way. Very near the right-hand edge, draw a line perpendicular to the axis, in each case to represent the film. Assume that all rays are focused on the film, so you need only draw the central ray to find an image. (This won't be exactly true because you have to take the objects fairly close in order to fit on the paper. It will be good enough, however.) (a) Suppose you have a lens with $f = 2.5$ cm. Draw it located that distance in front of the film. You will have two objects, each 16-cm high. Draw the first object, *A*, at a distance of 8 cm from the lens and the second object, *B*, located 4 cm farther from the lens. Using the central rays, draw the images and measure the sizes of them on the film. Calculate the ratio of the image sizes. (b) On your next sheet of paper, use twice as large a focal length: $f = 5$ cm. As before, you will draw two objects, each 16-cm high, but have, to keep the image of *A* equal to that in part (a), draw *A* twice as far away, at 16 cm from the lens. Keep the distance between *A* and *B* 4 cm, as before. That is, you've doubled your focal length, but moved back twice as far to keep the image of *A* unchanged. As before, find the images and measure the sizes of them on the film. Now calculate the ratio of the image sizes. (c) Discuss how the spacing between two objects (say people) would look if you had photographed them in the two cases.

PH3 An inexpensive way to get telephoto pictures is to use a teleconverter. This is an auxiliary diverging lens that you place *between* the camera and your normal lens. Suppose your normal lens has focal length of 50 mm, the teleconverter has focal length -50 mm, and the spacing between the two lenses is 25 mm. Draw a full-size ray tracing diagram to find the effective focal length of the combination, as follows. (a) Draw an incoming ray R

parallel to, and 2 cm above, the axis, and let it strike the first lens (your normal lens) at a point Q. Treat Q as an object for the second (teleconverter) lens and locate *its* image. You will then know what happens to any ray leaving Q that hits the teleconverter. In particular, you will know what happens to R after it is bent by your normal lens. Using a ruler and drawing very carefully, trace R through the teleconverter and find where it crosses the axis. Label this focus of the lens combination F. This is where the film should be. The barrel of the teleconverter is designed so the lens lies at the correct location. (b) Extend the ray you have just constructed, between the teleconverter and F, back in a straight line until it intersects the original ray R. You have now located the position of a single, simple lens that would have the same effect as the combination of normal lens plus teleconverter. Draw that lens and measure its focal length to get the effective focal length of the combination. What is it? By what *factor* has the teleconverter increased the focal length of your normal lens? (c) The combination of these two lenses thus behaves like a telephoto lens. Why should that come as no surprise to you?

PH4 Describe the development procedure for reversal film, from initial exposure to final transparency. What does each step in the process accomplish?

PH5 Relate the contrast of a film to properties and distribution of its silver halide crystals.

MATHEMATICAL PROBLEMS

PM1 Suppose you have an f/5.6 lens on your camera. If the lens has a 56-mm focal length, what is its effective diameter?

PM2 A lens of speed f/1.8 has an effective diameter of 28 mm. (a) What is its focal length? (b) If you want a lens of double that focal length, and the same speed, what effective diameter would that lens have to have?

PM3 A particular 120-mm focal length lens has an effective aperture of 7.5 mm (a) What is the f-number in this case? (b) If the aperture was 15 mm instead, what would be the f-number? (c) Suppose the speed of this lens is f/4. What is the effective diameter of the largest opening?

PM4 Extend Table 4.3 by supplying one additional f-number on each end of the table.

PM5 Extend Table 4.4 by supplying one additional equivalent exposure on each end of the table.

PM6 Assume the exposure time given in Table 2.1 for Tri-X film in the bright sun is exactly correct. (a) Use that exposure time and the ISO index to calculate the exposure time for Kodacolor. (b) The result you get in part (a) should be less than that given in the table. Why do you think we gave you the number we did?

PM7 What ASA index corresponds to (a) DIN 18, and (b) DIN 24?

PM8 What is the ISO index of a film that is four times as fast as ASA 100?

The Human Eye and Vision—I: Producing the Image

5.1
INTRODUCTION

Our eyes have evolved to deal with a remarkable variety of situations and tasks: from guiding us down a dark path on a moonless night, to helping us choose in the noonday sun which subtly colored mushroom is safe to eat, to gauging the distance to a tree branch as we swing through the air. The variety of our visual experience is immense: brightness, color, form, texture, depth, transparency, motion, size, and so on. The deep link between vision and knowledge is captured in our language: "See what we mean?"

The importance of the eye is reflected throughout our culture. For thousands of years, the human eye has been revered as an organ of mystery, imbued with power, not only to help us perceive the world, but also to charm or influence people and events. The pervasive legend of the evil eye reflects this fascination. The legend probably arose from the belief that there are rays emitted by the eye that can influence others as well as "touch" the world. In the first millennium B.C. in Mesopotamia, the legend involved certain demons, called *utukku*, who haunted and spooked deserts and graveyards and whose power lay in their glance, which could injure anyone unlucky enough to get too close. Saint Mark grouped the evil eye among the lowest of human traits: "adulteries, fornications, murders, thefts, covetousness, wickedness, deceit, lasciviousness, an evil eye, blasphemy." Marcus Terentius Varro (116–27 B.C.) claimed that the evil eye in women arose from unbridled passions. The Hindu god Shiva could consume objects in flames with a look from his third eye. A glance from an evil eye could ruin crops, cleave rocks, split precious stones, transmit disease (as Shakespeare wrote, "They have the plague and caught it of your eyes"), adversely affect sexual function, and rob Christians of faith. Death, too, is a common result of a glance from an evil eye. The beautiful slave girl Twaddud in *The Arabian Nights* attracted men and then "shot them down with the shafts she launched from her eyes." Two Moroccan proverbs quantify this: "The evil eye owns two-thirds of the graveyard" and "one-half of mankind die from the evil eye." Perhaps evil eyes were particularly virulent in third-century Babylon, where the Talmudic scholar Rab attributed 99 out of 100 deaths to the evil eye. Happily, modern adaptations of this mythology depict a powerful benefactor using his vision to further the cause of "Truth, Justice and the American Way" (Fig. 1.1).

The eyeball itself was thought to posess magical powers. Jerome Cavdan, a professor of medicine in the mid-sixteenth century, discussed the belief that by holding the eyeball of a black dog in your hand, you could prevent the neighborhood dogs from barking. Naturally, such eyeballs were prized by thieves. As late as the mid-1930s, some of the Pennsylvania Dutch believed that pinning the eye of a wolf to the inside of a sleeve could protect the wearer from accidents. (The method of pinning is unclear.) Today, psychiatrists treat ophthalmophobia—fear of being stared at.

But eyes need not be evil. For example, Cervantes calls them "those silent tongues of love." "Love had his dwelling in the subtle streams of her eyes," writes Chaucer in "Troilus and Criseyde." The Polish poet Daniel Naborowski expresses this more romantic view: "The single word 'eye' embraces all:/The torches, stars, suns, firmaments, and gods."

Here we shall be concerned not with the eye's power for good or evil, but rather with its power for vision. What does it do, and how does it do it? This concern with the physical mechanism of the eye led René Descartes, in the seventeenth century, to demonstrate and explain the image-forming properties of the eye, and it is this approach that we'll pursue. We'll begin by relating the eye to something familiar: the camera (Chapter 4). Though there are important similarities between the eye and camera, they do not tell the whole story. In this chapter we'll stress these similarities, then return to the eye in Chapter 7 to consider properties of the eye (and brain) that are rather different from those of the camera.

5.2
EYE AND CAMERA

There are a number of similarities between a human eyeball and a simple camera (Fig. 5.1). We've seen that a camera is a light-tight box with a lens system for forming a real image on the light-sensitive film. Stops and a shutter are used to control the amount and duration of the light admitted. The eye, too, is basically a light-tight box, whose outer walls are formed by the hard, white **sclera.*** There is a two-element lens system, consisting of the outer **cornea** and the inner **crystalline lens** or **eyelens.**† The lens system forms an inverted image on the **retina**‡ at the back of the eyeball. The colored **iris** corresponds to the diaphragm in a camera. Its **pupil,**§ the dark, circular hole through which the light enters, corresponds to the camera aperture. Though *not* analogous to a shutter, the eyelid *is* like a lens cover, protecting the cornea from dirt and objects. The eye or tear fluid cleans and moistens the cornea and can be compared to a lens cloth or brush.

The volume of the eye is not empty, as is that of the camera, but is filled with two transparent jelly-like liquids, the **vitreous humor** and **aqueous humor**‖ which provide nourishment to the eyelens and cornea. Additionally, their internal pressure helps to hold the shape of the eyeball. (Glaucoma is a

---

*Greek, *skleros*, hard. At the Battle of Bunker Hill, Colonel William Prescott is said to have shouted, "Don't shoot until you see their scleras."

†This convenient terminology comes from Frans Gerritsen's *The Art of Color.*

‡Latin *rete*, net—the nerve cells of the retina give the retina an appearance of a net.

§Latin *pupa*, doll. You see a tiny image of yourself in another's eye (you doll!).

‖Latin *vitreus*, glassy, and *aqueus*, water, plus *umor*, fluid. Actually, the humors are somewhat gelatinous, neither as dense as glass, nor as fluid as water.

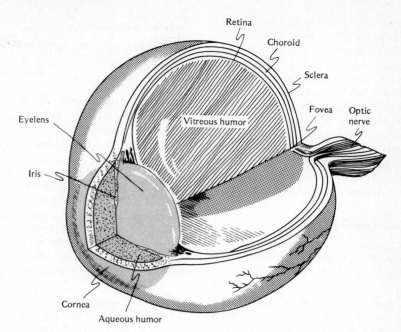

(a)

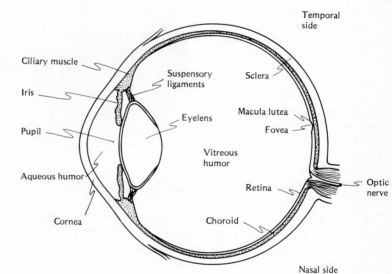

(b)

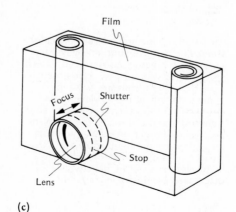

(c)

FIGURE 5.1

Human eyeball: **(a)** in three dimensions, **(b)** in two-dimensional cross section, and, for comparison, **(c)** a simple camera.

disease in which the pressure of these fluids is too high.) To see evidence of the fluid nature of these humors, stare at a large, bright, uniform area, such as a blank sheet of well-lit white paper or the sky on a cloudless day. Move your eyeball from side to side quickly, then hold it fixed. The moving dots and chains you see are shadows due to dead blood cells floating in your humors. The older you get, the more of these *floaters* you will have.

PONDER

Why are the humors, rather than blood vessels, used to nourish the cornea and eyelens?

The lack of blood vessels in the cornea and eyelens means that, when surgically transplanted, these elements are less likely to be rejected by the immune system, which resides in the blood. Nowadays, cornea transplants are performed routinely.

The "film" of the eye, the retina, contains over a hundred million light-sensitive cells, the *rods* and *cones*, packed together like the chambers of a honeycomb. Near the center of the retina—in a small region called the *fovea**—there are a great number of cones and no rods. This region is responsible for the most precise vision, and also for color vision (Plate 5.1 and Fig. 5.2). Away from the fovea rods predominate, and at the extremes of the retina, corresponding to peripheral vision, there are very few cones. The rods and cones are connected through a network of nerve cells in the retina to the *optic nerve.* There are only about one million nerve fibers in the eye's optic nerve; thus each such fiber is connected to many rods and cones.

There are no rods or cones at the point of the retina where the optic nerve leaves the eye, so you cannot see light that strikes this spot—the *blind spot.* You can find your blind spot by using Figure 5.3.

*Latin, depression, small pit. The fovea is a shallow depression in the retina.

PONDER

Is the blind spot on the part of the retina near the nose or near the ear?

Upon learning of the blind spot, Charles II is said to have amused himself by looking to the side and making the heads of his courtiers disappear.

As you might expect, the existence of a special region of sharp, detailed viewing near the center of the retina results in some significant differences in behavior of the eye as compared to the camera. A camera should be held fixed to produce an unblurred image, and then

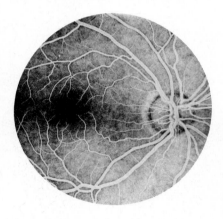

FIGURE 5.2

In this fluorescein angiograph, the blood vessels in the retina are made visible by injecting a fluorescent substance into the blood system. Note the fine network of blood vessels, except over the fovea.

FIGURE 5.3

Locate your blind spot. At a distance of about 25 cm, close your left eye and stare at the X with your right eye. Slowly move the tip of your pen across the page toward the right from the X. Keep looking directly at the X, viewing the pen with your peripheral vision. At about 8 cm to the right of the X, the tip of the pen will no longer be visible, when its image falls on the blind spot. You can "map out" the extent of the blind spot by placing dots on the page where the tip of your pen becomes invisible.

each part of the exposed film is equally sharp and detailed. But your eye must *scan*—move to point in different directions—so that different parts of the scene fall on your fovea, the small region of your sharpest vision. In this way, you can build up a sharp, detailed, mental picture of the world without having large numbers of optic nerve fibers attached to all parts of the retina. The efficiency of scanning is illustrated by the eye of the tiny marine organism *Copilia* (Fig. 5.4). Each of its eyes has only one light-sensitive cell, which scans across the focal plane of the lens in front of it every second or so. Rather than getting a complete picture of the world all at once, the *Copilia* brain (such as it is) receives a sequence of messages telling how the intensity of the light changes as the cell senses different directions. With this information, *Copilia* builds up

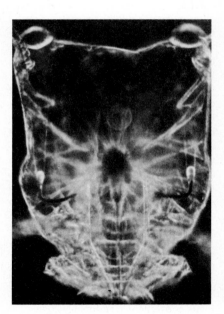

FIGURE 5.4

Photograph of the tiny (millimeter size) sea animal *Copilia*. Each eye contains an outer (corneal) lens and an inner lens attached to the photoreceptor. The inner lens and its associated receptor sweep side to side while sending signals through a single optic nerve to the brain.

its mental picture of the world. The wiring efficiency of the scanning process is the reason that TV cameras also scan. The scanning there is faster and more orderly—and done electronically, rather than optically—to produce a sequence of electronic signals that can be transmitted to your home TV or stored on tape.

In other animals, the nature of the region of sharp, detailed viewing in the retina depends on the needs and habits of the animal. Birds, for example, have a special problem; their eyes are set into the sides of their heads, giving them a maximum field of view. However, in addition to side vision, a bird must have sharp forward vision to see where it is flying. Many birds solve this problem by having two "central areas" (similar to our fovea) in each eye—one close to the optic axis for sharp side vision (recall how a pigeon looks at you), and one toward the back to provide sharp forward vision. Rabbits' eyes are different; they have a foveal "strip," which gives them excellent acuity where danger or predators are most likely to be—on the horizon.

A. Focusing and accommodation

The cornea and eyelens form a lens system similar to that of Figure 3.31, and produce a real, inverted image on the retina. (The first TRY IT tells you how to verify that it is inverted.) The amount of bending of light at each surface depends on the ratio of the indexes of refraction of the two media involved, in accordance with Snell's law. For this reason, most focusing occurs at the air-cornea surface ($n_{air} \simeq 1.000$, $n_{cornea} \simeq 1.376$), while less occurs at the surface of the eyelens, whose index of refraction is only slightly greater than that of the surrounding humors ($n_{eyelens}$ ranges from 1.386 near its surface to 1.406 near its center, whereas $n_{humors} \simeq 1.336$). Because n_{cornea} is very close to n_{water}, there is almost no bending of light at the cornea when the eye is underwater. Thus, fish have almost spherical eyelenses that do all

the focusing. The shape of their cornea is not important and differs in different fish. If an animal is to see both in air and in water, the cornea should be flat to provide no focusing for distant objects in either case. Crustaceans and some ducks have such eyes.

The anableps, a fish with a face that only its mother could love, solves the problem of simultaneous vision in air and water by a different technique (Fig. 5.5)—a bifocal eye!

FIGURE 5.5

(a) Photograph of an anableps. (b) The anableps eye at the water's surface, looking upward and downward simultaneously. Rays from the airborne prey are focused by both the air-cornea surface and the gently curved (long focal length) part of the eyelens. Rays from the underwater predator are not focused at the water-cornea surface but are focused by the sharply curved (short focal length) part of the eyelens.

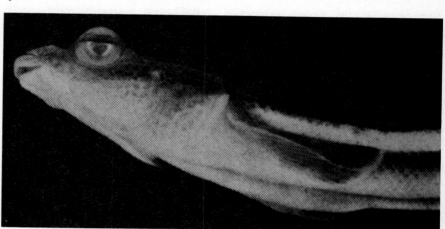

(a)

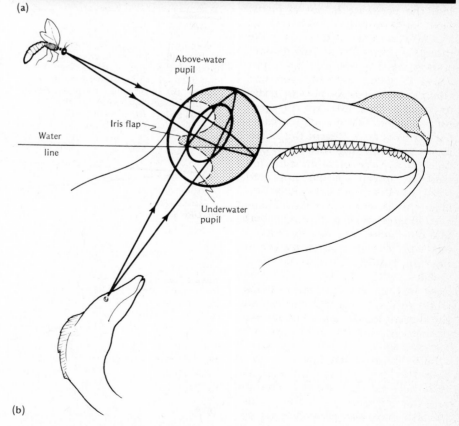

(b)

Only low-quality cameras have fixed lens systems; better cameras can be adjusted to focus on distant or on nearby objects. In the camera, adjusting the focus is achieved by changing the lens-film separation—increasing it for closer subjects (Sec. 4.2B). Our eyes do not change focus this way, but the eyes of many fish do.

Changing focus in a camera is necessary when the depth of field is limited. You can easily check that your eyes also have limited depth of field, and that you can change your focus for near and far objects. Hold up your thumb in front of a distant object such as a bookcase across your room. Focus on your thumb with one eye, closing the other. Without refocusing or pointing your eye in a different direction, notice that the bookcase is out of focus. Next, keep your thumb where it is, and focus on the bookcase; now the thumb is out of focus. This demonstrates that the eye has only a fair depth of field, and that you can change your focus for objects at different distances. (Making the aperture smaller, of course, increases the depth of field. Repeat the experiment while looking through a pinhole in a piece of aluminum foil held close to your eye. The improved focus and depth of field may be sufficient to give you a sharp image even if you remove your glasses.) You often make use of your limited depth of field when driving a car in the rain. You lean close to the windshield to make the spattered drops close to you, and thus out of focus while you are properly focused on the distant signs. Likewise, specks of dirt on your eyeglasses are not too noticeable.

Your ability to change focus depends on the fact that your eyelens is rather elastic. Instead of changing the lens-retina separation as the fish does, you change focus by *changing the focal length of your eyelens*—a process called **accommodation.** This is performed by the **ciliary muscles.** When these muscles are relaxed, the eyelens is pulled out by the **suspensory ligaments** and has the shape shown in Figure 5.6a—a long focal length for viewing distant objects. When the ciliary muscles are tense they form a smaller ring, thus releasing the tension in the ligaments and allowing the elastic eyelens to bulge into the shape that it prefers (Fig. 5.6b)—with a shorter focal length for nearby objects. (To see these changes in lens shape, refer to the second TRY IT.) Greater precision is possible by making a large adjustment in a weak element (the eyelens), rather than a small adjustment in a powerful element (the cornea)—just as you can set your UHF TV station more accurately by adjusting the *fine* tuning knob carefully. The strain you feel in your eyes after hours of doing close-up work is the fatigue of the ciliary muscles (and also, perhaps, the fatigue of muscles outside the eye tending to "cross" your eyes for close work—Sec. 8.3).

Normal eyes are able to accommodate for object distances between infinity and about 25 cm, but if the cornea bulges too much or too little, the eyelens may not be able to bring the image to a focus on the retina. For instance, a sharply bulging, short-focal-length cornea may be proper for focusing on nearby objects, but too strong for distant objects, no matter how relaxed the ciliary muscles are—a condition called **myopia.*** Similarly, too long a focal length produces **hyperopia.†** A person with either of these conditions may require eyeglasses or contact lenses to compensate (Sec. 6.2). (A more extreme method of correcting myopic vision involves surgically scoring the cornea surface. This process, radial keratotomy, relaxes the internal stress in the cornea, allowing it to assume a

---

*Greek, *muo*, shut or close, plus *ops*, eye. People with myopia squint when trying to see distant objects without glasses. By reducing the aperture, they increase the depth of field as with the pinhole.

†A contraction of "hypermetropia," from the Greek *hupermetros*, beyond measure plus *ops*. The hyperopic eye can focus on very distant objects.

FIGURE 5.6

Accommodation. **(a)** Relaxed ciliary muscles allow the suspensory ligaments to stretch the eyelens, which then has a long focal length for viewing distant objects. **(b)** Tense ciliary muscles release the suspensory ligaments and the eyelens bulges for viewing near objects.

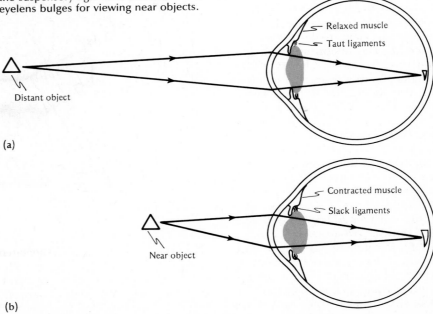

flatter shape. Other processes, suitable for both conditions, involve carving or molding the cornea into the desired shape.)

Occasionally, particularly as one gets older, the eyelens may develop an opaque white cloudiness—cataracts. Light can no longer reach the retina. In this case, the eyelens is surgically removed and the focusing power of it is replaced by glasses, a contact lens, or an artificial lens surgically implanted. While the eye then can see, it no longer has the ability to accommodate that it previously did.

First TRY IT

FOR SECTION 5.2A
The orientation of the retinal image

You can easily tell that the image on your retina is inverted by touching your eye. Open one eye and look at a piece of white paper. Use your little finger to touch gently the outside of the lid near the corner of the open eye, by the nose. You will see a dark spot off to the side toward your ear. The gentle finger pressure on the eye restricts blood flow in the retina at that position, and the retinal cells cannot respond. Thus, there is an apparent black spot. The spot you see is toward the side opposite the finger because the image on the retina is inverted; light from the side near your ear is imaged on the part of your retina near your nose.

If the image is inverted, why do you see the world as "upright"? In Chapter 7 we'll see that your sense of sight corresponds to complex patterns of activity in your brain—activity that is not simply related to the image on your retina itself. The relationship between your sensation of sight and this neural activity is, at best, problematic. It is meaningless to compare rightside up and upside down between material things, such as trees, and the immaterial, subjective sensation of your sight of the tree. What is important is that all your subjective sensations agree as to the location of objects. You need only bump into a few things for your brain to achieve this agreement. If you wear goggles that invert the retinal image from the usual orientation, you can learn to ride a bicycle, jump rope, play catch and so on—evidence of the remarkable ability of your brain to correlate sight with physical experience.

Second TRY IT

FOR SECTION 5.2A
Accommodation

With a candle, a dimly lit room, and an accommodating friend, you can study the action of the eyelens. Hold the lit candle about ⅓ meter from your friend's eye, slightly to the side of her direction of gaze. Look carefully at the reflections of the candle in her eye. You should see something like Figure 5.7. These are the **Purkinje images**. The first, which is the brightest, is due to the outer corneal surface. The image is erect, virtual, and smaller than the object. The third Purkinje image, due to the front of the eyelens, is also erect, though somewhat dimmer than the first. The fourth Purkinje image is due to the rear surface of the eyelens. (The second Purkinje image, due to the inner surface of the cornea, is probably too faint to be seen.)

PONDER

Why are the first and third Purkinje images erect (and virtual) while the fourth is inverted (and real)?

Have your friend accommodate by focusing on something close, say your ear. Note carefully the positions of the Purkinje images. Now have her focus on a distant object, without changing her direction of gaze. Only the fourth Purkinje image shifts significantly, because accommodation is achieved by changing the curvature of the rear surface of the eyelens (see Fig. 5.6).

FIGURE 5.7

The first, third, and fourth Purkinje images. The third image is blurred because it is reflected from the pebbly front surface of the eyelens. The fourth (inverted) Purkinje image moves during accommodation, implying that the rear surface of the eyelens changes shape.

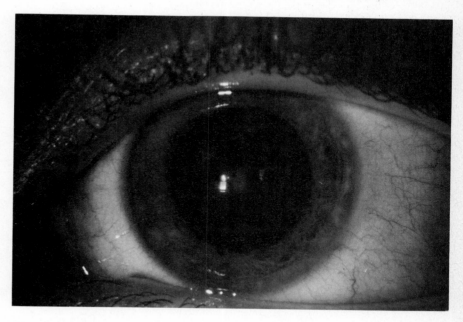

*B. Aberrations

As in any image-forming device, there are *aberrations* associated with the lens system of the eye. Nature has been very clever in minimizing these aberrations, and the few aberrations that are not well corrected turn out to be of little consequence.

The fovea is the only region where aberrations need to be reduced to a minimum, as it is the region of your most precise viewing. It lies nearly on the axis of the lens system of the eye, where the effects of *all* aberrations are smallest. Although points on the retina away from the fovea are more subject to the effects of aberration and the image is constantly degraded there, that is not very critical—you can always check any fuzzy image there by moving the fovea into the region of interest.

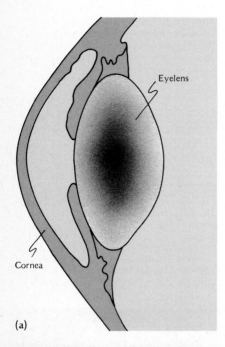

(a)

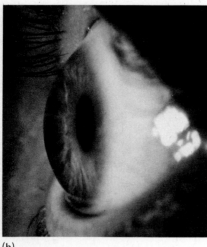

(b)

FIGURE 5.8

(a) The cornea is less curved at its periphery and thereby reduces spherical aberration. The index of refraction of the eyelens is lower near its surface than at its center, also helping to reduce spherical aberration. **(b)** Photograph of the cornea from the side.

Aberrations are also reduced in other ways. In the camera, *curvature of field* (Sec. 3.5D) is a problem as the image plane should lie on the flat film. In the eye, the retina is already curved, reducing this problem. The eye avoids *spherical aberration* (Sec. 3.5B) in several ways. First, the cornea, where most of the focusing is done, is not a spherical lens—it is flatter, less curved near its periphery than a spherical lens is (Fig. 5.8), so there is less bending of the peripheral rays, as desired. Furthermore, the eyelens has neither a spherical shape nor a uniform index of refraction. The eyelens is made in layers, like an onion, with the inner layers formed first and the outer layers added as the eye ages (Fig. 5.9). This makes the index of refraction lower at the periphery, so there is less focusing of the peripheral rays—less spherical aberration. We saw (Sec. 3.5D) that *distortion* is reduced when the stop is between the elements of a two-lens system; in the eye, the iris (stop) lies between the cornea and eyelens.

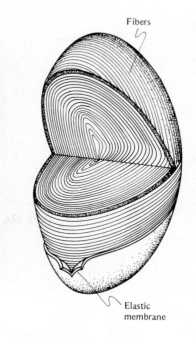

FIGURE 5.9

Eyelens consisting of layers of transparent fibers in clear elastic membrane. The lens of an 80 year old is $1\frac{1}{2}$ times as thick as that of a 20 year old.

The eye does not have an achromatic doublet to reduce *chromatic aberration* (Sec. 3.5A). Although the material of the lens system doesn't have much dispersion between the red and the blue-green regions of the spectrum, chromatic aberration is appreciable in the blue and ultraviolet (UV), so the eye tends to discard the latter wavelengths. The eyelens absorbs the UV as well as some blue. (With age, the eyelens becomes yellower—it absorbs more blue light. It has been suggested that most artists use less blue during their later years because they see less blue. However, if the eyelens is removed, say for cataracts, even ultraviolet light becomes visible.) Additionally, very little blue light is processed by the fovea where fine focus is most important. The fovea is populated by cones, which, taken as a group, are most sensitive to the yellow-green and somewhat less sensitive to the blue. On top of that (literally) lies the **macula lutea,*** a small, yellow-pigmented layer that covers the fovea, further rejecting blue light.

The corrections for chromatic aberration are not perfect, though, as you can easily see—"black lights" look fuzzy as a result of chromatic aberration. The eye focuses on the little amount of yellow-green light present in these UV lights, while the enormous amount of blue gets through the lens and macula lutea and is detected by the cones. Because it is out of focus, you see a blue fuzz around the bulb. Alternatively, you may notice a thin light line at the border between the bright red and blue regions of Plate 5.2. This line is not present in the painting or the reproduction and is due to chromatic aberration. Since your eye cannot focus on the red and blue simultaneously, at least one region is slightly blurred. Thus, there is a line on your retina receiving light from both red and blue regions; the result is a light line. Another example uses the meters on certain stereo equipment that have

---

*Latin *macula*, spot (cf immaculate = spotless), plus *luteus*, yellowish.

some figures glowing in red and others in blue. Focus on lit figures of one color in a dark room and you'll notice that the other figures are out of focus—they may pulsate, alternately in and out of focus.

Probably the most effective compensation for aberrations is the complex learning and processing done by the brain. For instance, if you wear goggles that have a great deal of chromatic aberration, you soon learn that the colored edges around objects are not real. Your brain compensates, and after a few days, you don't experience any colored edges. When the goggles are removed, however, you see colored edges even though they are not physically present in your retina. Another experiment uses glasses in which each lens is blue on the left half and yellow on the right half, with a vertical line dividing the two areas. After a while, you learn to ignore the blue and yellow half-worlds and sense no difference from your usual perception—you perceive no color or boundary. Similarly, millions of wearers of bifocal glasses learn to ignore the line that divides their field of view. These examples of effects more drastic than optical aberrations demonstrate the remarkable ability of your brain to compensate for unusual images on the retina.

C. The iris

Like the diaphragm in a camera, your iris opens and closes in response to the average ambient light level. (You can easily check this with the TRY IT.) When your iris is fully open, your eye has an f-number between f/2 and f/3 (compared with f/0.9 for the nocturnal cat, and f/4 for the diurnal pigeon). Your iris controls light admitted to your eye fairly rapidly, in about $\frac{1}{5}$ second, but the amount of light reduction is rather small, less than a factor of 20. The *main* function, therefore, cannot be to control the light intensity, since the range of light intensities to which you can reasonably respond is enormous (about a factor of 10^{13}). Instead, in bright con-

ditions your iris stops down to *reduce aberrations* and *increase the depth of field*. When you do close work, such as threading a needle, your iris tends to stop down. This increases your depth of field and hence allows you to focus on objects closer to you than you otherwise could focus on. It also diminishes the need to keep changing accommodation. As you grow older and your vision deteriorates, your iris will stop down under a greater variety of conditions to increase depth of field and reduce aberrations. Under low light levels, when your eye can't afford these luxuries, the iris opens up allowing more light in. This occurs at times when gathering light is more important than a sharp image. For example, in the dark it is more important that you be able to see *that* there is a tiger present, than being able to tell its sex. We have the same trade-off—between light gathering power and **resolution** (the ability to see fine detail)—that we saw in camera film (Sec. 4.7G) and that we'll see again and again.

TRY IT

To see some properties of the action of the iris, all you need is a flashlight and a friend (or a mirror). In a dim room, shine the flashlight into your friend's right eye and notice the contraction of his iris. Turn off the light and watch his iris open again. Without touching his eye, use a plastic ruler marked in millimeters to measure the diameter of his pupil in each case. Using the fact that the area of a disk is proportional to the square of the diameter, calculate the ratio of the pupil areas in the two cases. Now shine the light in his left eye while watching his right eye; his right iris responds. The brain causes the irises to operate together. Hold the flashlight about $\frac{1}{2}$ m directly in front of him, and notice the size of your friend's pupil. Now turn on an additional light, dimmer (or more distant) than the first, but somewhat to the side, and observe Fechner's paradox. Although there is more light entering your friend's eye, his pupil opens up

further! The size of the pupil is determined by the average light over the illuminated area of the retina. Even though there is more total light present in the latter, two-light case, it is spread out over more of the retina, so the average is lower (and the iris opens).

The size of the iris also depends on the state of accommodation. Under normal room illumination, have your friend first focus on something near and then on something far. Notice that the iris stops down when he focuses on the close object, even though there is the same average illumination. The smaller pupil increases the depth of field and allows him to see quite close objects in good focus.

5.3
THE RETINA

The retina is, in many ways, analogous to the film in a camera. Each receives a real image, the first step in a complicated process that results in either a photograph or our sense of sight. Figure 5.10 shows a schematic cross section of a human retina, with its three layers. The light sensitive layer consists of a dense array of rods and cones (the **photoreceptors**), which absorb the light. The **plexiform layer*** consists of several types of nerve cells

---

*From the Latin *plectere*, to braid, to plait. A plexus has come to mean a network.

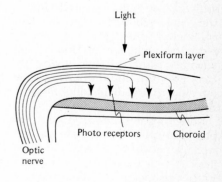

FIGURE 5.10

Layers of the human retina (at a point other than the fovea).

that process the signals generated by the rods and cones and relay them to the optic nerve. Finally, the **choroid** carries major blood vessels to nourish the retina and also acts as an antihalation backing, absorbing light so that it cannot reflect and strike the rods and cones on a second pass.

Our pupils look so dark because much of the light that enters them is absorbed by either the photoreceptors or the choroid. If a bright light is shined directly into the eye, though, some will come back out. This is why the eyes of subjects in some flash photographs appear red (Plate 5.3). Light from the flashbulb is focused on the retina by the lenses of the eye, and some of this light is retroreflected—a phenomenon similar to dew heiligenschein (Sec. 3.4B), only here the reflection occurs on the retina, inside the eyeball, rather than on the blade of grass, outside the dew drop. Because the light is reflected back along the direction from which it came, the effect is greatest with cameras that have the flashbulb near the camera lens.

Nocturnal animals like cats need to respond to low light levels. Rather than a choroid, then, a cat's eye has a reflective **tapetum lucidum.**[*] The light that misses the photoreceptors on the first pass is reflected by the tapetum and gets a second chance to be absorbed. This reflection slightly defocuses the light, so the resultant image is degraded, blurred. Again we see a trade-off of resolution for light sensitivity. The light retroreflected by the cat's yellowish tapetum isn't completely absorbed on its second pass through the retina. Some emerges from the eye, giving the cat's eyes the often sinister yellowish glow that your headlights sometimes reveal.

Notice in Figure 5.10 that the photoreceptors are *behind* the neural network in the plexiform layer. It is as if the retina (film) were in the eye backward! In normal film,

---

[*]Latin, luminous carpet. Crystals containing zinc cause the reflectivity.

the antihalation backing and the emulsion are usually kept separate so the antihalation backing can be removed easily during development. The photoreceptors of the retina, however, are living cells, which need the nourishment provided by the blood vessels in the choroid. Although the plexiform layer in front of the photoreceptors absorbs *some* light, this is minimized since its cells are transparent and require only a thin network of capillaries. In the critical, foveal region, there are no capillaries (see the TRY IT) and the plexiform layer is thinner. (Since they live at the bottom of the sea, where light is scarce, cephalopods—octopus, squid, cuttlefish—have their light-sensitive receptors in the front of their retinas.)

Another reason for having the receptors next to the "antihalation backing" is to diminish the chance of reflections from the intervening material, which would cause a loss of sharpness. For this reason, some special films have the antihalation backing between the emulsion and the base.

TRY IT

FOR SECTION 5.3
Seeing blood vessels, capillaries, and cells in the retina

Hold a small flashlight just above your closed eyelid and move the light back and forth slightly. The larger vessels on your retina's surface become visible as they cast a moving shadow on your retina. They appear like a fine network of rivers and tributaries. Notice that they lead away from the blind spot (as in Plate 5.1). Also note that no vessels cover your fovea, at the center of your field of view.

John Uri Lloyd, in his book Etidorhpa; or the End of the Earth describes a similar way to see these vessels using a candle:

Placing himself before the sashless window of the cabin, which opening appeared as a black space pictured against the night, the sage took the candle in his right hand, holding it so that the flame was just below the tip of his nose, and about six inches from his face. Then facing the open window he turned the pupils of his eyes upward, seeming to fix his gaze on the upper part of the open window space,

and then he slowly moved the candle transversely, backward and forward, across, in front of his face, keeping it in such position that the flickering flame made a parallel line with his eyes, and as just remarked, about six inches from his face, and just below the tip of his nose. . . .
"Try for yourself," quietly said my guide.
Placing myself in the position designated, I repeated the maneuver, when slowly a shadowy something seemed to be evolved out of the blank space behind me. It seemed to be as a gray veil, or like a corrugated sheet as thin as gauze, which as I gazed upon it and discovered its outline, became more apparent and real. Soon the convolutions assumed a more decided form, the gray matter was visible, filled with venations, first gray and then red, and as I became familiar with the sight, suddenly the convolutions of a brain in all its exactness, with a network of red blood venations, burst into existence.
I beheld a brain, a brain, a living brain, my own brain, and as an uncanny sensation possessed me I shudderingly stopped the motion of the candle, and in an instant the shadowy figure disappeared.

Lloyd's character, of course, can only be said to see his brain in the sense that, as Aristotle said, "The eye is an off-shoot of the brain."
To see the smaller, finer capillaries of your retina, close your left eye, hold a pinhole very close to your right eye, and through the hole view a uniform bright area, such as a well-lit unlined page or a cloudless sky. Now, without moving your eye, jiggle the pinhole with a small circular motion. The small squiggly lines that appear on the uniform bright area are the shadows of small blood vessels on your retina. Convince yourself that there are no vessels in front of your fovea.
You do not need any devices, just patience and care, to see the effects of blood cells nourishing your retina (these are not the same as the dead blood cells in the humors, the floaters). Simply gaze at the bright cloudless sky. If you're observant, you'll see tiny white dots wandering across your field of view at points away from the fovea. These are the effects of single blood cells as they pass through the small capillaries you saw earlier. (You won't see these cells on your fovea, which is nourished at the rear.) Notice that the motion of the cells is greatest a short time after each heart beat. This is particularly evident after exertion, such as jogging, when the heart is pumping very hard.

A. The rods and cones

The heart of the retina is the light-sensitive photoreceptor layer (Fig. 5.11). There are about 7 million

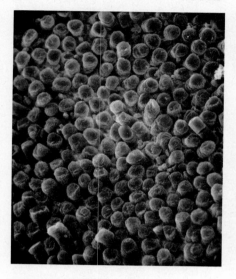

FIGURE 5.11

Electron micrograph of the back of a retina. The photoreceptors face you, and the light would strike them from the far side. The rods are big and blunt; the cones are smaller and pointy. Each cone is about 1 μm across, and each rod is about 5 μm across.

cones (sensitive to high light levels) and 120 million rods (sensitive to low light levels) in each retina, in an area of 5 cm². If you could image it sharply, a baseball nearly one mile away would produce an image that would cover just one cone. Figure 5.12 shows that the distribution of these cells across the retina is not uniform.

The relative numbers of rods and cones in the eyes of other animals differ from ours, depending on the animal's particular life style. Thus, in nocturnal animals, such as cats and rabbits, the photosensitive cells are all or mostly the highly sensitive rods, allowing the animal to see under low, night, light levels (though they cannot distinguish colors). Some diurnal animals, such as pigeons, turtles, and mongooses, have only cones.

FIGURE 5.12

Distribution of rods and cones along the equator of an eyeball. The fovea has no rods but many cones. Rods predominate in the periphery. The blind spot has no photoreceptors. Inset shows left eye viewed from above. (Check your answer to the PONDER of Sec. 5.2.)

*B. The mechanisms of light absorption

Figure 5.13 shows a schematic of a typical rod and cone. Each cell has four parts: the outer segment, the inner segment, the cell nucleus, and the synaptic ending. The shapes of the outer segments give the cells their names. Light is absorbed in the outer segment by the light-sensitive photochemical, which is manufactured in the inner segment. A complicated chemical process then generates an electrical signal that passes down the cell to the synaptic ending. At that point, another chemical (called a neurotransmitter) is released by the cell and passes across a gap, the **synapse,*** to the nerve endings in the plexiform layer. As with other cells, the nucleus produces nutrients and directs chemical reactions within the cell.

The outer segment of a rod looks similar to Figure 5.14. Each rod is filled with a stack of disks that contain the photochemical. Some of the photochemical is used up when it absorbs light. To replenish it, the inner segment produces new disks and photochemicals at the base of the outer segment. The new disks continuously push the older disks up to the end of the rod where they break off and are broken down in the choroid.

The rods' photochemical is made of a derivative of vitamin A plus a protein, opsin. Its purple color gives it its name, **rhodopsin,†** or visual purple. When visible light strikes a molecule of rhodopsin, it causes the electrons in the molecule to resonate and the molecule changes shape. In the process, an electrical signal is generated that becomes the neural signal to be transmitted to the rod synapse and, ultimately, to the brain. The molecule in its new shape is unstable, and it decays through a sequence of inter-

---

*Greek *syn*, with, plus *hapto*, join. The synapse is the place where one nerve cell joins with another.

†Greek *rhodon*, rose (cf rhododendron = rose tree), plus *opsis*, sight.

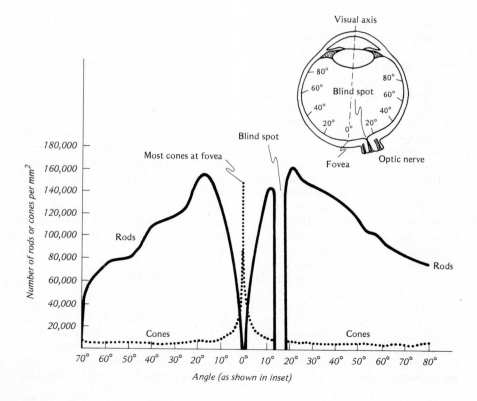

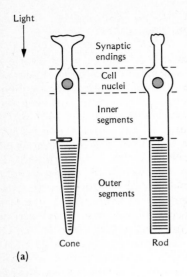

Light

Synaptic endings

Cell nuclei

Inner segments

Outer segments

Cone   Rod

(a)

FIGURE 5.13

Physiology of the photoreceptors:
(a) schematic, (b) actual photograph of rods.

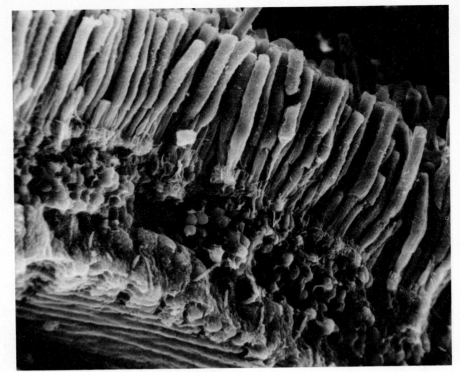

(b)

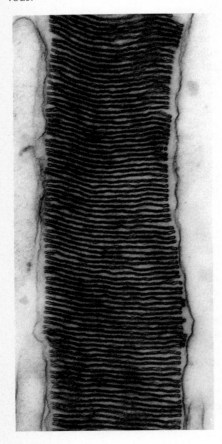

FIGURE 5.14

Electron micrograph of the outer segment of a cone.

mediate unstable molecules, finally yielding the original components, the vitamin A derivative and the opsin. In each successive step in the decay chain, the chemicals become lighter in color—as they no longer absorb visible light, they reflect all wavelengths and thus tend to look white—so the process is called **bleaching.** At the end of the bleaching process, enzymes reform the rhodopsin from its components. The entire process may be repeated many times, but ultimately the photochemicals must be replenished, so new disks are made.

Similar processes occur in the cones. One difference between rods and cones, however, involves the way in which they absorb light. Cone vision is involved when light is plentiful, so the cones can sacrifice some of the excess light in order to improve the quality of the image; the cones are less effective in absorbing the rays that strike them at large angles. Such rays may be due to light scattering within the eye and would degrade the image. Alternatively, they may be peripheral rays from the lens, which contribute most to aberrations. The re-

duced response to these undesirable rays—the **Stiles-Crawford effect**—is thought to be due to the fact that each cone acts as a light pipe, guiding the incident light down its body until it reaches the outer segment, where it is absorbed. Just as in fiber optics (Sec. 2.5B), the light is totally reflected from the inner sides of the cell. This mechanism allows light that is focused at the base of a cone to be absorbed throughout the entire volume of the light-sensitive outer segment. However, light reaching the retina at too large an angle strikes the inner sides of the cell at less than the critical angle, is not totally reflected, and thus is less likely to be absorbed. The eye discards this excess light in order to improve the sharpness of the image.

C. Processing time

The cells in the retina do not respond instantaneously to a flash of light. Further, their response neither ceases the instant the flash stops, nor lasts forever. The first effect, the delay, is called **latency** and the latter, **persistence of response.** Perhaps these effects may be made clearer by considering the following. It takes you only a moment to tell a child to clean his room. Undoubtedly, there is a delay (latency period) before he starts, and once he begins, he takes a long time to clean it (persistence). In vision, the intensity of a flash affects the latency and persistence periods. The response to a more intense flash is both prompter (shorter latency) and less persistent. When you *yell* at the child he may begin cleaning his room sooner and finish it faster (Fig. 5.15).

The persistence of the response can be considered analogous to the exposure time in a camera. If the shutter of a camera is open for a certain time, say $\frac{1}{2}$ second, anything moving in the field of view during that time appears blurred on the film—washed out if it is moving rapidly. When we look at moving objects, however, they generally don't look blurred even though our eyes remain open essentially continuously. This is due, in part, to the fact that receptors and connecting nerves remain active only for the persistence time (as opposed to film, which records the image as long as the shutter is open), so an object moving across our field of view is recorded only briefly by each receptor. The persistence time varies from about $\frac{1}{25}$ second at low in-

tensities to $\frac{1}{50}$ second at high intensities (Sec. 7.7A). This is sufficient, combined with the brain's ability to interpret motion, to allow us to see many moving objects clearly. (Not all, by any means, as you know if you've ever tried to read the label on a rotating phonograph record.) In a camera, of course, the exposure time not only helps stop the action, it also limits the amount of light during an exposure. Good cameras have exposure times ranging over a factor of a thousand, much greater than the factor of two in the persistence time of the human eye. Thus, this mechanism, like the iris, does not give the eye much control over the range of light intensities we experience in the world. In the next section we'll see how the eye *does* allow for the large range of light intensities.

D. Sensitivity

In Section 5.2C we found that the action of your iris (stop) is not sufficient to compensate for the enormous range of light intensities you experience in the world. Likewise, we saw that the processing time (Sec. 5.3C) can only vary by a factor of two and thus is also inadequate for this purpose. What else is there?

In the camera, you can use films of different speed—a fast film for dark conditions, a slower film when enough light is present. Changing a camera's film every time you walk from the noonday sun into a basement may be cumbersome and awkward, but in fact such a process

occurs in your eye—your retina automatically changes its sensitivity (its speed).

In a sense, the retina contains two types of film; one is made up of the system of cones and the other, the system of rods. The former is analogous to a slow, fine-grain, color film while the latter is analogous to a fast, coarse-grain, black and white film. The cones are sensitive to high levels of light—**photopic***—conditions—such as you're experiencing now. The rod system is sensitive to low light levels—**scotopic**†—conditions—such as occur at night in the forest. The rods "turn off" under photopic conditions; conversely, the cones do not respond under scotopic conditions.

Your resolution is not uniform across the retina. The cones are densely packed in the fovea to help produce your high resolution and color vision there. Away from the fovea there are few cones, so your peripheral color vision is quite poor. Cone density is not the complete story, though; your resolution is determined by how much information gets to your brain via the fibers of the optic nerve. Typically, just a few cones in the fovea are connected to the same optic nerve fiber, while several thousand rods in the periphery of the retina may be funneled into one fiber. This funneling implies that the retina must, in some way, process the im-

---

*Greek *phos*, light.

†Greek *skotos*, darkness.

FIGURE 5.15

(a) Latency. (b) Persistence of response.

(a)   (b)

age—less funneling in the fovea resulting in more fine details of the image being sent to the brain from this central region (see the TRY IT).

The rod system is sensitive under scotopic conditions, but alone it cannot mediate color perceptions. That is why everything appears black, white, and gray in low light levels—as John Heywood wrote, "When all candles bee out, all cats be grey." Lacking rods, your fovea doesn't work under scotopic conditions. To detect a dim spot of light, such as a dim star at night, you avert your vision slightly—look to the side of the star a bit—so that its image falls off the fovea and onto a region of many rods. Because many rods are "wired together" into a single optic nerve fiber, their sensitivity to low light levels is increased. Thus, as in film, the faster the "film" speed, the coarser the grain.

The two systems, rods and cones, provide two different "film speeds" to deal with different levels of light. Even within just one system, however, the eye adjusts its sensitivity as the light level varies. The changing of sensitivity of the retina in response to overall light level is called **adaptation** and can be illustrated by considering what you see when you walk from the midday sun into a dark theater. At first, it's too dark to see anything. You may accidentally bump into chairs or sit on an old lady's lap. Bit by bit, though, your retinas adapt to the lower light levels (scotopic conditions) and you can begin to see the people around you. Eventually you can make out your surroundings, but they appear only in black, white, and grays. You don't see much fine detail—you can't read your program. Throughout the period of adaptation, a constant intensity source, say an EXIT sign, appears brighter and brighter.

Let's be more specific. As a measure of the sensitivity of the retina, we can plot the intensity of a spot of light that is just barely detectable. This intensity is called the **threshold of detection,** or simply, threshold. In Figure 5.16, we plot your threshold as a function of the time you are in a dark room. The vertical axis denotes the intensity of

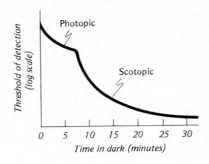

FIGURE 5.16

Dark adaptation. The threshold of detection (log scale) versus time in the dark. The first section (photopic) is due to the cones. You sense color at these thresholds. The second section (scotopic) is due to the rods. You see the world as black, white, and gray at these thresholds. Details of this curve will vary, depending upon the specific conditions and the individual subject.

the light at threshold. Values at the top of the graph mean your threshold is high, your sensitivity poor, as the source must be intense for you to see it. Conversely, values at the bottom of the graph mean that you are quite sensitive. We use a logarithmic scale (as in Fig. 4.42) because the range of your sensitivity is so great. The horizontal axis is the time in the dark, measured in minutes. The range of times of interest is not great, so we can use a regular (linear) scale for the time axis. On the left, the graph shows that just after you come into a dark room from a bright area, an intense light is needed for you to detect it. During the next few minutes your retina becomes more sensitive—the light intensity at threshold decreases, at first rapidly, and then more slowly. When shown the light, you can tell its color, a sure sign that your cones are still operating. About 5 to 10 minutes after you enter the dark room, the curve drops rapidly again, as you are rapidly getting more sensitive at this time. After this kink, you cannot tell the color of the light at threshold—you are now using rods only. (Of course, f you look at brighter light, such as from the stage, your cones respond and you can distinguish the colors.) The curve becomes nearly flat at

about 30 minutes. After a very long time in the dark, you can detect a light intensity equal to that of a candle viewed from 10 miles away! During World War II, blackouts required that *all* lights be extinguished; even a single lit cigarette could be seen by enemy pilots.

PONDER

If we repeat this experiment using a small light source and its light falls only on the fovea, no kink is found. Why? If the light falls only on the periphery, the threshold is very high until around 7 minutes, and then it joins the curve in the figure. Why?

When you go from a dark room to the bright outdoors, the process is reversed; at first you are "blinded" by the light. (This phenomenon is sometimes used in the theater with great dramatic effect. The opening scene of Mussorgsky's opera *Boris Godunov* is played in dim light, making the following scene—Boris's coronation—appear all the more dazzling.)

Thus, when you enter a dark room, your cone sensitivity begins to increase. When it can increase no more (the leveling off at around 5 minutes in Fig. 5.16), you switch to the rods. If necessary, their sensitivity will also increase. Using these two different systems, each with adjustable sensitivity, you automatically and continuously adjust your "film speed" until you have the appropriate one for the available light. While this is a slow mechanism compared to the response of the iris or the processing time, it has a much greater range available. In a camera, when you switch to faster film, you usually have to sacrifice grain size, the sharpness of the final picture. Further, the fastest films are black and white. Similarly, in the eye, the switch to rods gives us a coarser-grained, less-detailed, black and white view of the world. Whenever necessary, you trade sharpness of vision for sensitivity to light. Whenever possible, you trade back.

The switch from cones to rods during dark adaptation has another consequence. Figure 5.17 shows

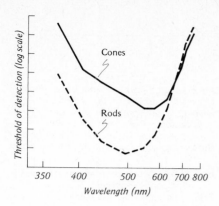

the relative thresholds for the rod and the cone systems versus wavelength of light. We see that, overall, the rods are more sensitive (have a lower threshold) than the cones, but there is another difference; the rods are most sensitive at the short wavelength end of the spectrum, the blue, while the cones are most sensitive at the long wavelength end, the red. Thus, at light levels high enough that the cones are operating, a given red object may appear brighter than a blue object. On the other hand, at light levels so low that only the rods operate, the same blue object may appear brighter than the red one. This change from the red appearing brighter at high light levels to the blue appearing brighter at low light levels is called the **Purkinje shift.*** You can easily experience this effect using Plate 5.2. Notice in bright light that the red region appears brighter than the blue; turn off the lights and wait about 10 minutes in the dim room and notice that the blue region appears brighter than the red.

This effect is utilized in situations where one must rapidly change from photopic vision to sco-

*After Johannes Purkinje (1789–1869), who first measured the dark adaptation curve (Fig. 5.16).

topic vision. For instance, astronomers may need their precise cone vision for such work as checking gauges, and shortly thereafter need their rod vision for work at the telescope. To allow for this rapid change, they use red lights to illuminate the observatory office building. The red light stimulates the cones preferentially, leaving the rods only slightly stimulated. In this way, the rods can stay more dark adapted while the cones are used for reading the gauges.

The adaptation of each of your eyes is independent of the other; one eye can be operating scotopically, while the other is operating photopically. When you're wakened in the middle of the night, keep one eye shut while you turn on the lights. Later, when you turn off the lights, open that eye (which is still dark adapted), and you can see clearly and avoid stumbling over the junk on your floor.

TRY IT

FOR SECTION 5.3D
Foveal versus peripheral viewing

The fovea is responsible for the sharp vision you need to read this book as well as for color vision. The peripheral parts of the retina provide neither of these. To demonstrate these facts, stare fixedly at one letter in the center of this page from about 25 cm away. Without moving your eye, try to read some other part of the page. If you don't move your eye, you may be surprised at the limited region of the page over which you can make out the words. Still keeping your eye fixed, ask a friend to hold up some fingers to your side and see if you can count them. These experiences should convince you that your sharp vision is only from the central, foveal, part of your retina.

To demonstrate the lack of color discrimination in the peripheral parts of your retina, again stare at one point. Have your friend hold a few colored pens or swatches of colored paper to your side while you attempt to judge the colors. Can you tell which color is which? On the other hand, if your friend holds up two pieces of paper, one black and one white, can you distinguish them? That is, can you distinguish the relative brightness of the light coming to the periphery of your retina?

SUMMARY

The compound lens **(cornea** plus **eyelens)** forms an image on the "film" **(retina)** of the eye, which contains a nonuniform distribution of the light-sensitive **rods** and **cones.** The diaphragm **(iris)** closes quickly to reduce aberrations and increase depth of field when enough light is present. The retina consists of a small area for precise vision (the **fovea,** containing cones), a broad area for more sensitive night vision (the peripheral parts of the retina, containing rods), and a region of no receptors (the **blind spot,** where the nerve cells leave the retina). This structure necessitates moving your eye as your view a scene **(scanning).** To change the focus of the eye as you look from distant to near objects, you **accommodate** (change the focal length of the eyelens). The fovea is responsible for the fine vision and, being near the axis of the eye, is least affected by aberrations. The brain's ability to learn is also crucial in overcoming aberrations.

The retina contains a light-sensitive layer of **photoreceptors,** a **plexiform layer** of nerve cells to carry the signals to the brain, and a dark **choroid** layer, which serves as an antihalation backing. In the photoreceptors, light affects the **photochemicals,** causing an electrical signal to pass along the cell to the **synapse,** which links the photoreceptor to subsequent nerve cells. The receptors take some time to start responding **(latency).** Most moving objects don't look blurred because, although the receptors keep responding **(persistence of reponse),** they do so for only a certain time. The retina's "speed" (sensitivity) is gradually changed automatically, depending on the overall light level. This process of **adaptation** changes the "film" from the fine-grained, color vision of the foveal cones in bright light **(photopic** vision) to the sensitive, coarse-grained, black and white **scotopic** vision of the peripheral rods in dim light.

PROBLEMS

P1 Compare and contrast the eye with the basic camera. What part of the eye corresponds to each major component in the camera?

P2 What is accommodation and how is it achieved?

P3 What are the layers in the retina (the "film")? Compare and contrast film and the human retina.

P4 Discuss how the eye and brain deal with the problems of aberrations. Pay particular attention to chromatic and spherical aberration, but discuss them all.

P5 Why do humans have choroids while nocturnal animals have tapeta?

P6 Explain in a brief sentence or two why a "black light" looks fuzzy.

P7 (a) What is the blind spot? (b) You don't "see" it as a dark spot in your field of view. What do you see? For example, draw a straight line. Draw another straight line parallel to the first, but shifted over by 2 mm, beginning even with the point where the first ends and continuing on. The two should look like one long line with a break in it. "Look" at the break with your blind spot. Can you tell there is a break in the line, or does it look like one long straight line? Try to make some other pattern which, when "looked" at with your blind spot, creates some interesting effect.

P8 In *The Divine Comedy*, Dante says, "a sudden flash blinds the eyes/so that it deprives them/of the sight of the clearest objects. . . ." Explain the phenomenon that Dante is talking about.

P9 What is the Stiles-Crawford effect? What is thought to be the mechanism for the effect? What benefit might this effect bring to human vision?

P10 In *The Girl in a Swing*, Richard Adams writes: "Look, everything's silver out there—the roses, the lupins, everything! Have you noticed, they've got no colour at all by moonlight?" Explain the phenomenon described.

P11 During the day, a certain red geranium appears brighter than a blue violet next to it. At dusk, though, the situation is reversed; the violet now appears brighter. Why?

P12 Isaac Newton, fascinated by the sun, viewed it through a focusing lens:
In a few hours I had brought my eyes to such a pass that I could look upon no bright object with neither eye but I saw the Sun before me, so that I durst neither write nor read but to recover the use of my eyes shut myself up in my chamber made dark three days together and used all means to divert my imagination from the Sun. For if I thought upon him I presently saw his picture though I was in the dark.
Explain what Newton's problem was and why it had come about.

HARDER PROBLEMS

PH1 Discuss the trade-off between sharpness of vision and sensitivity to low light levels in the eye. What mechanisms are responsible for this trade-off?

PH2 Suppose the threshold of detection experiment was done using a small light source. Sketch the analogous curve to that of Figure 5.16 for the case where the light from the test source falls only on: (a) the fovea, (b) the periphery of the retina.

PH3 Suppose the curve of Figure 5.16 had been obtained by measuring the threshold of detection of a red light (650 nm). What would the curve look like? (See Fig. 5.17.)

PH4 In "Stanzas on Death," Jean de Sponde writes:

My eyes, no longer cast your dazzled gaze
Upon the sparkling beams of fiery life;
Envelop, cloak yourself in darkness, eyes:
Your customary keenness not to dim,
For I shall make you see yet brighter lights,
But leaving night you shall see all the brighter.

What phenomenon of human vision is de Sponde referring to?

PH5 Hold a pinhole in a piece of aluminum foil about 3 to 5 cm from your eye and look through it at a bright light (e.g., a light bulb $\frac{1}{2}$ m away). Take an ordinary straight pin, stick it in an eraser on top of a pencil, and put it between your eye and the pinhole, moving the pin up and down so you can see the pinhead crossing in front of the pinhole. (a) Describe what you see. (b) Draw a picture showing the light bulb, the pinhole, the pin, your eye, and rays coming through the pinhole from the top and the bottom of the light bulb. Use this picture to explain what you see.

PH6 Hold an 8-cm object 30 cm from your eye. Notice that you must move your eye to view the entire object carefully. How could you use this observation to estimate the size of your fovea? Explain.

MATHEMATICAL PROBLEMS

PM1 If the (compound) lens of the human eye were taken out and examined, it would have a focal length of about 16 mm. (In the eye, however, there is not air behind the lens, but rather the vitreous humor with an index of refraction of about $\frac{4}{3}$, so the focal length of the lens *in* the eye is somewhat longer.) When dilated, the pupil diameter is about 4 mm. What is the speed of this lens (in air)?

Optical Instruments

6.1
INTRODUCTION

In the last two chapters, we discussed devices used for observing the world. The earliest such device was, of course, the eye; all you have to do is look. The camera allowed you to record an image on film, so you could look at it later. This chapter treats devices that improve the image to be detected. These not only allow you to see a better quality image, but to see images of objects that you otherwise couldn't see because they are too small, too far, or too transparent. As John Trumbull wrote, "But optics sharp it needs, I ween,/To see what is not to be seen."

6.2
SINGLE-LENS INSTRUMENTS

Several optical instruments use only one lens—the simple camera is an example we've already met. Here we'll see that the simple lens can solve a variety of optical problems.

A. Eyeglasses: spherical correction

Eyeglasses, or contact lenses, are used to improve the image on the retina when the lens system of the eye is defective. When the ciliary muscles of a normal eye are relaxed (we call it "relaxed eye" for brevity), its power is just enough to bring parallel rays from a distant object to a focus on the retina. We say that

the *far point* (the farthest object that can be clearly seen) of a normal eye is at infinity. In order to focus on closer objects, the ciliary muscles tense to shorten the focal length of the eyelens (accommodation). We call the closest point on which an eye can focus the *near point.* For a normal eye, this is taken to be 25 cm. (This distance actually varies from eye to eye, even for eyes in no need of correction. To locate your own near point, hold this page so near that it looks blurred, using only one eye. Then move the page away until the letters just come back into focus. The page is then at the near point for that eye. You should try this both with and without your glasses or contact lenses, if you wear them.)

As the normal eye accommodates between infinity and 25 cm, the power of the eyelens changes. We can think of the accommodated lens of the eye as being made of two *fictitious* lenses, one of which is the relaxed lens of the eye (focused on infinity) and a second lens that allows the eye to focus on a point only 25 cm away. The second fictitious lens must then accept rays from a point 25 cm away and convert them to parallel rays, so that the first lens can focus them onto the retina. Hence, the second lens must have a focal length of 25 cm = $\frac{1}{4}$ m. Its power of 4 diopters adds to that of the first lens (Sec. 3.4E), so during the process of accommodation the

power of the eye must change by 4 diopters. (The actual range of accommodation will vary from eye to eye. In youth it is as large as 14 diopters, corresponding to a near point of 7 cm, and it usually decreases with age.)

Many eyes, however, do not have the correct power in the relaxed state for the size of the eyeball (Sec. 5.2A). Figure 6.1 shows, all in the relaxed state, a normal eye, an eye with too much power *(myopia),* and an eye with too little power *(hyperopia),* all looking at an object located at infinity (parallel incident light). The images on the retinas of the defective eyes are blurred. The obvious cure is to put an appropriate lens in front of each defective eye, so as to increase the power of the hyperopic eye or to decrease the power of the myopic eye. If the defective eyes have the normal range of accommodation, both the near and the far point of vision will thereby be moved to the normal positions. Let's look at the way this is done.

Consider first the myopic eye. Its power in the relaxed state is already too large, and accommodation will only make it larger. Therefore, the unaided myopic eye will always see distant objects blurred; its far point

---

FIGURE 6.1

Relaxed eyes viewing a distant object, **(a)** normal, **(b)** myopic, **(c)** hyperopic.

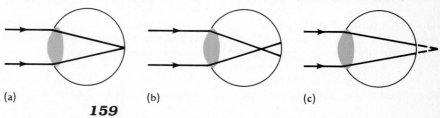

(a)   (b)   (c)

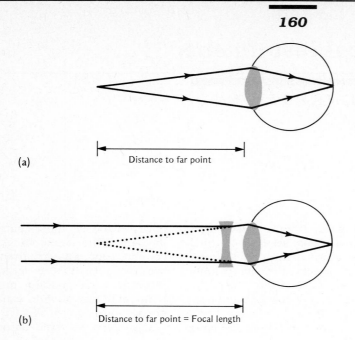

(a)

Distance to far point

(b)

Distance to far point = Focal length

is at some closer distance, say at 50 cm (Fig. 6.2a). (This close distance would correspond to the power of the eye's compound lens being only 3% too large.) If the eye's accommodation is normal (4 diopters), its near point is then closer than the normal near point (in this case it turns out to be 17 cm). Thus a myopic person has only a relatively small range of distances over which she can see clearly, but at the near point she can see a larger, more detailed image of the object than a normal viewer can at his near point—the myope has a kind of built-in magnifying glass. In fact, if your eye is normal and you want to know how the world looks to a myope, you need only look through a magnifying glass held close to your eye. Since magnifying glasses were not used as visual aids until the thirteenth century, it has been suggested that the artisans responsible for the delicate engraving found in jewelry and coins from well before that time were actually myopic. Being hereditary, this trait, as well as the tricks of the engraving trade, would have been handed down from one generation of artisans to the next.

The cure for a myopic eye is a diverging eyeglass, whose negative power decreases the eye's excessive power and moves the far point back

FIGURE 6.2

Relaxed myopic eye, **(a)** without glasses, **(b)** with glasses. The diverging lens is drawn separated from the eye to show that the rays leaving the glasses look as in **(a)**; but there need be no separation; the correction works also for contact lenses.

to infinity, where it belongs. Alternatively put, it bends parallel rays from distant objects so that they *appear* to be coming from a virtual image. If this image is located at the far point, the relaxed myopic eye can see it clearly (see Fig. 6.2b and compare with Fig. 3.23a). Hence, the lens must have a focal length equal to the distance to the far point of the relaxed myopic eye. For our example, where the far point is at 50 cm, the proper diverging eyeglass has $f = -50$ cm $= -\frac{1}{2}$ m, so the prescription for this lens would read -2 diopters. (The normal eye's compound lens has a power of about 60 diopters when relaxed. The myopic eye, in this case, had the excessive power of 62 diopters; the eyeglasses reduce it back to the normal value.)

Now consider the hyperopic eye. Its power is too *small* in the relaxed state—suppose it is only 57 diopters. This hyperope already has to accommodate (by 3 diopters) to the normal value of 60 diopters in order

to see distant objects (Fig. 6.3a). However, the very highest power he can reach by accommodation is 61 diopters. Since this is only 1 diopter above that necessary to see distant objects, his near point is at 1 m (see Figs. 6.3b and c). To see objects that close he must use all his accommodation ability; therefore, he cannot focus on anything closer.

The cure for this eye is a converging eyeglass that raises the power by 3 diopters to reach the normal 64 diopters when fully accommodated. That is, the prescription would read +3 diopters. Consider the effect of such an eyeglass lens on an object at the normal near point of 25 cm. It will make a virtual image at 1 m, where the hyperopic eye can see it. (See Fig. 6.3d and try the construction by ray tracing.)

As the accommodated, uncorrected hyperopic eye can focus on distant objects, hyperopic people were once called "farsighted." Similarly, the extra strong myopic eye, when fully accommodated, can focus on objects closer than the normal 25 cm and was therefore called "nearsighted." These names, however, are misleading if the eye can't accommodate properly, as often happens as you grow older. The aging process adds flatter layers to the eyelens (see Fig. 5.9). These layers not only reduce the eyelens' power (reducing any myopia that may be present!), but also separate the inner core from the humors that bathe the eyelens. This older inner core then tends to become less flexible and pliant, and the eye does not *accommodate* as well. This condition, called **presbyopia,** * prevents an otherwise normal eye from focusing on near objects. In his poem "Typical Optical," John Updike writes;

In the days of my youth 'mid many a caper
I drew with my nose a mere inch from the paper,
but now that I'm older and life has grown hard
I find I can't focus inside of a yard.

*Greek *presbus,* old man.

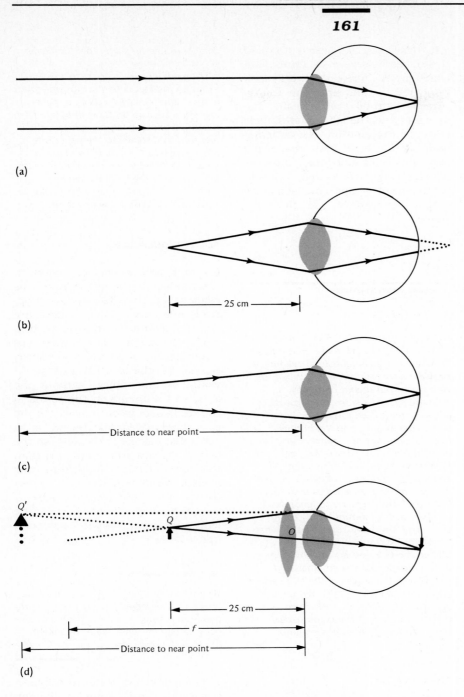

(a)

(b)

25 cm

(c)

Distance to near point

(d)

Q'

Q

O

25 cm

f

Distance to near point

FIGURE 6.3

The hyperopic eye, **(a)** partially accommodated, **(b)** fully accommodated, object at 25 cm, **(c)** fully accommodated, object at near point, **(d)** corrected, fully accommodated, object at 25 cm appears to be at near point.

The aged eye may then require one eyeglass prescription for distant viewing and another for close-up. The two lenses are often combined into a **bifocal** lens, with the close-up lens in the lower half of the frame (since you generally look downward toward close objects). Bifocals can now be made in which the lenses in the two halves gradually fade into one another, thus eliminating the conspicuous line across the lens.

Convex lenses for the hyperopic eye were in use at the end of the thirteenth century, but it was a

century and a half later before concave lenses appeared for the more common myopic eye. It was another century or so before anybody understood how these lenses worked to assist the eye.

B. Eyeglasses: cylindrical correction

Another common eye defect, **astigmatism,** occurs when the cornea is *not* spherical but is more curved in one plane than in another. That is, the focal length of the astigmatic eye is different for rays in one plane than for those in a perpendicular plane. It cannot, for example, focus simultaneously on both the horizontal and vertical glazing bars of a window. Figure 6.4 provides a sim-

FIGURE 6.4

Test for astigmatism. Close one eye and view this figure through the other eye (without glasses or contact lenses). Hold the figure sufficiently close to the eye so that all lines look blurred. Then gradually move the figure away until one set of lines comes into focus, with the rest blurred. (If two adjacent sets come into focus together, rotate the figure a little until only one is in focus. If all sets come into focus together, you don't have astigmatism.) You have now found the near point for a line in the direction of the lines that are in focus. Move the pattern away further until the lines perpendicular to the first set come into focus. (The first set may or may not remain in focus.) This is the near point for a line perpendicular to the original set. The different near points mean that your eye has a different focal length for lines parallel to and perpendicular to the original set. Try the procedure again with your glasses or contact lenses to see if your astigmatism is corrected.

ple test for astigmatism in your eyes. (The difference in focal length in the defect astigmatism occurs all over the field of view; it should not be confused with the off-axis aberration astigmatism, Sec. 3.5D.)

PONDER

The anamorphic lenses described in Section 4.4B are thick lenses. Why can't a thin astigmatic lens be used as an anamorphic lens?

The way to correct astigmatism is to use a lens that converges (or diverges) rays in one plane, while not affecting rays in the perpendicular plane. Such a lens is a **cylindrical lens** (Fig. 6.5), which is curved in one direction but not in the perpendicular direction, as if cut out of a cylinder of glass (see the TRY IT). The ophthalmologist will then prescribe a certain amount of cylindrical curvature to your glasses, specifying the orientation along which the cylinder is to lie. Such astigmatism may be concurrent with an overall myopia, say. In that case, your prescription may have a spherical component of −2.5 diopters and a cylindrical component of −0.75 diopters at a 10° orientation.

FIGURE 6.5

Cylindrical lens. Rays 1 and 2, in the horizontal plane, are made to converge. Rays 3 and 4, in the vertical plane, are essentially unaffected.

It has been suggested that El Greco had severe astigmatism, which would make the world look elongated to him, and that this is why he painted the elongated figures for which he is famous. This is really no explanation. His astigmatism would have also made his paintings look elongated, and his figures would have looked to him to be the same shape as his models only if they were, in fact, the same shape.

TRY IT

FOR SECTION 6.2B
A cylindrical lens

As described in the first TRY IT for Section 3.4A, you can make a cylindrical lens by filling a straight-sided cylindrical jar with water. Fill the jar about half full with water, put the lid on tightly, and hold the jar on its side. The top surface of the water is then the flat side of a plano-convex lens, and the bottom (curved) surface is the convex side. Hold the lens about ½ m above this page, and look down through it at the print. Keeping the lens still, rotate the page. Also try this with a pencil as the object, instead of this page. Try to ignore any wobbles due to nonuniformity of the glass. You now have some idea of what an astigmatic eye sees.

Hold the lens about ½ m above a piece of graph paper (or any crosshatched paper) with the length of the jar oriented parallel to one set of lines. With your eyes close to the lens, look down through the lens at the graph paper. What happens to the lines running

parallel to the jar? To the perpendicular set of lines? Why?

Use your lens to try to focus a line image of the sun (when it is overhead) or a high lightbulb onto a piece of paper. Move the lens up or down until you have made the line as sharp as you can, then measure its distance to the paper (the focal length of the cylindrical lens) in meters. Determine the power of the lens in diopters. If the jar had a larger diameter, would the power be greater or smaller? Find another jar and try it.

*C. Contact lenses

Contact lenses correct vision in essentially the same manner as glasses, but they do have a few optical differences. These lenses are worn in contact with the cornea, hence the name. The eye then sees the image at the *same angle* (determined by the central ray) as it would see the object without the lens, so the image on the retina is the *same size* in either case. In the case with eyeglasses the eye is at some distance from the lens, so the image appears a different size than the object. (Hold a lens from your glasses, or a contact lens, a few inches from your eye, and look at an object that extends beyond the lens, so you can compare object and image sizes.)

PONDER

Why does the converging lens cause the image to look larger? Try ray tracing using two lenses, one to represent the converging lens, and one to represent the lens of your eye.

This magnification (for the hyperopic eye) and shrinking (for the myopic eye) is generally insignificant and rarely a reason to get contact lenses. (When you first put on strong glasses the world may appear smaller or larger and seem to shift as you move your eyes. However, your brain soon compensates for this effect.)

Whereas the orientation of the astigmatic correction in an eyeglass is important, a *spherical* contact lens can correct for astigmatism even

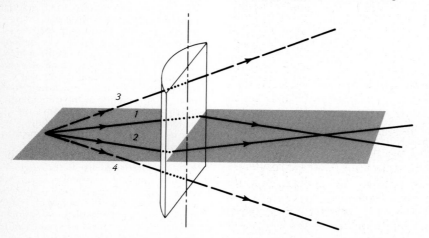

without remaining in a single orientation. The reason for this is that the contact lens, the cornea, and the tear fluid that fills the space between them all have about the same index of refraction. The only bending of light, therefore, occurs at the front surface of the contact lens (Fig. 6.6). Since this surface is spherical, there will be no astigmatism. (Soft contact lenses tend to be too flexible to correct astigmatism, but hard contact lenses do it quite well.)

Since your contact lens moves when your eye moves, you can't look through different parts of the lens

for reading and for distant viewing as easily as you can with bifocal glasses. However, when you look down sufficiently, the motion of the contact lens is stopped by the lower lid, so the contact lens slips on the cornea. Bifocal contact lenses take advantage of this fact. The outer part of the lens, which is located over the pupil when you look down, has a different curvature or index of refraction than the center of the lens (which is normally over the pupil, and which will return to that position when you look up, in order to resume its tight fit to your cornea).

D. The magnifying glass

We have argued that if you place your eye immediately behind a converging lens (which, we'll see, is a magnifying glass), the image appears the same size as the object did without the lens (see Fig. 6.3d). One often sees movie detectives holding the magnifying glass at arm's length. In that case the image does appear larger than the object. Nevertheless, this is not the best way to hold the magnifying glass. To get the *largest* image on your retina, you should place your eye close to the magnifying glass, at which distance the magnifying glass does *not* magnify. The expla-

nation of this paradoxical statement depends on the fact that the magnifying glass enables you to bring objects very close to your eye and still keep them in *focus*. The objects look big because, being close, they subtend a large angle. Without the magnifying glass, you would have to move the objects farther away in order to focus sharply on them and thus would have a smaller image on your retina.

Let's see how this works. In Figure 6.7a, the object is located at the normal eye's near point, 25 cm. The image on the retina is shown, its location determined by the central ray alone (since we know the eye will focus it on the retina). In Figure 6.7b, a larger retinal image is obtained with a magnifying glass (of focal length less than 25 cm). Here the object is located at the focal point of the lens, and consequently the rays leaving the magnifying glass are parallel to each other as they enter the eye. Such rays are focused by the relaxed eye. Again, the central ray locates the image on the retina. Because the object is closer, the central ray makes a bigger angle with the axis in Figure 6.7b than in Figure 6.7a, so the retinal image is larger. The magnifying glass has done its job. (The TRY IT suggests several ways to make a magnifying glass.)

The **magnification**, or **magnifying power** of a lens is usually given as the ratio of the image sizes in these two situations (i.e., with and without the magnifying glass), which is equal to the ratio of the object distances:

Magnifying power = $^{25}\!/_f$ (f in cm)

For example, if $f = 100$ mm $= 10$ cm, the magnifying lens has magnifying power of $\frac{25}{10} = 2.5$. This is written as 2.5X (and read as "2.5 power").

You can get somewhat greater magnification if you place the object closer than the focal point, at the point where it forms a virtual image 25 cm from the eye (Fig. 6.7c). Your eye can focus on this virtual image by accommodating, which is more tiring than viewing with parallel

---

FIGURE 6.6

(a) A contact lens on the cornea with tear fluid in between. (b) Photograph of a contact lens.

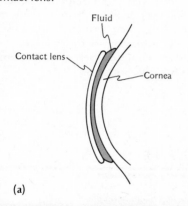

Fluid

Contact lens

Cornea

(a)

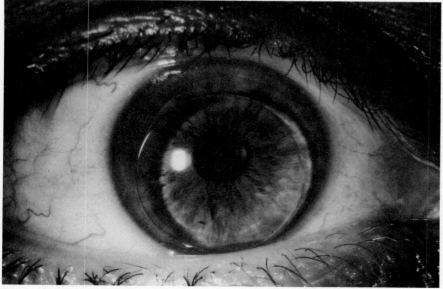

(b)

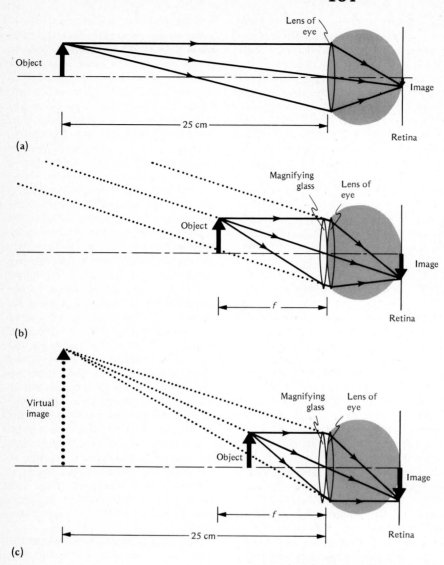

(a)

(b)

(c)

FIGURE 6.7

(a) Unaided normal eye, object at near point. (b) Eye with magnifying glass, image at infinity. (c) Eye with magnifying glass, image at 25 cm. Note the relative sizes of the images on the retinas. The compound lens of the eye is treated as a thin lens in this figure.

rays, but does give a bigger retinal image than that obtained in Figure 6.7b. The magnification is then larger (by 1) than the nominal magnifying power given by the formula above. (So our 2.5X magnifier actually magnifies 3.5 times, used this way. Of course, you want the eye as close to the lens as possible

to make it as close to the virtual image as possible.)

The same principle can be applied to a camera for close-up photography. A close-up lens is nothing but a magnifying glass. Placed in front of the camera lens, it allows the camera to focus on very close objects. Just as with the eye, the camera actually focuses on the virtual image produced by the magnifying glass. For example, if you place the object at the focal point of the magnifying glass, you should set the camera to focus at infinity.

In principle, by using a very short focal length lens, you can make the image on your retina (or film) as

large as you like. Generally, however, the aberrations become too large for magnifications much beyond 5X. We've seen that, assuming enough light is available, we can reduce aberrations by stopping down or using a smaller diameter lens. This is just what van Leeuwenhoek (1632–1723) did with his pioneering "microscopes"; his lenses were pinhead size, ground with incredible skill, and had focal lengths as short as $1\frac{1}{4}$ mm, giving a magnifying power of 200! With good lighting and careful mounting of such a tiny lens and the specimen, he could see spermatozoa and other "wee animals." He opened up the world of microscopy with a "microscope" whose entire size was about two inches. Modern microscopes are bigger and compound—they use more than one lens. It is interesting to note that the compound microscope was invented before Leeuwenhoek was born, but his lenses were so far superior to the competition that he could do better with a simple magnifying glass.

TRY IT

FOR SECTION 6.2D
A water magnifying glass

If you have a spherical glass bottle, you can make a thick lens by filling it with water (Secs. 3.4A and 3.4B), or make a plano-convex lens as in the TRY IT for Section 6.2B. (One of the attractions of a wine glass, besides its content, is that the spherical lens formed by the wine in the round-bottom glass creates on the table cloth images of nearby candles. The next time you're served wine, while pretending to judge its color like a true connoisseur, look through the wine at the inverted image of the room.) Spherical glass-bottle lenses were used to concentrate the light of a candle on fine work, such as lace making (Fig. 6.8). Such a "lace-makers' condenser" was described by John White in 1651 under the heading "How to make a glorious light with a candle like sunshine." Round bottles of water, left in an open window, have been known to start fires when the

FIGURE 6.8

Lacemakers' condenser, capable of focusing the light of one candle on the work of four different lacemakers.

sun moved into the right position. In Jules Verne's Mysterious Island, *the heroes, who are stranded on a desert island without matches (of course!), make a lens by holding water between the glasses from two pocket watches. The lens is then used as a burning glass to start a fire.*

A water droplet that makes a nice lens can be formed with an eyedropper. If you hold the drop in front of a printed page, you should be able to measure its focal length—roughly the greatest distance from the page that you can hold the drop and still see an erect image. (You may have to examine the image with another magnifying glass.) Your drop will probably be bigger than Leeuwenhoek's lenses, so you'll get some idea of the viewing problems he had. (You can also measure the focal length by the method of the TRY IT for Sec. 3.4B.)

To hold a slightly larger droplet, pierce a piece of aluminum foil with the tip of a pencil and rotate the pencil, making a reasonably round hole about 5 mm in diameter. Pour a little water over the hole in the foil to form a droplet of water across it that can be used as a magnifying glass. (Alternatively, a wire loop of the same diameter can be used.) Looking through the droplet, you can estimate its

focal length and magnification power. The focal length should be quite small (about 1.5 cm, which means a 16X magnifier).

A way to make a larger water lens is to freeze the water in the bottom of a spherically shaped bowl. The water will freeze from the outside inward, and concentrate the less easily frozen impurities on the inside, as well as developing bumps and cracks as it freezes, leaving you with a rather foggy lens that scatters the light instead of focusing it. If you freeze the water slowly, and take it out of the freezer before the last bit of water is frozen, you can get a reasonably clear lens. In the short time before your ice lens drips all over everything, you should be able to verify that it is a magnifying glass and make some estimates of its magnifying power and focal length. It is very hard to get any kind of decent image from an ice lens, but you may be able to concentrate the sun's rays enough to use the lens as a burning glass, as Richard Adams describes in The Girl in a Swing:

"You don't know about the ice-burn? I'll tell you. Sometimes, in the North, in winter, the ice forms right across the curved top of a hill. Then when the sun shines, the ice becomes like a magnifying glass, so that the sun burns off all the grass and heather underneath. Later, the ice melts and all through spring the hill's bare until the grass grows again. . . Don't you see, the ice is what burns—the last thing you'd expect to burn anything, and yet it does?"

6.3
COMPOUND MICROSCOPES

The **compound microscope** uses a magnifying glass, not to look directly at the object, but rather to look at a real image created by another lens. Consider the simplified compound microscope shown in Figure 6.9a. (The TRY IT at the end of Sec. 6.4B tells you how to make one.) The first lens, called the **objective** because it is closest to the object PQ, forms a real image $P'Q'$, the *intermediate image*. The viewer looks at this real image with a magnifying glass, called the **eyepiece** or **ocular.** For viewing with relaxed eye muscles, the rays leaving the eyepiece should be parallel to one another. This means that the eyepiece should be located at its focal distance, f_e, from the intermediate image (see the figure and the important note in its caption). In order to eliminate any stray light, the two lenses are encased in the ends of a tube. Sometimes a calibrated length scale, called a **graticule,** is drawn on a thin glass disk located in the plane of the intermediate image. This allows measurement of the size of objects observed in the microscope.

Both lenses, then, provide magnification. As with any magnifying glass, the shorter the focal length f_e, the greater the magnification of the eyepiece. The magnification of the objective is determined by the intermediate image size, $P'Q'$, compared to the object size, PQ. The *location* of the intermediate image is fixed because it must be at the focal point of the eyepiece, and the spacing between the lenses is usually fixed at about 18 cm so the viewer's eye can be at a convenient height above the work table. You can see by examining the central ray 2 that a larger image at that location means the object must be closer to the objective. This can be accomplished with a smaller objective focal length, f_o (Fig. 6.9b). That is, for larger magnification we want both f_o and f_e smaller.

For the same diameter, a smaller

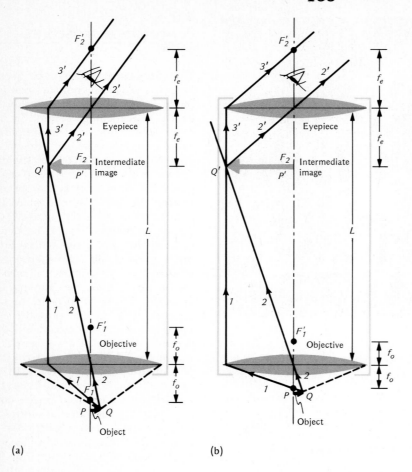

(a)   (b)

FIGURE 6.9

The compound microscope, **(a)** with larger objective focal length, **(b)** with smaller objective focal length. The magnification is greater in **(b)**. Rays from Q between the dashed lines strike the objective lens. Note: This figure uses our ray tracing convention that the intermediate image serves as a source of new rays, not necessarily extensions of the rays converging to it. Thus, for example, ray 2' must actually originate at Q and pass through the objective before reaching Q'. If the objective is sufficiently large, such a ray will, in fact, exist. Since we are only interested in locating images, or, in this case, the *direction* of the beam emerging from the eyepiece, we don't really care whether ray 2' actually exists. We simply treat it as originating at Q'—we treat the intermediate image as a new object. (A field lens would cause ray 2 to become ray 2'—see Sec. 6.6A.)

focal length means a smaller f-number and, consequently, a greater light-gathering ability. This is seen by examining the cone of rays from the object point Q that enter the objective lens (cf Figs. 6.9a versus 6.9b). Small f-number also means small depth of field. This sometimes is advantageous. It allows you to examine different layers of a transparent object by focusing on the depth of interest. You do, however, lose the three-dimensionality of the object, as you can only examine one plane at a time.

*A. Dark-field and oil-immersion microscopy

Because the object must be close to a small-focal-length objective, very little space is available for the object and for auxiliary equipment useful for special techniques. For example, in **dark-field microscopy** the ob-

ject is illuminated only from the sides. One sees in the microscope, then, those parts of the object that scatter light (such as cell walls) against a dark background. Adequate space is necessary for proper illumination.

On the other hand, when we want to get as much light as we can through a microscope, even that which has left the object at a large angle to the microscope's axis, the space between object and lens must be small. This creates a new problem; light rays bend as they leave the cover glass of the microscope slide and enter the air space (Fig. 6.10a), and rays at different angles bend different amounts. Thus the inner rays, which leave the cover glass in a direction almost perpendicular to the glass, are bent very little and appear to come from a point close to the actual object. The outer rays, leaving at a greater angle, are bent more and appear to come from a different, higher, location. These rays and all the rays in between thus appear to come from a range of depths, giving a blurred image. (The flat cover plate thus behaves as a lens with spherical aberration!) In the **oil-immersion microscope** this problem is solved by filling the air space with oil that has an index of refraction close to that of glass—it is as if the object were embedded in the glass of the lens. Thus, the light from the object is refracted only by the second surface of the lens (Fig. 6.10b), eliminating a blurring that would otherwise occur when the lens is close to the object.

B. Scanning microscopes

The technique of *scanning* is familiar from a variety of examples. In Section 5.2 we discussed how the human eye scans when viewing the world. Another example is the focal-plane shutter (Sec. 4.5A), where the moving slot exposes the film strip by strip, rather than all at once. Similarly, in television an entire picture is not flashed on the screen all at one time. Rather, individual points on the fluorescent screen are

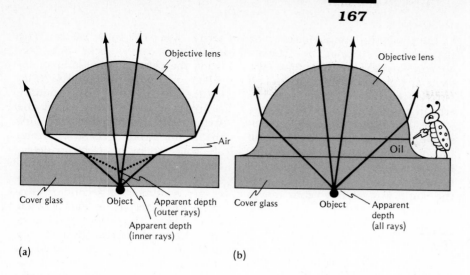

FIGURE 6.10

The oil immersion microscope objective, **(a)** without oil, **(b)** with oil.

---

successively hit by a beam of electrons, making a tiny spot whose brightness depends on the number of electrons hitting it. This number, and hence the brightness, are controlled by the input signal to the TV tube. The electron beam is made to scan (move rapidly) across the screen many times, each time starting on a lower line, until the screen is covered by a set of lines or **raster.** Each point in the raster then has a controlled brightness, so the entire screen forms a picture. (Also see the TRY IT for Sec. 4.5A.) A similar scanning technique can be applied to microscopy.

An example of a scanning microscope is the **flying-spot microscope,** which uses two TV tubes and a compound microscope run backward (Fig. 6.11). On the first TV tube, a uniform bright spot moves over the raster. The face of this tube is placed where your eye normally goes on the microscope, and the backward microscope focuses this moving bright spot down into a tiny microspot on the object to be studied. As the electron beam scans the first TV tube, the microspot scans the object. On the other side of the object is a single photocell (similar to the light meter in your camera) that measures the intensity of light transmitted from the

microspot through the object. This intensity varies depending on how opaque the part of the object is that is struck by the microspot. The output of the photocell controls the intensity of the electron beam of the second TV tube. This beam scans in

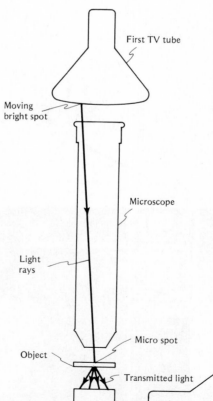

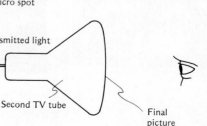

step with the first TV tube (and thus with the microspot) and, therefore, forms a TV image that is bright where the object is transparent and dark where it is opaque. Whatever area is scanned by the microspot will fill the final TV image, so if this area is made very small (by using a powerful backward microscope), the magnification can be very large. Notice that all the focusing occurs in the light that *illuminates* the object—there is no focusing of the light transmitted by the object.

Like the simple pinhole camera (Sec. 2.2B), even the best visible-light microscopes have limitations due to the wavelength of the light, as mentioned in Section 2.2B. The essential point is that you cannot focus light into a spot much smaller than the wavelength of that light. Consequently, to examine object details finer than about 500 nm, you must use something other than visible light. **Ultraviolet microscopy** therefore uses light of about half the wavelength of visible light to achieve sharper images. Even better than using UV, for this purpose, is to use electrons. A typical beam of electrons that one may work with behaves as if the electrons had a wavelength less than 0.01 nm! (See Sec. 15.2C.) Thus, they can be focused to a much smaller spot than visible light can. Electrons are focused by magnetic fields, which can be made by running currents through wire loops—choosing these loops appropriately we can get them to exert forces and focus the charged electrons just as lenses focus light. (These loops are called magnetic lenses.)

FIGURE 6.11

The flying-spot microscope.

Like light, electrons can be transmitted through or reflected from an object. In the **transmission electron microscope,** electrons are sent through an object and then focused on a fluorescent screen to form an enlarged image, in much the same way that a slide projector produces an enlarged image. The electrons are capable of distinguishing objects only 0.2 to 0.3 nm apart—a distance almost as small as the size of an atom! However, as the electrons cannot be focused sharply unless they all have the same energy, only thin samples can be used lest the electrons lose energy in passing through the sample.

The **scanning electron microscope** is the electron analogue of the optical flying-spot microscope. Electrons are focused to a very fine beam that moves across the sample the same way as the beam moves across the TV tube raster. The scanning beam knocks out electrons that were in the sample, so-called secondary electrons (Fig. 6.12). The number of secondary electrons emitted depends on the structure of the sample at the point the beam strikes. These electrons are collected by an electron detector. As in the flying-spot microscope, no further focusing is necessary, so almost all secondary electrons are detected no matter what their energy. The output of the electron detector controls the intensity of the beam in a TV tube, just as the photocell of the flying-spot microscope does. Although magnifications of 50,000X are attainable this way, magnifications of only several hundred to several thousand are also commonly used. This is because the electron microscope, with its very narrow beam (effectively, a very small aperture), has much greater depth of field than ordinary light microscopes, making the three-dimensionality of objects visible in a manner not otherwise available (see Fig. 5.11).

The major problem with electron microscopes is that they require elaborate sample preparation in order to get sufficient numbers of secondary electrons. One technique requires coating the sample with metal. Another involves creating a three-dimensional image of the object in plastic with the help of x-rays and an etching procedure. While cumbersome, these procedures have helped make the microscopic world visible in an incredibly realistic fashion.

6.4
TELESCOPES

Like any device designed to be used by the eye, the telescope produces outgoing parallel rays of light, which can be focused on the retina by the relaxed eye. Since telescopes are made for viewing distant objects, the incoming rays are also parallel to one another (or nearly so). Hence, a telescope must take

FIGURE 6.12

(a) Diagram of a scanning electron microscope. **(b)** Photograph of a scanning electron microscope.

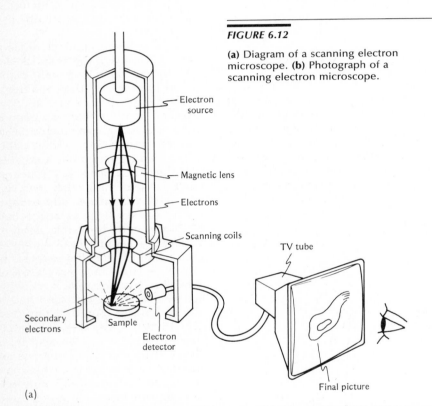

(a)

(b)

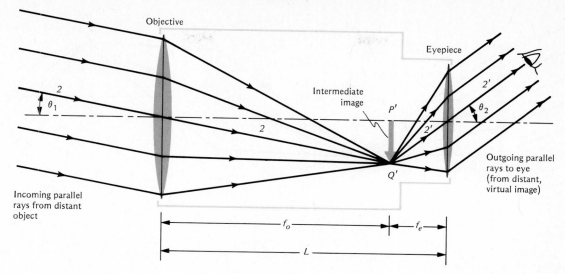

FIGURE 6.13

Principle of an astronomical telescope. This figure uses our ray tracing convention (see caption Fig. 6.9) whereby the intermediate image serves as a source of new rays, not necessarily extensions of the rays converging to it. This is achieved physically by a field lens (Sec. 6.6A).

incoming parallel rays and convert them to outgoing parallel rays. All it can do, therefore, is change the angle and density of these rays. It is the change of angle that constitutes the *magnification*. The increase in density of the rays increases the *brightness* of the image. (The TRY IT at the end of Sec. 6.4B gives several examples of telescopes.)

A. The refracting astronomical telescope

Consider a telescope made of two converging lenses (Fig. 6.13). As in the microscope, the two lenses are called the **objective** and the **eyepiece.** The incoming parallel rays are focused in the second focal plane of the objective. The intermediate image produced there must also lie in the first focal plane of the eyepiece in order to have parallel outgoing rays. Thus the spacing between the lenses, L, must equal the sum of the two focal lengths, $f_o + f_e$. Magnification results if the rays leave the telescope at a steeper

angle than they entered. An object, such as the moon, which to the naked eye may be spread out over an angle θ_1, now appears to be spread out over the larger angle θ_2.

The magnification of a telescope is large if the intermediate image is large and if this image is viewed through an eyepiece (magnifying glass) of large magnification. Examining Figure 6.13, we see that the intermediate image is larger the farther it is from the objective, that is, the larger f_o. As usual, an eyepiece having shorter focal length, f_e, also gives greater magnification. In fact (see Appendix J):

Telescope magnification = $-f_o/f_e$

The negative sign in this expression means that if both focal lengths are positive (as in Fig. 6.13), the image is inverted; the eye must look down when the incoming rays come from above the axis (as shown).

However, increasing the intermediate image's distance from the objective spreads the incoming light over the larger intermediate image, which thus becomes dimmer. This is unfortunate because a telescope is often used to look at objects that are dim to begin with, making light-gathering ability important. In order for a longer focal length objective to produce the same brightness as a shorter one, we must have a larger diameter in the former, proportional to the larger focal length. (That is, we get the same bright-

ness if we keep the f-number the same. Most astronomical telescopes of this type operate between f/12 and f/15.) For this reason, a telescope is usually specified, not by the focal length, but rather by the diameter of its objective. For example, we speak of a 6-inch telescope when we refer to one with a 6-inch objective diameter. The largest objective *lens* is in the 40-inch Yerkes Observatory telescope in Williams Bay, Wisconsin.

B. Terrestrial telescopes

For terrestrial purposes, such as looking for whales or spying on your neighbor, the inverted image of the refracting astronomical telescope is rather inconvenient. One way to make a telescope that produces an erect image is simply to insert, between the objective and the eyepiece, an erecting system, which inverts the intermediate image $P'Q'$, by creating another intermediate image $P''Q''$ (Fig. 6.14). If the erecting system is a simple converging lens, the telescope becomes rather long (of the type favored by pirates), but it does produce an erect image for you to view.

An erecting system that allows a shorter telescope is the combination of prisms shown in Figure 2.56c. This not only erects the image without taking up much space, but the long, multiply reflected path

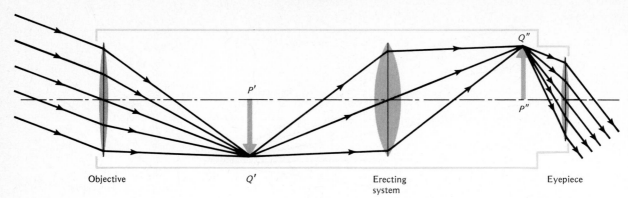

Objective   Q'   Erecting system   Eyepiece

FIGURE 6.14

Principle of a terrestrial telescope, which uses an erecting system. This figure uses our ray tracing convention (see caption Fig. 6.9) whereby the intermediate image serves as a source of new rays, not necessarily extensions of the rays converging to it.

through the prisms allows the objective and eyepiece to be placed close together. Such a system is used in *prism binoculars*. Binoculars are specified by both their magnification and objective diameter. Thus a 7 X 35 binocular has magnification 7 and objective diameter 35 mm.

An alternative approach avoids the erecting system by using a *diverging* eyepiece, as in the **Galilean telescope** (Fig. 6.15). Its advantage is that the distance between the lenses, L, is actually less than f_o. This is because L is still given by $f_o + f_e$, and f_e is negative here. However, the *field of view* is *limited*, so this technique is used in

small, inexpensive opera glasses (but it was good enough to enable Galileo to discover four of Jupiter's moons). The telephoto lens (Fig. 4.15) can be thought of as a Galilean telescope where the eyepiece and the camera's normal lens have been combined.

PONDER

Would a "telescope" with a diverging objective (negative f_o) and converging eyepiece magnify?

If you are hyperopic, you can make a Galilean telescope with a single lens. Your hyperopic eye, having less power than a normal eye, can be thought of as consisting of a normal lens and a diverging lens. The latter behaves as the eyepiece of a Galilean telescope, and

FIGURE 6.15

Galilean telescope.

you need only hold a magnifying glass (the objective) at the proper distance in front of your eye to give yourself "telescopic vision."

Looked through backward, a Galilean telescope presents a tiny, wide-angle view of the world. Therefore, backward Galilean telescopes are used as peepholes in doors so, without opening the door, you can see the salesman who rang your bell. Should the salesman put his eye to his side of the peephole, the narrow field of view would show him only the telescopic image of the fly on your opposite wall.

TRY IT

FOR SECTIONS 6.2D, 6.3, 6.4A, AND 6.4B
Optical instruments

You can verify some of the basic ideas used in the construction of many of the optical instruments described in this chapter with about four inexpensive lenses (purchased, e.g., from Edmund Scientific Co., Barrington, NJ 08007). You will need two converging lenses of reasonably short focal length (f ≈ 5 cm is convenient, if possible the two should be slightly different) and one diverging lens of short focal length (f ≈ −5 cm). You'll also need a longer focal length lens (f ≈ 20 cm). At least one of the f ≈ 5-cm lenses and the f ≈ 20-cm lens should have a diameter no smaller than 2 or 3 cm, and the larger the better.

(a) The Magnifying Glass. Use one of the f ≈ 5-cm lenses as a magnifying glass. To determine its magnifying power, first measure the height of some object, such as the print in a newspaper. Then cut the newspaper so the print comes to the edge of the paper. Next,

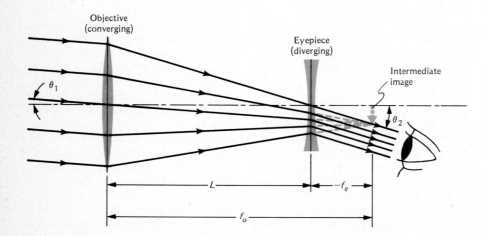

Objective (converging)

Eyepiece (diverging)

Intermediate image

θ_1

θ_2

L

$-f_e$

f_o

lay a ruler on a table and locate your eye above the ruler, so that the ruler is at your near point (about 25 cm away). Keeping your eye in this position, hold the lens up to your eye and bring the paper with the print at its edge close enough to the lens to be in focus. By looking simultaneously through the lens at the print and past the lens at the more distant ruler, you should be able to measure how big the print appears to be (through the lens) as measured on the ruler (without the lens). Divide this apparent size of the print (with the lens) by its actual, originally measured size to estimate the magnification of the lens. Does your result agree, approximately, with the expression given in Section 6.2D? If your two f ≃ 5-cm lenses are slightly different, compare them to each other. Also compare them to the f ≃ 20-cm lens. Which gives the most magnification?

You can get the same magnification with a pinhole as you can with a magnifying glass. Make a small pinhole in a piece of aluminum foil and, holding it over your eye, bring your eye close to the printed lines. Compare the focus with and without the pinhole, when it is about 5 cm away from the print. Now compare the pinhole at that distance with the f ≃ 5-cm lens. Is the magnification the same? What difference do you observe? What is the advantage of the lens?

(b) The Compound Microscope. Use the f ≃ 5-cm lens with the larger diameter as the objective, and the other f ≃ 5-cm lens as the eyepiece. With one hand, hold the objective steadily at a little more than its focal length from a well-illuminated object (say a printed letter). While holding the eyepiece up to one eye with your other hand, move it and your head closer to or farther from the objective until the magnified image of the printed letter comes into focus. Notice the small angle of view—a clear area within the (blurred) image of the objective lens. Is the magnified image erect or inverted? How does it compare in size to that obtained with just one of these lenses?

(c) The Refracting Astronomical Telescope. Use one of the f ≃ 5-cm lenses as the eyepiece, and the f ≃ 20-cm lens as the objective, separated by about 25 cm, with the eyepiece close to your eye. It may help to brace your hand against something, say the window frame. Notice the orientation and size of the image and the field of view and compare them to those obtained with the Galilean telescope, below. With a third hand you can try inserting the other

f ≃ 5-cm lens about 5 cm in front of the eyepiece as a field lens (Sec.6.6A).

(d) The Galilean Telescope. Here use the f ≃ −5-cm lens as the eyepiece, and the f ≃ 20-cm lens as the objective. The spacing between the lenses should be about 20 cm − 5 cm = 15 cm. Try to look at the same object as you did with the refracting astronomical telescope. The image should now be erect. How do the magnifications compare in the two cases? How about the angle of view? Reverse the positions of the two lenses and look through them. Notice the wide angle of view, useful for a peephole in a door. (Of course, unless you live in a castle, you probably don't have doors 15 cm thick. Actual peepholes use shorter focal length lenses.)

C. Reflecting telescopes

The preceding telescopes all use objective *lenses* (so-called **refracting telescopes**). Large diameter lenses, however, create numerous problems. Their aberrations (particularly chromatic) are difficult to correct and they tend to sag under their own weight because they can only be supported by their edges. A technique that avoids all these difficulties is to use a concave, focusing *mirror* as the objective (the so-called **primary mirror**). Because no refraction is involved, mirrors have no chromatic aberration. They can be supported over their entire back, and because they are front-surface mirrors, the back of the glass can even be honeycombed to reduce the weight. A smaller, **secondary mirror** is normally used to enable you to look through the telescope without getting your head in the way of the incoming beam. There are a variety of such **reflecting telescopes,** a few of which are pictured in Figure 6.16. In the very large reflecting telescopes, the astronomer (or her film) can actually sit at the focal point of the primary mirror (Fig. 6.17). The largest reflecting telescope is in Zelenchukskaya, Soviet Union, with a 6-meter objective.

A new technique to avoid the weight and cost of a single large mirror is to use a number of smaller mirrors in concert. (While we have called this technique new, it is sim-

ilar to the one ascribed to Archimedes. See Sec. 3.3D.) The light collected by each of the separate mirrors is focused to a common image. The mirrors' sensitive alignment is accomplished by means of laser beams that accurately detect each mirror's position, computers that rapidly calculate any required position change, and motors that continually realign the mirrors. The *Multiple Mirror Telescope* on Mount Hopkins, Arizona, has six 1.8-m mirrors that combined have the light gathering ability of about one 4.5-m mirror.

*D. Catadioptric telescopes

From the seventeenth century, when all the telescopes we've discussed were first developed, various aberrations have constituted a major limitation in the construction of large-aperture telescopes. In 1930 the first **Schmidt telescope** was built, using a new approach to avoid this limitation. Its objective incorporates both reflecting and refracting elements, and is therefore *catadioptric.†* Its design is based on the fact that a concave spherical mirror with a stop in a plane passing through its center of curvature has no off-axis aberrations other than curvature of field. This is true because any line passing through the center of curvature makes a perfectly good axis, so there can be no "off-axis." To correct for spherical aberration, Bernhard Schmidt (who once claimed that his best ideas came to him as he awoke from a drunken stupor) introduced a **correction plate**—a specially shaped piece of glass that passes through the center of curvature and causes additional conver-

---

†This word is a combination of the words catoptric, which means involving reflection, and dioptric, which means involving refraction. The former is from the Greek *katoptron*, mirror, while the latter ultimately comes from the Greek *dia*, through. Catadioptric lenses are available for 35-mm cameras. They are not inexpensive.

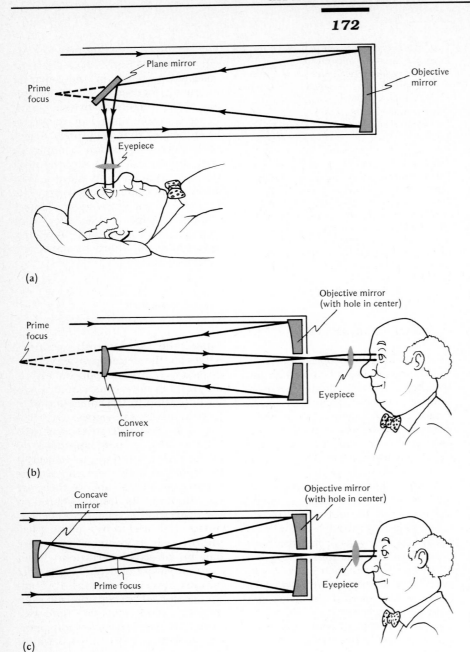

(a)

(b)

(c)

FIGURE 6.16

Reflecting telescopes: **(a)** Newtonian telescope, typically has speeds between f/4 and f/8, **(b)** Cassegrain telescope, the most compact reflecting telescope, with typical speeds between f/7 and f/12, **(c)** Gregorian telescope, longer and less aberration-free than the Cassegrain, but produces an erect image.

gence near its center and divergence farther out (Fig. 6.18a). It compensates for the spherical aberration of the mirror without introducing appreciable chromatic aberration. This approach has allowed the construction of telescopes with apertures as large as f/0.6.

As with many of our sophisticated instrumental designs, Nature anticipated the catadioptric principle, equipping each of the tasty scallop's 60 eyes with such a lens (Fig. 6.18b). The back of each lens carries a spherical mirror (see Sec. 12.3C), while the cornea is converging near its center and diverging toward its periphery. After reflection, the light is focused on the light-sensitive cells in the curved front sur-

face of the fovea (which lies too close to the eyelens for the light to be focused on it prior to reflection).

*6.5
SCHLIEREN PHOTOGRAPHY

So far we have been describing optical instruments that increase the *size* of the image on the retina or film, to help us see small or distant objects. Another kind of object difficult to see is a transparent object. Examples include the shock waves that a bullet produces as it passes through air, the convection currents of hot air rising above a flame, or clear biological material (which must also be magnified). These objects refract light but do not affect its intensity. In order to see them clearly, you need a device that produces intensity variations whenever the index of refraction varies. Without the **schlieren**† device used to make the photograph of Figure 6.19, you would not see the density variations of the transparent air.

Suppose we try to take a close-up photograph of a transparent object, say a slab of undisturbed air, using an ordinary camera (Figure 6.20a). The transparent object is illuminated from behind by parallel light. The light remains parallel as it passes through and thus is brought to a focus at the focal point of the lens, F'. From there, this light continues, diverging, until it strikes a large area of the film. So we get uniform illumination—a picture of clear, undisturbed air as it should be. But now suppose that in the center of the object, P, there is some air that has been disturbed by heat or pressure, giving it a slightly different index of refraction than the undisturbed air. Light passing through P is then bent a little, as if it originated at a source *at P*. Any of this light that gets to the film lies within the cone shown in Figure 6.20a and is brought to a focus on the film at P', making P' brighter. The trouble is that, if the object is

---

†German, streaks or striations.

FIGURE 6.17

Photograph of the cage at the primary focus of the Mt. Palomar telescope.

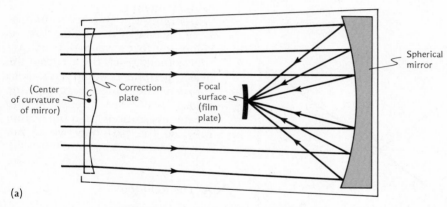

(a)

(Center of curvature of mirror) — Correction plate — Focal surface (film plate) — Spherical mirror

(b)

FIGURE 6.18

(a) The Schmidt telescope, and **(b)** a scallop's eye.

really transparent, the same amount of light was taken *out* of the original beam at P, making P' darker. Both processes together leave the intensity at P' nearly unchanged, so the film will not show much of an image of the disturbance—the disturbance is invisible.

To create a visible image of the disturbance, we prevent one of these contributions to the intensity at P' from actually reaching the film (Fig. 6.20b). An obstacle or stop, usually in the form of a knife edge, is slid into the beam at the focal point, F', so it just blocks all the rays that are *not* bent by the disturbance. However, whenever the change of index of refraction in the object is such as to refract the light upward at P, this bent light will reach the film at P', so there will be a corresponding bright spot on the film against an otherwise dark background. (A variation is to put only a *small* obstacle at F', so that light that is bent downward or sideways will also illuminate the film. This is a modification of the *dark-field* technique described in Sec. 6.3A.)

The basic schlieren idea may be incorporated in wind tunnels for aerodynamic analyses, in microscopes to produce enlarged schlieren images of otherwise transparent cellular material, and in a variety of other applications. So sensitive are the sophisticated versions of this device, that it is common practice to align them by using, as an object, the streams of warm air rising from the operator's hand.

*6.6
FIELD OF VIEW

The discussion of Section 6.5 shows the importance of considering not only the location of the image but also the stops that limit the beam. A stop at one lens of a multilens instrument may prevent a full cone of rays from reaching subsequent lenses, restricting the field of view. For example, when discussing the microscope (Fig. 6.9) and the refracting astronomical telescope

FIGURE 6.19

Schlieren photograph of shock waves in air that moves at supersonic speed past a sphere. The horizontal rod above the sphere measures the pressure.

FIGURE 6.20

Principle of schlieren photography of a transparent object: **(a)** without knife edge, **(b)** with knife edge.

(Fig. 6.13) we emphasized that ray 2' was just an artifact introduced to locate the final emerging rays. There actually would *not* be such a ray because the objective was *not sufficiently large* to pass it. If we extend ray 2' backward, we see that it doesn't intersect the objective. (The objective lens is thus acting as a *stop* that limits the beam.) Conversely, very few of the rays that the objective sends to Q' reach the eye because the eyepiece is *not sufficiently large* to pass them and refract them to the eye. (The eyepiece thus also acts as a *stop.*) To the eye, then, the point Q appears much dimmer than a point on the axis, P. The object does not appear uniformly bright, but gradually shades off toward the periphery. This effect is called **vignetting** (Fig. 6.21a), and while it is occasionally attractive in, say, photographic portraits, it is generally undesirable in microscopes, telescopes, and other optical instruments.

Enlarging the eyepiece would enlarge the field of view by moving the vignetting further out; but the rays that actually reach the eye from the field's periphery would only pass through the outer parts of the lenses. Such rays are subject to greater aberrations than the central rays, so the periphery would not only be darkened but also blurred.

A. The field lens

We redirect the rays from the periphery of the field of view so they pass through the central region of the eyepiece by inserting another lens, called a **field lens,** at the in-

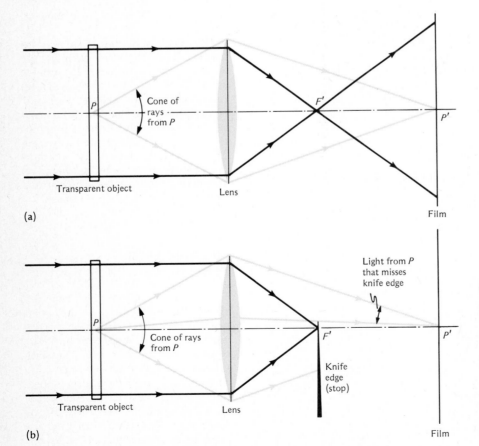

FIGURE 6.21

Principle of the field lens. **(a)** Without field lens the final image is vignetted, only the heavy part of the arrow $P''Q''$ is significantly illuminated, the intensity falling to zero at Q''. (See this effect in Fig. 4.5c.) **(b)** With a field lens the entire field is well illuminated, because the field lens bends the entire cone of rays $AQ'B$ into $CQ'D$, so they can reach the eyepiece and the eye. **(c)** and **(d)** Photographs of the eye's view in the two cases, respectively.

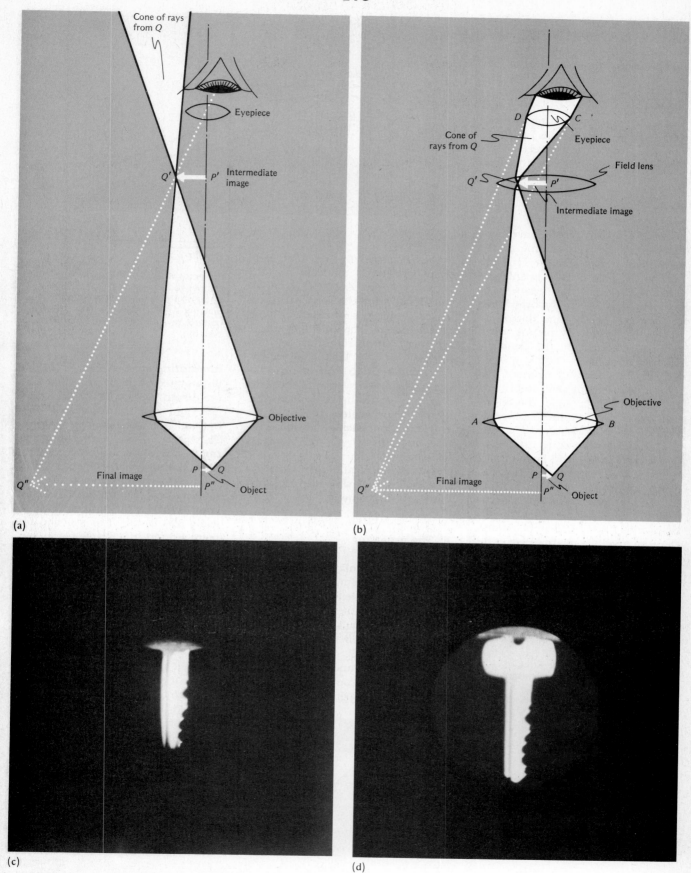

Cone of rays
from Q

Eyepiece

Q'   P'   Intermediate
image

Objective

Final image

Q"   P   Q
      P"   Object

(a)

Cone of
rays from Q

D   C   Eyepiece

Field lens

Q'   P'

Intermediate
image

Objective

A   B

Final image

Q"   P   Q
      P"   Object

(b)

(c)

(d)

termediate image $P'Q'$. The field lens is chosen to image the objective lens at the eyepiece. This is illustrated in Figure 6.21b for the case of the microscope. Since the field lens lies right *at* the intermediate image $P'Q'$, it doesn't modify it or its subsequent image—the final image still resides at $P''Q''$. However, the rays between these images *are* modified by the field lens (cf Figs. 6.21a and b) so that all rays striking it pass through the eyepiece. Hence, the final image is brighter and more evenly illuminated. Rays that would have been blocked by the stop at one or the other lens now reach the final image, so the field of view is enlarged (Figs. 6.21c and d). Because the field lens only affects the brightness of the final image but is not involved in the actual imaging, aberrations produced by this lens are not very important.

An actual optical instrument may have a field lens at each intermediate image. For example, the terrestrial telescope of Figure 6.14 would require two field lenses, at $P'Q'$ and at $P''Q''$. By repeating the sequence of erecting lens - field lens several times, such a telescope can be made quite long. The ability to look through a long, relatively narrow tube but with an ample field of view is useful in several applications. One example is a **cystoscope,** a device that allows a doctor to examine your insides. (The cystoscope's rigidity makes the fiber optics tech-

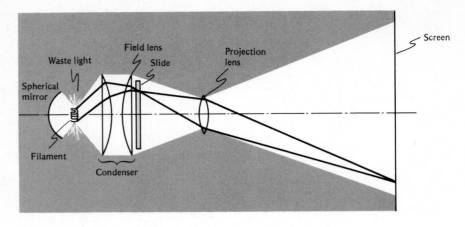

nique described in Sec. 2.5B generally preferred, at least by the patient.) A cystoscope (Fig. 6.22) will usually have a long string of lenses, with every other lens serving as a field lens. The only light losses are then due to reflection and absorption at the lenses. Typically, there will be a 45° mirror near the beginning of the system to allow the doctor to see to the sides as she examines your insides. Another example is a **periscope,** a similar device, but with two 45° mirrors at its ends (see the TRY IT for Sec. 2.4C). Many other optical instruments use field lenses, and we'll next examine one such instrument.

B. The projector

To project an enlarged image of a slide, or movie film, we need only illuminate the slide and place it close to the focal point of a converging lens. The lens will then image the slide on a distant screen, as noted in Section 3.4C. Aside from the problems of advancing the slide or film, the main complication of the **projector** is associated with the **il-**

FIGURE 6.23

Projector with illuminating system. Solid outline shows a cone of rays from one point on the slide, defined by the imaging system. White region shows the cone of rays defined by the illuminating system.

luminating system (Fig. 6.23). The illuminated slide is projected onto the distant screen by the projection lens, and a sharp image is obtained by adjusting that lens' position. This constitutes the **imaging system.** To achieve uniform illumination, several other elements are introduced, which constitute the illuminating system. The projection lamp, which is the light source (typically 500 watts), has a coiled filament wound in a single plane, parallel to the slide. This filament lies at the center of a spherical mirror, located behind the lamp (or inside it). The purpose of the mirror is to reflect any backward-directed light, which would otherwise be lost, and image it in the plane of the filament. Maximum illumination will occur if the reflected images of the filament coils lie between the actual coils, so that the reflected light isn't blocked by the coils themselves. This combination of the glowing filament coils and their images then forms a fairly uniform source of light, directed forward from the filament plane (see the TRY IT). Between this plane and the slide lies the **condenser,** which consists of two lenses. Because the filament lies at the focal plane of the first lens, that lens produces a parallel

FIGURE 6.22

Erecting lenses *(E)* and field lenses *(F)* that may be part of a long string of lenses in a cystoscope or a periscope. Note that the full cone of rays striking the first lens emerges from the system.

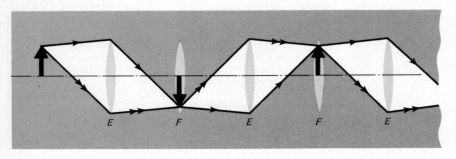

beam of light, which illuminates the slide uniformly (Fig. 6.24). The second lens is the field lens. Because it lies quite close to the slide, it does not significantly affect the uniform light intensity near the slide, but it does focus this light to a point close to the projection lens. By imaging the filament on the projection lens, the field lens assures that the maximum amount of light gets from the filament through the projection lens and on to the screen.

The two simultaneously coexisting systems, for illumination and for imaging, are designed to intermesh and yet to exist almost independently of one another—the projection lens having little effect on the illuminating system, while the field lens doesn't affect the imaging system.

Some slide projectors have a zoom lens (Sec. 4.4C). This allows you to vary the distance between the projection lens and the slide, while still keeping the image in focus on the screen. The closer the projection lens is to the slide, the larger the image on the screen. You can see this by looking at the central rays from the top and bottom of the slide in Figure 6.25. Without such a lens, the image size is fixed by the distance to the screen. The zoom lens thus provides an extra degree of flexibility in the slide projector.

Overhead projectors, commonly found in lecture halls, are similar to slide projectors. Since the transparency to be projected is quite large, a large field lens is necessary, usually a Fresnel lens (Sec. 3.4D). Its rings can be seen just below the transparency. (Being so close to the transparency, they can easily be focused on the screen.) Since you want to have a horizontal transparency and a vertical image, these overhead projectors have a (somewhat adjust-

(a)

(b)

(c)

FIGURE 6.24

Image thrown by projector **(a)** with all lenses in place, **(b)** with the field lens removed, and **(c)** with the entire condenser removed. Note the "keystoning" of the image.

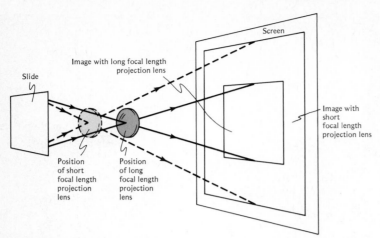

Screen

Image with long focal length projection lens

Slide

Image with short focal length projection lens

Position of short focal length projection lens

Position of long focal length projection lens

FIGURE 6.25

Zoom projection lens.

able) 45° mirror either just after the projection lens or between its two elements.

Enlargers are basically projectors equipped with a variable stop and exposure timing to control the amount of light projected on the print paper. The illuminating system may use a condenser or it may consist of a diffuse, uniform light source made, say, by inserting a piece of ground glass between the lamp and the negative (the "slide"). In the latter case most of the light misses the projection lens, but not much light intensity is needed to expose the paper.

TRY IT

FOR SECTION 6.6B
The home slide projector

If you have a slide (or movie) projector, examine it (carefully!) to see the various elements described in Section 6.6B. With the projector turned off, you should be able to remove the various covers that house the lamp and the condenser. Examine the lamp's filament. (If you have to remove the lamp, it is a good idea to use a handkerchief. This is particularly important if you have a quartz-halogen lamp. Any grease or damp spots that you leave on the lamp may result in uneven heating when the lamp is later turned on, which may cause the glass to crack.) What kind of lamp is it (see Sec. 1.4C)? What is the shape of the filament? Notice whether the lamp socket requires the filament to be oriented in any particular way. Notice the curved mirror behind the lamp (built into the lamp in some cases) and, in front of the lamp, the two-lens condenser system (which might include a heat absorbing—that is, infrared absorbing—glass plate). If the filament lies in a plane, it should be parallel to the slide, and the mirror should image the filament in the spaces between the coils. To verify this, and also that the field lens images both the filament and its reflected image at the projection lens, replace the lamp and remove the projection lens by turning the focusing knob until the projection lens moves sufficiently forward so that you can gently pull it out. Now hold a piece of white paper at about the original location of the projection lens, and turn the lamp on. Look at the piece of paper from the front, rather than at the lamp directly. The filament should be imaged there, but you will not be able to make out much detail because the coils and their interlaced reflected image make a fairly uniform source. The image you see, therefore, should be a moderately uniform bright spot.

While you have the projection lens out, you may want to measure its focal length. You can then measure the lens' diameter and compute its f-number, which you can check against the manufacturer's marking.

SUMMARY

An eye that has too strong a lens system **(myopic)** requires a *diverging* (negative power) **eyeglass** lens, which puts the **far point** properly at infinity. A **hyperopic** eye requires a *converging* lens to move the **near point** back to the normal *25 cm*. **Bifocal** lenses have two parts, each a lens of different focal length. **Astigmatism** in vision (unequal focal lengths along different meridians) is corrected by lenses having a *cylindrical* component. A **magnifying glass** (converging lens) forms a *virtual image*, at a point on which you can focus, of a much closer object. The **eyepiece** of a **compound microscope** acts like a magnifying glass for the real image formed by the **objective lens**. **Scanning microscopes** link a scanning electronic display with a scanning spot of illumination. **Telescopes** accept parallel light and produce a parallel beam at a different direction. **Refracting astronomical telescopes** produce an inverted image. **Terrestrial telescopes** provide erect images by means of a prism system or by a diverging eyepiece **(Galilean telescope)**. **Reflecting telescopes** use *concave mirrors* to collect and focus light. **Catadioptric telescopes** (such as the **Schmidt telescope**) combine reflecting and refracting elements. A **schlieren** system uses a knife edge at the focal point to block undeviated light, while permitting light that has been bent to strike the screen. Otherwise invisible objects are thereby made visible. A **field lens** preserves the field of view, making the image uniformly illuminated (without **vignetting**) while not changing the image-forming properties of the system, an important consideration in **projectors, enlargers,** and most optical instruments.

PROBLEMS

P1 State whether the following people have normal, myopic, hyperopic, or presbyopic vision: (a) Someone with eyeglasses of strength 3 diopters. (b) Someone with eyeglasses of strength −3 diopters. (c) Someone with bifocals.

P2 Repeat Problem P1 for the following people: (a) Someone with a near point of 1 m and no accommodation problems. (b) Someone with a near point of 25 cm and a far point of infinity. (c) Someone with a near point of 17 cm and a far point of 50 cm. (d) Someone with a near point of 50 cm and a far point of 1 m.

P3 People who have their eyelenses removed because of cataracts usually make up for this loss of focusing by wearing contact lenses. In addition, they often have to wear glasses for reading. Why?

P4 When trying to measure the power of her eyeglasses, a student discovers that the focal length she measures depends on how she uses the light source. The light source is a long straight filament. When it is held horizontally she measures a different focal length than when it is held vertically. What can she conclude about her eyes, and why do eyeglasses with these properties help?

P5 (a) Which lens focal length, of the following, would be best to use for a magnifying glass, assuming the only concern was to get the maximum magnification: $f = 2$ cm, $f = 8$ cm, $f = -6$ cm, $f = 5$ cm? (b) Why might one of the other lenses enable you to see more detail?

P6 You have one each of the following lenses, all of large diameter: $f = 50$ mm, $f = 100$ mm, $f = 200$ mm, $f = 25$ mm, $f = -50$ mm. Which of them would you use, and why, (ignoring aberrations) if you were making: (a) A magnifying glass? (b) A compound microscope? (c) An astronomical telescope? (d) A Galilean telescope?

P7 What does the correction plate in a Schmidt telescope correct?

P8 An opaque projector is similar to the overhead projector used in lecture halls, but is designed to project opaque material (for example, a page in a book) onto a screen. The basic idea is to shine sufficient light on the book to make the book a bright source, which can be projected by a lens. (a) Draw a diagram showing a design for a simple opaque projector. As in the overhead projector, the object (book) should be horizontal and the image (on the screen) vertical.

(b) Indicate, with an arrow in your diagram, the location of the top of the book if its image is to be rightside up on the screen.

P9 In the viewing system of an SLR camera, the camera lens images the scene on the ground-glass viewing screen, which is then viewed through a small magnifying lens in the viewing window. (The mirror and pentaprism need not concern us here.) Discuss the function of a field lens in this system. Where would you put it and what would it do? (See Sec. 4.2C.)

P10 The figure shows an object PQ, an intermediate image $P'Q'$, and a final image $P''Q''$ formed by two identical lenses L_1 and L_2. Redraw the figure. (a) Draw the cone of rays from P to P' to P''. (b) Draw the cone of rays from Q to Q'. How much of it goes on to Q''? (c) Pick an intermediate point M on the object and (in another color) draw the cone of rays from M to its intermediate image M'. Indicate, by shading that cone, which part of it reaches the final image M''. (d) To the right of M'', locate an eye E_1 that can see P'' but not M''. (e) Locate an eye E_2 that can see both P'' and M''. (f) Is there any location for an eye E_3 to see Q''?

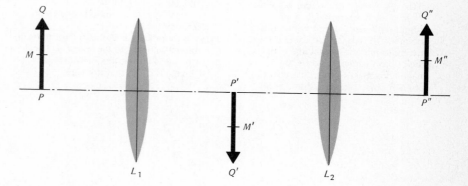

HARDER PROBLEMS

PH1 Draw an eye with a rather bad myopic condition in the following way. Represent the compound lens system of the eye by a single thin lens. Represent the retina by a plane 3 cm behind the lens. When the eye muscles are relaxed, the focal point of this myopic lens is at F'_{MR}, 2.5 cm behind the lens, instead of on the retina. Your job is to prescribe a contact lens that will correct this condition, as follows: (a) Draw a (not too large or too small) image on the retina, where a distant object *should* be focused. (The exact size doesn't matter, as you can check later by repeating the construction with a different size image.) (b) Use ray tracing and the known focal point of the relaxed myopic eye F'_{MR} to work backward and construct where an object would have to be in order that this relaxed myopic eye would focus it on the retina. Mark the point where this object would touch the axis with a P. (c) Measure the distance in centimeters between P and the lens. What is this distance? (d) The contact lens to correct this myopia should be such that rays from very far away (i.e.,

parallel rays) are made to appear to come from P. What is the focal length of the contact lens that will do this? (Note particularly the sign of the focal length, i.e., whether this is a converging or a diverging lens.) (e) What would the power of such a contact lens be (in diopters)?

PH2 Measure your range of accommodation in diopters and describe how you did it.

PH3 A certain myopic eye (when tensed) has a focal length $f_{eye} = 2.45$ cm, while the distance between the lens and retina is 3.00 cm. Eyeglasses with focal length $f_{glasses} = -20$ cm (power $= -5$ diopters) are worn 1.5 cm in front of the eye. The eye views an object PQ that is 8 cm high and 30 cm in front of it. Make a one-half scale diagram carefully, locating the focal points of the eyeglasses and the eye. Draw the retina flat. Use ray tracing to construct the image on the retina using the two-step procedure: (a) Construct the virtual image of the object PQ due to the eyeglass lens, ignoring the eye. Label this virtual image $P'Q'$. (b) Treat $P'Q'$ as the new object, ignore the eyeglasses now, and construct the image of $P'Q'$ due to the lens system of the eye. (c) Repeat steps (a) and (b) but with the eyeglass lens 3 cm in front of the eye. (d) Compare the sizes of the images on the retina obtained in the two cases, and also compare them with the image that would have been obtained without eyeglasses. Does the image size depend on the location of the eyeglasses? Is the image obtained with these diverging eyeglasses larger or smaller than that obtained without eyeglasses?

PH4 Magnifying glasses can sometimes produce unusual perspective. For example, it is possible to make the *farther* of two identical objects look *larger* (instead of the usual perspective where more distant objects look smaller). Show this effect as follows: (a) Draw a converging lens with a 20-mm focal length (the magnifying glass). To the left of the lens draw two objects, each 10 mm high. One should be located 5 mm from the lens, the other 10 mm farther. (b) Use ray tracing to find the virtual image of each object as produced by the magnifying glass. Use a different color for each object. (c) At a distance 60

mm to the right of the lens, draw a pinhole to represent the pupil of the eye. (You can draw a lens there if you wish but the lens will focus the rays on the retina, so you will only need ray 2, the same ray that goes through the pinhole.) Draw a "retina" (a straight line) 30 mm to the right of the pinhole (or even farther if you have room on the paper). Draw a ray from the tip of the virtual image through the pinhole to the retina, for each virtual image (still using two colors). (d) Draw, each in its own color, the final images on the retina. Which is larger, the image of the closer or more distant object? (e) Without the magnifying glass, which object would produce a larger image on the retina? (If you have a magnifying glass, you can confirm your results. Look through the magnifying glass at a match box or other small rectangular object held lengthwise. Hold the lens at arm's length and touching one end of the match box.)

PH5 (a) to (e) Repeat Problem P10, parts (a) to (e) with a field lens present. Locate this field lens at the intermediate image $P'Q'$ so that it images

L_1 on L_2. (f) Locate an eye E_3 that can see the entire final image $P''Q''$.

PH6 A better field lens than the one in Problem PH5 would image L_1 on the pupil of the observing eye. Why would that be better?

MATHEMATICAL PROBLEMS

PM1 A certain person has an uncorrected near point of 50 cm. What would be his eyeglass prescription?

PM2 A certain person is wearing eyeglasses of strength -3 diopters. (a) What is the farthest she can see clearly without eyeglasses? (b) What is the closest she can see with eyeglasses, assuming her range of accommodation is normal?

PM3 A certain magnifying glass has a focal length of 5 cm. (a) What is its magnification? (b) What does that mean? Show how the lens should be held for best effect.

PM4 Using the two lenses of Problem P6 that give the greatest magnification for an astronomical telescope, calculate: (a) the magnification of that telescope; (b) the length of that telescope, from objective to eyepiece.

PM5 Repeat Problem PM4 for a Galilean telescope.

The Human Eye and Vision—II: Processing the Image

CHAPTER 7

7.1
INTRODUCTION

The formation of an optical image on the retina (Chapter 5) is but the first stage in vision. Our analogy between the camera and eye is useful up to that point, but it breaks down for subsequent, higher visual processes. Cameras merely record an image; a human can distinguish one object from another, recognize and interpret scenes, and so on. And, of course, you don't take your eye's "film" to the drugstore to be developed. So to understand how you see, we want to consider what happens in the visual system *after* light is absorbed by the photoreceptors. As the physiologist Ewald Hering wrote:

The whole visual world and its content is a creation of our inner eye, as we may call the neural visual system . . . , in contrast to the dioptric mechanism, which may be designated the outer eye.

When discussing the "outer eye," we've been concerned with an image. But that is the *only* optical image formed in the visual system. While there is some correspondence between points in your field of view and the part of your brain that responds to them, there is no little TV screen inside your head, no "little person" in your brain who looks at an image back there. The information sent from your eyes to your brain is coded in the activity of millions of transmitting nerve cells (Fig. 7.1). This, along with the activity of billions of nerve cells within your brain, forms a **symbolic representation** of the scene. The word "calf," and the sound "kav" bear no

FIGURE 7.1

Electrical activity in a nerve leading from a cat's eye to its brain, as recorded by a tiny microelectrode placed in the nerve. The pulses are of equal amplitude, only the frequency of the pulses changes— the more pulses per second, the larger the signal relayed by the nerve cell.

direct similarity to a young cow (or the part of a leg, for that matter), but are just symbolic representations. Nor do the cells in your brain that respond as you look at a calf lie in a calf-shaped region, but they represent a calf symbolically. Understanding the visual system means understanding how the symbols of the "inner eye" are constructed; how information from the retinal image is combined or mixed to yield information in a different form.

Unlike photographic film, where the information is processed by development that produces one optical image, visual processing consists of sending symbolic information to the brain about various features (lines, edges, brightness, color, motion, etc.) that tend to be analyzed independently of each other. A subsystem of the visual system that responds preferentially to such a feature is called a **channel.** Each channel processes specialized information. We can compare the operation of the visual system to the way news is reported. A major event may involve sports, political, business, and society news, and those aspects of the event will reach the newspa-

per through separate channels—the separate reporters and editors for those particular features.

Like the newspaper, the visual system only reports the exciting news, the *changes* that it sees. The brain doesn't want to be cluttered with bits of information that, say, record that one point on the margin of this page looks like all the rest, any more than a newspaper wants to be cluttered with "dog bites man" stories. It is the changes that are of value—just as the newspaper reports only "man bites dog" stories, your visual system responds only to something new, the print.

How the "inner eye" processes information—emphasizing changes and sorting the information into its various channels—and the resultant effects on the way we perceive the visible world, is the subject of this chapter.

7.2
OVERVIEW OF THE HUMAN VISUAL SYSTEM

How does the information get from the rods and cones to the higher processing centers of the brain? Figure 7.2 shows the neural connections within the retina itself. The light-sensitive rods and cones (the **photoreceptors**) are connected to **bipolar** cells, which in turn are connected to **ganglion** cells. The latter send the information toward the brain. But even before the information leaves the retina, there is

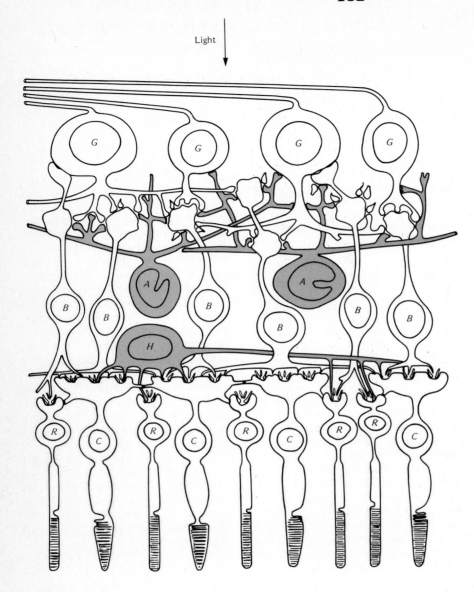

Light

FIGURE 7.2

Neural connections in a human retina. The photoreceptors (R, rods; C, cones) are connected to bipolar (B) cells. These in turn are connected to ganglion (G) cells, which lead to the brain. Sideways connections are made through the horizontal (H) and amacrine (A) cells.

ample opportunity for processing. Notice that a given bipolar cell connects to more than one photoreceptor (in fact, to as many as a thousand) and that a given photoreceptor connects to more than one bipolar cell. Further, there is a complicated cross-linking layer of **horizontal** cells, which provides sideways connections among photoreceptors. The bipolar cells are connected to the ganglion cells through another layer of cross-linking cells, the **amacrine** cells. Here again, each cell is connected to numerous other cells. These sideways connections allow information from adjacent parts of the retina to be compared and analyzed before being sent on by the ganglion cells. It is this analysis within the retina that reduces the information gathered from 10^8 photoreceptors so it can be transmitted by only 10^6 ganglion cells.

All these ganglion cells from each eye are bundled together to form the **optic nerve,** which leads to the **optic chiasma*** (Fig. 7.3). Here the ganglion cells separate; from *each* eye, roughly half of them *cross* to the other side of the head, while the rest stay on the *same* side of the head as that eye. In this way, the bundles of nerves that continue to each side of the brain contain ganglion cells from both eyes. In fact, the division of the ganglion cells at the optic chiasma corresponds to a division of the field of view into a right and a left half. Information from the right field of view of *both* eyes is conveyed by ganglion cells to the left side of the brain, while information from the left field of view of both eyes is similarly conveyed to the right side of the brain. Each half of the brain communicates with the other sufficiently, so there

---

*From the Greek letter chi (χ), which is in the shape of a cross or X. The Greek spelling of "Christ" begins with a chi, which is the root of our abbreviation, Xmas.

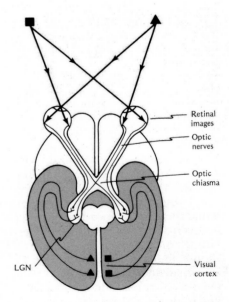

Retinal images

Optic nerves

Optic chiasma

LGN

Visual cortex

FIGURE 7.3

Overview of the human visual system. Light focused on the photoreceptors of the retina initiates a sequence of nerve firings in the nerve cells of the retina, leading to the ganglion cells, to the lateral geniculate nucleus, and to the visual cortex.

is no break seen in the middle of the field of view. (This arrangement of connections facilitates comparison of the images from your two eyes, which is important in your ability to see spatial depth—Chapter 8.)

Most of the ganglion cells lead to the **lateral geniculate nucleus (LGN),** often considered a sort of relay station, where the ganglions connect with the nerve cells that lead to the **visual cortex†** (at the back of the brain), the part responsible for vision. (About ⅕ of the ganglion cells do not go to the LGN, but to other parts of the brain that control reflex eye movement and pupil size.)

7.3
ELEMENTARY LIGHTNESS PERCEPTION

We saw in Chapter 5 how the sensitivity of the retina changes automatically according to the overall light level. The TRY IT (after Sec. 7.3C) tells how you can verify this. Perhaps surprisingly, the brain isn't very interested in the overall illumination—the overall light level is less important than the local *variations* in it.

A. Brightness and lightness

We've spoken of changes in **brightness** when there were changes in light intensity coming from a scene or from an isolated light source such as a flashlight. Brightness ranges from dark and dim up to bright and dazzling. Thus, you say a sunny day is *bright,* as is a strong searchlight. A flashlight with old batteries emits a *dim* light.

Your sensation of brightness also depends on your state of adaptation, as we saw in Section 5.3D. Thus, your bedroom illuminated by a single desk lamp may appear

---

*Latin for cap. The cortex is the gray outer layer of the brain.

bright when you wake up in the middle of the night. On the other hand, the same bedroom may appear dim when you come in from the bright sunshine.

In addition to the brightness of the light, you also have perceptions about the *surfaces* that reflect the light. In a given scene, a handkerchief appears *white,* a black cat appears *black,* no matter how bright the scene is. When speaking about the appearance of individual surfaces, we use the term **lightness.** Lightness is a sensation that ranges from black and dark gray up to light gray and white.

By and large, your sensation of the lightness of a surface depends *neither* on your state of adaptation *nor* on the overall illumination. The handkerchief appears white and the cat black whether you wake up in the middle of the night or come in from the bright outdoors, and in your dim bedroom as well as out in the bright sunlight.

B. Lightness constancy

If your sensation of lightness corresponded to the absolute light intensity striking your retina, you would continually notice changing lightnesses (and colors, see Chapter 10) whenever the sun slipped behind a cloud. However, rather than the absolute light intensity, it is the *relative* intensity coming from each surface that determines whether a surface appears white or black. The cat appears black because it reflects *less* light than other objects near it, while the handkerchief reflecting *more* light than do surrounding objects appears white. You respond to the *ratio* of the light from an object compared to that from its surroundings, and that ratio stays constant even though the overall illumination or your state of adaptation may change. This type of response leads to **lightness constancy:** all objects appear (for the most part) to maintain their familiar lightnesses as the lighting changes. The handkerchief always appears white even though a light meter might measure less light

coming from it at night than from a sunlit black cat. Since you respond to the *relative* light intensity throughout a scene, you never try to blow your nose on a black cat.

C. Weber's law

That the *ratio* of light intensities is important to your brain can be seen from Figure 7.4. (The first TRY IT gives another check on this.) Notice that steps of equal ratio of light intensity correspond to equal steps in lightness. That is, unlike those of Figure 7.4a, the steps of Figure 7.4b look like equal steps. Equal steps in lightness arise from steps of equal *ratio* of light intensity. This result is called **Weber's law.** We might say that the visual system works on a logarithmic scale (see Appendix I).

Weber's law also applies to your sensation of brightness—your perception of luminous objects. In this context, it explains why you don't see stars in the daytime. During the day, the ratio of the intensity of the star to the intensity of the sky is so

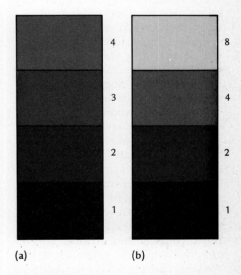

(a)   (b)

FIGURE 7.4

Weber's law. **(a)** Reflected light intensity increases by steps of an equal *amount* (1, 2, 3, 4, . . .)—a linear scale. **(b)** Reflected light intensity increases by steps of an equal *ratio* (1, 2, 4, 8, . . .)—a logarithmic scale. Steps of an equal ratio, **(b),** are equal steps in lightness.

small that you can't see the star. During the night, the intensity coming from the *sky* is small, so the ratio of the intensity of the star to that of the sky is then large. Hence, most stars are visible only at night, but not during the day.

PONDER

How does this Weber's law argument apply to the spy behind the half-silvered mirror of Section 2.3E?

Now, Weber's law has its limitations. If, for a given state of adaptation, the intensity of a region is extremely high, there will come a point when your visual system is pushed to its limits and cannot respond to increased light any further. Increasing the light intensity then does not result in an increase in the brightness, and so Weber's law ceases to operate. At the other extreme, the law breaks down as well. If we decrease the light intensity by a certain ratio again and again, there will come a point where the light intensity is so low that it is swamped by the "noise" in the visual system itself. This noise may be in the form of random chemical reactions in the nerves and synapses, pressure fluctuations, cosmic rays striking the nerves, and so on. Then, decreasing the light intensity further does *not* decrease the sensation of brightness. The second TRY IT shows how you can experience the effects of such noise.

Your visual system's response is thus similar to the behavior of the H & D response curves of film (Fig. 4.42). Over an extended, intermediate range of light intensities, the responses of both your eye and the film depend on the intensity, but at the extremes, they do not. The region between these extremes, where Weber's law holds, corresponds to the *latitude* of your eye for a particular state of adaptation. For higher or lower average light intensities, your eye changes its *speed* by adapting to the lighting conditions. In this way you can extract useful lightness information throughout an extremely large range of light intensities.

TRY IT

FOR SECTION 7.3
Uniform fields of light

Cut a table tennis ball in half, take one of its hemispheres that doesn't have any writing on it and cup it over one eye. Hold it in place like a monocle, without using your hands. Keep your other eye closed while your friend shines bright light, such as from a slide projector, directly toward your covered eye. Keep that eye open and pointed directly forward, toward the projector. After a few seconds, the inside of the ball will appear a uniform gray—not white. Have your friend place a filter (a lens from sunglasses will do) in the light beam to reduce the light striking the hemisphere. After a few seconds, the field will again appear a uniform gray. (Try colored filters, as well.)

The plastic scatters the light, making it uniformly bright throughout your field of view. The hemisphere is too close to be in focus, so you can't see subtle variations due to the texture of the plastic. You have verified that uniform fields (within a large range of intensities) appear the same—gray. Any nonuniformity, such as a pencil mark on the ball, destroys the effect. Try it!

First TRY IT

FOR SECTION 7.3C
Weber's law and its limitations

Look at a bright scene first with your naked eye and then through your sunglasses. Does the contrast change? Since the sunglasses cut out about half the light to your eye from both the dark and the bright parts of the scene, the ratio of the light intensities from different parts of the scene remains the same.

Now look through several layers of sunglasses (two pairs of polarizing sunglasses held almost at right angles will do nicely—see Chapter 13). What has happened to the contrast in the darker parts of the scene? Compare your observations to the H & D curve of film (Fig. 4.42).

Second TRY IT

FOR SECTION 7.3C
Fun with phosphenes

Phosphenes* are patterns of light, dark, and (often) color that you can see when no light is present, if your eyeballs experience pressure. We used phosphenes in the first TRY IT for Section 5.2A to verify that the image on your retina is inverted. Some of the phosphenes described here require greater and more prolonged pressure, so be extremely careful when generating them. Do not push on your eyeball longer than a minute at a time. If you experience any pain, stop immediately.

In a dark room, after your eyes have become dark adapted, close both eyes and cup your hand over your left eye, avoiding pressure. With the heel of your right hand, carefully push on your closed right eyelid for several seconds. Try to keep the pressure evenly distributed over the eyelid. You may see any one of a number of patterns: pulsing blobs, flashes, lines, dots, checkerboards, and so forth. The pressure restricts your retina's blood supply and may affect its nerves too, both of which play important and complicated roles in your visual perception.

Patterns can be seen even in total darkness and without pressure. While drifting off to sleep at night, notice vague, fuzzy patches of light that drift, fade, and are blocked by subsequent patches. After half an hour in total darkness, the sensitivity of your visual system has been turned up so high that it responds to the "noise" in it, letting you see this "prisoner's cinema."

7.4
RETINAL PROCESSING I: LATERAL INHIBITION

We've noted that your brain is less interested in the overall illumination than it is in the *relative* light intensity throughout the scene. How does your brain compare the light striking different cones, while

---

*Greek phos, light, plus phainein, to show.

ignoring the overall light level? In Figure 7.2 we saw that there is ample opportunity for this comparison *within the retina;* horizontal and amacrine cells, by providing *lateral** connections, permit the signal from one region of the retina to be modified because of the illumination of a neighboring region. *Increased* illumination of one region of the retina *diminishes* the signal to the brain from a *neighboring* region. This makes the signal sent to the brain relatively insensitive to overall illumination changes, but very responsive to differences in light striking the two regions. This ability of one part of the retina to inhibit the signal from another is called **lateral inhibition.**

A. Mechanism of lightness constancy

Let's see how lateral inhibition assists lightness constancy. Suppose two receptors of the retina, A and B (Fig. 7.5), receive some level of illumination and send an appropriate signal to the brain. What happens if the overall illumination is increased? The signals from each of the receptors tend to increase, but so does their ability to inhibit each

---

*Latin, *latus*, side.

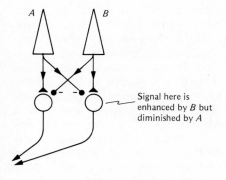

FIGURE 7.5

Two cones, A and B, and lateral neural connections (highly schematic). Inhibitory connections, indicated by minus signs (−), cause the signals from each cone to be diminished when nearby cones are stimulated.

other. The net result is that the overall increase in illumination is largely ignored. As an analogy, imagine a couple in bed under a two-part electric blanket having separate controls. Each control normally determines the warmth on the corresponding side of the bed. Suppose (for whatever reason) the controls have been interchanged; *he* gets *her* control, and she gets his. If the bedroom becomes colder, each person turns up the current and stays warm and comfy despite the change in room temperature. Similarly, the signal to the brain stays at the same level if the illumination at *both A and B* is increased.

But what happens if the illumination at *only A* is increased? The enhanced signal from A *further* inhibits that from B. However, because B doesn't receive any extra illumination, its inhibitory effect on A remains unchanged. The effect of the increased illumination at A is thus an *increased* signal from the more highly stimulated A and a *decreased* signal from the more inhibited B. The brain is made more aware of this difference in illumination at A and B than it would be without the lateral inhibition. The result is that an **edge,** where the light intensity changes rapidly from brighter to darker, is made more noticeable *(edge enhancement),* while an overall illumination change is not so apparent. Think of what happens in the bedroom scenario if it gets colder on only *one* side of the bed (because it is closer to an open window, say). He (near the cold window) turns *up* his control, which makes it warm for *her.* She, consequently, turns *down* her control, making it yet *colder* for him. Thus, a small difference between the room temperatures on the two sides of the bed is enhanced—he gets extra cold, and she gets extra hot.

By responding to *variations,* such as edges, a pattern can be processed more efficiently than if each single point were recorded. It is easier to describe a black square as four lines, black on one side, white on the other, than to state, point by point, which places are black and

which are white. Further, edges convey a lot of information; we can easily recognize figures given just their outlines (see Fig. 2.1c). Artists often employ outlines around their figures, and your visual system concentrates on edges for the same reason—to achieve efficiency in the storage and transmission of information.

The reliance of your visual system on ratios and edge information not only makes information storage and transmission more efficient, it also allows you to see the same lightness distribution under different conditions of illumination, that is, it accounts for lightness constancy. Thus, when you look at a given scene, the information that one area appears twice as light as its neighbor comes from the edge between them and is the same no matter how bright the overall illumination. Even if the scene is illuminated nonuniformly, say from a nearby source on the right, you still get the same information at the edge—just at the edge one region is twice as bright as the other—so you still see the same relative lightness (Fig. 7.6). By depending on ratios and edge information, your visual system is able to concentrate on the scene itself and not on the possibly capricious illumination.

B. Simultaneous lightness contrast

We can use our knowledge of lateral inhibition to explain many illusions associated with lightness. Notice that the small gray rectangle on the right in Figure 7.7 appears darker than the one on the left. Actually, the two reflect equal amounts of light; the rectangle on the right appears darker because it is surrounded by a white region, while the one on the left appears lighter because it is surrounded by a black region. This effect—**simultaneous lightness contrast**—in which the lightness of an area is influenced by neighboring regions, is easy to understand using lateral inhibition. The parts of your retina responding

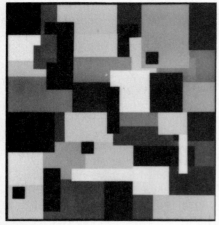

(a)

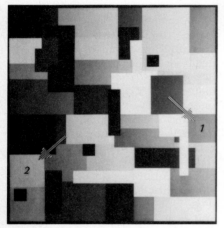

(b)

FIGURE 7.6

(a) Black, white, and gray paper rectangles photographed with uniform illumination. (b) The same rectangles illuminated by a lamp near the right. The edge information causes you to see rectangle *2* as lighter than *1*, as in part (a), even though the illumination was adjusted so the same amount of light comes to your eye from the same size area in each rectangle. To verify this, cut holes in a piece of paper to form a mask so that only the central regions of the two rectangles show.

to the gray rectangle on the right are surrounded by white regions and so, via lateral inhibition, respond *less* than they would otherwise. There is no such lateral inhibition to reduce the response for the rectangle on the left. The result

FIGURE 7.7

Simultaneous lightness contrast. The two gray rectangles reflect equal intensities but are unequal in lightness because they are surrounded by regions of different light intensity. (You can check that the small rectangles are objectively equal by cutting two holes in a piece of paper and covering the figure so that only the two gray regions show through.)

is that the gray rectangle on the right appears darker gray than the one on the left. Artists have long known this effect. For example, El Greco surrounds his white figures with gray or black areas to make the figures appear lighter, almost luminous. As Charles W. Stork wrote: "White is the skimming gull on the somber green of the fir trees;/Black is the soaring gull on a snowy glimmer of cloud."

Simultaneous lightness contrast is also relevant in picture framing. A picture surrounded by a white frame appears a darker gray and therefore perhaps dull and lifeless. (See the first TRY IT.) You can revitalize such a picture by viewing it through a tube made by your curled hand, thereby obscuring the frame. Sometimes a white frame is appropriate, for instance to make a dark photograph of underground coal miners more striking. For more standard black and white photo-

graphs, however, a frame of intermediate lightness may be best.

The apparent nonuniformity seen in each band of Figure 7.8 is again due to lateral inhibition. A region of your retina responding to the right side of a band is inhibited by the neighboring band that reflects more light. A retinal region responding to the left side of a band, however, is less inhibited, as the neighboring band here is darker. The result is that the left side of each band appears lighter than the right. A related effect is demonstrated in Plate 7.1. (See also the second TRY IT.) The pointillist artist Georges Seurat incorporated simultaneous lightness contrast into his paintings, compounding the effect (Plate 7.2).

Lateral inhibition is also responsible for the Hermann grid illusion (Fig. 7.9). How can we explain the illusory gray spots at the intersections of the horizontal and vertical white bands? Consider two regions of the retina, one viewing an intersection of a white horizontal and vertical band, *a*, and the other viewing a white band away from such an intersection, *b*. Whereas the two regions themselves receive the same amount of light, the situation in their neighboring regions is different. Since there is light coming from four sides of *a*, but from only two sides of *b*, the retinal

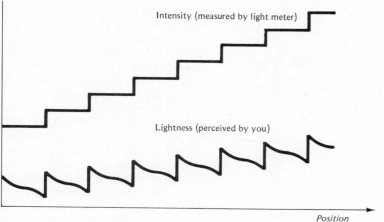

Intensity (measured by light meter)

Lightness (perceived by you)

Position

ing body, the surrounding parts are brightly illuminated and consist of shiny metallic objects, white cloth, etc. The parts the magician wishes to conceal are black, in front of a black background, usually drapery, and appear even darker to the viewer, dazzled by the rest of the display. The eye cannot make out the details in the darker parts. This technique was used in the nineteenth century by a Dr. Lynn, who produced half a woman, suspended in midair, on the stage of the Folies Bergères, a theater better known for displaying rather more of a woman.

First TRY IT

FOR SECTION 7.4B
Simultaneous contrast

When can a white region not appear white? When it is adjacent to a brighter region.

In a dimly lit room, hold this book vertically at arm's length with a desk lamp (turned off) visible above the top of the pages. Notice that the pages appear white. With the book in the same position, turn on the desk lamp so that light shines in your eyes. Now the pages appear light gray. Compare this effect with that in Figure 7.7.

FIGURE 7.8

Each vertical band has equal light intensity across its width. (Use two pieces of white paper to obscure neighboring bands to check this.) Notice, though, that the right side of each bar appears darker than the left side—a consequence of lateral inhibition.

region viewing *a* is more inhibited than that viewing *b*—*a* thus appears darker than *b*. We thus see dark spots at the intersections of the white bands but not at the points away from the intersections.

Various magic tricks depend on simultaneous lightness contrast. In order to conceal parts of the apparatus, say the supports of the float-

FIGURE 7.9

Hermann grid. Stare at the tiny black dot at the center of the figure and notice that illusory dark spots apear at the other intersections of the vertical and horizontal white bands. These spots

result from lateral inhibition. (The effect is greater in your peripheral vision, where lateral inhibition acts over greater distances.)

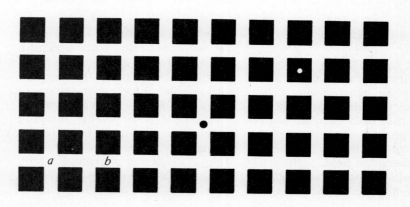

Second TRY IT

FOR SECTION 7.4B
Lateral inhibition and shadows

Shadows produced using uncovered, dual-tube fluorescent lights can reveal the effects of lateral inhibition. Use a piece of white paper as a screen, and another to cast a shadow. Hold the screen paper horizontally, below the lights. Hold the other piece of paper, with one edge parallel to the two fluorescent tubes, about 5 cm above the screen (Fig. 7.10).

Now look carefully on the screen at the shadow of this edge. In region A, the screen receives light from both tubes and is therefore bright. Region B receives light from only one tube—a penumbra. Region C is in total shadow (ignore any scattered light). Thus, the light reflected from the screen has an intensity profile similar to that found in Figure 7.8. Notice that (as in Fig. 7.8), there is edge enhancement at each shadow edge.

FIGURE 7.10

Arrangement for seeing simultaneous contrast in shadows of edge *a*. To see Mach bands, look at the shadows of edge *b*.

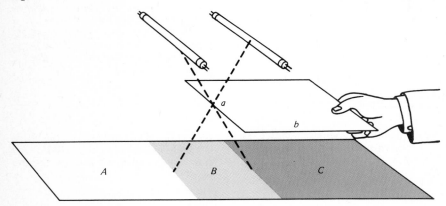

(Now, however, the intensity distribution is due to variations in illumination and not in reflectance.)

A related effect, **Mach bands,** *can be seen by looking at the shadow in the vicinity of a blocking edge oriented perpendicularly to the lights. There is, as before, a region illuminated by the full extent of the tubes and there is a full shadow well beneath the blocking piece of paper. However, between these regions the light intensity varies gradually. (Why?) Notice an illusory bright line along the edge of the bright region. This is one Mach band. The other is an illusory dark band along the edge of the region of the full shadow.*

PONDER

Draw the light intensity versus position on the screen for the Mach bands. Considering lateral inhibition, sketch the lightness versus position. Do your perceptions of the Mach bands agree with your sketch?

FIGURE 7.11

Schematic receptive field for a human ganglion cell. When light strikes any of the cones in the center of the receptive field, the ganglion is excited (denoted by +'s); when light strikes any of the cones in the surround, the ganglion is inhibited (denoted by −'s).

C. Receptive fields

We have said that the response of each point in the retina is influenced by the light striking neighboring regions. In Figure 7.2 you can see that each ganglion cell receives signals from an *area* of the retina. We call this area the ganglion's **receptive field.** Light striking anywhere within the ganglion's receptive field can change that ganglion's response, but light striking other parts cannot. Figure 7.11 shows schematically the receptive field of a typical ganglion cell. Light striking the *center* of the receptive field *excites* the ganglion (increases its rate of firing). Light striking any noncentral point of the receptive field (the *surround*) *inhibits* the ganglion (decreases its rate of firing). This **center-surround** organization, then, produces the lateral inhibition we've been discussing.

In Figure 7.12 we show several stimuli and the ganglion responses. Part *a* shows that when a uniform gray area is imaged on the receptive field, a small "background" response results. When, additionally, bright light strikes the receptors in the center of the receptive field, *b*, a large response occurs. This is the excitation we would expect in the absence of lateral inhibition. If, instead, the bright light strikes the surround of the receptive field, *c*, the response is *decreased*. When bright light strikes *both* center and surround, *d*, there is virtually no response above the background activity—the effects of excitation of the center and inhibition of the surround balance. If a bright bar strikes the center of the receptive field and flanking dark bars fall on the surround, *e*, a very large response occurs—even larger than in *b*. The extra dark bars on the surround reduce the inhibition, making the total response greater. If a very finely spaced pattern of bright and dark bars is used, *f*, there is virtually no response above background. Here the center gets white and black (no net excitation), and the surround also gets both white and black (no net inhibition).

D. Processing edges

The processing of visual information within the retina, designed to decrease the effects of uniform changes in illumination and to enhance the effects of edges, suggests that we rely on edges for information about the uniform regions be-

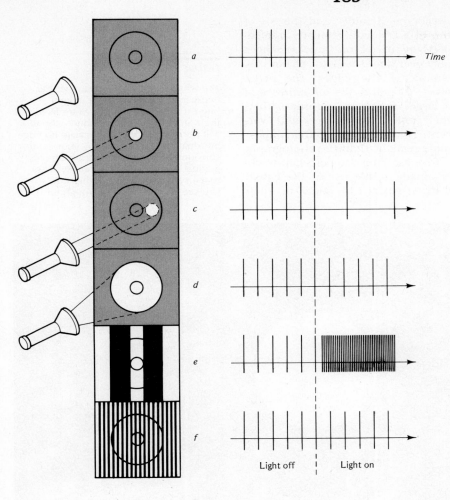

FIGURE 7.12

Response of a cat ganglion cell to various patterns of light (measured using tiny microelectrodes as in Fig. 7.1).

tween them. We can illustrate this with the **Craik-O'Brien illusion.** Even though the extreme right and left of Figure 7.13a are equally bright, the entire right side *appears* a darker gray than the left. The illusion is due to the form of the distribution of the reflected light near the central edge between the two regions. As with Seurat's painting (Plate 7.2), one side of the edge (the right) is made more reflecting than the extremes of the display, while the other side is made less reflecting. Center-surround receptive fields that view the edge tell you that the right region is lighter than the left. Because the intensity changes so gradually away from the edge, your center-surrounds do not pick up the change (compare Figs. 7.12a and d). Your visual system then **fills in** the regions, taking its clue from the behavior it notices at the edge, so the entire right field appears lighter than the left. When you cover the center line in the figure with a pencil, you lose the edge information, so the (subjective) lightness of the sides more closely corresponds to the (objective) light intensity distribution.

By using this illusion to enhance the edges in their pictures, painters can "fool" the observer into seeing one side of an edge as much lighter than the other, even though the objective difference is small. This is one advantage the painter has over the photographer, who cannot enhance the edges without a major ef-

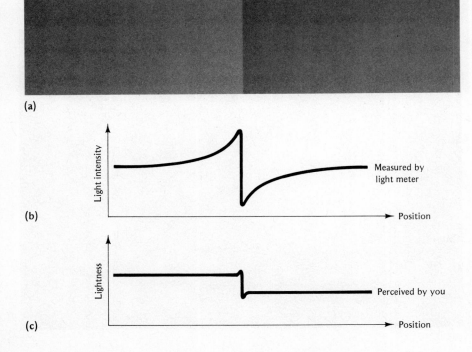

(a)

(b) Light intensity — Position — Measured by light meter

(c) Lightness — Position — Perceived by you

FIGURE 7.13

(a) The Craik-O'Brien illusion. The light intensity on the extreme right and left sides of **(a)** are equal, but the right side appears darker because of the way the edge is formed. **(b)** The actual reflected light intensity across the display in **(a)**, assuming uniform illumination. Cover the central, vertical line in **(a)** by placing a pencil over it and notice that the two halves there now appear equally light. **(c)** Lightness across the display in **(a)**.

fort. The range of light intensities in most scenes is much greater than can be reproduced in either a painting or a photograph. (One part of a sunlit scene may be a million times brighter than another. Indoors, the range between the brightest and darkest parts of the scene may be greater than a factor of a thousand. The best white on a photograph, however, is only about fifteen times as reflecting as the darkest black—the difference be- tween the shoulder and the toe of the H & D curve. Similarly, the reflectivity of oil on canvas ranges only over about a factor of ten.) By enhancing the edges, the artist gives her paints the *appearance* of a larger range of brightness than their reflectivity alone would warrant. The painting may then appear more realistic than the photograph.

Chinese Ting Yao porcelain workers made subtle use of the Craik-O'Brien effect. In some of their

(a)

(b)

(a)

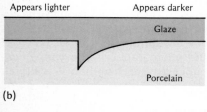

(b)

FIGURE 7.14

(a) A Chinese Ting Yao porcelain saucer (ca 1000 A.D.), an early, artistic example of the Craik-O'Brien illusion. **(b)** The shape of the groove cut by the artist. After glazing, the light reflected by the porcelain has an intensity distribution given by the same shape, because the thicker the depth of the glaze filling the deep notch, the more light is absorbed.

FIGURE 7.15

The white disk in **(a)** is clearly visible because of the sharp edge in light intensity at the perimeter. Stare fixedly at the X in **(b)**. Within 20 seconds, the disk will normally disappear because no edge information is picked up by your center-surround receptive fields from the gradual change in light intensity. Your visual system then *fills in* the disk, so you see a uniform gray throughout.

plates, designs were created by special grooves cut in the porcelain (Fig. 7.14). The plate was then coated with a layer of semitransparent glaze. The amount of light reflected at any point from the porcelain beneath this glaze is determined by the thickness of the absorbing glaze layer there—the thinner the layer, the more light is reflected—so the intensity of the reflected light matches the contour of the grooves. Since the variation of light intensity away from the sharp edge is so gradual as to be barely noticed, your visual system infers from the sharp "edge" in light intensity that the entire region on one side of a groove is lighter than the region on the other. Thus, the flower appears lighter than its background.

Figures 7.13 and 7.14 show that regions reflecting objectively equal light intensities can differ in lightness. Conversely, by drastically *reducing* edge information, we can make two objectively *unequal* regions appear equal, as in Figure 7.15.

7.5
RETINAL PROCESSING II: NEGATIVE AFTERIMAGES

We've seen that the sensitivity of a region of your retina is decreased if a neighboring region is simultaneously exposed to a bright light—simultaneous lightness contrast. Similarly, the sensitivity of a given region of your retina decreases *after* it is exposed to a bright light for a period of *time*—**successive lightness contrast.** Prolonged stimulation **adapts,** or desensitizes, part of your retina so that it has a weaker response to subsequent stimulation. This is the principle behind standard **negative afterimages.** During the period of adaptation to Figure 7.16, the parts of your retina corresponding to the cat are active and thus become desensitized. The parts of your retina corresponding to the outside of the cat are not so desensitized. Later, when you look

FIGURE 7.16

Standard negative afterimage. Stare for at least one minute at the black dot in the center of the cat. Then, look at the dot on the right. (It may take a moment to see the effect.)

at the white region on the right, the desensitized parts of your retina cannot respond well to the white light, while the rest of your retina can, so you see a dark cat on a white background. As you move your eyes about. the afterimage follows along, always at the desensitized region of the retina. The afterimage may remain for 30 seconds or longer, depending on how long you adapt and on the light conditions. (Rapid blinking helps preserve the afterimage.)

Besides prolonged stimulation, brief intense stimulation can also evoke a negative afterimage, as you know if you've ever been the victim of flash photography. The dark spot that follows your gaze is at the desensitized region of your retina that was stimulated by the flash and, hence, has a much higher threshold than the unstimulated region. As Dante wrote:

. . . a sudden flash blinds the eyes, so that it deprives them of the sight of the clearest objects . . .

This idea finds application in a personal defense weapon that produces an extremely intense burst of light designed to "blind" a (dark adapted) mugger, much as the photographer (Jimmy Stewart) used his flash bulbs to blind an assailant in Alfred Hitchcock's film "Rear Window."

Negative afterimages stay with the eye that was adapted; they *do not* **transfer** (to the opposite eye). Close your right eye and adapt to

Figure 7.16 with your left eye. You'll see an afterimage only if your left eye looks at the white region; your right eye doesn't show the effect. (Similarly, at the end of Sec. 5.3D you saw that there was no transfer of dark adaptation.) Such lack of transfer helps determine where in the visual system the effect arises. Thus, for the simple lightness perception we've discussed here, the processing must occur *before* information from the two eyes comes together at the LGN—that is, in the retina. You may be surprised to find that there are afterimage effects that *do* transfer (Sec. 7.8B).

Negative afterimages are an indication of another means by which (useless) constant stimulation is ignored, while *changes* are not. The desensitizing of your retina, which produces negative afterimages, results in a decreased signal to your brain from a region receiving constant stimulation. Regions receiving varying stimulation are not so desensitized and, thus, continue to signal variations to your brain.

*7.6
EYE MOVEMENTS

The illusion of Figure 7.15 works while you keep your eyes fixed (pointed at the *X*)—the variations in light intensity across the figure are too gradual to be noticed. However, when you allow your eyes to scan the page, you see the disk at the left as (correctly) lighter than its background—your **eye movements** translate the variation in stimulation across space into a variation of stimulation *in time*. If this variation in time is large enough, you

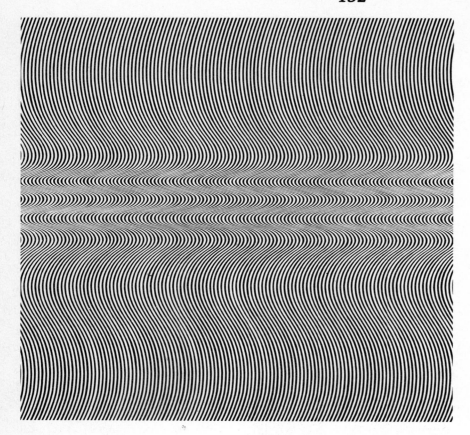

FIGURE 7.17

Bridget Riley, "Current." As your eye scans, image and afterimage superpose to give a lively, shimmering, apparently moving figure. The perception is not unlike that resulting from looking at a rippling flow of water, hence the painting's name. Try holding your eye fixed—the shimmering persists. (1964. Synthetic polymer paint on composition board. 148 × 149 cm. Collection, The Museum of Modern Art, New York.)

notice it. Thus, scanning plays an additional role to that described in Section 5.2, where we saw that the visual system builds up a detailed mental image of the world by scanning the fovea throughout the field of view.

Even when you think you're holding your eye fixed, you are not. You may have noticed, as you stared at the dot in the center of the cat (Fig. 7.16), that the outline of the cat seemed to "shimmer." This is due to small, involuntary eye movements that displace the edge of the afterimage with respect to the pattern at which you are staring. For another demonstration, stare at the small black dot in the center of Figure 7.9 (as you did to see the illusory gray spots), but now stare for at least a minute to create a strong afterimage. (As with the cat, the edges should appear to shimmer.) Then look over to the tiny white dot at the upper right of the figure. The afterimage, fixed to your retina, will appear to jiggle and dance back and forth as a result of your involuntary eye movements, while the image of the page appears properly fixed and unmoving.

Several op artists employ these effects of eye movements and afterimages. The pattern in Figure 7.17 seems to shimmer and flow no matter how hard you try to hold your eye still—a testament to the relentless eye movements. In Figure 7.18, you see the same effect from a slightly different angle. Small eye movements along a direction 45° to the *right* of vertical make *some* of the areas of the painting shimmer (such as the small central region), but not others. This is because the

afterimages on the lines coincide with the lines themselves only for those lines tipped in the direction of the eye motion. Eye movements along a line tipped 45° to the *left* of vertical result in shimmering of *other* parts of the painting. Random eye movements result in a lively interplay of areas shimmering in alternation. (Involuntary changes in accommodation may also play a significant role in the shimmering in both these paintings.)

You can eliminate the effects of eye movements in the following way. Stay in the dark for at least 5 minutes. Then, illuminate either of the last two figures with a brief, intense flash of light, such as from a camera's flash unit. The combination of your sensitive (dark adapted) retina and the intense light will produce an afterimage that will remain long enough for you to see and study it in the dark. Notice that no shimmering or apparent motion results, as there can be no superposition of image and afterimage.

Eye movements fall into three broad classes: ***drifts, tremors,*** and ***saccades.*** Drifts are slow, smooth movements of about a minute of arc ($\frac{1}{60}$ of a degree) per second. Concurrent with the drifts are small but rapid jittery motions called tremors, typically a quarter of a minute of arc at a frequency of about 50 Hz. Saccades* are sharp, abrupt movements, which may occur as frequently as 4 times per second. The size of a saccade is often about 5 to 10 minutes of arc. As you read this sentence, your eye skips to groups of words, and back to the beginning of each line by means of saccades.

A. Retinal stabilization

It has been found that eye movements are necessary for your vision by seeing what happens when their effects are cancelled. We can remove the *effects* of your eye movements if we can assure that an image always falls on the same spot on your retina, no matter how your eye moves.

*Old French *saquer,* to pull.

FIGURE 7.18

Reginald Neal, "Square of Three—Yellow and Black."

This can be done, for example, by reflecting the image to your eye with a mirror whose orientation is continually changed so as to compensate for your eye movements. Such **retinal stabilization** reveals a curious effect; within a few seconds, you find that the image fades to a uniform gray and thus becomes invisible.

A ganglion cell desensitizes within a few seconds under the constant stimulation accompanying retinal stabilization, even if there is a light-intensity edge in the cell's receptive field. Usually (that is, without retinal stabilization), each ganglion's response changes rapidly as eye movements sweep its receptive field across different edges in the field of view; but under retinal stabilization, this does not occur. For example, the shadows of the blood vessels that lie over the retina are effectively stabilized and thus invisible. Only through tricks similar to those described in the TRY IT for Section 5.3 can these shadows be made to move a bit and thus become visible. The fading of stabilized images can be thought of as ganglion desensitizing or as cancellation of image by afterimage, but the upshot is the same; *change* in stimulation is required for vision. No change, no vision.

When you look at a large piece of blank white paper, the center of your field of view is, effectively, retinally stabilized. That is, even when your eye moves slightly, the same image (a pure white field) reaches the central part of your retina. Why, then, doesn't the piece of paper fade at the center? The answer, as in the Craik-O'Brien illusion, involves *edges*. Your brain receives the information about the edges of the paper and fills in (much as it fills in across your blind spot) throughout the area of the paper.

Without the edge information the image fades, as you saw in the TRY IT for Section 7.3.

An interesting example of the dependence of vision on change occurred when a subject viewed his friend's face under retinal stabilization. Within a few seconds, the image of the face disappeared. Then, the friend smiled. Such motion could not be retinally stabilized by the mirror, and so the mouth was visible for a few seconds before it faded again. For a few seconds, then, only the smiling lips were visible to the subject. The effect was described as "like the smile of the Cheshire cat in *Alice in Wonderland.*"

7.7
TEMPORAL RESPONSE

Your retina, including its ganglion cells, responds only if there is a change in stimulation *in time*, such as a light turning on and off or your eye scanning across an edge. The way your eye responds in time has a number of interesting consequences.

A. Positive afterimages

The response of your visual system to a brief flash of light is both delayed **(latency)** and of longer duration **(persistence)** than the brief flash itself (Sect. 5.3C). Here we'll deal primarily with persistence, which causes you to see the flash as lasting longer than the actual flash, thus generating **positive afterimages** (white where there was white in the stimulus, black where you saw black). These may last as long as $\frac{1}{20}$ sec at low ambient light levels but are shorter at high light levels. (The first TRY IT tells how to prolong them.)

If two images are presented in rapid succession or alternation (such that the time between presentations is less that the persistence time), the images will appear as one because your response is too slow to separate and distinguish them. You can easily verify this using the second or third TRY IT.

When a rapid sequence of images is shown, each of which differs only slightly from the ones coming directly before and after it, you see a continuous *motion*. The images arrive in too quick a succession to be individually processed. If the change from one image to the next is small, the visual system fills in, and you perceive one image that moves. Slowly flip the edges of pages 11 through 81 while watching the figure at the bottom. When the pages are flipped slowly, a rough, jerky "motion" results. When the pages are flipped quickly, you experience an illusion of continuous motion. When stationary light bulbs on movie marquees and advertising signs are flashed in the proper, rapid sequence, the illusion of continuous motion can be quite compelling. In "Love Happy," Harpo Marx escapes the bad guys by sitting on a "moving" flying horse in an advertising billboard (flashed neon lights)—and is carried along with it!

This motion effect is the basis for movies and TV. A sequence of images is presented in quick succession so that the illusion of continuous motion results. (See the fourth and fifth TRY IT's.) The persistence of vision can be as short as $\frac{1}{50}$ second under high light levels. American television pictures are broadcast at the (slower) rate of one frame each $\frac{1}{30}$ second, so you'd expect that the screen would appear to flicker. This problem is avoided by scanning the horizontal lines of the TV screen in the order 1,3,5,7, . . . 2,4,6,8, . . . ; that is, two frames, each covering the screen, are interleaved. In this way, the TV screen is essentially filled each $\frac{1}{60}$ second and so flicker is avoided.

Movie frames are projected at one frame each $\frac{1}{24}$ second, and again, one would experience flicker were it not for a special shutter in the projector that projects each frame three times. Thus, effectively, 72 frames are projected every second—a rate sufficient to avoid flicker. If successive images don't differ too drastically, the illusion of continuous motion results. Older projectors did not have the special shutter and so movies of yore flickered and were called "the flicks."

First TRY IT

FOR SECTION 7.7A
Positive afterimages

Face a window through which bright sky is visible. Close your eyes and gently cover them with your hands for at least 30 seconds, blocking as much light as possible. Then, take you hands away and open your eyes for about three seconds, looking at an intersection of the window's glazing bars against the bright sky. Be sure to keep looking at one point for the three seconds. Then, close and cover your eyes as before. Notice that you experience a positive afterimage. Observe it for as long as it lasts (perhaps 10 to 15 seconds), noticing any color changes that occur (see Sec. 10.7B).

By closing and covering your eyes and thereby lowering the light striking them, you have caused their sensitivity to increase. The light from the brief exposure strikes these highly sensitive eyes. Returning them to the dark increases their persistence time, so they remain excited and you see the afterimage for a long time.

Second TRY IT

FOR SECTION 7.7A
Thaumatrope

*The **thaumatrope*** employs persistence of vision (positive afterimages) to create a single picture from two individual pictures flashed in quick succession. Cut a square piece of white cardboard, about 4 cm on a side, and punch four small holes, two each on opposite sides, as shown in Figure 7.19. Draw two pictures (or paste photos), such as a bird cage and a bird, on opposite sides of the cardboard. (How should they be oriented?) Twirl the card between two long rubber bands, as shown. The persistence of your vision will make you see the bird in the cage. With a little*

---

*Greek *thauma*, marvel, plus *tropos*, turn.

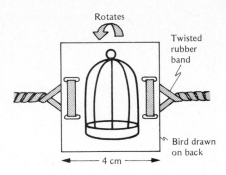

FIGURE 7.19

Design for a thaumatrope.

practice, you can "pump" the card by pulling on the rubber bands at the proper moments (an example of resonance!), so you can keep the bird from escaping for as long as you like.

Third TRY IT

FOR SECTION 7.7A
The ghost in the window

You can also use persistence of vision to conjure up a "ghost" and make its image hover in midair. In a dark room at night, place a slide of your ghost (any portrait) in a slide projector aimed out an open window. Hold a piece of paper a short distance in front of the window as a temporary screen, and focus the image of the slide on the screen. Then, take the paper away. All the light from the projector should go out the window so the room remains dark.

Now, conjure up the ghost. Sweep a long stick rapidly up and down just in front of the window, in the place where the slide is focused.

Fourth TRY IT

FOR SECTION 7.7A
Zoetrope

*The **zoetrope*** or wheel of life is the precursor to the modern cinema and graced the drawing rooms of many a Victorian home. You can make one using a cylindrical oatmeal box, pencil, thumbtack, and some paper. Cut the oatmeal box to a height of about 10 cm*

---

*Greek *zoe*, life.

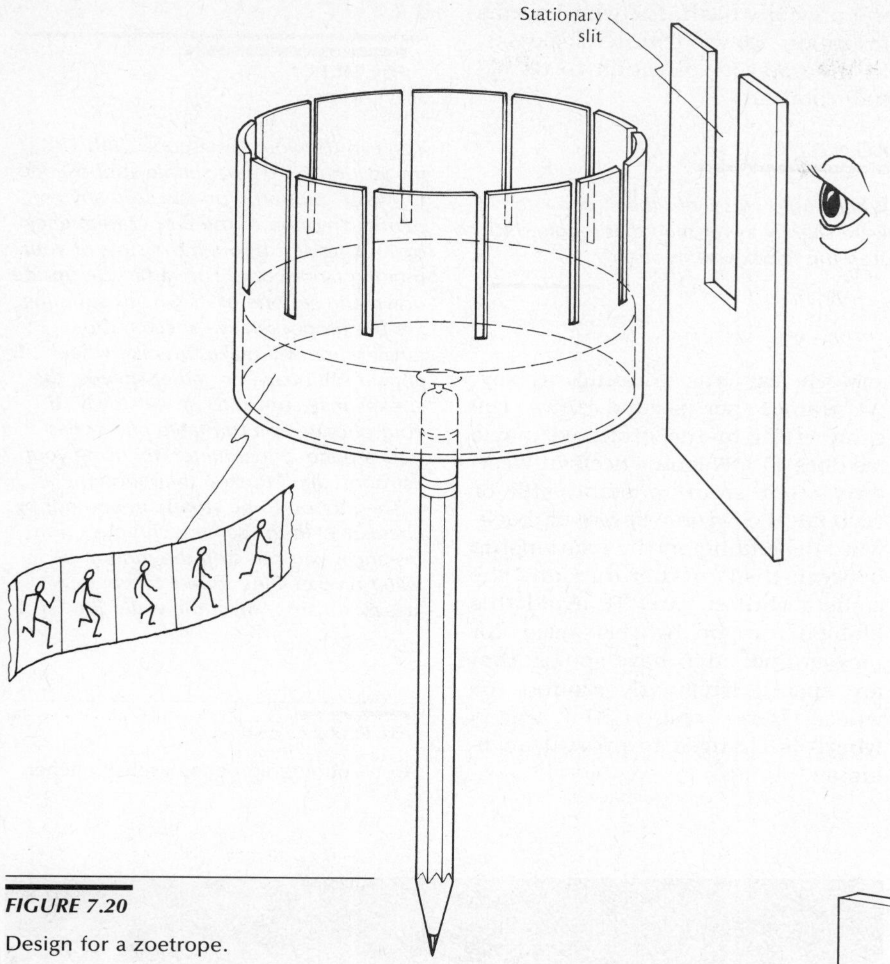

Stationary
slit

FOR SECTION 7.7A
Fantascope

*The **fantascope** or **phenakistoscope***
*operates on the same principle as the
zoetrope. Cut 12 radial slots in the
perimeter of a flat, circular disk about 30
cm across (Fig. 7.21). Draw 12 pictures (as
for the zoetrope), and place them
between the slots on one side of the
disk. Push a thumbtack through the
center of the disk into the eraser of a
pencil. The pencil should be on the side
opposite the pictures. Face a mirror and
hold the pencil so that the disk is
between you and the mirror. Look
through one of the slots at the reflection
of the pictures. Spin the disk.*

*Greek *phenakistes*, deceiver.*

FIGURE 7.20

Design for a zoetrope.

FIGURE 7.21

Design for a fantascope.

*and cut 12 vertical slots in the side,
evenly spaced around the perimeter, as
shown in Figure 7.20. On a strip of paper
whose length is equal to that perimeter,
draw 12 simple figures, such as stick
men, that differ slightly from one
drawing to the next. (Or use the January
1980 issue of* Playboy *for a somewhat
racier set of pictures.) Place the strip of
paper inside the box around the side
walls, just below the slots. Push the
thumbtack through the center of the
bottom of the box, into the pencil's
eraser. Look through the slots as you
spin the box, and see your stick men
come to life. For best results, view with
one eye from a distance of about 25 cm
and through a stationary slit (made from
paper, as shown in the figure), so your
eye sees only one moving slit of the
zoetrope at a time. If you make a
cylinder out of stiff paper, about 50 cm
around, and cut 20 evenly spaced slots,
you can get good results by placing it on
a phonograph turntable.*

*B. Stroboscopes

The persistence of vision can be used to make repetitive motion appear "stopped" or frozen. For example, suppose a bicycle is turned upside down and one of its wheels set in rapid rotation. The spokes will appear blurred under normal illumination—they rotate too quickly to be seen individually. Suppose, however, that the only illumination comes from a **stroboscope,** or strobe—a source of a regular sequence of bright flashes of light. If the strobe flashes once for each revolution of the bike wheel, the wheel will appear stopped. Every time a particular spoke gets to the vertical position, say, the flash is emitted. Because there is no illumination between flashes, that spoke always appears vertical and thus stopped. Likewise, all other spokes will appear stopped at their respective positions.

The flashes may be too infrequent, however, for the wheel to appear visible continuously, and you'll notice the flickering of the strobe light. This problem can be surmounted by increasing the frequency of strobe flashes so that every time *any* spoke is in the vertical position, a flash is emitted. Because the spokes look identical, the wheel again appears stopped, without flicker. The TRY IT shows how to make and use your own stroboscope.

What happens in our bike wheel example if the flashes are not at the exact frequency that stops the motion? Suppose that the frequency of the flashes is a bit too *high* and the wheel is rotating clockwise. Assume that at some instant a strobe flash illuminates the wheel when a given spoke is in the vertical position. Just *before* the next spoke gets to that vertical position, the next flash of light is emitted (since the strobe frequency is too high). Thus, it appears as if the spoke that was originally vertical has moved a bit *counter*clockwise. Similarly, each successive flash reveals each spoke displaced slightly counterclockwise from where another spoke was on

the previous flash; the wheel seems to rotate slowly counterclockwise, in the opposite direction to its actual rotation.

PONDER

What would you see if the strobe rate were slightly lower *than that required to stop the (clockwise) motion?*

Movie cameras running at, say, 24 frames per second show the same effect in the projected movie (as does TV). Wagon wheels in Westerns often seem to stand still or turn at the wrong speed or backward depending on the relationship between their rotation rate and the camera shutter rate. To avoid this illusion, wagon wheels used for movies now often have spokes that are spaced irregularly around the wheel. Here, reality (the wagon wheel) is changed to prevent an illusion.

TRY IT

FOR SECTION 7.7B
Stroboscopes

You can use your fantascope (fifth TRY IT for Section 7.7A) as a simple stroboscope to "stop" motion. You can stop any periodic motion of the proper frequency by looking at it through the slots of your rotating fantascope. Turn a bicycle upside down and set one of its wheels spinning. For the proper relative speeds, the fanatascope will make the bike wheel appear stationary. For other speeds, the spokes may appear to rotate slowly. If your phonograph turntable platter has dots around its perimeter, try using your "stroboscope" to stop their motion.

The motion to be stopped need not be circular. With the same techniques, for instance, you can stop the motion of water drops falling from a faucet in a steady stream. Adjust the water flow so a

FIGURE 7.22

Strobe photograph of a baseball pitcher.

smooth stream emerges from the faucet, and then "breaks" into droplets. If you spin your fantascope at the right speed, you can see individual water droplets "frozen" in midair.

Intermittent illumination (just like intermittent viewing) can also stop motion. Strobe lights having adjustable flash rate are sold for use in discos and other places. If you have access to such a stroboscope, use it instead of your fantascope to stop motion. With all other lights off, adjust the flash rate to stop the motion of the bike spokes. Next, increase the flash rate slightly. Do the spokes then seem to rotate forward or backward?

With some care, you can photograph the stopped motion. Your exposure meter can be trusted if the flash rate is sufficiently high that the meter reading is constant. Otherwise, first double the flash rate so that the rate is high enough and find an exposure setting. Then, halve the flash rate to its proper value and set the exposure setting to compensate: double the exposure time or decrease the aperture setting by one f-stop marking. Bracket the exposure by trying several f-stop settings above and below this one to ensure that at least one of your photographs is properly exposed.

You can photograph nonrepetitive motion, too (Fig. 7.22). Be sure to use a dark background; otherwise your subject, which in a given position receives only one flash, will be darker than the fixed background, which receives all the flashes while the shutter is open. The exposure time should be long enough so that several flashes are included, and the aperture opened wide enough so that each flash gives a proper exposure. This latter requirement is a bit tricky. As described above, find an exposure time and f-stop as if you were going to photograph a repetitive motion. Knowing (or guessing) the flash rate, determine the number of flashes that will occur during the exposure time you set. Open the aperture accordingly: the more flashes that will occur, the wider the aperture. For example, if eight flashes occur during your exposure time, you must lower the f-number by three f-stops, so that eight times as much light enters the camera from each flash. As above, bracket your f-stop settings.

*7.8
CHANNELS: SPATIAL FREQUENCY AND TILT

We've seen that your ganglion cells are not excited very much by either a very gradual change in light intensity across the field of view (Fig. 7.15b), or by a set of very narrow, closely spaced vertical bars (Fig. 7.12f). We can make this quantitative by studying your response to visual **gratings,** sets of parallel bright and dark bars. In Figure 7.23, the horizontal axis is **spatial frequency**—the number of light and dark bars of the grating per degree of visual angle—and ranges from low spatial frequency (broad, spread out bars) to high spatial frequency (narrow, close bars). The vertical axis is (grating) **contrast,** which is

FIGURE 7.23

Measure your contrast sensitivity function (CSF). (The bars that you see will roughly correspond to the marked spatial frequencies, in cycles per degree, if you view the figure at arms length.) The contrast is large across the bottom of the figure and small along the top. The envelope of the visible part of the figure marks your CSF. (A more accurate CSF is obtained by separate measurements at each spatial frequency.)

proportional to the difference between the light intensities from the lightest and the darkest parts of the grating. Where the contrast is very low, there is virtually no difference between the lightest and darkest parts of the grating, so you cannot see the grating there. Where the contrast is high, there is a large difference between the light intensities from the lightest and darkest parts, and the grating is more easily seen.

A. Contrast sensitivity function

Roughly speaking, the boundary between the regions of Figure 7.23 where you can see the grating and where you cannot is called your **contrast sensitivity function (CSF).** As we might deduce from Figure 7.12, you are not sensitive to very low-spatial-frequency gratings (Fig. 7.12d) or to very high-spatial-frequency gratings (Fig. 7.12f) but are fairly sensitive to intermediate frequencies (Fig. 7.12e). Thus, a high contrast (near the bottom of the figure) is required for you to see low- and high-spatial-frequency gratings, while you can still see the intermediate grating even when it is of low contrast (near the top of the figure).

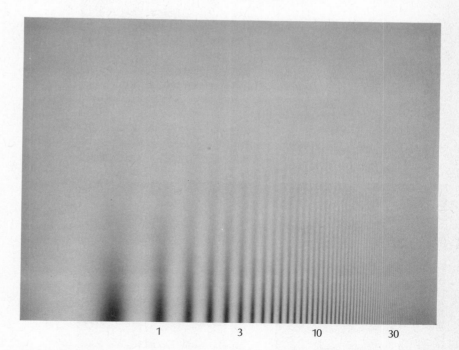

| 1 | 3 | 10 | 30 |

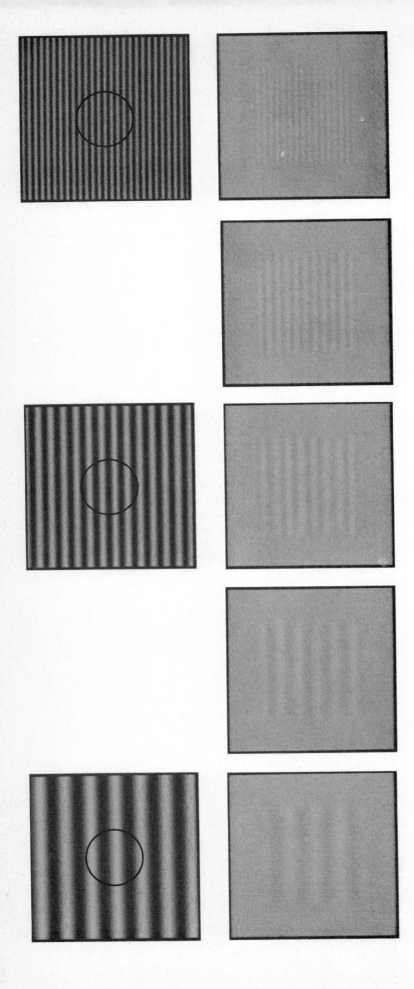

Suppose you try to desensitize the mechanism that responds to gratings. If you adapt to a grating having a certain spatial frequency, you *might* expect that your CSF would be lower, overall, when you look again at any pattern of bright and dark bars—this would happen if you had only one mechanism that responded to all spatial frequencies. When you do this experiment, however, sensitivity is reduced *only* for those spatial frequencies *near* that of the adaptation grating. It's like the negative afterimage (Sec. 7.5), where the desensitizing occurred only on the area of your retina exposed to bright light. Unlike that case, however, you move your eyes back and forth across the grating during adaptation to ensure that no standard negative afterimage is formed here. This type of adaptation thus involves both a level of your visual system *beyond* the photoreceptors and mechanisms that each respond to only a limited range of spatial frequencies.

You can use Figure 7.24 to desensitize such a mechanism of your visual system—a subsystem that responds primarily to low-spatial-frequency gratings, say, but not to high-spatial-frequency gratings. Low-spatial-frequency gratings then require more contrast than normal in order to be seen. Your sensitivity to higher-spatial-frequency gratings, however, is not affected. After the effects of the last experiment have subsided (about ten minutes), adapt to a different grating on the left of the figure and notice that your sensitivity to only *that* spatial frequency is affected. In short, the CSF after adaptation to a grating of a single spatial frequency is affected only *near that spatial frequency.*

FIGURE 7.24

Spatial-frequency adaptation. The five test gratings along the right each have a low contrast, so they are barely visible. Adapt to a high-contrast grating at the left by allowing your eye to move around inside one of the circles. After two minutes, look at each of the test gratings. You'll find that the test grating having the same frequency as the one you adapted to will be invisible, while the others remain visible.

FIGURE 7.25

Normal CSF (solid line) and CSF after prolonged adaptation to a grating having the spatial frequency marked (dotted line). The spatial frequencies marked L, M, and H correspond to those of the left gratings in Figure 7.24 viewed from 1¾m.

As Figure 7.25 shows, it has a dip in a region centered on the adaptation spatial frequency but is unchanged at other spatial frequencies.

B. Channels

We can describe the above effects by saying that adaptation to different spatial-frequency gratings desensitizes different *channels*. As mentioned, a channel is a subsystem of a visual system that responds preferentially to one type of stimulus rather than another. It is an abstraction by your visual system of some attribute of the stimulus—in this case, spatial frequency or, roughly, size. Thus, there are channels identified with low, with intermediate, and with high spatial frequencies. These channels naturally result from the center-surround receptive fields of Section 7.4C. If the excitatory center just "fits" into the retinal image of a bright bar, the receptive field will be most responsive to gratings of that size and less responsive to others (Fig. 7.12). Large center-surround receptive fields will respond most to low spatial frequencies, small ones most to high spatial frequencies. Receptive fields do, in fact, come in the necessary variety of sizes. (We'll see shortly, though, that there is more processing necessary to account for *all* the

properties of spatial-frequency channels.)

Adapting a particular channel can also affect your **percept,** that is, your subjective sense of *what* you see when a grating (or other stimulus) is clearly visible. For the channels we've been describing, the percept corresponds to bars of a certain size, separated by a certain amount. This percept is determined by a comparison of the responses in different channels. Normally, when you look at a grating of intermediate spatial frequency, your intermediate channels respond a great

FIGURE 7.26

Successive size contrast. For two minutes, direct your eye toward the short horizontal bar on the left, thereby adapting the top of your field of view to a low spatial frequency, and the bottom of your field of view to a high spatial frequency. Keep your eye moving along the bar to avoid standard afterimages. Then, look at the dot between the central gratings and notice that the grating on the top right appears higher in spatial frequency (bars seem narrower) than the one below it. (The analogous experiment involving orientation can be done using the gratings on the right.)

deal, while your higher and lower channels do not respond much. Your percept corresponds to an intermediate spatial frequency because most of your response occurs in your intermediate-spatial-frequency channels.

However, if first you adapt to a high-spatial-frequency grating, your high channels become desensitized. This creates an imbalance if you now look at a clearly visible *intermediate*-spatial-frequency grating; your low channels have not been desensitized and thus respond as usual, so there is *more* activity in your *low channels* compared to that in your high channels. The channel of peak response then corresponds to a *lower* spatial frequency than before adaptation. Thus, the intermediate-spatial-frequency grating will *appear lower* in spatial frequency than the grating objectively is; the bars will appear spread apart—your percept of that grating has changed. You can verify this effect using the four vertical gratings on the left part of Figure 7.26.

Although more subtle than the **size aftereffect** (or *successive* size contrast) just described, **simultaneous size contrast** results when gratings of one spatial frequency are surrounded by gratings of a different spatial frequency, as in Figure 7.27. This is loosely analogous to the simultaneous lightness contrast we saw in Figure 7.6, where light in the surround of a receptive field tended to decrease the activity

FIGURE 7.27

Simultaneous size contrast. The two central gratings have equal spatial frequency, but appear unequal because of their surrounding gratings.

in its center. Here, a low-spatial-frequency surround decreases the activity of the low channels responding at the center, making the center appear higher in spatial frequency. As with lightness, the visual system also emphasizes *differences* in spatial frequency or size.

But no size contrast occurs in Figure 7.28, where the surround grating has a very different orientation than the center. This shows that there is another attribute (besides spatial frequency) that is associated with these channels: **orientation.**

The center-surround receptive fields respond equally well to a grating of *any* orientation (of a given spatial frequency), so differences in sensitivity based on *orientation* cannot be due to the ganglion cells alone. Instead, these channels correspond more closely to subsequent nerve cells, each connected (perhaps indirectly) to *many* ganglion cells whose receptive fields lie roughly along a line (Fig. 7.29). The cell that adds, or **pools,** the information from the ganglion cells would then respond differently for gratings of different orientation. The stimulus that excites this pooling cell most is one that excites

FIGURE 7.28

Lack of size contrast. The two central gratings, having equal spatial frequency, now appear equal in spatial frequency because the orientation of the surrounding gratings differs from that of the central gratings.

each of its pooled ganglion cells a great deal. But such a stimulus has light striking the centers of each of the ganglion cells and an absence of light on the surrounds—that is, a bar or grating oriented along the line of ganglion cells. A grating of a different orientation would not excite this pooling cell much at all.

Because the channels have orientation preference, the spatial-frequency (size) illusions we saw before have counterparts in orientation or tilt, as you can easily check.

PONDER

Use Figures 7.30 and 7.31 and the four gratings on the right part of Figure 7.26 to demonstrate orientation illusions analogous to those of spatial frequency (size) in Figures 7.24, 7.26, and 7.27. In the orientation illusions, what attribute of the stimulus is analogous to spatial frequency in the size illusions? What is analogous to the percept of size or bar spacing?

Unlike standard negative afterimages, the size and tilt adaptation and aftereffects do *transfer* from one eye to the other. Using a longer adaptation time, repeat any of the above experiments, adapting with the left eye (right eye closed) and testing with the right (left eye closed). The effects are reduced somewhat from the earlier versions

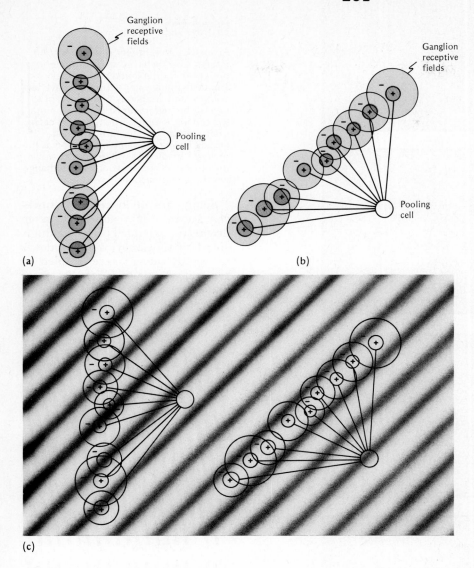

(a)

(b)

(c)

FIGURE 7.29

A possible "wiring" of ganglion cells to a subsequent nerve cell for **(a)** a vertical tilt channel, and **(b)** a 45° tilt channel. **(c)** A 45° grating of the right spatial frequency will excite the 45° channel but not the vertical channel.

but should be clearly visible. Transfer can only result if the pooling cells of a particular channel occur at a point *after* information from your two eyes comes together, such as in your visual cortex. Indeed, pooling cells with the appropriate size and tilt response have been found in the visual cortex, at least in monkeys.

*7.9
OTHER CHANNELS

Your visual system has many other channels beside those for spatial frequency and tilt. Let's briefly consider a few more.

If you look at a nearby hill after watching a waterfall for a while, the hill appears to drift upward slowly—without getting anywhere. This is an example of the **waterfall illusion** or **motion aftereffect.** Similarly, if you've been looking out of your window on a moving train that then stops in the station, the platform appears to drift forward slowly.

There seem to be channels sensi-

tive to **motion** in various directions, just as there are channels sensitive to gratings of various orientations. Normally, when you view a hill, there is a balance between responses in all of your motion channels—your upward and downward motion channels have the same activity in them. Your percept, then, is of a stationary hill. However, after adaptation to the waterfall, your channel that responds to downward motion is desensitized. When you then view the hill, there is an imbalance in response—more response in the upward motion channel than the downward motion channel, so you see the hill as drifting slowly upward.

PONDER

Which motion channels are relevant to the train example? Which are adapted, and how can that explain the aftereffect?

Likewise, when jogging, or driving a car, the world appears to rush toward you. When you stop, the world seems to recede slowly. This is another motion aftereffect, only here the channels are not simple motion channels, but **looming channels**—channels most stimulated by simultaneous motion of parts of the scene away from your point of fixation. (Movie makers reverse this; they excite your looming channels by making the image on the screen expand away from a central point, and this gives you the illusion that you are moving toward that point. This is part of the thrill associated, for example, with Han Solo's daring escapes into hyperspace in "Star Wars"—you can almost feel yourself thrown back against your theater seat.)

PONDER

An engineer looking out the front of a moving train has her looming channels stimulated, while a passenger looking out the side has his motion channels stimulated. What is the motion of the image on each of their retinas? What are the aftereffects seen by the engineer and the passenger when the train stops?

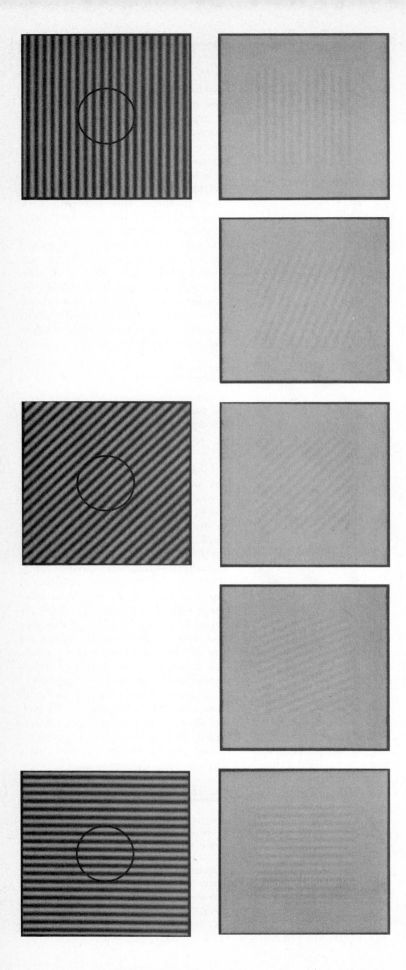

FIGURE 7.30

Adaptation to different orientations. All figures have the same spatial frequency, but differ in orientation and contrast. The test gratings along the right have low contrast and are barely visible. Adapt to one of the high-contrast gratings on the left for two minutes. Then, look at each of the test gratings in turn. The test grating having the same orientation as the one you adapted to will disappear, while the others will not. Tip your head to change which grating becomes invisible.

Adaptation of looming channels can cause traffic accidents; after driving for extended periods at high speed on a highway, you slow down to exit. Because you've adapted to rapid looming motion, the intermediate speed seems slower (and safer) than it really is (compare the effect in Fig. 7.26).

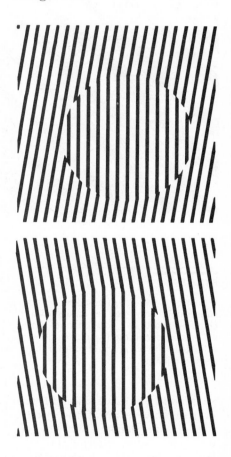

FIGURE 7.31

Orientation contrast. The apparent orientation of the center (vertical) gratings is affected by the presence of their surrounding gratings.

Using the TRY IT you can experience an aftereffect associated with spiral motion; in Chapter 10 we'll see channels associated with color, and there are yet other channels of your visual system.

FOR SECTION 7.9
Spiral aftereffect

Photocopy (or trace or redraw) the spiral in Figure 7.32. Place it on a phonograph turntable, and set it rotating at 45 rpm (33⅓ rpm will work, but not as well). Stare at the center for 2 to 3 minutes. You are adapting a clockwise rotation channel, as well as a looming channel. Now stop the turntable to see the aftereffect—apparent counterclockwise rotation. If you look at your friend's face immediately after adaptation, you'll be in for a shock.

FIGURE 7.32

Spiral for generating aftereffect.

*7.10
MACHINE VISION

Another approach toward understanding vision is to *make* a machine that "sees"—the more we can duplicate features of the visual system, the more insight we gain about mechanisms of vision. The simplest such machine consists of a *source of light* and a *photocell* (a device to convert light into electrical signals—Sec. 15.2B). The photocell can tell whether light strikes it or not and can be wired, for example, to ring a bell whenever it doesn't "see" any light. Such an "electric eye" can be used in a burglar alarm, or in a device to ring a door bell when a customer enters a store or to count items on a conveyor belt. Although the processing here is trivial compared to that in humans, it shows that a vision system can convey information without conveying an optical image (here, by ringing the bell) and need process only some attributes or features of the scene (here, just the overall light striking the photocell).

A. Template matching

A more elaborate system compares the light at each point of an optical image to that at the corresponding point of a standard, stored image. For instance, microelectronic chips (consisting of many tiny electrical circuits) must be carefully inspected to see that all the circuits are properly connected without breaks. For years, bored humans peered through microscopes to inspect the chips. Now, a TV camera peers through the microscope and sends a signal to a computer that compares the image with one stored electronically in memory. If the two match point for point, the chip is OK; if there is a break in any of the connections, the images do not match and the chip is rejected. The machines that read the bank-account numbers on your checks operate on the same **template matching** principle, except there are more stored templates, one for each numeral and special symbol. Each character is compared to each of the stored templates and the machine finds which template matches best, thereby "reading" that character. The choice of typeface used on checks is a compromise between one that is easily read by humans and one in which each symbol is most easily discriminated by such a machine.

Another use of template matching is to read the universal product code (UPC) symbols, which consist of closely spaced white and black bands and are found on many store items. The UPC symbol is read by machine so that pricing and inventory are faster and more free from errors. Each digit (often written in human-readable form below the bars) has a machine-readable form consisting of two dark bars. The width and spacing of the bars is unique for each digit. The checkout person sweeps a light wand (a small light source) across the symbol. Light reflected from the symbol is converted by a photocell on the light

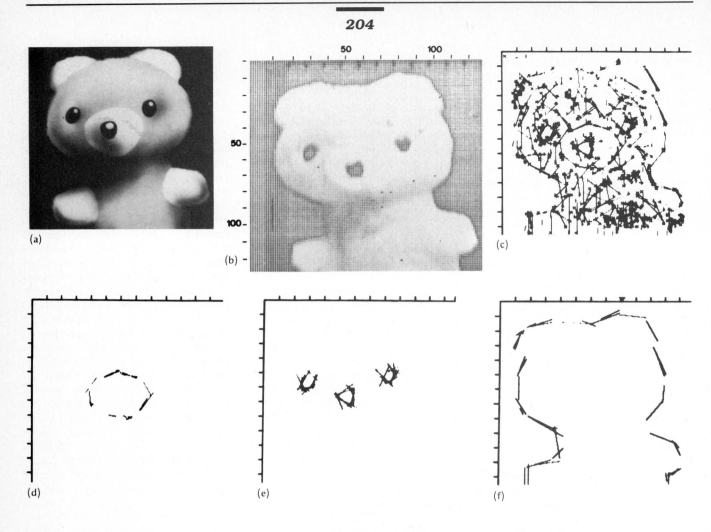

FIGURE 7.33

Processing stages in one computer vision system. **(a)** Original scene. **(b)** Pictorial representation of the gray-level description of the scene, in which individual discrete points in the computer are assigned the average brightness of a corresponding small region of the scene. In the picture the dot size is determined by the light intensity assigned to that point. **(c)** Schematic representation of the output of various channels. **(d)**, **(e)**, and **(f)** Groupings of features into structures.

wand into an electrical signal, which is then sent to a central computer. As the light scans the symbol, the *spatial* variation in the symbol pattern is converted into a *time* variation in the photocell signal. Template matching is used— but here, *in time*—to compare the measured sequence to those stored in the machine. A modification of the light wand, found in supermarkets, consists of a glass window over which the checkout person draws your cereal box and its UPC symbol. Here the light source is a laser swept across the window from below, while a photocell beneath the window detects the reflected light.

B. Channel approaches

In complicated applications, such as factory assembly lines or robots, the objects to be identified do not have a simple, expected form suitable for template matching. Also, the objects may be at different distances and orientations and the lighting may not be uniform. Under such circumstances, more sophisticated techniques are used, including some based on human visual processing schemes.

Figure 7.33 shows the stages in one such program. The image of the original scene (a teddy bear in Fig. 7.33a) is first converted to its **gray level description.** This consists of dividing the image into little picture elements **(pixels).** For each pixel, a number representing the brightness of the scene at the corresponding point is stored in the computer. Figure 7.33b is a pictorial representation of the gray level description. (The machine doesn't need such pictorial representations, only the stored numbers. There is no little person in the machine to look at

the picture, any more than there is one in your brain.) Then, for each point in the scene, the computer calculates the response in different "channels" that have both spatial frequency and orientation properties. A pictorial representation of the output of these channels is shown in Figure 7.33c. For instance, at the upper right of the figure, the output is large for channels oriented at about 15° to the left of vertical. This is signified in the representation by a line at that orientation. (Actually, the computer program calculates and stores more information, such as the sharpness of an edge, the direction across an edge from light to dark, the relative brightness, and so on.)

The next step is to "connect the lines" to identify separate structures (eyes, nose, . . .) within the bear. This is somewhat like connecting the dots in a child's puzzle, only here no one has numbered the lines giving the order they are to be linked. Although this problem may appear simple to us (after all, we already know the structure represents a bear), much work in computer vision is centered on different mathematical techniques for deciding how such features should be grouped. Figures 7.33d, e, and f show the groupings found by one computer program.

Further stages in processing, in which the computer interprets the structures and puts them together to recognize the bear, are so far only partially successful. Some day perhaps you may make a machine so sophisticated that it can look at an original Cézanne and tell it from a good forgery.

SUMMARY

The optical image on the retina is **processed** to form a symbolic, neural representation of the scene. In order that such processing be most sensitive to *changes* (both across *space*, and in *time*) in the light intensity of the retinal image, *ratios* of light intensity are fundamental to the visual system—as exemplified by **Weber's law.** Without changes, such as those produced in space by **edges** and in time by **eye movements (drifts, tremors,** and **saccades),** vision cannot occur, as is demonstrated with **retinal stabilization.** To process spatial changes, the retina employs **lateral inhibition** (in the form of **center-surround receptive fields),** which accounts for **lightness constancy, simultaneous lightness contrast,** and **edge enhancement.**

Because of **persistence** in the visual system, there are **positive afterimages.** Rapidly presented images that vary slightly merge into the appearance of a single *moving* image. This fact is basic to movies and TV. Likewise, because of **persistence of vision,** a **stroboscope** can appear to **stop motion.** It does this if its periodic flash occurs whenever a moving object (or one just like it) arrives at a given position.

Information from ganglion cells is **pooled** by subsequent nerve cells, thereby forming **channels** that respond preferentially to a set of lines **(grating)** of a particular **spatial frequency** and **orientation.** These channels can be desensitized (analogous to standard **negative afterimage** and **successive lightness contrast)** and are responsible for **size** and **orientation adaptation** and **illusion** effects. Such effects **transfer** (to the other eye), indicating that the pooling cells are in the **visual cortex** rather than in the retina. Other channels, such as those associated with **motion** and **looming,** explain related aftereffects and illusions. Artificial vision systems abstract certain appropriate features from the scene and use **template matching** or other, more advanced schemes such as those of channels to identify some useful attribute of the image.

PROBLEMS

P1 Why is it advantageous for the visual system to process information from the photoreceptors before it reaches the brain? How is this processing *un*like that used in photography?

P2 You're sitting under a shade tree on a bright, sunny day, reading this book. You see a black dog that is walking in the sunshine. The dog sends more light to your eyes than does the page of this book, yet the pages appear white, while the dog black, even when you view them at the same time. Explain.

P3 Even though sunspots are about 4000°K and emit lots of light, they appear as dark spots on the sun's surface. Why?

P4 What is Weber's law and why is it useful? How might you demonstrate its limitations?

P5 Weber's law applies to hearing. Use the law to explain why the difference in loudness between 50 and 51 musicians playing simultaneously does not seem as great as the difference between two and three musicians.

P6 A certain chandelier has three symmetrically placed 100-watt bulbs, each of which can be separately controlled. As you switch the bulbs on, the brightness of the light seems to increase quite a bit when you go from one bulb lit to two bulbs lit, but from two to three bulbs, the brightness does not seem to increase nearly as much. Explain why the (subjective) sensation does not increase in proportion to the wattage, or intensity of the light emitted by the chandelier.

P7 In some buildings, there are long, dimly lit corridors with windows to the outdoors only at each end. In such corridors, the lighting makes it difficult to identify someone approaching you. Why?

P8 One set of gray papers reflects light having intensities 1, 3, 5, 7, 9, ... Another set reflects light having intensities 1, 3, 9, 27, 81, ... Which set appears to have equal steps of lightness?

P9 A lit lighthouse light is very visible from many miles away during the night but not during the day. Why?

P10 What is lightness constancy? How can it be explained?

P11 A light object looks even lighter to you when it is surrounded by a dark background. This is due to: (a) accommodation, (b) binocular disparity, (c) polaroid sunglasses, (d) simultaneous lightness contrast, (e) dispersion. (Choose one.)

P12 Explain how lateral inhibition contributes to the Hermann grid illusion and the Craik-O'Brien illusion.

P13 How can lateral inhibition be used to explain edge enhancement?

P14 To "clean" a coffee stain from a rug, smart homemakers concentrate on the *edge* of the stain, smoothing and blending the stain gradually into the rug. Why might this help make the stain less visible than if the stain were cleaned uniformly, but perhaps not thoroughly, overall?

P15 (a) Give two examples to illustrate that the visual system's processing is preoccupied with *changes* within a stimulus. (b) State two advantages of this kind of processing.

P16 Give two reasons why eye movements are beneficial.

P17 (a) Approximately how rapidly must a flashing image repeat for it to appear continuous? (b) Describe how TV avoids flicker. (c) Describe how movies avoid flicker.

P18 You have no screen, but your slide projector projects a slide through a stationary wheel on an inverted bicycle. Although the slide is focused onto the spokes, it is impossible for you to tell what the image is because you can't see enough of the image. What is the simplest way to make the entire image visible? What accounts for this effect?

P19 A device for checking the speed of a turntable relies on a stroboscope effect. Explain how it works.

P20 You hold a comb at arm's length oriented such that you can see all its teeth. Consider the teeth as a crude grating. When you bring the comb closer to your eye, the apparent spatial frequency of this grating: (a) increases, (b) decreases, (c) remains constant. (Choose one.)

P21 (a) What does it mean to say that a particular aftereffect *transfers*? (b) Give an example of an afterimage that transfers. (c) Give an example of one that does not transfer. (d) How does the transfer clue help to determine which stage in the visual system is responsible for a particular effect?

P22 (a) Superman flies above a field, looking down on it. Are his motion channels or his looming channels most excited? (b) While viewing the wall of a building, he flies directly toward it. Now which channels are most excited?

P23 State the fundamental principle of vision to which Thomas Middleton refers in the following words from his seventeenth-century play "Women Beware Women": "As good be blind and have no use of sight,/As look on one thing still. What's the eyes' treasure,/but change of objects?"

HARDER PROBLEMS

PH1 Consider a single receptive field. (a) Why won't a grating of very low spatial frequency excite this receptive field much? (b) Why won't a grating of very high spatial frequency excite it?

PH2 The figure represents light patterns falling on a receptive field of the type shown in Figure 7.11. As in Figure 7.12, there is some background response in the ganglion cell when overall light of intermediate intensity strikes the entire receptive field. For each pattern, state whether the response in the ganglion cell is greater than, less than, or nearly equal the background response.

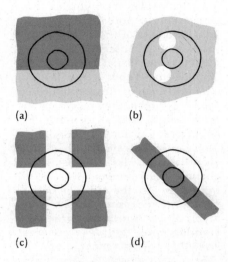

(a) (b)

(c) (d)

PH3 Repeat PH2 for the light patterns shown in the figure here.

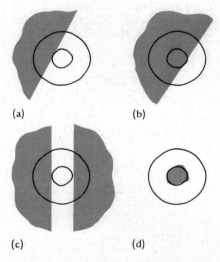

(a) (b)

(c) (d)

PH4 Why does the CSF fall off at very high spatial frequencies? Give a few reasons that might limit the fineness of very small receptive fields.

PH5 A stream of water produces drops that strike a pool of water every $\frac{1}{100}$ second. (a) If this scene is illuminated solely by a strobe light flashing at 100 Hz, what will you see? (b) If the strobe is set to 99 Hz, what will you see? (c) Repeat for 101 Hz.

PH6 A movie projector can be used as a strobe light. Unlike a normal strobe, however, the projector's "flashes" are fairly long, whereas the dark interruptions of the light are brief. (a) A bicycle wheel illuminated by this projector rotates in front of a black screen. When the wheel rotates at a certain rate, you see what appear to be stationary *dark* spokes separated by light regions. Explain. (b) The black screen is now replaced by a white screen, on which the wheel casts a shadow. What does the shadow of the rotating wheel look like?

PH7 Suppose you look for two or three minutes at a grating that is tipped to the left by 10° from vertical. Then, you look at a high contrast (i.e., easily visible), vertical grating of the same spatial frequency. (a) Does this grating appear to be vertical, tipped to the right, or tipped to the left? (b) Explain this effect using the channels described in Section 7.8B.

PH8 (a) In the manner of Figure 7.29, draw a series of receptive fields and a pooling cell connected such that the pooling cell responds well to a low-spatial-frequency horizontal grating. (b) Repeat as in (a) for a pooling cell that responds well to a high-spatial-frequency grating tipped 60° to the right of vertical. (c) Will the pooling cell of part (a) respond well to the grating that excites the pooling cell of (b)?

PH9 A few cats have been raised in "strobe" environments, in which the only illumination comes from a strobe light flashing once per second or so. It was later found that these cats lacked visual cells that respond to motion. Why might you expect such a result?

PH10 Suppose you had to make a machine that could recognize your mother's face. Would a template matching scheme work well? Why or why not?

MATHEMATICAL PROBLEMS

PM1 The light reflected from region A has *intensity* 2 (in some units), while that from region B has intensity 4. The light reflected from region C has intensity 3. (a) What should be the *intensity* of light reflected from region D to ensure that the difference in *lightness* between C and D is the same as that between A and B? (b) Repeat as in part (a), only now assume that the light reflected from A is 1, while that from B and C is unchanged.

PM2 Consider three regions, A, B, and C. The light reflected from A has intensity 1 (in some units), while that from region C has intensity 16. What should be the intensity of the light reflected from B so that the lightness of B appears midway between that of A and C?

Binocular Vision and the Perception of Depth

CHAPTER 8

8.1
INTRODUCTION

Up to now we've been concerned with monocular* optics; the cameras we considered had one (possibly compound) lens, the properties of vision we described required only one eye, and the optical instruments we discussed were mostly designed for the use of one eye. We now want to acknowledge that we, like many other animals, have two eyes and ask what might be the advantage of this **binocular**† arrangement.

One advantage of the second eye is that it provides us with an increased *field of view*. Close one eye and you immediately notice that part of the scene previously visible is no longer in your field of view. Many animals, such as fish or rabbits, have eyes set in opposite sides of their heads, each eye providing a separate view of the world. Together, two such eyes, with practically no overlap in their fields of view, enable the animal to see a sweeping panorama. (The field of view of a rabbit is 360°, allowing it to see all around without turning its head, with only 24° overlap between the two eyes.)

However, your two eyes, placed in the front of your head, have fields of view that overlap considerably. (Your field of view is 208°, with 130° overlap.) Your eyes thus provide slightly different views of almost the same scene. Close one eye and hold

up your thumb at arm's length. Now open that eye and close the other. Notice that your thumb appears to move against the background as you alternate eyes—each eye sees a slightly different view.

Nevertheless, normally you see only one view—somehow your brain combines the two images into one view of the world. In Figure 7.3, we showed the neural connections that allow this mixing of the signals from your two eyes. Notice that a given side of the brain gets signals from the corresponding points of each eye, allowing it to combine them into one view of the world. (There are occasions, however, when your brain isn't able to provide a single, smoothly combined, view, and you become aware of the two images produced by your two eyes. Hold up a finger in your field of view, fairly close to your eyes, and

focus on a more distant object. You will notice two somewhat transparent fingers in front of the distant object. Alternately, focus on the finger, and you can see two images of the distant object. The TRY IT offers another example.)

Why bother with two images of the same scene? These slightly different images enable you to gauge the **depth** of a *three-dimensional* scene. Among animals, predators (such as cats) have their two eyes in front, with overlapping fields of view, to enable them to judge accurately the distance to their prey. On the other hand, animals (such as

---

FIGURE 8.1

Edward Hicks' "The Peaceable Kingdom" shows both predators and prey. They are easily distinguishable by the location of their eyes.

---

*Greek *monos*, single, plus Latin *oculus*, eye.

†Latin *bini*, a pair.

rabbits or deer) who are likely to be someone else's dinner have non-overlapping fields of view to give them the wide angle of view best for detecting predators (Fig. 8.1). Similarly, animals that leap about the branches of a tree (such as squirrels and our simian ancestors) must be able to gauge the depth of those branches, and correspondingly have two eyes in the front.

The brain, as we'll see, uses many **cues** in its determination of depth. Some require signals from two eyes, but others do not. These latter include the cues that artists rely on when they convey a feeling of depth in two-dimensional pictures. It is also possible to play off one cue against the other—to create scenes that provide conflicting visual depth cues. By examining the resulting illusions, we can learn about the way the brain processes the various cues in arriving at its depth determinations.

Let's examine a number of the techniques by which we visually fathom the depths of the world around us. We'll begin by separating the depth cues that can be used in a painting from those normally unavailable to the artist. Imagine that while you sleep an artist or photographer has made an extremely realistic picture of the view from your bedroom window and pasted it to the outside of your window so when you awake you see the picture (Fig. 8.2). How can you tell whether you are looking at such a painting or at the actual scene outside your window? In the next few sections, we'll discuss several techniques for distinguishing the two alternatives.

FIGURE 8.2

René Magritte, "The Human Condition I." How can we visually distinguish between the artist's rendering of the outside world and the world itself?

FOR SECTION 8.1
Two eyes provide two views

Hold one end of a string against your upper lip and pull the other end straight in front of you. You will see not one but two strings stretching out in front of you and crossing. These correspond to the images from your two eyes, as you can readily confirm by alternately closing each eye. The point at which they cross is the point on the string at which you "aim" your eyes (the point toward which your eyes converge). Try looking at different points of the string, beginning close to you and moving away, and notice how the cross-over point moves away from you as you do this. (Having a friend slide a finger along the string may enhance this effect.)

8.2
ACCOMMODATION

Just as you can measure the distance to an object by focusing your camera on it and noting the lens' position, the amount of *accommodation* necessary to focus your eye on an object tells you the object's distance. If you see an object clearly while your eyes are relaxed, you know that it is far from you. If, however, you must tense your ciliary muscles to make the object come into focus, then the object must be closer. Thus when looking out your window at the actual street scene, you would accommodate differently for objects at different distances in the scene. The artist, attempting to simulate that scene with a picture, must decide what is in focus and what is not. Once the artist makes that decision, no amount of accommodation on your part will sharpen the focus of an object painted out of focus.

If you were a chameleon, you'd rely very heavily on the technique of accommodation to gauge the distance to flying insects as you flicked them with your tongue. If you cover one eye of a chameleon, he maintains his high degree of accuracy in fly-flicking—binocular vision is not important here. However, give a chameleon a pair of glasses that change the amount of accommodation necessary and his tongue flaps futilely at the fleeing fly.

Humans, on the other hand, make little use of this technique, possibly because our potential meals are generally more than a tongue's distance away. Accommodation as a way of determining depth is at best only useful for close objects. If our mischievous artist confines herself to a distant scene and if you cannot get too close to the window, then both the painting and the actual scene would be in focus to your relaxed eye and you wouldn't be able to distinguish the one from the other by accommodation.

8.3
CONVERGENCE

If you look at a near object, the angle between your two eyes' direc- tions of gaze is bigger than if you look at a distant object (Fig. 8.3). This angle is called the angle of **convergence** of your eyes. (For an object at the normal near point, 25 cm, the angle of convergence takes its maximum value of about 15°. It is only 1° for an object about 4 m from your eyes.) If your brain keeps track of the convergence of your eyes, it can determine the distance to the object that your eyes are viewing by using the rangefinder principle (Sec. 4.2D). As you scan a painted picture, your convergence, like your accommodation, remains unchanged, because all the objects lie in the plane of the picture. However, when you view objects at different depths in an actual scene, your convergence changes.

Like accommodation, convergence is most effective for determining depth in nearby scenes, but you make relatively little use of it. That you make *some* use of convergence, however, can be seen from the TRY IT.

TRY IT

FOR SECTION 8.3
Convergence and depth

Look at a distant street lamp. Cross your eyes (say by looking at your finger, which you hold in front of and just below the light). Notice that the light appears closer (and smaller) than before—the more your eyes converge, the closer. This is true even though you see two images, but it may be easier if you position yourself so that only one eye can see the lamp.

8.4
PARALLAX

To gauge depth, you rely much more heavily on the fact that your view is different from different positions—the phenomenon of **parallax.*** With one eye closed, hold your thumb a few centimeters from this page, so it blocks your view of the word "parallax." By moving your head, you can change your view sufficiently so that you can see the word. This is possible only because your thumb and the word are at different distances from your eye. So, your view of different objects changes as you move, according to their distance from you (Fig. 8.4).

No matter how carefully our artist simulates other depth cues, as long as she confines herself to a flat canvas, she cannot overcome parallax. You need only move your head and compare the relative positions of the distant scene and the window glazing bars to determine if she has tried to trick you. Similarly, the painted finger of Lord Kitchener (Fig. 8.5) always points at you, no matter where you move with respect to the picture, but an actual finger points in one direction—when you move out of that direction, it no longer points at you. Because your view of Lord Kitchener's finger doesn't change, your brain may interpret the image as if he were rotating as you walk by, so as always to point at you. This provides the recruiting poster with a rather personal touch. (The FOCUS ON X-Ray

---

*Greek *parallaxis*, change.

FIGURE 8.3

The angle of convergence, ϕ, is **(a)** large for near objects, and **(b)** small for distant objects.

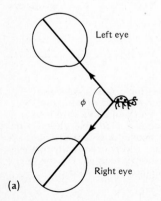

(a)

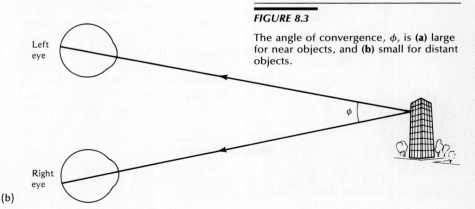

(b)

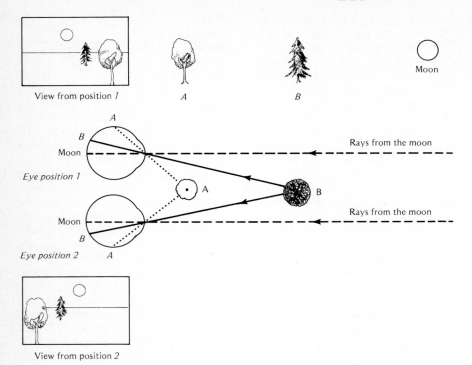

View from position *1*

A

B

Moon

View from position *2*

FIGURE 8.4

The locations of the images on the retina change as the eye moves from position *1* to position *2*. The location of the retinal image of the nearest object, *A*, changes the most—from one extreme to the other. The image of a more distant object, *B*, moves less, while the parallel rays of the very distant moon are always imaged on the same place on the retina. This is why the moon seems to follow you as you look at it, say, from the window of a moving car. The closer the object is to you, the more it appears to move in the direction opposite to your motion.

FIGURE 8.5

Alfred Leete's 1914 recruiting poster of Lord Kitchener, a British hero of the Boer War.

Tomography gives an application of parallax.)

8.5
BINOCULAR DISPARITY

Those of us blessed with two eyes need not move in order to gain the benefits of parallax for gauging depth. The two eyes, separated by about $6\frac{1}{2}$ cm and with significantly overlapping fields of view, see slightly different views of any object they look at. This difference between the views of the two eyes **(binocular disparity)** thus provides a way of determining the distance of the object in sight. Your brain attempts to reconcile the two views by ascribing the difference to depth. You "see" one three-dimensional view of the world, rather than two two-dimensional views. If the images are too diverse, your brain cannot **fuse** them, and you get double vision.

This happens sometimes when you're drunk or injured and your eyes don't properly converge to a single object. (As Shakespeare put it, "Methinks I see these things with parted eye, when everything seems double.")

Consider the two eyes viewing the cube in Figure 8.6. The views seen by the two eyes, shown in the figure, are slightly different. A common feature, say the front edge *aA*, is imaged at corresponding points of the two retinas. Another feature, say the left edge *dD*, is imaged at locations on the two retinas that do *not* correspond to each other. Some cells in the visual cortex of the brain respond strongest when a common feature occurs at corresponding points of the retinas. Other cells have their strongest responses at particular *differences* between the locations of the two

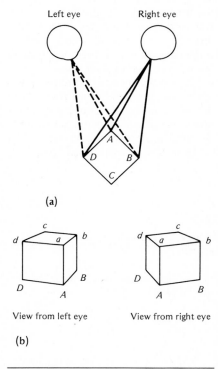

(a)

View from left eye   View from right eye

(b)

FIGURE 8.6

The two eyes looking at a cube see slightly different views of the cube. **(a)** View of eyes and cube. **(b)** Views seen by each eye. Note that the edge *da* subtends a smaller angle for the *right* eye than does *ab*, and consequently that eye sees *da* as shorter than *ab*. For the left eye, the situation is reversed.

retinal stimulations. Thus, one set of such cells in the visual cortex responds strongest to aA, while a different set responds strongest to dD. This difference in response leads to the viewer's perception that these features lie at different depths. (See the TRY IT's.) Under some circumstances, two objects will be seen in depth if the separations of their images on the two retinas differ by as little as 1 μm—less than the diameter of a cone!

Of course, binocular disparity is useful for depth determination only if your two eyes see different images. A horizontal clothes line looks the same to both eyes, so when you want to know where it is, it helps to tilt your head in order to introduce some disparity. (Objects that repeat themselves, such as bars on a cage, can offer confusing binocular disparity. If your eyes fuse the wrong bars together, the cage appears at a different distance than it should.) Distant scenes also present essentially the same view to your two eyes, so, like the other techniques discussed so far, binocular disparity is of little use for such scenes. But for relatively close objects it is an extremely effective way of gauging depth. In fact, by presenting two different views to your two eyes, it is possible to fool your brain into believing that there is depth even when there actually isn't. A variety of optical instruments, devices, and toys are based on this idea.

First TRY IT

FOR SECTION 8.5
Depth and chromatic aberration

The lenses of your eyes have some chromatic aberration—they bend blue light more than red. If you cover half your pupil, the resultant half lens acts like a prism, shifting the blue retinal image slightly away from the red image. By arranging that this shift is in opposite directions for the two eyes, you can produce binocular disparity—color differences can therefore be translated into differences in apparent depth.

In order to eliminate other depth cues as much as possible, you should look at blue and red patches separated and surrounded by black or white. Blue and red squares on one side of a Rubik's Cube can be used, as can a picture of an American flag on a white background. With pieces of cardboard or stiff paper, cover the outer half of each pupil while you look at the colored patches. Notice which color appears closer, then suddenly remove the cardboard pieces and notice the change. Now cut a piece of cardboard so it fits between your pupils, covering the inner half of each. (If you cut one of the vertical sides of the cardboard at an angle, you can get a good fit without precision measurements by raising or lowering the cardboard until it just doesn't block your view.) Look at the patches again, and notice that the depth has reversed from the previous case.

PONDER

Why did you need to use the cards?

Considering your half lenses as prisms, draw ray diagrams to convince yourself that the dispersion of the "prisms" is responsible for the effects you saw.

Second TRY IT

FOR SECTION 8.5
Increase your binocular disparity

The amount of depth perceived depends on the binocular disparity between your two views. The wider the separation of the points of view, the greater the apparent depth. Thus, stereoscopic aerial photographs are often taken from the two wing tips of the airplane. If you find the world too shallow, you can use the periscope described in the TRY IT for Section 2.4C to increase your binocular disparity. Hold the periscope horizontally and look through it with one eye while also looking with the other, unaided eye at the same object. A short periscope works best; about 6 cm will double your eye separation. The difference in apparent depth is most obvious if, after looking through it, you quickly remove the periscope while continuing to look at the scene.

Just as your brain interprets nerve signals from both of your eyes to produce the sensation of depth under normal viewing conditions, it can interpret signals accompanying afterimages with similar results. The persistence of positive afterimages also permits you to achieve an enhanced depth. Use the technique described in the first TRY IT for Section 7.7A, but here, since each eye is separately exposed to the window scene, care must be taken that the eyes are pointed in the same direction. After covering both eyes with your hands for 30 seconds, expose your right eye. To be sure that it is aimed properly, momentarily squint and point your eye at a distant point, such as the top of a distant tree. Then open your right eye wide for three seconds, keeping it pointed at the treetop. Next close and cover your right eye. Immediately move your head a few centimeters to the left and similarly expose your left eye to the scene, pointing it at the same treetop. Expose this eye for only two seconds. Close and cover the left eye (the right should already be closed and covered) and watch the stereo afterimage develop. Compare the depth in the afterimage to the actual depth.

To appreciate fully this "superstereo," compare it to reduced stereo, achieved by moving your head to the right between the exposure of your eyes in the order described above. The reduced effective separation between the eyes gives a flatter view. Moving your head about 13 cm to the right may result in a pseudoscopic view (Secs. 8.5A and 8.6H).

A. The stereoscope and related optical instruments and toys

How can we present two different pictures to the two eyes, so that each eye sees only the image intended for it? A simple technique is that used in the nineteenth-century **stereoscope.*** Here two photographs taken from slightly different angles are placed side by side. You view them through special lens-prism combinations, one for each eye (Fig. 8.7). The lenses produce distant virtual images of the photographs, while the prisms cause these virtual images to appear at the same place. Thus, the lenses assure the proper accommodation

---

*Greek *stereos*, solid.

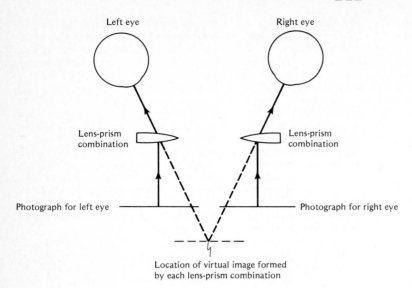

Left eye   Right eye

Lens-prism combination   Lens-prism combination

Photograph for left eye   Photograph for right eye

Location of virtual image formed by each lens-prism combination

(a)

(b)

FIGURE 8.7

(a) The principle of the stereoscope. The amount of depth perceived depends on the disparity between the two photographs. **(b)** A stereoscope from the 1880s viewing a stereo pair of photographs of Indians.

and the prisms the proper convergence for an object at one, average distance. But the binocular disparity resulting from differences in the photographs often produces quite a realistic impression of depth. (The modern "Viewmaster," used for scenic views or as a toy, is just an inexpensive version of Hermann von Helmholtz's 1866 stereoscope.)

With a little effort, you can obtain depth from stereoscopic pairs of pictures without paraphernalia. (That is, most of you can. About 2% of the population lack the stereoscopic vision to see depth from stereo pairs, even if they have two good eyes.) Figure 8.8 shows a stereoscopic pair. If you hold a piece of cardboard between your eyes so it blocks each eye's view of the other eye's photograph (being careful to avoid shadows), you may be able to fuse the two pictures even though the convergence and accommodation is wrong. It may take a little time and concentration before the images fuse. If you are able to do this, you can then use this talent to pick out small differences in otherwise identical photographs, since only the different parts will appear in depth. This depth effect has been used to notice changes in the position of stars. Two photographs of the night sky, taken at different times, are viewed stereoscopically. Those stars that have moved appear in depth. (A better way is to view the two photographs in rapid succession, repeatedly, and then look for the stars that appear to move—Sec. 7.7A.)

Stereo pairs of pictures may be obtained in a variety of ways. Cameras have been made with side-by-side lenses that produce two photographs simultaneously. You can make stereo pairs with an ordinary camera by photographing a still scene, then moving the camera to one side, and taking another picture. Stereo pairs are produced on the scanning electron microscope by taking a picture, tipping the microscopic sample slightly, and taking another picture. You can even make drawings, such as Figure 8.6b. (See the first TRY IT.)

What happens if you interchange the two pictures, say of Figure 8.6b, so the right eye views the left eye's picture and vice versa? The left eye will then see edge *da* as shorter than *ab*, while the right eye sees the reverse. This is just what these eyes would see if they viewed the three-dimensional object shown in Figure 8.9. That is, instead of the original front edge *aA* appearing closer to the viewer, it appears *farther* than the side edges *dD* and *bB*. Such a view, where the parts of the original object that came forward

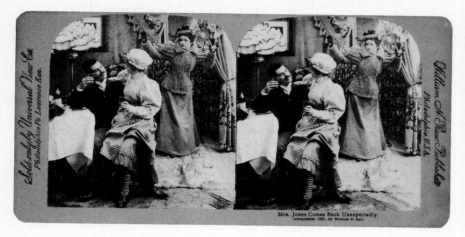

FIGURE 8.8

Stereoscopic photograph: "Mrs. Jones Comes Back Unexpectedly" (1897).

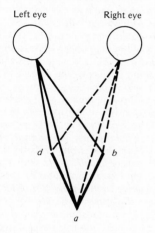

Left eye   Right eye

d   *b*

a

FIGURE 8.9

The two eyes looking at this object would have views corresponding to those shown in Figure 8.6b, but with the views interchanged. The edge *da* now subtends a smaller angle for the *left* eye, so that eye sees *da* as shorter than *ab*.

appear to go back and vice versa, is called **pseudoscopic.***

Various other optical devices that offer separate images to each eye

---

*The name (Greek *pseudes*, false) given by Sir Charles Wheatstone who, in 1838, first constructed a stereoscope. The first TRY IT shows how to reproduce his model.

present binocular disparity and hence a three-dimensional view to the observer. (The second TRY IT gives an unusual example.) The most common such instrument is binoculars—actually binocular telescopes, one for each eye. Binocular microscopes are more complicated. Because the image produced by a microscope is usually *inverted* (Fig. 6.9), simply providing each eye with its own microscope directed at the same object would produce a pseudoscopic view.

PONDER

Why does each eye viewing an inverted image result in a pseudoscopic view? A lens inverts the image (turns it upside down) by reversing both up-down and left-right. Draw these reversals separately for the two views of Figure 8.6b, first a left-right reversal and then an up-down reversal of that. A mirror may help in each case. Which reversal causes the stereo pair to result in a pseudoscopic image?

For this reason, some binocular microscopes do *not* offer a stereo view. They simply have a binocular eyepiece so that you can look at the *same* image with both eyes. Other binocular microscopes do offer a proper stereo view by inserting an *erecting system* in each of two microscopes. This may be a set of Porro prisms (Fig. 2.56c), as in binoculars, or some other system involving a reflection.

Another way of providing separate images to the two eyes from two-dimensional pictures does not require that the two pictures be separated. The pictures are superimposed, and the viewer is given special eyeglasses that allow each eye to see only the image meant for it. This may be done in two ways: using *color* or using *polarized light* (Sec. 1.3B and Chapter 13). Consider first the use of color. Here the two eyes' views are printed in different colors, say red and green, as in Plate 8.1. (Such a picture is called an **anaglyph.***) The viewer is given a different colored filter for each eye. If the filter is red, the green printing appears black, while the red printing is almost invisible. If the filter is green, the reverse is true. Thus, the eye with the red filter sees only the green picture, while the eye with the green filter sees only the red picture. Despite the difference in colors, the images can usually be fused into a three-dimensional view. (To produce three-dimensional shadows this way, see the third TRY IT.) This technique was used in early "3-D" movies, but because the color is used for the stereo effect the resultant three-dimensional view is not in full color. Full-colored 3-D movies can be achieved by using polarized light. The idea here is to project the two images with light of different polarization, say one with vertically polarized light and one with horizontally polarized light. The viewer looks through polarizing filters. The filter over one eye allows only the vertically polarized light to pass, while the filter over the other eye, oriented perpendicularly to the first, passes only the horizontally polarized light—but each "in living color." (Actually, the polarization is usually at 45°, one running between upper right and lower left, and the other between upper left and lower right.)

It is even possible to make stereo pairs of pictures where each member of the pair, by itself, doesn't

---

*Greek *ana*, upward, plus *gluphein*, carve. The term was first used for low relief carvings.

FIGURE 8.10

The eye and brain can pick out the dog in this seemingly random array of dots.

look like anything recognizable, but when the pair is properly viewed, a three-dimensional figure becomes apparent. This can be done with patterns of random dots. The eye and brain are quite good at finding order and sorting out familiar patterns when presented with a seemingly random array of dots (Fig. 8.10). (See the fourth TRY IT for other examples.) If we present to the two eyes arrays of dots that, viewed monocularly, have no pattern but do have a definite stereoscopic relation with one another, the arrays will exhibit a pattern of depth when viewed properly binocularly. We can make such arrays by the method shown in Figure 8.11a. To construct the left eye's view, we need only make a rectangle (R) cov-

ered with some pattern. To construct the right eye's view, we simply cut out a square (S) from the center of the left eye's view, slide it over, and fill the gap produced (Y) with more pattern. Any pattern will do, including a random array of dots, providing there *is* some pattern that the brain can correlate between the two views. (A solid color would not work because we are not going to draw any boundaries on S.) With random dots, each eye's view shows no sign of the square S when viewed monocularly. Nevertheless, by comparing the two apparently random patterns, the brain can notice the correlation (some of the dots are moved with respect to the others) and attribute this correlation to depth (Fig. 8.11b and Plate 8.2). These random dot stereograms, developed by Bela Julesz, have been extremely useful in sorting out which states of visual processing occur before stereoscopic fusion and which after.

FIGURE 8.11

Constructing a stereogram. **(a)** The views of the two eyes. The left eye sees X but not Y. Everything else in the rectangle R is the same for the two views. Properly viewed stereoscopically, the two eyes will fuse these views into a three-dimensional image even if R and S are covered with random arrays of dots. **(b)** The three-dimensional view is that of a square S floating above the rectangle R. For the left eye, S obscures the region Y of R. For the right eye, S obscures the region X of R.

First TRY IT

FOR SECTION 8.5A
Constructing and viewing stereo pictures

Stereo pictures may be made and viewed in a variety of ways. For viewing with a stereoscope, you need two views from slightly separated positions. To do this photographically, choose a scene that has objects at a variety of distances from the camera. Get the greatest possible depth of field by using the largest f-number on your camera. It is best to rest your camera on a flat surface, so you can carefully slide it to one side for the second picture. The distance between camera positions depends on the way the pictures are to be viewed and whether you want normal or exaggerated depth. If the object is distant (beyond 2 m) and the pictures are to be viewed through magnifying glasses (as in most stereoscopes) so the image you see is distant, then you'll want to slide the camera sideways about 6½ cm, the distance between your pupils. (The focal length of the magnifying glasses should equal that of the camera lens so you see the same perspective the camera did.) Try taking several pictures at various separations—the greater the separation the greater will be the apparent depth, but if the separation is too large, you won't be able to fuse the two pictures. Very distant shots, from airplanes or tall buildings, are usually taken with a separation between pictures of about $\frac{1}{50}$ the distance to the subject in order to increase the apparent depth. For close-up photographs, you should rotate the camera as you slide it, so it is always directed at exactly the same point on the object. For very close objects, you'll want to slide the camera less than 6½ cm. (For magnifications greater than one, a good rule of thumb is to divide 6½ cm by

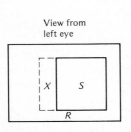

(a)

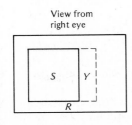

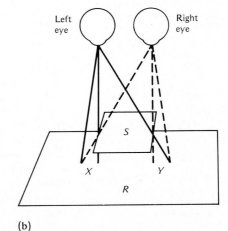

(b)

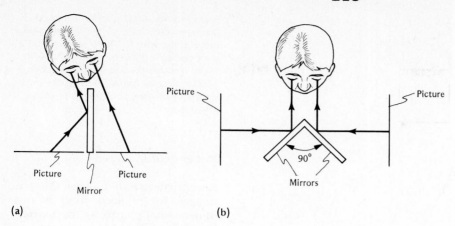

(a) (b)

FIGURE 8.12

(a) Single-mirror stereoscope.
(b) Wheatstone's stereoscope.

the magnification.) Be sure to record which picture was taken from the left side and which from the right.

The stereo pairs may be viewed without a stereoscope by the method described in Section 8.5A. Fusing the images may be easier if the views are crossed—that is, the photograph for the left eye is placed on the right. (Looking at a pin held in front of you may assist you in crossing your eyes.) In another technique, you print one of the stereo pair of photographs left-right reversed, and instead of using a piece of cardboard between your eyes, you use a mirror (Fig. 8.12a). Alternatively, you can use Wheatstone's 1838 design to construct your own stereoscope. You will need only two small plane mirrors and some way of supporting them (chewing gum, if all else fails). Position the two mirrors at an angle of 90°, with the reflecting surfaces facing outward (Fig. 8.12b). Looking into the mirrors from the vertex of the angle, the left eye will see a view to the left and the right eye a view to the right. Position the photographs parallel to and facing each other and far enough from your eyes so that you can focus on them (25 cm, including the reflection). If you have two identical converging lenses, you can hold them in front of your eyes and place the photographs at the focal plane of each lens. Since there are mirrors (which reverse the images), you should print the pictures left-right reversed or you'll get a pseudoscopic view.

You can also draw stereo pairs for your stereoscope. This may be done in the same way as Figure 8.6 was drawn. If you plan to view the pictures at 25 cm, you

should draw a diagram like Figure 8.6a, but full scale, with the object at 25 cm from the eye, and the eyes separated by 6½ cm. To obtain the proper horizontal distances for your version of Figure 8.6b, draw a horizontal line through the object in your version of Figure 8.6a and measure the distances between the locations where rays cross that line. For example, the horizontal distance between points c and d for the left eye would be the distance, along that line, between the rays that go from those points to the left eye. Alternatively, if you have a sufficiently simple real object and some artistic talent, you can directly draw the pictures as seen by each of your eyes.

Very simple figures can be drawn without artistic talent. For example, two circles, one inside the other, can be drawn with a compass. Make the outer circle about 6 cm across and the inner circle 2 cm across. For one eye's view, the center of the inner circle should be a few millimeters off center to the right, and for the other eye's view, it should be off center the same distance to the left. The resultant stereo pair can be viewed in all the usual ways. The single-mirror technique of Figure 8.12a is particularly simple here, since for this case you draw the inner circle off center in the same direction for both eyes. (Why?)

Instead of a stereoscope, you can use colored filters and make anaglyphs. The colored filters can be the cheap plastic kind. You can make a photographic anaglyph by taking a double exposure on color film, holding first one filter in front of the lens, and then the other, and moving the camera to the side between exposures. Alternatively, you can draw an anaglyph by one of the above methods, using colored pens, pencils, or crayons, and drawing the two pictures one on top of the other. The colors with which you draw should be matched to

the filters. You want colors that are easily seen through one filter but are invisible through the other.

You can also use polarized light (Chapter 13) to project stereo pictures. You'll need four seperate polarizing filters and two projectors, as well as a "metallic" projection screen that does not destroy the polarization. You can use polarizing sunglasses, but you'll either need four pairs, or you'll have to break them in half. (If you still have your glasses from a 3-D movie, you can get away with two pairs of them, only one of which you need break.) Take two stereo pictures, as above, to make color slides. Tape a polarizing filter over the lens of each projector and project the two slides, with their images the same size and overlapping as much as possible. View the screen with a polarizing filter over each eye. These should be oriented perpendicularly to each other (i.e., so you can't see through them when they are held one behind the other). Rotate the filters over the projectors until each eye can see only the image from the projector meant for that eye.

Second TRY IT

FOR SECTION 8.5A
The dark axle

You'll need a metal ring 6 cm across or larger (like those used in macrame). In a dim room, illuminate a table top at a low angle, say with a flashlight. Spin the ring as you might spin a coin—it will look like a transparent sphere spinning around a dark axle. The spherical appearance is due to persistence of your vision in the dim light; the dark axle comes about because the ring looks extra dark when its front hides its back. This occurs for each eye when the ring is lined up with that eye. Convince yourself that the resulting binocular disparity locates the axle in the center of the sphere.

Third TRY IT

FOR SECTION 8.5A
Three-dimensional shadows

For this you'll need two lamps that throw good sharp shadows, a red and a green light bulb, a white bed sheet, and a red and a green filter. (The filters should be colors that make the biggest difference when held in front of the two bulbs.)

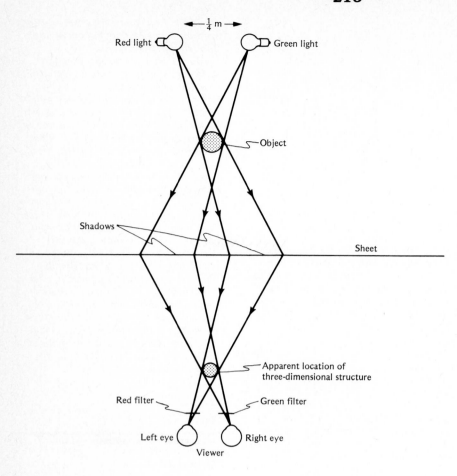

FIGURE 8.13

Set-up for three-dimensional shadows.

Arrange them as in Figure 8.13. With all other lights off, an object behind the screen will cast two shadows on the sheet, one due to each light. Viewed through the filters, the shadows will fuse into one three-dimensional shadow, apparently in front of the screen. A good object to use is a person. As he walks back from the screen toward the lamps, his shadow will appear to move forward from the screen toward the viewer.

Fourth TRY IT

FOR SECTION 8.5A
Order out of chaos

Excellent random arrays of dots can be easily found on your TV screen if you simply tune to a nonbroadcasting channel, so the screen is filled with "snow." This consists of randomly placed

bright spots which, because the TV scans through all 525 lines in $\frac{1}{30}$ sec (in North America), change completely 30 times a second. Therefore, the dots seem to jump about at random, as the eye associates motion to the disappearance of a dot at one point and the subsequent appearance of another dot at a nearby location. Thus, the snow seems to consist of randomly placed dots randomly jumping about—a truly chaotic scene. It is, however, quite easy to "train" these dots to behave in a more orderly fashion, thanks to the ability of your brain to organize the visual information presented to it. All you need do is give your brain the appropriate hints.

Form a narrow channel between two pieces of paper held against the TV screen, separated by a few centimeters. Notice that the dots now seem to flow along the channel. Now make a loop out of wire or string, about 5 to 6 cm across, and hold it against the screen. Notice how the dots jump around within and without the loop, but do not seem to cross the loop. This remains true even if

you move the loop across the screen—the dots seem lassoed by the loop and move along with it. There seems to be a void immediately behind the loop as it slides through the sea of jittering dots. You can form a furrow through the dots simply by moving your fingertip across the screen.

B. Lenticular screens

Another device that allows different images to be seen from a two-dimensional picture is the **lenticular screen.** Consider a sheet of cylindrical lenses (called **lenticules**) resting on a flat picture at the focal plane of the lenses (Fig. 8.14). (Such a sheet can easily and cheaply be made of plastic.) Light originating or scattered from any point in the picture plane is made parallel by the lenticules. The direction in which the light emerges from the lenticules depends on the location of the source in the picture plane. Light from a point slightly to the left of the axis of any of the lenses will emerge directed toward the right. Thus, light from all the points labeled R in the figure will emerge headed to the right, as shown. Similarly, light from the points labeled L will emerge directed to the left. This means that an observer located at O_R, looking slightly to the left, will see only the points R, while an observer at O_L, looking slightly to the right, will see only the points L.

If the picture plane consists of alternating strips of two different pictures, one at the points R and the other at the points L, then you see two different pictures, depending on the angle at which you view the lenticular screen. It may seem a bit tedious to cut a picture into little strips and paste them on the picture plane, but the strips can be made quite simply photographically by using the lenticular screen itself. Projecting a picture from O_R toward the points R will only expose the points R. Another picture can be projected from O_L toward the points L, only exposing those points. In fact, it is possible to get several more pictures between those

FIGURE 8.14

The lenticular screen. A sheet of cylindrical lenses is backed by a picture, lying at the focal plane of the lenses. Light originating (or scattered) from the points labeled R is made parallel by the lenses and sent to the upper right. Light from the points labeled L is made parallel and sent to the upper left. If we place two different pictures, intermeshed, at the points R and L, we see the R picture only if we look from O_R, the L picture only if we look from O_L.

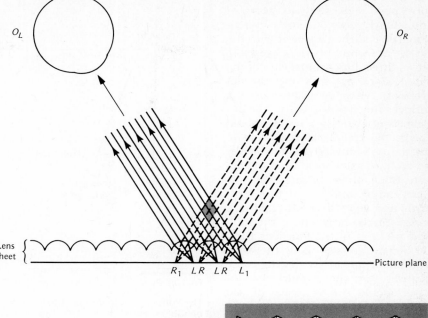

points, each corresponding to a different direction of exposure, and later of viewing. The resultant device, which presents a different picture to you as you move past it, changing your angle of view, is sometimes seen in advertisements, picture postcards, and even political campaign buttons. It is reminiscent of the nineteenth-century puzzle pictures that were painted on a series of wedges (Fig. 8.15). The artist Yaacov Agam has made modern abstract (usually geometrical) pictures on such wedges, so that the viewer sees several different patterns as she moves about. He has also used lenticular screens similarly to present a series of different images to the viewer. (The latter he modestly calls "Agamographs.")

FIGURE 8.15

(a) By painting one picture on the R faces of the wedge and a different one on the L faces, an artist can present two pictures. Viewed from the right, the observer sees only the R picture; from the left, the observer sees the L picture. **(b)** Two views of such a Victorian puzzle picture. **(c)** Another puzzle picture that shows three views. Two views are drawn on opposite sides of the parallel slats.

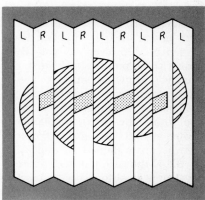

(a)

(b)

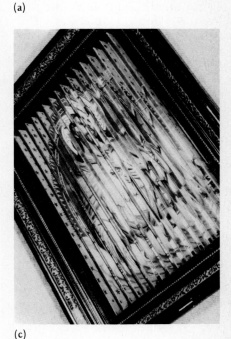

(c)

If the *R* pictures and *L* pictures correspond respectively to the right- and left-eye views of a three-dimensional object, then a viewer with his left eye at O_L and his right eye at O_R sees a three-dimensional view of the object. Each eye sees only the appropriate view necessary for the correct binocular disparity. The angle between the two views should not be too great, corresponding to the natural convergence the viewer would have when looking at the object. Such three-dimensional pictures can be found on picture postcards and sometimes on children's books. Stereo prints from the multiple pictures of the four-lens Nimslo camera are also made this way. It is important when viewing these pictures that the cylindrical lenticules should run vertically. If you rotate the picture so they run horizontally, both your eyes see the same image and the picture loses its depth. (However, if you then slowly tip the picture backward, the image you see will change.)

The picture may even appear in front of the lenticular screen. Suppose that the picture plane is painted a solid color, say red, except for two points (labled R_1 and L_1 in Fig. 8.14) that are painted green. If a viewer is positioned with his left eye at O_L and his right eye at O_R, both eyes see the same thing, red, except at the point where the rays from R_1 and L_1 intersect. This point, shaded in the figure, appears green. Since this green point lies in front of the picture, the viewer sees a green spot floating in front of a red background. This effect has been used by the artist Frank Bunts, whose dizzying pictures may have spots floating as much as a foot in front of the picture plane. It is well suited to producing three-dimensional geometrical patterns, and simplified versions are sometimes found in toys and novelties.

*C. The Pulfrich phenomenon

Another phenomenon that produces a stereoscopic effect by pre-

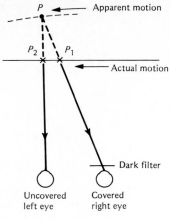

(a)

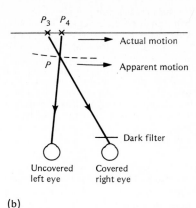

(b)

FIGURE 8.16

The Pulfrich pendulum. **(a)** The pendulum moves to the left. The covered right eye sends a delayed signal to the brain, so the viewer "sees" the pendulum at P_1 with that eye at the same time that she "sees" it at P_2 with the uncovered left eye. Her convergence and binocular disparity then tell her that the pendulum is at P, behind the actual plane of motion of the pendulum.
(b) The pendulum has now reversed its direction. Again, the right eye "sees" the pendulum at the delayed location, P_3, which is now to the left of the location at which the left eye "sees" the pendulum, P_4. The point P, on which both eyes agree, is now in front of the actual pendulum plane.

senting different views to the two eyes relies on the fact that the latency time of the eye (Sec. 5.3C) varies with the intensity of light. In dimmer light, the retinal cells respond more slowly, as they must wait until there is sufficient exposure. Thus, an eye that is covered with a dark filter will respond more slowly to the same subject than one that is not.

Suppose you put a dark filter over one eye and look binocularly at a pendulum swinging in a plane perpendicular to your direction of vision (Fig. 8.16 and the TRY IT). Your brain gets the information about the location of the pendulum from the covered eye a little later than from the uncovered one. Thus, your two eyes locate the moving

pendulum at two different positions (your covered eye "sees" the pendulum slightly in the past, compared to the uncovered eye). When the pendulum swings toward the side of your uncovered eye (Fig. 8.16a), the views of your two eyes, extended backward, converge behind the plane of the pendulum. The convergence and binocular disparity information lead you to "see" the pendulum behind its actual position. When the pendulum reverses its direction (Fig. 8.16b), its apparent path is in front of its actual path. Thus, the pendulum appears to swing out of its actual plane, in an ellipse.

This effect is named for Carl Pulfrich, who never saw it. He had been blind in one eye for sixteen years before the phenomenon was first noticed as a nuisance in an optical instrument that he designed. His colleague Ferdinand Fertsch actually suggested the explanation. The phenomenon was once used in diagnosing syphilis. Currently, it is used for the differential diagnosis of optic neuritis occuring in multiple sclerosis. In this case, the difference in response time of the two eyes is actual, so the patient sees the effect without any filter.

FOR SECTION 8.5C
The Pulfrich pendulum and a variation

You will need a white object (a plastic tea cup works well), a piece of string about a meter long, and a dark filter (as from a pair of sunglasses or any reasonably dark piece of plastic). In a well-lit room, have a friend stand a few meters from you and swing the object, on the end of the string, back and forth perpendicularly to your line of sight. With the filter in front of one eye, use both eyes to view the swinging pendulum. What happens if you move the filter to the other eye? Notice how the size of the ellipse in which the pendulum apparently swings changes as you change the distance to your friend or the darkness of the filter. Increase the pendulum speed by shortening the string (see the TRY IT for Sec. 1.2B) and note the changes in apparent motion. Can you explain these changes?

Notice also that your brain is not a slave to geometrical reasoning. Our arguments would imply that the swinging pendulum should appear to move in a perfect elliptical orbit. Your brain, however, won't believe that the string passes through your friend's legs, so the apparent orbit is somewhat flattened in back. The pendulum never appears to pass behind your friend. By suitably locating other obstacles in the apparent path of the pendulum, you can make it appear to move in an extremely strange orbit (see Sec. 8.6G).

There is a variation on this effect that can be observed on your TV screen. Tune to some nonbroadcasting channel, so all you see is "snow"—a screen full of random dots that keeps changing. The dots appear to move from one position to another. This apparent motion can also serve as a source for the Pulfrich effect. Cover one eye with a filter, as above, and view the snow on the screen. The jittering dots now appear to move in a more coherent fashion. While still jittering, the dots appear to lie in two planes, one slightly in front of the TV screen, and the other behind, and these two planes appear to move sideways in opposite directions. (Some people see a whole range of depths rather than just two. Since there is a whole range of apparent motion of the dots on the screen, this range of depths shouldn't be surprising.) Move the filter to the other eye. Also try tilting your head.

8.6
THREE DIMENSIONS VERSUS TWO DIMENSIONS, AND THE AMBIGUOUS DEPTH CUES

None of the depth cues considered so far (accommodation, convergence, and parallax as manifested either through motion or binocular disparity) can be produced in a two-dimensional picture. The artist (photographer, movie maker, etc.) nevertheless manages to create the illusion of a three-dimensional scene by using a host of other depth cues. These depth cues are all intrinsically *ambiguous*, they allow us to interpret them in a variety of ways, and the interpretation that we choose depends on what is most likely to occur. Convergence, for example, is unambiguous—when our eyes converge more, we are looking at a closer object. However, the apparent smallness of objects is an ambiguous cue to their distance—it is possible that they really are smaller, that the people the lecturer believes to be at the back of the lecture hall are actually, however improbably, small people in small chairs, suspended directly above the people she believes to be in the front of the hall.

The skilled artist makes careful use of these other depth cues to convey to us the impression of depth in his picture. If well done, the artwork can be quite convincing. Mastering the art of conveying three-dimensionality in a two-dimensional picture is by no means easy—much great art either doesn't succeed, makes no attempt to do so, or deliberately avoids these cues. Artists in the eighteenth century trying to learn to convey depth, found it helpful to study images produced in the camera obscura, for these were automatically two-dimensional reductions of the three-dimensional world. The advent of photography in the nineteenth century offered a more permanent two-dimensional reduction, which could be examined by the artist at his leisure, and many artists had collec-

tions of photographic studies of the subjects they painted.

But an image drawn on a two-dimensional surface need not actually be a true reduction of a three-dimensional object. Much art makes no attempt to represent three-dimensionality; an unskilled artist may represent a three-dimensional object badly due to his misuse of the depth cues; or conversely, a skilled artist may, by careful use of the depth cues, present a two-dimensional image that conveys depth to our eyes and yet could not actually exist as a three-dimensional object. A series of examples of this type of illusion is shown in Figure 8.17. In each of them, small regions are faithful representations of possible three-dimensional objects but are put together in a manner that, though possible in two dimensions, would be inconsistent in three. In the next sections, we will discuss the **ambiguous depth cues** on which artists rely. Our general depth perception is due to a combination of all the cues, and we are capable of suppressing any of them if it requires an interpretation that our brain somehow considers unlikely.

A. Size

Distant objects appear *smaller*. Figure 8.18 shows, by ray tracing, that the more distant of two identical objects produces a smaller retinal image. We often use our knowledge of the size of familiar objects to gauge their distance. Some cameras rely on this for focusing (Fig. 4.6b). If, however, the size of the object is not what we think it is, we can be fooled in our estimates of distance. Many special effects in movies are produced by taking close shots of miniature models to give the illusion of more distant views of full-sized objects (see the FOCUS ON Special Effects in the Movies). Architects use this effect to make buildings appear taller by having smaller windows on the higher floors (Fig. 8.18c). In Japanese gardens, bushes and trees with larger,

FIGURE 8.17

Impossible objects: various versions of illusions that contain depth cues designed to mislead the eye. The cues convey depth in a manner that contradicts the structure of the objects, which can only be two-dimensional. **(a)** The apparent construction of an impossible triangle out of three solid blocks. **(b)** A mechanical and an architectural example of the same impossible object. **(c)** "Waterfall" by M. C. Escher. **(d)** Frontispiece in *Dr. Brook Taylor's Method of Perspective Made Easy*, Book I, by Joshua Kirby (1754). The artist, William Hogarth, wrote, "Whoever makes a design, without the knowledge of perspective, will be liable to such absurdities, as are shown in this frontispiece." **(e)** A crate for shipping optical illusions.

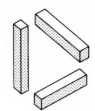
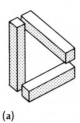
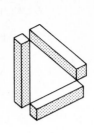
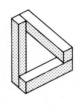

(a)

(b)

(c)

(d)

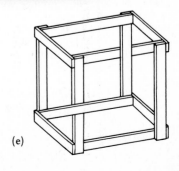

(e)

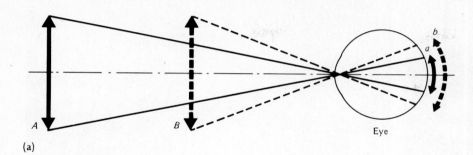

(a)

FIGURE 8.18

(a) Objects *A* and *B* are the same size but *B*, being closer, subtends a greater angle. The image *b* of the closer object is therefore larger than the image *a* of the more distant one. As usual, we have used only the central rays because we know that the lens of the eye will focus all the rays on the retina. **(b)** Roger van der Weyden's "Saint George and the Dragon" has smaller figures in the background. (Earlier art used smaller figures for less important people, no matter where they were located.) **(c)** Half-timbered buildings in Strasbourg.

(b)

(c)

broad leaves are planted near the house, and smaller-leaved plants are used in the background, so the garden looks larger than it is. ***Trompe l'oeil**** artists have used the same effect by constructing miniature scenes in a box that, when viewed through a peephole (to eliminate parallax and convergence), gave the appearance of full-

*French, deceives the eye.

sized scenes to the unsuspecting eye. (The first recorded such peepshow was by Brunelleschi. His model of the Florence baptistry was made even more realistic through the use of a sky of burnished silver that reflected the actual sky with its moving clouds.) Nature also is involved in this sort of deception. The pupa of the butterfly *Spalgis epius* mimics the face of the *rhesus macaque* monkey of its native India. Even though the pupa is only 5 mm

long, the disguise seems to scare off hungry birds who apparently think that it is a distant, larger, and inedible monkey rather than a nearby digestible pupa.

B. Geometrical perspective

Parallel receding lines appear as if they are coming together. This ***perspective*** effect is easily seen with

(a)

(b)

FIGURE 8.19

(a) Photograph of seemingly convergent railroad tracks. (b) Photograph of crepuscular rays of the setting sun (sometimes called Rays of Buddha).

FIGURE 8.20

(a) Same tracks as in Figure 8.19a, photographed in the opposite direction. (b) Photograph of anticrepuscular rays. The sun is behind the camera.

(a)

(b)

railroad tracks. Imagine that *A* and *B* in Figure 8.18 are railroad ties separating the parallel rails. The image of the more distant tie is smaller (that is, *a* is smaller than *b*), so the rails, which are attached to the ties, must appear closer together at *A* than at *B* (Fig. 8.19a). Just like these rails, the rays from the sun that reach the earth are essentially parallel. When these rays pass through clouds, the clouds cast dark shadows, allowing us to see these parallel rays as lines in the sky. Perspective makes these rays appear as if they come from one point (like rays from a candle flame) and spread out, even though they are parallel (Fig. 8.19b) just like the railroad tracks. We can check this by noticing that the parallel railroad tracks appear to come together in *both* directions (compare Fig. 8.20a to Fig. 8.19a). This must also be true for the sun's rays. If we look *away* from the sun, and if we are lucky, we should be able to see the sun's rays coming together toward a point in the sky opposite to the sun. Such so-called anticrepuscular rays are shown in Figure 8.20b.

Like size, perspective as a depth cue is ambiguous. In Figure 8.21, we see a number of objects all of which cast the same image on the

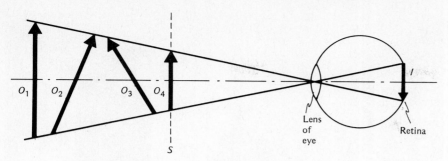

FIGURE 8.21

The ambiguity of perspective. Objects O_1, O_2, O_3, and O_4 all cast the same image I on the retina. A two-dimensional image on a screen at S would also create the same image I.

retina. This ambiguity allows for a certain amount of deception. Examples are the perspective boxes made by artists in the seventeenth century. The insides of these boxes were painted in such a way that, when viewed through an appropriately placed peephole, the perspective was just right to make the view look like a realistic, full-sized room. Of course, when viewed from some other point, the paintings appeared severely deformed, like slant anamorphic art (Sec. 3.3C). The corners of the "room" may not even coincide with the actual corners of the box. On a larger scale, in the apartments of St. Ignatius in Rome, the ceiling and walls of the corridor were painted with careful perspective to make them appear quite a different shape than they actually were. Again, this is only convincing when viewed from one point—from another point the corridor appears strangely distorted. In the auditorium of the Imperial Theatre in Vienna, the ceiling is painted so that, when viewed from the orchestra seats, it appears that there is an extra balcony. More recently, the artist Jan Beutener has constructed "The Room," which appears like a perfectly normal room when viewed through an appropriate peephole but is in fact a fantastic array of the components—for example, a coat apparently draped over a chair is actually suspended on wire while the chair is partly painted on the floor.

The experiments of Adelbert Ames take advantage of the ambiguity of perspective. The Ames room is a rather distorted room that is designed with such a perspective that the view through one peephole has the appearance of a more familiar, rectangular room (Fig. 8.22). A person in a more distant part of the room will, as always, appear smaller than he does when closer, but so convincing is the illusion of rectangularity that as the person walks about the room he appears to the peephole viewer to be growing and shrinking rather than approaching and receding (Fig.

8.23). Figure 8.24 shows the same idea in reverse. The converging lines appear as a perspective view of receding lines, causing us to interpret what are actually identical objects at the same distance (the distance to the picture) as being different size objects at different distances.

Tricks of perspective are often used in architecture. Some skyscrapers get narrower toward the top. The resultant converging lines make them appear even taller. This idea has been used since the Greeks built the 10-m high columns of the Parthenon with about an 8-cm inward tilt. More purposely deceptive is the Palazzo Spada in Rome (Fig. 8.25). A colonnade there is made to appear longer than it actually is, by being smaller at the far end. Its entrance is about 6×3 m, but the passageway diminishes to about $2\frac{1}{2} \times 1$ m. An apparently full-size statue that is actually less than 1 m high is located at the far end to enhance the illusion. This type of false perspective is commonly used by stage set designers in the theater to give the illusion of a larger scene than would otherwise fit on the stage. The same idea is used to con-

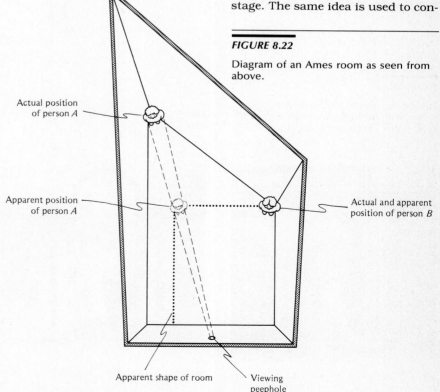

FIGURE 8.22

Diagram of an Ames room as seen from above.

Actual position of person A

Apparent position of person A

Actual and apparent position of person B

Apparent shape of room

Viewing peephole

FIGURE 8.23

A photograph of three normal-size people in an Ames room. The room is not rectangular; the apparently smallest person is actually most distant.

ceal hidden compartments in magicians' equipment—the narrower, shorter inside of a box is made to appear as wide and deep as the outside by converging lines drawn on the inside.

To convey depth accurately, artists carefully study perspective (Fig. 8.26). By the high Renaissance, they had become adept at drawing rather odd-shaped objects in accurate perspective. Perspective was used not only to convey depth but for artistic purposes. The lines in da Vinci's "Last Supper" (Fig. 8.26c) converge toward the head of Christ, bringing the viewer's eye to the most important part of the pic-

FIGURE 8.24

Perspective illusion. These three figures, which are actually the same size, appear to be of different sizes because the converging lines cause us to interpret the figures as being at different distances.

FIGURE 8.25

Palazzo Spada, Rome. **(a)** View through colonnade. **(b)** Views from other end of colonnade, showing the deception.

(a)

(b)

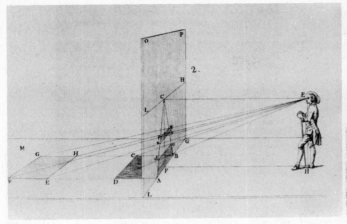

(a)

(b)

(c)

(d)

ture. Notice in the Panini painting (Fig. 8.26d) that the vertical lines do *not* converge, as if the artist had a PC lens (Sec. 4.3B). We are more accustomed to seeing horizontal receding lines converge than vertical lines, possibly because our view is generally horizontal—as discussed in Section 4.3B, if we keep our film (or retina) vertical, vertical lines will not converge on it. Since we view the *picture* at eye level, converging vertical lines in it would make it appear that the building was falling away from us (Fig. 4.11a).

There are times when the artist may chose to draw a less extreme perspective than a camera would capture so that the painting will look more "realistic." Consider a basketball and a tennis ball (Fig. 8.27a). Accurately tracing the rays, as in Figure 8.26a, would result in the views from *P* and *Q* shown in

Figure 8.27b. However, we rarely "see" a tennis ball as big as it appears in *Q*'s view—we know that it is smaller than a basketball. (Imagine the basketball were a face and the tennis ball its nose—or examine the photograph in Fig. 8.27c which, while accurate, appears distorted.) Both artists and photographers avoid this extreme perspective when making realistic portraits. To make her subject's face fill the frame, the photographer will use a telephoto lens rather than move closer to the subject. (Portraits with pleasing perspective can be taken with a 100-mm lens on a 35-mm camera. Also see Appendix G.)

To see an artist's "corrections" to perspective, examine the Delacroix "Odalisque" and the photographic study found in his album (Figs. 8.27d and e). Notice that he has

FIGURE 8.26

The rules of perspective. **(a)** From *Dr. Brook Taylor's Method of Perspective Made Easy*, Book II, by Joshua Kirby. **(b)** A. Dürer, "Draftsman Drawing a Lute." In each case, the draftsman determines where the light rays from the object intersect a chosen plane. **(c)** Leonardo da Vinci, "The Last Supper." **(d)** Panini, "Interior of the Pantheon."

elongated the foreshortened thigh, twisted the foot, and moved the leg. While the resultant position is actually more uncomfortable and precariously balanced, it appears more natural (even more so in 1857 when viewers were less accustomed to such photographs). As Delacroix

FIGURE 8.27

(a) Side view of tennis ball and basketball. (b) Views of balls as seen from *P* and *Q*. (c) View from close up, appears distorted. (d) Photograph of nude found in Delacroix's album. (e) "Odalisque" by Delacroix (1857).

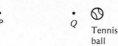

(a)

(b)

(c)

(d)

(e)

wrote, "the daguerreotype [is] only a reflection of the real, only a copy, in some ways false because it is so exact . . . . The eye corrects," and, "The obstinate realist will therefore rectify in his picture this authentic perspective which falsifies appearance by being correct."

What about the artist who is not an "obstinate realist"? De Chirico uses a very inconsistent perspective to imbue his picture with a dreamlike, menacing quality (Fig. 8.28a). John Grazier gives a twisted, dizzying effect by changing perspective throughout his picture (Fig. 8.28b). Henri Matisse deliberately avoids the appearance of depth and attempts to produce a flat surface.

FIGURE 8.28

(a) Giorgio di Chirico, "Mystery and Melancholy of a Street." (b) (facing page) John Grazier, "Dark Rooms." (c) Henri Matisse, "The Painter's Family." Matisse specifically seeks to avoid the feeling of depth in this painting.

(a)

(b)

(c)

His inconsistent perspective is part of his effort to force your eye back to the picture plane (Fig. 8.28c).

*C. Variations in brightness (shadows)

The way an artist varies the *brightness* of a picture significantly affects the depth we perceive in it. Three levels of brightness convey three layers of depth in Hiroshige's print (Fig. 8.29a). The introduction of *shadows* produces a more gradual rounding (Fig. 8.29b). Thus, to give substance to the billiard balls, pool halls typically have highly directional lighting—usually a small lamp, directly over the table, which gives good shadows. Similarly, *highlights* introduce rounding. Matisse creates apparently rounded glasses and bottles by painting nothing but highlights and shadows (Fig. 8.29c). Without painted shadows, pictures often look flat. Thus, when Matisse wants a flat picture (Fig. 8.28c), he omits all shadows. Printing photographs on low-contrast paper has this effect.

Shadows are particularly impor-

(a)

(b)

FIGURE 8.29

(a) Hiroshige, "Miyanokoshi." (b) The circle becomes a sphere when shadows are added. (c) Matisse, "The Lunch Table."

(c)

(a)

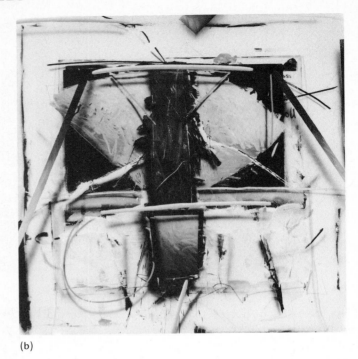

(b)

FIGURE 8.30

(a) Yrjö Edelman, "Packed Landscape (verso)." (b) Richard A. Johnson, "Blue Sundae."

tant when there are only small variations in depth. They convey the *texture* of material to the viewer. There is a Roman floor mosaic that depicts scraps of food and their shadows. When the lighting is right, it appears as if the floor hadn't been swept. The modern *trompe l'oeil* artist Yrjö Edelmann makes paintings of brown wrapping paper apparently taped to the picture (Fig. 8.30a). The shadows are captured so carefully that you must bring your eye quite close to his picture before you can convince yourself that they are not collages of actual wrapping paper bits taped to the background. Similarly, a number of abstract illusionists (Ceravolo, Havard, Johnson, Lembeck, etc.), by deft use of shadows, make it appear that there is a streak of paint on the glass over their painting, rather than in the picture plane (Fig. 8.30b). The temptation

FIGURE 8.31

The Helmholtz illusion. (a) The small white square appears larger than the small black square. (b) The lighter magazine cover appears larger than the dark one.

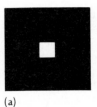

(a)

(b)

to touch the picture to determine the depth of the streak is almost irresistible.

Brightness combines with size, by virtue of the *Helmholtz illusion* (Fig. 8.31), to convey depth. In this illusion, a lighter figure appears larger than an otherwise identical darker one. Because closer objects appear larger, the lighter figure may also appear closer to us. (Magicians take advantage of the Helmholtz illusion. A subject is placed into a fairly bright box and disappears—concealed in a hidden compartment, the outside of which is painted black to appear small. The apparent smallness of this region of the apparatus, combined with the incredible ability of people to squeeze into small spaces, convinces the audience that there is no room to conceal the subject.)

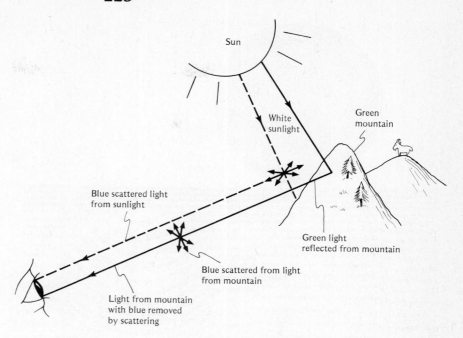

*D. Variations in color

Artists also convey the roundedness of an object by variations in *color* that indicate the different light falling on different surfaces. A color print seems to have more depth than the identical picture printed in black and white. Shadows may even be conveyed without variations in brightness—Renoir made shadows with color changes alone. Even abstract uses of color can create the impression of depth (Plate 8.3).

Distant landscapes tend to lose their color contrasts—the colors get duller, less pure. Distant mountains appear blue (Plate 8.4) due to the blueness of the intervening air (Sec. 13.2). How blue a distant scene appears depends on the relative amount of light coming from the intervening air (which provides the blue) and from the distant scene itself (which tends to have the blue removed from it due to scattering in the air)—see Figure 8.32. Haze or smog in the air can introduce gray or brown. Thus, the longer the space of intervening air, the more the color of the distant scene can be changed, and the change in color between nearby and distant scenes conveys depth information to you.

FIGURE 8.32

The intervening air can make a mountain appear more or less blue, depending on the relative amounts of light from the sun that strike the mountain and the intervening air.

Chromatic aberration sometimes creates apparent edges between regions of different colors (Plate 5.2). These edges, seen in many examples of op art, may then convey the impression that the two colors are at different levels. The fact that a red area appears closer than an otherwise identical blue area has also been attributed to chromatic aberration—the red, being focused less than the blue, requires additional accommodation. Additionally, because the fovea is slightly off the optic axis of the eye, chromatic aberration acts in opposite directions in your two eyes, creating binocular disparity that is different for red and blue.

*E. Variations in sharpness

Distant objects also may appear *fuzzier*, less sharply focused. Their images are smaller on your retina and the lack of infinitely perfect focusing is more apparent on these smaller images. Thus, the artist will convey a feeling of depth by a loss of detail in distant objects. A shallow depth of field in a photograph also allows the viewer to distinguish depth planes. To enhance the apparent size of a garden, Japanese gardners design more detail in the nearby plantings and greater simplicity farther away.

Loss of brightness, color saturation, and sharpness together are sometimes called **aerial** or **atmospheric perspective** because these effects are a result of the scattering of light from the air between the distant scene and the observer. On exceptionally clear days the intervening air blurs distant objects less, and they appear closer than usual.

*F. Patterns

An abstract *pattern*, by recalling perspective and shading effects, may create the feeling of depth. Simply by diverting straight lines, Vasarely conveys the impression of indentations and elevations (Fig. 8.33). Notice how Matisse's conflicting array of patterns tends to flat-

FIGURE 8.33

Victor Vasarely, "Ondho."

ten the picture in Figure 8.28c. In other works, Matisse blends the figure and background by painting, for example, the identical pattern on a table cloth and on the wallpaper behind it.

*G. Overlay (interposition)

We perceive one object to be farther than another if the second object *blocks our view* of the first. So obvious is this fact that otherwise flat pictures take on some depth through the use of overlay (Fig. 8.34a). Yet even as compelling a depth cue as this is ambiguous.

The apparently more distant object may in fact be the closer but cut in such a shape that it fully reveals the apparently closer object. It is the unlikeliness of that situation that makes this cue so convincing. Thus, in Figure 8.34b, Heracles and the bull appear in front of the background on both sides of the Greek amphora, despite the fact that on the one side they are dark with a light background and on the other the coloring is reversed. It is not reasonable to interpret the scene as a Heracles- and bull-shaped hole in a wall (but see Fig. 8.34c).

*H. Previous knowledge

Finally, we must remember that you look at any scene or image with all

your *experience* stored in your brain, and your brain, in interpreting an image, compares it with its *previous knowledge.* It is the fact that alternate interpretations are so unlikely—that is, go against your previous experience—that makes the ambiguous depth cues generally useful. An interpretation that goes against common experience may be suppressed, even when all depth cues argue for it. Thus, both faces in Figure 8.35 appear to bulge out toward you, like all other faces you've seen. Indeed, you even see a real, three-dimensional face mold this way. The "Haunted Mansion" in Disneyland, California, contains such a hollow mold, which appears to bulge toward you. You might think that walking past the mold would cause parallax to give the game away, even if all the other depth cues don't. Those parts that are really closest to you then appear to move the fastest and in the direction opposite to your motion. The illusion is so convincing, however, that rather than accept an inside-out face, your brain decides that it is the back of the face that is rapidly moving opposite to you, that the nose (actually farthest from you) is moving with you, and thus that this strange (but convex) face is rotating and keeps looking at you as you walk by it. This in spite of the fact that every depth cue, other than your familiarity with normal faces, should make you perceive an inside-out face.

When you look at a two-dimensional picture, you bring your experience that it represents a three-dimensional subject and you interpret it as such. As soon as you recognize the three-dimensional interpretation, the picture takes on the corresponding depth. In Figure 8.2, the scene through the window appears distant until you notice the painting on the easel. When that happens, that painting leaps forward from the background. Figure 8.36 is a flat, abstract drawing that suddenly acquires depth when it is recognized as a picture of a giraffe passing a window. Clearly the giraffe is behind the plane of the window. Anyone who has watched an

(a)

(b)

FIGURE 8.34

(a) Detail from a stained-glass window in the Washington National Cathedral.
(b) Two sides of the Andokides Amphora, Heracles and the Cretan Bull.
(c) Dali, "Old Age, Adolescence, and Infancy (The Three Ages)."

FIGURE 8.35

The left figure is a model of a face. The right figure is a concave mold of that model—an inside-out face. Nevertheless, both faces appear to bulge out, like normal faces. Notice that the illumination appears to come from opposite directions, to be consistent with this appearance.

artist sketching has experienced the sudden depth that appears in the drawing when recognition of the content of the picture sets in. An artist may convey a rounded figure with a few deftly placed lines

(c)

FIGURE 8.36

The figure is a flat, abstract pattern until you recognize it. Then it acquires depth.

(Fig. 8.37) because you recognize the figure and ascribe to it the depth you associate with it.

When a figure contains two equally likely interpretations, you cannot

FIGURE 8.37

Henri Matisse, "Seated Nude, Back Turned."

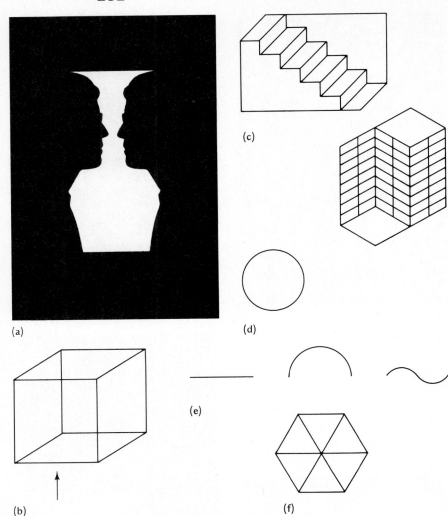

FIGURE 8.38

(a) An oscillating figure illusion. Is it an upside-down goblet or two faces (see Fig. 2.1c)? (b) The Necker cube. Is the edge nearest the arrow on the back face or on the front face? (c) Other oscillating illusions: Schröder's staircase and Thiéry's figure. (d) The circle is seen to represent a flat disk. That is, you imagine that, seen from the side, it looks like the first figure of (e), rather than one of the other two. (f) The Necker cube viewed along the main diagonal looks two dimensional.

perceive both simultaneously, but interpret it first one way and then the other, oscillating back and forth. It is as if there were inhibitory neural connections that prevent one interpretation while the other is in effect, but that eventually fatigue, allowing the second interpretation which in turn inhibits the first. Does Figure 8.38a show an upside-down goblet or two faces? Your interpretation will depend on whether you ascribe the boundary to the dark area (the faces) or the white area (the goblet), since your brain stores boundary information, as we've seen. But as the interpretation oscillates, so does the depth—if it is a goblet, that area is in the foreground; if two faces, the reverse is true. A similar situation occurs with the Necker cube (Fig. 8.38b)—which face of the cube is closer? In Schröder's staircase and in Thiéry's figure (Fig. 8.38c)—are you looking up or down at these figures? So appealing are these oscillating figures that numerous artists have incorporated them into paintings, draw-

ings, and even patchwork quilts. In fact, the confusion of planes fundamental to cubism bears a strong dependence on that found in the Necker cube.

Generally, your brain tries to make the *simplest* interpretation. If

the images received by your two eyes are different, your brain would rather ascribe the difference to depth than to a complicated system of lenses, prisms, and stereo pairs. If you see a circle (Fig. 8.38d), your brain assumes it represents a flat disk rather than one of the other possible shapes shown in Figure 8.38e. A complicated two-dimensional figure may be simply described if interpreted as a picture of a three-dimensional figure, so your brain seeks to make such an interpretation, even when there is no unique interpretation as in these oscillating figures and in the "impossible objects" of Figure 8.17. Note, however, that a simple enough two-dimensional figure does not have depth ascribed to it (Fig. 8.38f).

Your brain, then, takes the two-dimensional images on the retina and imparts three-dimensional form to them according to your experience. Thus, a table top, except when viewed from directly above, makes a rather odd shape on your retina—a trapezoid if seen from the center of one side, almost any four-sided figure if viewed from an appropriate point. However, the table top always appears rectangular. It is not very likely that the table top would change its shape as you move about it, so your brain ascribes the shape changes on your retina to your changes in viewing position, and the table appears always to have the same rectangular shape. This phenomenon, known as **shape constancy,** allows you to identify your changing perceptions with constant objects in the external world. Similarly, the people in the back of the hall, referred to at the beginning of Section 8.6, do not appear to be tiny people—the lecturer

FIGURE 8.39

Faceless face by David Suter. What do you perceive and what do you actually see?

recognizes them as normal-size, distant people who do not shrink or grow as they recede or approach. This phenomenon, **size constancy,** also helps you in interpreting the world, although like many other brain processes, it can be suppressed in the face of conflicting evidence (Fig. 8.23).

Since previous experience plays so significant a role in depth perception, it should come as no surprise that the effectiveness of different depth cues varies with culture. One culture may be much more successful than another in picking out the figure from the ground, interpreting perspective, being fooled by illusions, or indeed in recognizing three-dimensionality in two-dimensional drawings. The world that you see is actually a result of a combination of the light that enters your eyes at a given instant and all the experience and organization already present in your brain. As Sartre said, "I always perceive more and otherwise than I see" (Fig. 8.39).

SUMMARY

Our perception of **depth**—the distance to the objects we see—depends on a number of **depth cues,** only some of which rely on the fact that we have two eyes. We make little use of the cue of **accommodation,** how much we must tense our ciliary muscles to bring an object into focus. Nor do we make much use of **convergence** information, the relative direction of our two eyes when they both view the same object. Much more important is **parallax,** the difference in scene obtained from different viewing positions—as we move to the left, say, nearby objects appear to move to the right, compared to distant objects. We can use parallax without moving, because we have two eyes with overlapping fields of view; the **binocular disparity** resulting from the two different views of the same object provides a significant depth cue if the object is not too distant. If our two eyes are presented with different views, we can be fooled into perceiving depth. This can be done in various ways. **Stereoscopes** provide the two views by means of separate pictures presented to separate lenses (a prism aids convergence). **Anaglyphs** offer the two views in separate colors, which each eye then sees through different filters. **Lenticular screens** use an array of small cylindrical lenses that present the different views to different viewing directions. Because the eye's latency time depends on the brightness of the incident light, a dark filter over one eye causes an object moving in a plane to appear to be moving in an ellipse—the **Pulfrich phenomenon.**

None of the above depth cues are available to the painter of a picture. Such an artist relies on more **ambiguous depth cues**—the apparent **size** of an object depends on its distance; **perspective** causes receding lines to appear to come together; **variations in brightness** and **shadows** give a clue to the roundedness and depth of an object, as do **variations in color; variations in sharpness** also give a depth cue because fine detail cannot be seen at great distances, and because of the **aerial** or **atmospheric perspective** produced by the air between the viewer and distant objects; the **patterns** we see may imply depth; close objects obscure distant ones **(overlay** or **interposition);** and finally our **previous knowledge** plays an important role in our perception of depth. When we view a given object from various angles and distances, the two-dimensional images on our retina change. Nevertheless, our brains provide us with a stable picture of the object **(shape constancy** and **size constancy).**

PROBLEMS

P1 Suppose object *A* is at a distance of 25 cm from your eye, while object *B* is 25 m from your eye. How can you use your accommodation to tell you something about their relative distance from you?

P2 Repeat P1, using your convergence to determine the relative distance.

P3 Suppose that you close one eye, and see two objects, *C* and *D*. As you move your head toward the right, *C* appears to move toward the left compared to *D*. Which object is closer?

P4 Suppose you looked at yourself in the mirror, and your two eyes saw the two views shown in the figure. Which view did your left eye see?

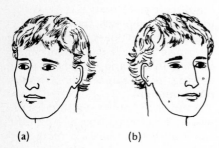

(a)   (b)

P5 Consider the lenticular screen shown in Figure 8.14. If you wanted a green spot to appear to float *behind* the picture plane, you would *not* paint the point R_1 and L_1 green. Redraw the figure and indicate which points you *would* paint green, and where the green spot would then appear.

P6 In "Cymbeline" Shakespeare writes:
I would have broke mine eye-strings, cracked them, but
To look upon him, till the diminution
Of space had pointed him sharp as my needle:
Nay, follow'd him, till he had melted from the smallness of a gnat to air;
To what depth cue is he referring?

P7 In Figures 8.19a and 8.20a, the distant railroad ties not only appear shorter, but they also appear closer together than the nearby ties. Demonstrate this effect of perspective by ray tracing on the figure, redrawn to scale as indicated: (a) Using only the central ray for each tie, locate the

position of its image on your retina. Do this carefully with a straightedge, for the two farthest ties and for the two closest ties. (b) Which pair of ties appears to be most widely spaced?

P8 Why does a landscape often appear to have less depth on an overcast day?

P9 Astronauts on the moon found it hard to judge distances. List several reasons why, taking into account the unfamiliar and peculiar landscape, the fact that craters come in all sizes, and the absence of atmosphere.

P10 Why do Figures 8.30a and b look less convincing than the description in the text would lead you to believe?

P11 The depth perception cue employed in the old-fashioned stereoscopes but *not* in plain photographs is: (a) variation in color, (b) variation in sharpness, (c) binocular disparity, (d) previous knowledge. (Choose one.)

P12 In *To the Lighthouse*, Virginia Woolf writes: "For sometimes quite close to shore, the Lighthouse looked this morning in the haze an enormous distance away." To what depth cue is she referring?

P13 (a) In the figure, which ring is closer, the one on the left or the one on the right? Why is there no "correct" answer for that question? (b) Why does your perception of the rings vacillate?

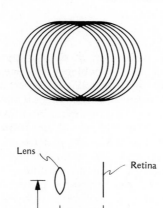

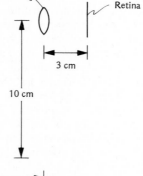

Lens

Retina

3 cm

10 cm

HARDER PROBLEMS

PH1 Ophthalmologists test your convergence with a stepped series of prisms. Placed in front of one eye, a prism shifts the apparent location of a spot of light, as seen by that eye only. The stronger the prism, the more your two eyes must converge in order to "see" only one spot. Show how you must increase your convergence when the prism is present by drawing one or more top view diagrams, showing the path of the light rays to your eyes.

PH2 A jet pilot says he saw a nearby, small, strange luminous object that appeared to move with him as his plane sped up, slowed, climbed, and dove. He claims that the object was a UFO with an unseemly interest in his plane. A passenger, however, says that the pilot was mistaken—the object was actually quite far away. Assuming the passenger is correct, explain the apparent motion of the object.

PH3 Manet's painting "The Dead Toreador" shows a full-length picture of the toreador lying, apparently, almost parallel to the plane of the picture. As you walk toward the side of the painting, however, the body appears to rotate until, when you look at the painting at a glancing angle, the body appears to lie almost perpendicular to the picture plane. Explain why this happens.

PH4 You look at a cage with a number of identical vertical bars. By mistake, your eyes fuse the wrong two bars—that is, your left eye is looking at one bar, call it *A*, but your right eye is looking at the bar to the *left* of *A*. (a) Does the cage appear to be closer or farther than it actually is? (b) Draw a diagram to prove your answer.

PH5 Two eyes in the figure, attached to the same brain, are looking through a stereoscope at a pair of pictures of two objects. The picture seen by the right eye has the objects at A_R and B_R. The picture seen by the left eye has them at A_L and B_L. The prisms merely bend the rays. Where does the brain locate the binocular images of these two objects? Redraw the figure and draw whatever lines you need to locate the images. Label them *A* and *B*.

1 cm   Ties

20 cm

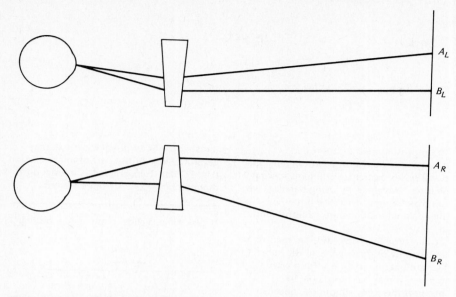

additional converging lens that both spreads the light over the retina, thus illuminating it, and creates a real image of the retina outside the eye. The ophthalmologist can then examine this real image closely with a stereo microscope. The essential parts of this stereo ophthalmoscope are shown in the first figure. In the second figure we simplify this by drawing one "effective lens" to represent the total combination of the converging lens, the cornea, and the eyelens. The focal point of this combination, F_{eff}, is shown. An arrow represents the retina. (a) Redraw the second figure and show, by drawing the appropriate rays, how the point source illuminates the retina. (b) Using a different color, construct the retina's real image, which is created by the effective lens.

PH6 Examine Figure 8.8 carefully and find two objects, one closer than the other. Explain how parallax tells you which is closer, referring to details in the picture.

PH7 If you put a stereocard (containing a stereo pair of pictures) into your stereoscope upside down, what happens to the apparent depth?

PH8 Suppose you tried the Pulfrich pendulum with a small pendulum swinging toward and away from you, on a line that intersects your head right between your eyes. Redraw Figure 8.16 for that case and describe the motion you would see.

PH9 Neutral-density filters are colorless, gray filters used for diminishing the amount of light without introducing any color. How can a series of such filters be used to diagnose the severity of a patient's optic neuritis? (See Sec. 8.5C.)

PH10 If your ophthalmologist wishes to examine your retina in stereo, he must both illuminate it uniformly and obtain two different views of it. The compound lens of your eye, however, tries to focus any incoming light to a point on your retina and also prevents him from getting very close. The solution to this is to use an

MATHEMATICAL PROBLEMS

PM1 A 35-mm picture is enlarged by a magnification $m = 3$ (to about a wallet-sized picture). From what distance should you view it so that it will appear with correct perspective if: (a) it was taken with a normal, $f_c = 50$-mm, lens, (b) it was taken with a telephoto, $f_c = 100$-mm, lens, (c) it was taken with a wide angle, $f_c = 25$-mm, lens? (d) Which of these distances are reasonable viewing distances?

PM2 You wish to view an 8×10-inch enlargement of a 35-mm picture ($m \simeq 7$) from a distance of 1 m. (a) What focal length lens, f_c, should the picture have been taken with if, at this viewing distance, the picture is to appear with correct perspective? (b) If, instead, you wish to view the picture from your near point, 25 cm, what focal length lens should have been used?

PM3 A 35-mm slide, taken with a telephoto lens ($f_c = 200$ mm), is projected to a screen by a standard projector ($f_p = 100$ mm) from a distance of 5 m. How far from the screen should the viewers sit so that the slide will appear in correct perspective?

PM4 A 35-mm slide is taken with a weak telephoto lens ($f_c = 85$ mm). (a) If this original slide is viewed through a magnifying glass, what focal length should the magnifying glass have so the picture appears with correct perspective? What is the magnifying power of that lens? (b) The slide is now enlarged to a 5×7-inch print. At what distance should the print be viewed in order to see correct perspective?

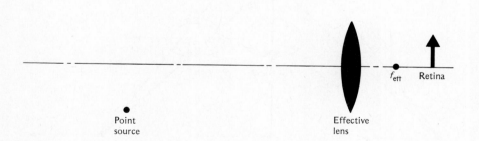

X-ray tomography

The representation of a three-dimensional object on a two-dimensional plane poses particular problems in x-ray pictures. However, by using parallax, we can make x-rays reveal the three-dimensional relationships of bones and organs.

In an ordinary x-ray, the film records the shadow of those parts of the subject that are opaque to x-rays. Even with training, it may be impossible to interpret this x-ray silhouette, because

the shadow of an opaque part, B, may get in the way of the part of interest, A (Fig. FO.4a). **Tomography*** is a procedure that combines x-rays from different

---

*Greek *tome*, a slice, plus *graphia*, writing.

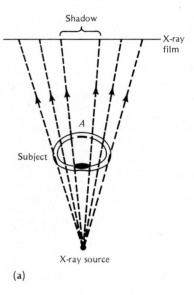

(a)

FIGURE FO.4

(a) Ordinary x-ray picture. The shadows of the two opaque objects A and B, overlap. **(b)** Tomograph. The subject moves to the right with speed v, while the x-ray source moves with twice that speed, $2v$. Opaque object A, which lies midway between the film and the source, always casts a shadow at the same location, A'. Opaque object B, which lies at a different depth, casts a shadow that moves across the film. The figure is drawn for three locations *(1, 2, and 3)* of the source, the subject, and the shadow B'. (In practice, the film and source move while the subject does not.)

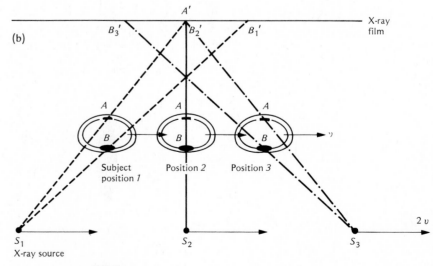

locations in order to isolate the shadow of objects that are located on one particular plane or section through the subject's body. Thus, it allows us to record the shadow of object A in the figure, while disregarding objects, such as B, that lie in front of or behind A.

Consider the example shown in Figure FO.4b. Here the subject is moved to the right, past the film, at a given speed, while the x-ray source is moved twice as fast in the same direction.

PONDER

In practice, the subject is stationary, and the film and source move in opposite directions at the same speed. Why do these two procedures give the same result?

Initially, the subject is located at position 1, the source at S_1, and the shadows cast by A and B are at A' and B_1'. As the subject moves to position 2, and the source to S_2, the shadow of A remains at A', while that of B moves to B_2'. At a later time, the subject is at position 3, the source is at S_3, B's shadow has moved to B_3', but A's shadow still is at A'. Thus, the shadow of A (and anything else opaque lying at the same depth) becomes a sharp image on the film, but the shadow of B, being spread across the film, appears only as a weak, overall haze. To obtain an x-ray photograph of a *different* depth layer, for instance that at which B lies, we need only change the speed at which, say, we move the source.

Instead of recording the image on film, a *scanning* technique (Sec. 6.3B) can be used. An x-ray detector records the attenuation of x-ray beams sent through the subject in various directions. A computer correlates the data for all the directions and reconstructs the image of any desired section. This **computerized tomography** (**CT** or **CAT***) scanner is useful for such difficult diagnoses as tumors in the brain (Plate FO.1).

Once the data are in the computer in the form of numbers, they can be used for **image enhancement,** that is, the modification of the constructed image in such a way as to increase certain aspects (such as contrast) or enhance certain features (such as edges). The speed, contrast, and latitude of the imaging system (Sec. 4.7G) can be adjusted by the computer, perhaps many times within the same picture. Further, in addition to enhancing edges as in Section 7.4D (by increasing the contrast across them), the computer can give a smoother picture (by averaging the signal over a small region or over a sequence of successive pictures). It is also possible to eliminate unwanted features, such as overlying bone structure, by combining several pictures. For example, to study blood flow, a substance that is opaque to x-rays, such as iodine, may be injected into the blood vessels. The x-ray absorption in iodine depends strongly on the x-ray frequency, while that in bone does not. Therefore, two pictures are taken using x-rays of different frequencies, and one picture is subtracted from the other. This emphasizes the *differences* between the pictures (the iodine-carrying blood), while suppressing the similarities (the bones). All these ideas are quite similar to processes that are performed by the neural computer used in human vision— the comparison of signals at different frequencies, for example, is analogous to color vision (Sec. 10.4).

---

*The A stands for axial; the patient is often rotated about an axis. Of course, there is more than one way to scan a cat.

237

Color

9.1
INTRODUCTION

So far, when we have talked about *color* we have usually been talking about *wavelength* of visible light. Although different wavelengths are associated with different colors, equating color and wavelength can often be very misleading—our intuitive idea of color is quite different from the result of a precise wavelength determination. In fact, while we associate a given wavelength with a particular color, we may nevertheless see that color when that wavelength is absent or not see it when the wavelength is present. Indeed, we see colors that are not even in the spectrum of visible light—that are not associated with any wavelength. We'll see that there is more to color than meets the eye, and, just as accurately, there is less to it than meets the eye.

Color is different for different people (Fig. 9.1). Consider the color green; it may evoke the freshness and life of lush vegetation or the decay of mold and slime. One diction-ary gives one meaning as "fresh, youthful, vigorous," and the next meaning as "pale and sickly." We think of green as implying fertility (a green thumb), seasickness, money, gullibility, strangeness (little green men), immaturity, jealousy, poison, or life. Our individual perceptions of green vary widely. Thus, one of the authors, whose color vision is deficient, can barely distinguish green from gray, while the painter Kandinsky looks at green and sees "a fat, healthy, immovably resting cow, capable only of eternal rumination, while dull bovine eyes gaze forth vacantly into the world."

In this chapter, we'll discuss the important facts about color, on which we can generally agree. In Chapter 10, we'll try to understand how color phenomena are dependent on our perceptual mechanisms.

9.2
COLOR VERSUS WAVELENGTH, AND NONSPECTRAL COLORS

When we break up white light into its component wavelengths, say by means of a prism as in Plate 2.1a, we see all the colors of the spectrum spread out according to the wavelength of the light. Is there a simple relation between the colors and the wavelengths? If you ask a collection of observers to locate a given color in the spectrum, there will be general, but not total, agreement. Most observers will identify blue with a wavelength between 455 and 485 nm, green with a region between 500 and 550 nm, yellow between 570 and 590 nm, and red with a wavelength somewhere above 625 nm. Thus the colors are not generally identified with a unique wavelength. Further, the identification depends somewhat on the intensity of the light—a wavelength that appears somewhat red at low intensity may seem orange as the intensity is increased. And the naming of colors is hardly an exact science. For instance, **monochromatic*** or **spectral** light (that is, ideally, light consisting of only one wavelength) with $\lambda = 600$ nm has been identified as orange chrome, golden poppy, spectrum orange, bitter sweet orange, oriental red, saturn red, cadmium red orange, red or-

---

*Greek *monos*, single, plus *chroma*, color.

FIGURE 9.1

ange, and yet other names in a list that continues to grow as advertisers take over the English language.

If you compare the monochromatic colors in a good spectrum with the colors you normally see around you, you'll find that most colors you see don't lie in the spectrum. For example, none of the colors in Plate 8.4 is identical to any of the colors seen in a spectrum cast by a prism. A few examples of nonspectral colors are purple, pink, brown, silver, fluorescent red, and iridescent green. Where do these other colors come from? How do we get the sensation of all these colors that do not appear in the spectrum?

9.3

THE INTENSITY-DISTRIBUTION CURVE AND THE CLASSIFICATION OF COLORS

Most colors around us are *not* monochromatic, but instead contain a *distribution of wavelengths.* Suppose we reflect white light from one of the greenish parts of Plate 8.4 and break up the reflected light with a prism. If we then measure the intensity at each visible wavelength and plot the result, we will get an **intensity-distribution curve** somewhat like the solid curve in Figure 9.2. Although there is a predominance of green light, there is also a little bit of every other visible wavelength present. Looking at the greenish part of the plate, then, your eye receives this entire distribution of different wavelengths.

When you simultaneously play two different notes on a piano, you usually hear the two separate notes. But when you shine two different wavelengths onto the same place on a screen, your eye doesn't separate the resulting light into two colors; rather you see some sort of *mixture.* How can your *sensations* of these mixtures be characterized? For example, consider what happens if we shine a little extra red light along with the greenish light of Fig. 9.2:

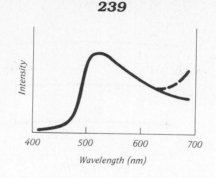

FIGURE 9.2

Intensity-distribution curve. Solid line: the intensity of light, at each visible wavelength, obtained when white light is reflected from a greenish region of Plate 8.4. Dashed line: the same light with a little extra red light mixed in.

the color mixture is changed—indeed, there are an infinite number of other ways that we could modify the intensity-distribution curve, and most people can distinguish a million or so such different color mixtures. Recognizing this huge variety of colors, the author Nabokov felt that a phrase like "the sky is blue" conveyed little information. Thus in *Speak, Memory* he attempts to express the variety of the color blue with such terms as "blue-white," "misty-blue," "purplish-blue," "silvery blue," "cobalt blue," "indigo blue," "azure," "china-blue," "dove-blue," "crystal blue," and "ice-bright." These are certainly more poetic than intensity-distribution curves, but may not convey the same image to every reader. What we want is a straightforward way of classifying all the distinguishable color sensations that is simpler than giving the intensity at each visible wavelength, but still contains all the necessary information.

We'll see that rather than specifying *all* the numbers in the intensity-distribution curve, for many purposes it is sufficient to specify only *three* numbers. Your eye cannot extract all the information contained in the curve by looking at the light; many lights with very different curves appear the same to you. The three qualities of the colored light that determine how the light

appears are hue, saturation, and brightness.

Hue corresponds to the *main color* or *color name;* it is what distinguishes one spectral color from another. For example, all yellows differ in hue from all blues, regardless of any other possible similarities. Hue is specified by the *dominant wavelength* in an intensity-distribution curve. Thus, the greatest intensities in the light represented by Figure 9.2 lie between 500 and 530 nm, so the hue of that light is some sort of green. (The term dominant wavelength is defined more precisely in Sec. 9.4C— the "dominant wavelength" may actually not be present, even though the color looks like light of that wavelength.)

Saturation corresponds to the *purity* of the color—a very saturated color generally has almost all its intensity fairly close to the dominant wavelength, while an unsaturated color would have contributions from many other wavelengths. The monochromatic, spectral colors have the highest saturation. White light, which generally consists of all wavelengths with no dominant one, is completely *un*saturated (Fig. 9.3a). (White is often thought of symbolically as the essence of purity— brides wear it. From our point of view, it is the *least* pure color you can get.) Other colors may be thought of as a mixture of white light with a saturated color. Figure 9.3b shows the intensity-distribution curve of a saturated red light. A mixture of that saturated red with white (Fig. 9.3c) gives a desaturated red, or pink. The saturation is a measure of how much dominant wavelength there is compared to the amount of white mixed in. Most objects around you have unsaturated colors because their dominant hue comes from absorption, which takes place somewhat below their surface, but there is also some surface reflection, which is independent of wavelength and thus mixes in some white. (See the TRY IT.) Artists use the terms *chroma* or (unfortunately) *intensity* to mean something similar to saturation. (Because the definitions come from

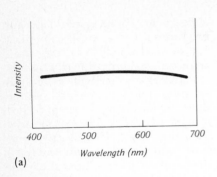

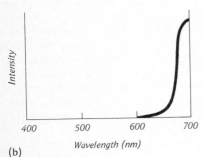

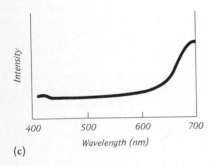

(a)

(b)

(c)

FIGURE 9.3

Saturation. **(a)** White light is completely unsaturated. **(b)** A saturated red light. **(c)** A less saturated red light—pink.

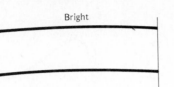

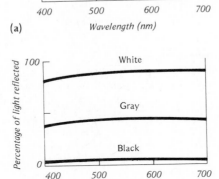

(a)

(b)

FIGURE 9.4

(a) Brightness of a light. The intensity-distribution curves for three different brightnesses. **(b)** Lightness of a surface. The curves correspond to the percentage of incident light reflected at each wavelength.

different fields, the meanings are slightly different.)

The third attribute of color depends on whether we're talking about colored lights or colored surfaces. **Brightness** refers to the sensation of overall intensity of a *light*, ranging from dark, through dim, to bright, and dazzling (Fig. 9.4a, see also Sec. 7.3A). **Lightness,** on the other hand, is related to the percentage of incident light reflected by a *surface* and refers to the *blackness, grayness, or whiteness* of a color (Fig. 9.4b).

We know how to change the brightness of a light (even a colored one); we need only adjust the intensity of the source. To change the lightness of a surface you must make it less reflecting, say by lightly rubbing pencil all over it. The surface will then appear grayer—you have decreased its lightness (Fig. 9.4b). Thus, the more reflecting the surface (in the visible), the lighter it is. We often think of lightness as being related to the total amount of light reflected at all visible wavelengths—but we must be careful. We can change the amount of reflected light in two ways. One is to make the surface *less reflecting*, say by the method just described. This indeed lowers the lightness. The other is to change the overall *illumination*—the incident light. But if you shine less light on a piece of white paper, it doesn't appear gray, it continues to look white. That is, the lightness of any surface is nearly independent of the overall illumination (lightness constancy—Sec. 7.3B).

To apply these ideas to color, take an orange surface and blacken it, lowering each point on the intensity-distribution curve of its reflected light by a factor of two, perhaps, without changing the relative heights. You have then decreased the surface's lightness without changing its hue. The surface will then look *brown,* providing you compare it to, say, the original orange, so you can tell that it is the amount of reflected light, not the illumination, that has changed. Thus, surprisingly, the sensation of lightness corresponds more closely to the surface's ability to reflect light than to properties of the light actually entering your eye from that surface. (The artist's term *value* is somewhat like lightness, but may also involve saturation.)

Colors may be arranged according to their hue, saturation, and lightness in a **color tree.** There are several different schemes, but they are all similar to those shown in Plate 9.1 and Figure 9.5. The "trunk" of the tree consists of the completely unsaturated colors—it ranges from black at the bottom, through grays, to white at the top. That is, the *height* is a measure of *lightness.* Out from the trunk, you find different hues in different directions— the *hue* varies *around* the tree.

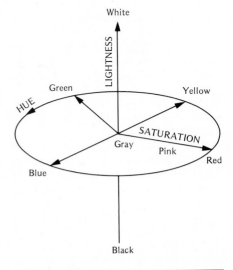

FIGURE 9.5

Schematic drawing of a color tree (compare with Plate 9.1).

Finally, as you move *away* from the trunk in some direction, the color becomes more *saturated*. For example, moving out in the red direction, you go from gray through pink, and only reach a saturated red at the tip of that "branch." Such standard color trees, or atlases, are of great importance to artists, paint manufacturers, printers, and anyone who must reproduce colors reliably, because any surfaces with the *same hue, saturation, and lightness* will *look the same color,* no matter what the intensity-distribution curves of their reflected light happen to be. Similarly, colored lights with the *same hue, saturation, and brightness* will *look the same color,* no matter what their intensity-distribution curves happen to be. As we've said, your eye cannot detect the wavelength distribution—rather it is sensitive to these three attributes. Since a given sensation can be produced by a variety of different stimuli, the color tree tells us very little directly about the physical processes that occur in color perception. We can get such information by exploring the various ways we can mix colors that appear identical to the eye.

TRY IT

FOR SECTION 9.3
Surface reflections and saturation

To see the effects of surface reflections on the saturation of colors, examine a shiny red apple or a piece of colored plastic in a room with several lights. Notice that no matter how intense the red of the apple, or how saturated the color of the plastic, the highlights (surface reflections of the lights) still appear white—completely unsaturated. View the apple outside on an overcast day, when there are no highlights. Does this affect the saturation? Crumple a piece of colored paper and compare the saturation of a part of the paper that is partially shaded with an adjacent part that is not.

9.4
COLOR MIXING BY ADDITION

One way to learn something about our perception of color is to start with *lights* for which we know both the intensity-distribution curves and the sensations that these lights produce, and then to ask what sensations are produced by *combinations* of these lights. In this section, we'll discuss combining two lights by shining them at the same spot on a **white screen,** a screen that diffusely reflects all visible wavelengths equally well. The light reflected to your eyes from that spot then contains both lights—it is an **additive mixture;** at every wavelength the intensities of the two lights add, so the combined intensity-distrubution curve is the *sum* of the two individual curves. (We'll refer to other techniques for mixing colors later.)

If we shine white light on the screen, that portion of the screen looks white to us. If, instead, we reflect the white light off a silver object and onto the screen, the screen again looks white, not silver. Had we used a copper object instead of the silver, the light on the screen would look orange; a fluorescent red object would reflect light to the screen that looks no different than ordinary red light, and so on. Thus, by concentrating on these lights, rather than on the objects themselves, we reduce the gamut of colors; but we're still left with the pure colors of the spectrum, washed-out colors, purples, and a host of other nonspectral colors, which remain for us to analyze.

A. The simple
additive rules

Suppose the intensity-distribution curves of a blue, a green, and a red light look like those shown in Figures 9.6a, b, and c. What will the combination of green plus red look like? Adding the two curves gives the result shown in Figure 9.6d—a rather broad, flat-topped curve with no particular wavelength dominating, centered about $\lambda = 575$ to 600 nm, that is, in the yellow. We cannot yet tell by looking at this curve what the sensation produced by the corresponding light will be, but if you actually do the experiment you find that the resultant light does look *yellow.* This is a fairly strange result; in no way does yellow appear to be a mixture of green and red. Thus we see that your eye can interpret rather different intensity distributions as the same color—the mixture of Figure 9.6d looks very much like the monochromatic yellow of Figure 9.6g. Indeed, you don't have to use the broad colors of Figures 9.6b and c. If you mix roughly equal amounts of monochromatic green and monochromatic red (Fig. 9.6h), the result *also* looks yellow, even though the spectral yellow ($\lambda = 580$ nm) is completely absent.

What about other color combinations? Mixing blue and green lights gives a broad peak centered in the blue-green, a color called **cyan*** (Fig. 9.6e). This is not too surprising, but again the result looks nearly the same whether we use broad or narrow blues and greens.

Mixing equal amounts of blue and red gives a double-humped distribution (Fig. 9.6f)—*no* one wavelength dominates. The sensation one gets is **magenta,†** a kind of purple. (Again, we must face the problem of color names. In this book, we'll reserve the name violet for the shortest wavelength visible light, the name purple for combinations of short and long wavelength visible lights, and the name magenta for this particular purple.) Now this is something new; purples aren't in the spectrum at all—a new hue! Shakespeare tells us: "To . . . add another hue/Unto the rainbow . . ./Is wasteful and ridiculous excess," and yet your eye does just that, producing a purple not present in any rainbow.

---

*Greek *kuanos,* dark blue(!).

†Neither Greek nor Latin, but the name of a town in northern Italy.

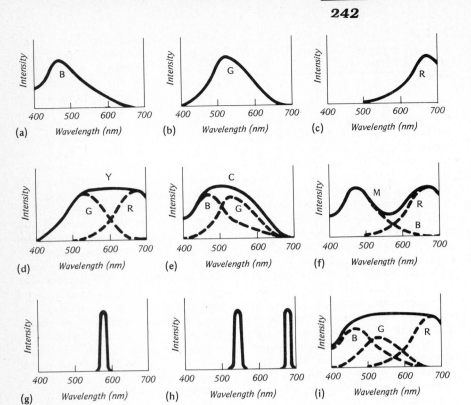

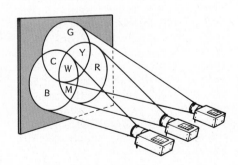

about the details of the distribution curves, is what makes additive mixtures so appealing to physicists and is the reason that we can get away with talking about only three properties of a colored light.

B. Complementary colors

We can use the rules summarized in Figure 9.7 to discover some new additive combinations. We've seen that adding all three colors, blue, green, and red (with proper intensities), gives white:

$$B + G + R \equiv W^*$$

But green and red together give yellow:

$$G + R \equiv Y$$

Hence, combining these two results:

$$B + Y \equiv W$$

blue and yellow lights added together give white. Two colors that, when added together, give white are called ***complementary colors.*** Thus *blue and yellow* are colors complementary to each other. Similarly, because we can think of cyan as being the same as blue and green mixed together:

$$C \equiv B + G$$

we can write:

$$C + R \equiv W$$

—*cyan and red* are complementary. Likewise:

$$M + G \equiv W$$

—*magenta and green* are complementary. (Note: these are not the complementary pairs your art teacher taught you.)

Again, it does not matter to your

FIGURE 9.6

Additive color mixing. Intensity-distribution curves of **(a)** blue (B), **(b)** green (G), and **(c)** red (R) lights. Intensity-distribution curves of the additive mixtures (in equal amounts) of **(d)** G + R ≡ Yellow (Y), **(e)** B + G ≡ Cyan (C), and **(f)** B + R ≡ Magenta (M). Intensity-distribution curves of **(g)** monochromatic yellow and **(h)** a yellow made of an additive mixture of monochromatic green plus monochromatic red. **(i)** Intensity-distribution curves of the additive mixture of B + G + R ≡ White (W).

look the same, they will give identical results in an *additive* mixture with another color. This great simplicity, which allows us to forget

FIGURE 9.7

Simple additive mixing rules. The drawing shows three partially overlapping light beams, which combine additively. B = Blue, G = Green, R = Red, Y = Yellow, C = Cyan, M = Magenta, and W = White:

$$G + R \equiv Y$$
$$B + G \equiv C$$
$$B + R \equiv M$$
$$B + G + R \equiv W$$

Finally, if you add all three colors (blue, green, and red) together in the proper proportions, you get the flat intensity-distribution curve shown in Figure 9.6i. This appears *white*, as we would expect from Figure 9.3a.

Figure 9.7 summarizes the rules we have developed so far for additive mixing of colored lights. The marvelous thing about these rules is that they are valid no matter what the intensity distributions of the three colors are. If two colors

*We follow the standard notation of the C.I.E. (*Commission Internationale de l'Eclairage,* the International Commission on Illumination) and use ≡ to mean "looks the same as," rather than = which would imply equality. Two colors that look the same need not be physically equal (see Fig. 9.8). These equations are valid only for the proper intensity balance of the colors being added.

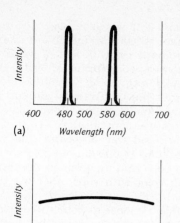

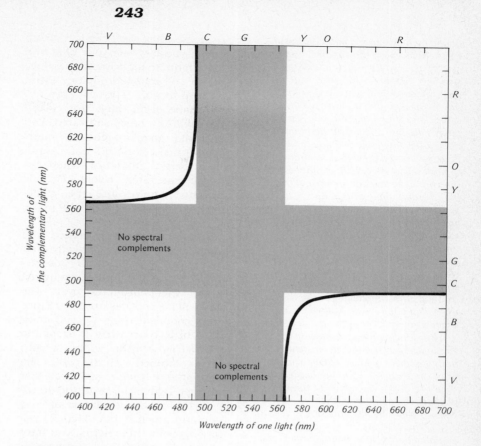

FIGURE 9.8

The lights with these two intensity-distribution curves look alike to your eye, even though one has only two wavelengths present while the other has all visible wavelengths.
(a) Monochromatic blue plus monochromatic yellow. **(b) Broad-band white** (all visible wavelengths).

FIGURE 9.9

Wavelengths of complementary pairs of monochromatic colors. To find the complement of a given wavelength, say λ = 600 nm, draw a horizontal line from the 600-nm mark on the vertical axis. Find the point where this line intersects the curve and drop a vertical line from that point to the horizontal axis to determine the wavelength of the complement, λ = 489 nm. Thus, the complement of orange (600 nm) is bluish cyan (489 nm). (V = Violet, B = Blue, C = Cyan, G = Green, Y = Yellow, O = Orange, and R = Red.) The curves depend somewhat on the observer and on the choice of white.

eye whether the intensity distributions are broad or monochromatic; if two colors are complementary, their additive mixture is white. So if we add monochromatic yellow (λ = 580 nm) to monochromatic blue (λ = 480 nm), the result will look just as white as ordinary white light, which has all the wavelengths in it. That is, to your eye, the lights represented by the two intensity-distribution curves of Figure 9.8 look the *same!* Two such colors, which look alike even though they have different intensity-distribution curves, are called **metamers.***

We can plot the combinations of monochromatic colors that are complementary (Fig. 9.9). Notice that there are no complementary monochromatic colors for the greens (roughly between 495 and 565 nm). This is because, as we already know, the complement of green is the nonspectral magenta—a double-

---

*Greek *meta*, with, plus *meros*, part.

humped intensity distribution, a combination of red and blue. This gives us a way to describe the nonspectral purple hues by wavelength, namely by the wavelength of their complements. For example, we can speak of a magenta 535c, meaning a color that gives white when added to 535-nm green.

The TRY IT uses negative afterimages to show you that complementary colors are intimately involved with the inner workings of your eye.

boundary between red and cyan, your eye tremors cause the afterimage of the cyan to overlap the red, and vice versa (see Sec. 7.6). Since the afterimage of cyan is already red, when it is seen on the red background it enhances the saturation of the red. Thus you see intense flashes of color at the boundary. (Kelly uses green rather than cyan. Examine the red-green border in the plate carefully. The effect here is different from the chromatic aberration effect at the red-blue border, referred to in Sec. 5.2B. The eye has very little dispersion between red and green.) Larry Poons has painted huge canvases of one color, scattered throughout with dots of another color. If you are fortunate enough to see such a work, stare at one dot for a half-minute and then let your eye roam over the painting. The afterimages of the dots roam with your eye and provide an exciting interplay with the painted dots.

Try producing the complementary afterimage in different circumstances. For example, place a small rectangular piece of colored paper on a white background, stare at a dot in the center of the colored paper for half a minute, then rotate the colored paper by 90°, and look at the dot again. The colored paper and its complementary afterimage form a cross. Notice that where they cross, the colored paper looks very desaturated.

C. Chromaticity diagrams

We have seen that different additive combinations of blue, green, and red lights produce yellow, cyan, magenta, and white. This suggests that we might be able to produce *any* color as an additive mixture of three properly chosen lights. A way of testing this idea may be thought of as a game. You choose three different colored lights and we pick an arbitrary fourth colored light. You then mix your three lights, adjusting their relative amounts, to see if you can make a mixture that matches our light in hue, saturation, and brightness. Suppose you pick three pure, monochromatic colors, one each in the blue, green, and red parts of the spectrum. Can you then match *any* color that we pick?

Well, almost any. Although this is your best choice, and there are many colors you *can* match, there are some that you can't. For instance, you can use your blue and green lights to make cyan, but if we choose a monochromatic cyan, you're in trouble. The best cyan you can make from blue and green lights will not be as saturated. You can make good matches to many unsaturated colors, but you cannot make perfect matches when we choose saturated colors. Strictly speaking, you've lost the game. However, if we change the rules slightly, you *can* win.

Although you can't match the monochromatic cyan perfectly, you can convert it to something you can match; if you add a little of your red to our monochromatic cyan, the cyan will become desaturated. Since red is complementary to cyan, mixing a little red with a little of the cyan gives a little white, which, when combined with the rest of the cyan, gives a desaturated cyan. You can then match the resulting desaturated cyan with a mixture of your blue and green lights. That is, using only your three lights, you have produced a match:

$$\text{blue} + \text{green} \equiv \begin{array}{c}\text{mono-}\\ \text{chromatic}\\ \text{cyan}\end{array} + \begin{array}{c}\text{a}\\ \text{little}\\ \text{red}\end{array}$$

If we treated this as a mathematical equation with an equals sign, we could rewrite it as:

$$\text{blue} + \text{green} - \begin{array}{c}\text{a}\\ \text{little}\\ \text{red}\end{array} \equiv \begin{array}{c}\text{mono-}\\ \text{chromatic}\\ \text{cyan}\end{array}$$

That is, if we allow you *negative* amounts, you can match our monochromatic cyan. In fact, if we allow negative contributions, you can match *any* color we choose, no matter how saturated—you can always win the game. (Remember, a "negative contribution" is just a shorthand way of saying that you mix some of one of *your* colors with *our* color and *then* match the resultant color with your other two colors.)

If you choose as your three colors a monochromatic red with λ = 650 nm, a monochromatic green with λ = 530 nm, and a monochromatic blue with λ = 460 nm, the relative amounts you will need to match any monochromatic color we choose are

---

FIGURE 9.10

The relative amounts of your three colors (460-nm blue, 530-nm green, and 650-nm red) needed to match any monochromatic (spectral) color that we choose. Notice that the required amounts of red and green colors are zero at 460 nm. This is because you can match our 460 nm using only your 460-nm blue. The relative amount of the blue, then, is 100% at that point. Similarly, the blue and red amounts vanish at 530 nm, while the blue and green vanish at 650 nm. (For historical reasons, "equal amounts of blue, green, and red" means that the intensity ratios of blue/green/red are about 1.3/1.0/1.8. The curves shown here and in succeeding figures are standardized; the actual data vary somewhat with observer, intensity of light, etc.)

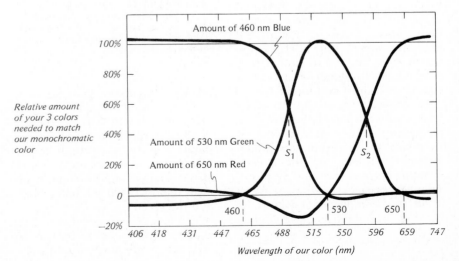

given in Figure 9.10. For example, if we choose a yellow with λ = 570 nm, you must add a little blue to it, and then you can match the resultant color with green and red. (The curves indicate 36% red, 66% green, and 2% negative blue will match this particular yellow.) Several features of Figure 9.10 should be noted. The curves are made at a standard amount of our monochromatic color. They will vary somewhat as we change the amount of that color. Further, the curves give the *relative* amounts—the fractions of the total. The three relative amounts must add up to 1 at each wavelength. This means that if we know any two of the curves, we can figure out the third. (For example, knowing that you need 36% red and 66% green to match yellow tells you that you need a negative 2% blue, since 36% + 66% − 2% = 100%.)

Figure 9.11 shows another way of plotting the relative amounts of 650-nm red and 530-nm green needed to match any color, and from these we can figure out the relative amount of the 460-nm blue. Here, points on the horseshoe-shaped curve correspond to the lo-cation of the spectral colors. For example, reading down from 570 (our yellow) on the curve, we see that the relative amount of red needed is 0.36 or 36%. Reading across tells us that the relative amount of green needed is 66%. (The fact that these two add up to more than 100%, i.e., that the 570-point lies outside the triangle formed by your three colors, means that a negative amount of one color is needed, in this case −2% blue.) Thus this curve contains all the information of Figure 9.10. Why go to this trouble to re-draw Figure 9.10 when the result just looks like a distorted color circle? Because on this new diagram

---

FIGURE 9.11

The relative amounts of 650-nm red and 530-nm green that, along with 460-nm blue, are needed to match any color. The horseshoe curve shows the locations of the spectral color to be matched. To use this, pick a spectral color to be matched, find its location on the horseshoe curve, and read off (by reading down) the amount of red and (by reading across) the amount of green. The amount of blue is determined by subtracting the sum of the green and red from 1.00.

the rules for *additive mixtures* are easily expressed:

Additive mixtures of any two colors lie on a *straight line* connecting those colors.

For example, any mixture of 650-nm red and 530-nm green lies on the diagonal straight line between red and green, inside the horseshoe. Further, the point *midway* between the red and green points corresponds to a *mixture of equal amounts* of the two colors. A point on the line that lies closer to the red has *proportionately* more red.

We know that any yellow made from red and green is slightly desaturated. Thus, the regions *inside the horseshoe* must correspond to the *unsaturated colors*. We also know that mixing red and blue gives the purples, so the region connecting *across the bottom* of the horseshoe must be the location of the *purples*.

Suppose you start with 570-nm yellow and add 460-nm blue. The resultant color lies along the straight line connecting those two points. A little blue changes the pure yellow to one just inside the horseshoe—a slightly desaturated yellow. A lot of blue moves the color close to the blue point—a desaturated blue. Somewhere between must lie the point corresponding to *white*, since blue and yellow are *complementary*—a properly balanced mixture of the two makes white. Hence white lies toward the middle of the diagram.

Knowing the location of the white point, you can find the *complement* of any color by the following rule:

The complement of any color is found by extending a straight line from that color through white and to the *opposite side* of the horseshoe.

For example, a line from 650-nm red through white intersects the horseshoe at around 490 nm, cyan—the complement of red.

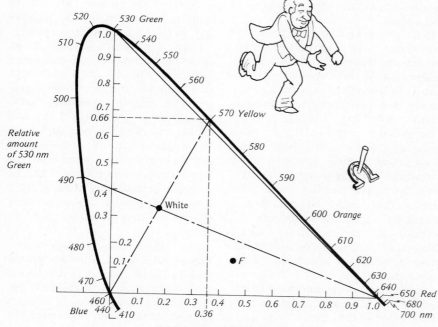

Relative amount of 650 nm Red

The white point lies approximately where the three colors, blue, green, and red, all contribute equally. Its exact location depends on which white we choose as our standard white (e.g., daylight or a particular artificial light source). The arbitrariness of our choice of what we call "white" is illustrated in the poem of Jan Andrzej Morsztyn, "On His Mistress":

White the polished marble of Carrara,
  White the milk from cow sheds
    fetched in baskets,
White the swan and covered with
  white plumage,
  White the pearl unspoiled by
    frequent stringing,
White the snow, fresh fallen, still
  untrampled,
  White the lily plucked in all its
    freshness,
But whiter still my lady's face and
  neck
  Than marble, milk, swan, pearl,
    fresh snow, and lily.

From this we can deduce that his mistress was Caucasian (although she sounds more like a polar bear), but it's hard to agree on what is really meant by the term "white," whether we're talking about surfaces or lights. Because of the ambiguity of the term, a **standard white illuminant** must be chosen. A variety of such illuminants is often used, and when precision is important we must specify the standard. The region in the chromaticity diagram near this white point (whatever the standard) contains the very desaturated colors. As we get farther from it, and closer to the horseshoe, the colors become more and more saturated.

If you want to know the hue of a color F on the diagram, draw a straight line from white, through F, to the spectral horseshoe. The intersection of that line with the horseshoe is the location of the corresponding saturated color—the hue or **dominant wavelength**—because F may be thought of as a mixture of white and that spectral color. If the hue of F is *purple*, extend the line *backward* through white. It will intersect the horseshoe in the green, and give you the

location of the *complementary* wavelength, which defines a purple hue (Sec. 9.4B).

Using the three colors, 650-nm red, 530-nm green, and 460-nm blue, you can make any color inside the large triangle without using negative contributions. Thus, if you wish to make the color F, start with red, add a little green (thus moving along the diagonal from red toward green), and then add some blue (thus moving from the red-green line toward the blue point). Proper amounts of green and blue will then get you to F. You can now see why your best choices for the color-matching game are blue, green, and red. These three colors give about as big a triangle as can fit inside the horseshoe, so they give the greatest variety of colors without negative contributions—that is, the greatest variety of colors that can be physically achieved with three lights. That is why *blue, green, and red* are called the **additive primary colors.** Had you chosen, say, blue, *yellow*, and red, you would not be able make as many colors; the triangle formed from these three points is smaller. For this reason, the phosphors used in color TV (which, we'll see, works by additive mixture) are near the additive primaries—as saturated as phosphors will allow.

Of course, even using the primary colors, you can't match other *spectral* colors. However, for most purposes that is not important, because most colors occurring in nature are unsaturated, as we've noted before.

Nevertheless, the idea of not being able to match the most saturated colors without negative contributions goes against the grain. What we would like are three primaries that could be mixed together, in positive amounts only, to produce all the spectral colors. Clearly the triangle formed from these primaries would have to be *bigger* than the spectral horseshoe, so these primaries would have to lie *outside* the horseshoe. That is not physically possible, but it is mathematically. We can invent three such

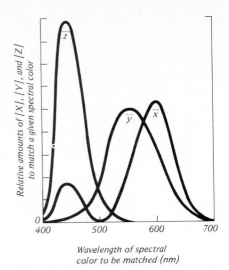

FIGURE 9.12

Tristimulus values: $\bar{x}$, the relative amount of [X], $\bar{y}$, the relative amount of [Y], and $\bar{z}$, the relative amount of [Z], needed to match a given spectral color. The [X], [Y], and [Z] are imaginary primaries, and the curve was not measured but rather was derived mathematically from Figure 9.10 and the definitions of the imaginary primaries.

imaginary colors (called [X], [Y], and [Z]) by making suitable combinations of the red, green, and blue we've been using. (For example, the imaginary [X] would consist of roughly 150% red and a negative 50% green—a kind a "super-red.")

Even if we can't actually make these new primaries, we can figure out from Figures 9.10 and 9.11 and from our choice of new primaries what the corresponding curves would be had we used [X], [Y], and [Z] instead of red, green, and blue. Redrawing Figure 9.10 using the new primaries (and somewhat different units) gives Figure 9.12, the so-called **tristimulus values.** That they are all positive means we have chosen our new primaries appropriately.

The figure that corresponds to Figure 9.11, using the new primaries, is Figure 9.13. It gives the relative amount, *x*, of primary [X] and the relative amount, *y*, of primary [Y] needed to match any color. The two numbers, *x* and *y*, are called

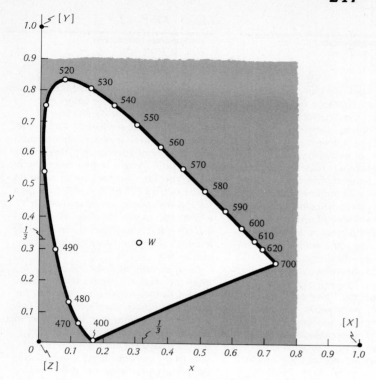

FIGURE 9.13

C.I.E. chromaticity diagram.

the color's **chromaticity.** (As with Fig. 9.11, the relative amount of primary [Z] can be calculated from x and y, since they must all add up to 1.) The units of x, y, and z are chosen so that white occurs in the vicinity of the "equal energy" point at $x = \frac{1}{3}$, $y = \frac{1}{3}$, and consequently $z = \frac{1}{3}$, with the exact location depending on the standard chosen. (These are different units from those used in Fig. 9.11.) Aside from these details, this C.I.E. **chromaticity diagram** has all the same properties as that of Figure 9.11; additive mixtures are found by drawing straight lines (the more of a color mixed in, the closer the mixture is to that point), complementary colors are found by extending a line from the color in question through white to the opposite side of the spectral horseshoe, the saturated colors are on the perimeter of the horseshoe, the unsaturated ones are on the inside, and the saturated purples are on

the line closing the bottom of the horseshoe (Plate 9.2).

Besides giving us precise rules of additive mixing (rather than the simplified version of Fig. 9.7), the chromaticity diagram gives us a precise way of specifying colors; we need only provide two numbers, the chromaticity, x and y. (We must also, of course, specify the lightness of the color. A chromaticity diagram is drawn for a given lightness, and the identification of points on the particular diagram with specific colors will depend on that lightness.) We don't have to go very far out on a limb to find uses for chromaticity coordinates; horticulturists use them to monitor the redness of apples. Also, we will be able to describe the results of say, mixing certain pigments, by drawing the path in the chromaticity diagram that the color of the mixture follows as the ingredients are changed (even though that color mixture is not additive). We'll make use of this diagram throughout our discussion of color.

There are many different techniques for mixing colors additively, and the chromaticity diagram rules apply to all of them.

A. Simple addition

Simple additive mixing occurs whenever two different light sources illuminate the same region, for example, when light from the blue sky coming through your window and yellowish light from your incandescent lamp both fall on this page. Since both lights enter your eye together, they add to each other. Such mixtures are extensively used in stage lighting. Here a number of different colored lights may be aimed at the stage, and their relative intensities may be varied throughout the performance. The first color photographs were projected using three projectors in register, one for each of the additive primaries. The same technique is still used in large-screen color television projection (such as is found in sports arenas and home projection systems). There the TV projector consists of three different mirrors or lenses that project enlarged images, in register, of three small very bright picture tubes, each of which contains the TV image in one of the three additive primary colors.

B. Partitive mixing

Another way of enabling lights of different colors to mix is to place small separate sources close to each other. If your eye cannot see them as separate sources, their colors will mix additively. This type of additive mixing is called **partitive** mixing. This is the way an ordinary color TV picture is produced. There is one picture tube with three electron guns in it. The electrons from different guns are directed to different points on the screen by a mask with many holes. The screen consists of dots of three different phosphors, each of which will produce one of the additive primary colors (Plate 9.3). The dots are arranged so that each type can be hit through the mask by only one of the guns. Each gun makes a complete TV picture in one of the primary colors, but the pictures are tightly interlaced on the screen, so you don't see them individually. Instead you see their additive mixture—the full color picture. While this is less cumbersome than using three different projectors, it means that you get less light from a given color because each primary color, at best, comes from only one third of the screen. (This is the reason most large-screen TV projection systems use three different picture tubes.)

Partitive mixing was also used by the pointillist painters, who put small dabs of different color paint near each other (see Plate 7.2). Up close you see just a confusion of differently colored dots, but if you step back (or take off your glasses) these colors fuse to give a more uniform color, corresponding to the additive mixture of the different colors of the dots. (Interestingly enough, the colors mix even though you can still make out the individual dots—see Sec. 10.6C.) You can see the same effect if you look at a mosaic or at a stained glass window from different distances or at a tree in autumn color, where the individual leaves may be green or red, but the tree looks brownish yellow from a distance. Textiles often achieve their colors by partitive mixing, using a tight weave of two different colors in the warp and weft. Early color film also used partitive mixing to achieve its color (see Sec. 11.3). You can study some of these examples using the procedures suggested in the TRY IT.

TRY IT

FOR SECTION 9.5B
Partitive mixing

Examine the screen of a color TV with a magnifying glass. First look before, then after, turning on the TV. Notice the array of colored phosphors. Different manufacturers use different arrays; some have the three colors intermixed in triangular arrays, others have the dots of a given color arranged in vertical lines, which alternate with the lines of other colors. Notice that most colors of the TV image involve all three phosphors.

Find a channel projecting in black and white, or turn down the color control, and move as close to the screen as necessary to resolve (without the magnifying glass) the separate colors (all three phosphors must give off light to produce white). Now move your head back until the colors fuse into white. Notice then that you can still see the individual dots (or lines of dots) even though you cannot resolve the individual colors.

Repeat this last procedure using the pointillist painting of Plate 7.2. Notice the colors of the individual dots when you are close. As you move back, notice how they fuse. If you are careful, you should be able to check the additive mixing rules.

Using your magnifying glass, examine the way the colors of textiles actually come from the different colors of the individual fibers.

*C. Other ways

Another additive way of mixing colors is to put them "near each other" in *time*—that is, change the colors so rapidly that the positive afterimage of one mixes additively in your eye with the image of the next. The most familiar example is a rapidly rotating wheel with different color segments: when it is standing still, you see differently colored wedges; when it is rotating, its color looks uniform and is the additive mixture of the individual colors. (See the first TRY IT.)

It's also possible to mix colors additively further along toward the brain by doing it *binocularly*—exposing each eye to a different color *may* result in an additive mixture of these colors. (See the second TRY IT.)

First TRY IT

FOR SECTION 9.5C
The color wheel

A disk with sectors of different colors can be rotated rapidly so the colors mix additively. You'll need some way of spinning the disk sufficiently fast so the colors fuse—even a 78 rpm phonograph turntable is not fast enough. You may be able to do it using a simple mounting on which you can spin the disk by hand (as in the fifth TRY IT for Sec. 7.7A), or on a toy top with some sort of spindle on which you can put the disk. More effective, if it's available, is the shank of a motor, such as on an electric fan or power drill.

Cut a number of disks out of differently colored paper. Also cut one cardboard disk to serve as a backing for the colored disks. Each disk should have a hole in the center that just fits over the spindle or shank. The colored disks should have a slit cut from the central hole to the outside, along a radius. This will allow you to slide the disks over each other, overlapping the different colors as much or as little as you like. Place the cardboard disk and then two colored disks, say red and green, on the spindle or shank. Start with half of each color showing. It is a good idea to fasten them down with tape, paper clips, a plastic disk, or a nut if the spindle or shaft is threaded. Take care that the overlapping edges are trailing (not leading) as the disks rotate. The colors will be best if you shine a good strong light at the rotating disks. They'll be even better if you use fluorescent colors, which you can cut out of the black-light posters you've outgrown. (If you use a fluorescent light, you may get interesting strobe effects—Sec. 7.7B.)

Notice the color mixture of the red-green combination. Vary the relative amounts of red and green, and trace out the path of the resultant mixture colors on the chromaticity diagram of Plate 9.2. Try it with other colors. See if you can find colors that form complementary pairs. (They will combine to form gray. Why not white?) Also try to make a white or gray mixture using three colors. When you try to make a mixture of a particular color, and your first attempt produces the wrong color, use your knowledge of the chromaticity diagram to figure out which color should show more, and which less.

Second TRY IT

FOR SECTION 9.5C
Binocular additive color mixing

If one color strikes one eye and a different color strikes the other, you will either get fusion, and see an additive mixture of the two colors, or binocular rivalry, and see the two colors alternately. There are two simple ways to do the experiment. If you have some colored filters (pieces of colored, transparent plastic, or cellophane), simply take two different filters, say red and green, hold one filter over each eye, and look at a plain white background—a bright white light (not the sun!) works best.

Alternatively, hold two colored pieces of paper about $\frac{1}{3}$ m in front of you, separated by a little less than a centimeter. Look at a distant object through the space between the papers. Each of your eyes will then form an image of the nearby colored papers, and, because your convergence is determined by the distant object, these images will overlap. In the region of overlap, you may get the additive mixture of the two colors. Whether you do or not depends somewhat on the colors involved and whether there is structure in the background.

If you achieve mixing in either of these ways, your brain will have concluded that, for example, there must be yellow at a place from which it received signals due to both red and green light, even though neither the red and green light, nor the signals produced by them, ever mixed prior to your brain.

9.6
COLOR MIXING BY SUBTRACTION

When you combine two colored lights, you get an additive mixture—the contributions from the two lights add. To make the original colored lights in the first place, you might shine two beams of white light through colored filters. But what happens if you *combine the filters* themselves, that is, place them back to back and shine one light through the combination at a white screen? Using the red and green filters, you see very little or nothing on the screen; you do not get the additive mixture, yellow. This is because a *filter doesn't add* color to the light that shines through it; rather it transmits different amounts of the wavelengths already contained in the incident light. The red filter transmits only red light—it absorbs blue and green light. The green filter transmits only green light—it absorbs red and blue light. In particular, the green filter absorbs any red light previously transmitted by the red filter. So by the time the light has passed through *both* filters, everything has been absorbed and nothing gets to the screen. Combining the filters this way illustrates **subtractive mixing,** so called because it is easiest to analyze the process involved by considering what each filter *takes away* from the original light beam.

The subtractive *process* is quite common. When we say an opaque object has a particular color, we generally mean that, when white light shines on it, the object *absorbs (subtracts) certain wavelengths* from the white light, *reflecting the rest* to our eyes. The color of the object is determined by the reflected part of the incident light, the part that the object doesn't absorb. Thus, if we know what is taken away from white light, we know the resulting color. For example, a robin's egg is blue because the shell absorbs green and red light, reflecting only the blue.

A. The simple subtractive rules

Filters are characterized by their **transmittance curves**—the curves that show what fraction of the incident light is transmitted at each wavelength. When you send *broadband white* light through a filter, the intensity-distribution curve of the transmitted light is the same as the filter's transmittance curve. It is important to understand that a filter cannot change the wavelengths contained in the light, just the intensities.

In order to determine the result of mixing two colors subtractively, you must know the transmittance curves of the filters involved. (The TRY IT shows how to estimate the transmittance curve of a filter.) In general, the subtractive rules are complicated. However, for **ideal filters,** which transmit 100% at some wavelengths and 0% at all others, the laws of subtractive mixing are fairly simple. (Figs. 9.14a, b, and c show transmittance curves for ideal filters. Compare them with transmittance curves of more realistic filters, which may look like Figs. 9.6a, b, and c.) If you use these three ideal filters for subtractive mixing, any two of them will result in black (no light passing through). (Why?)

For subtractive mixing, it is more interesting to consider filters that pass the *complements* of the additive primaries. That is, yellow (made additively of green plus red) is the complement of blue, so it may be made subtractively by *absorbing* the blue in white light:

$$Y \equiv G + R \equiv W - B$$

Similarly for magenta and cyan:

$$M \equiv B + R \equiv W - G$$
$$C \equiv B + G \equiv W - R$$

Transmittance curves for filters of these three **subtractive primaries** *(yellow, magenta, and cyan)* are shown in Figures 9.14e, f, and g. If white light shines through both a yellow and a cyan filter, the yellow subtracts the blue and the cyan subtracts the red, so the combination subtracts both blue and red, leaving only green. The results for all subtractive combinations of these subtractive primaries are summarized in Figure 9.15. Notice that the triple combination of all three filters subtracts blue, green, and red, leaving nothing—black. Contrast that with the *additive* case (Fig. 9.7) where, when everything is added together, you get white. (Recall that, for the additive case, we found it surprising that green and red gave yellow, since yellow doesn't look like reddish green. Here we should similarly be surprised that cyan and magenta give blue—blue doesn't look like cyanish magenta.)

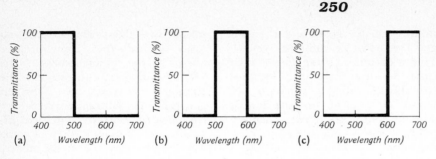

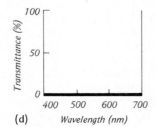

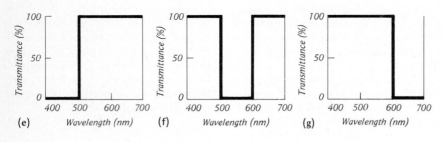

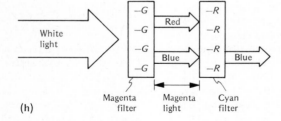

FIGURE 9.14

Ideal filters. Transmittance curves of
(a) an ideal blue filter, **(b)** an ideal green
filter, and **(c)** an ideal red filter.
(d) Result of a subtractive mixture of any
two of **(a)**, **(b)**, and **(c)**. Also shown are
transmittance curves of filters that
transmit the same light as one gets from
additive mixtures of **(a)**, **(b)**, and **(c)**: **(e)**
ideal yellow filter (green plus red ≡ white
minus blue), **(f)** ideal magenta filter (blue
plus red ≡ white minus green), and **(g)**
ideal cyan filter (blue plus green ≡ white
minus red). Notice that **(a)**, **(b)**, and **(c)**
also give the results of *subtractive*
mixtures of **(e)**, **(f)**, and **(g)**. For
example, a subtractive mixture of **(f)** and
(g) gives **(a)**, as shown schematically in **(h)**.

FIGURE 9.15

Simple subtractive mixing rules. The
drawing shows the effect on white light
of three partially overlapping broad-band
filters, which produce subtractive mixing:

A subtractive mixture of C and M ≡ B.
A subtractive mixture of C and Y ≡ G.
A subtractive mixture of Y and M ≡ R.
A subtractive mixture of C, Y, and M ≡ Bk
(Black).

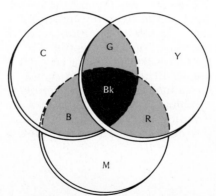

TRY IT

FOR SECTION 9.6A
Making a spectrum
and transmittance
and reflectance curves

1. Making a spectrum

*If you can make a spectrum of the visible
colors, you will be able to try a number
of the effects associated with color. The
spectrum in Plate 2.1a will not often be
adequate, since it consists of only the
three dyes used in printing it. Neither
will a rainbow do, since there the
different spectral colors overlap (Sec.
2.6B). To make a good spectrum you will
need a slide projector and either a prism
or a diffraction grating (see Sec. 12.3A).
Prisms can sometimes be found hanging
from chandeliers, or the corner of a fish
tank can be used, and both prisms and
gratings may be purchased cheaply. An
adequate substitute for a grating is an LP
record.*

*Instead of putting a slide in your
projector, insert a blank cardboard frame
covered with aluminum foil, in which
you have cut a short slit using a sharp
razor blade. Focus the image of the slit at
about the distance where you will project
the spectrum (onto a white screen, wall,
or ceiling). Then place the prism or
grating in front of, and close to, the
projection lens. You should see a
spectrum located some distance away
from where the slit image originally was.
In the case of the diffraction grating or
the record, you'll see the white slit, but
off to the side you should see the
spectrum. If you use an LP record,
choose one with only one or two broad
bands on a side (a symphony works
well), and hold it at a slant so the beam
strikes it at a grazing angle and is
reflected to your screen. Rotate the
prism, grating, or record to get the best
colors, and then adjust the focus to get
the sharpest image.*

*For further experiments using the
spectrum, you'll want to secure the
prism, grating, or record so that it
doesn't shake. Chewing gum works
adequately, but be sure to keep it away
from the grooves of the record.*

2. Transmittance curves
of filters and dyes

*Once you've made your spectrum, you
need only hold a filter in the light beam
to estimate the filter's transmittance
curve. Let the filter block only half of the
beam, so you can see both the original
and the filtered spectrum. By comparing
the two spectra, you can see, for
example, that most of the red, yellow,*

and green, but very little blue, passes through a yellow filter. Try to draw the transmittance curve by estimating how much light of each color passes through the filter.

The filter can be a piece of transparent colored cellophane, colored glass, etc. You can also use dyes. Put a little food coloring in water in a clear glass bottle with straight sides, so all the light passes through the same thickness of colored water. Try other clear liquids you may have, such as apple juice, shampoo, etc.

3. Reflectance curves

In the same way, you can estimate how much of a given color is reflected by colored objects by holding them in your spectrum. Try using pieces of colored paper, comparing how the colors of your spectrum look on the paper and on the adjacent white screen. Also try dyed textiles.

B. Subtractive mixture laws for realistic filters and dyes

The subtractive rules summarized in Figure 9.15 are only approximate—strictly speaking they apply only to the ideal filters of Figure 9.14. They would also apply to ideal transparent **dyes.** Dyes are substances that absorb certain parts of the spectrum. Mixing transparent dyes also results in a subtractive color mixture. In general, there are no simple rules that allow you to predict the result of a subtractive mixture of *real* dyes or filters, if you know only the colors (i.e., chromaticities) of the constituents you are mixing. For such *subtractive mixing*, the result depends on the details of the intensity-distribution curves (that is, on the transmittance curves of the filters or dyes)— a subtractive mixture of, say, a particular yellow and cyan may look very different than one of a different yellow and cyan, even though the two original pairs of colors may look the same.

To illustrate how complicated even a simple case can be, consider a filter or dye with the transmittance curve shown in Figure 9.16a.

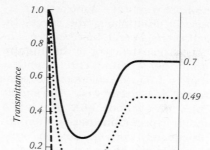

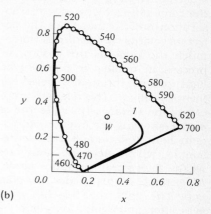

(a)   (b)

FIGURE 9.16

Subtractive mixing of a color with itself. **(a)** Transmittance curve of one filter (solid line); of two identical such filters, one behind the other (dotted line); and of many identical such filters, one behind the other (dashed line). Here, instead of giving the percentage of incident light transmitted at each wavelength, we give the *fraction* transmitted. That is, we've simply changed from percentage to decimal equivalent. **(b)** Path, in the chromaticity diagram, of the color of the light transmitted as more and more filters are used and the intensity of the incident light is increased proportionally. One filter gives a desaturated orange (the point marked *1*). Several filters give an unsaturated purple. Many filters result in a monochromatic violet.

Most of the transmission lies in the region between 550 and 700 nm, giving an unsaturated orange; the little bit of complementary blue (violet) at 400 nm only desaturates the orange further. What happens if we put one such filter behind another (or double the concentration of the dye)—that is, if we subtractively mix this color with *itself*? (Subtractively mixing the *ideal* filters of Fig. 9.14 with themselves has no effect on the color—100% of the yellow passes through the filter of Fig. 9.14e, no matter how many such filters the light encounters.) Using the transmittance for one filter, shown in Figure 9.16a, we can figure out the transmittance of two filters, as follows. We see, for example, that 0.7 (70%) of the 600-nm orange light passes through the first filter. When this orange light

reaches the second, 0.7 of *it* will pass through. That is, $0.7 \times 0.7 = 0.49$ (49%) of the orange passes through both filters. We see, then, that the transmittance of two filters together is found by *multiplying* the individual transmittance curves—for each wavelength, you multiply the transmittances of each filter. Figure 9.16a shows the resulting transmittance of the two identical filters—a slightly more saturated orange. But suppose we increase the number of filter layers to ten, or the dye concentration tenfold. The transmittance in the wide orange range will then be:

$$0.7 \times 0.7 \times \cdots \times 0.7$$
$$= (0.7)^{10} = 0.03$$

that is, hardly any orange (3%) is transmitted. However, the transmittance at 400 nm is:

$$1.0 \times 1.0 \times \cdots \times 1.0$$
$$= (1.0)^{10} = 1.0$$

so *all* of the 400-nm violet light still gets through the layers of filters. So *this* is the dominant wavelength of our ten-filter combination—the light passing through ten of these orange filters looks violet! (Of course, much less total light gets through ten filters than through one, so the violet is darker than the original orange. We can make up for this by increasing the intensity of the incident white light.) Figure 9.16b shows how the color changes as more and more filters are used. (The TRY IT gives other examples.)

Subtractive mixing of a color with itself can also occur due to *multiple reflection.* If light is reflected from the same object repeatedly, the absorption process that gives the object its color is repeated each time—much the same as the absorption process in a filter is repeated as light passes through a series of identical filters. The relevant property of the surface is its **reflectance curve** (showing the fraction of incident light *reflected* at each wavelength), which plays the same role as the filter's transmittance curve. An example of this occurs in a gold cup. The light that reaches your eyes after undergoing only one reflection from the cup looks yellow (golden). But the light reaching your eyes from inside the cup may have undergone several reflections from the walls and bottom, and takes on a warmer tone with each reflection. The inside of the cup consequently looks more orange, in some cases almost scarlet. Similarly, the inside of a pink rose looks a more saturated red, and the colors in the folds of drapery may appear more intense.

The story becomes even more complicated if we examine the subtractive mixing of two *different* colors. (Fig. 9.17 gives a somewhat pathological example of a blue and a yellow dye that, when mixed, give almost any color *other* than green.) The moral of all this is that, if we know the transmittance (or reflectance) curves in full detail, we can

find the result of a subtractive mixture by multiplying the curves together. In general, however, we cannot predict the color of the mixture from the separate chromaticities, that is, from the colors of the separate dyes as seen by eye. The simple rules of Section 9.6A give us a rough expectation that is often approximately correct, but on occasion they can be very misleading.

FIGURE 9.17

Subtractive color mixing of two different dyes at various concentrations. **(a)** Transmittance of *(1)* blue and *(2)* yellow dyes at unit concentrations, and *(3)* a one-to-one mixture of the two dyes, also at unit concentration. **(b)** Chromaticity paths as the concentration of mixtures of the two dyes is increased. At very low concentration the dyes are almost transparent, so the illuminating white light passes through unchanged *(W)*. As the concentration is increased, the color becomes more saturated, ultimately becoming red at high concentrations. The path between white and red depends on the ratios of the concentrations of the two dyes. Shown are a one-to-one mixture (1:1), a three-blue-to-one-yellow mixture (3:1), and a one-blue-to-three-yellow mixture (1:3). The points marked *b* and *y* are the colors of the unit concentration dyes shown in **(a).** Thus, appropriate mixtures of these yellow and blue dyes result in almost any color in the lower half of the chromaticity diagram, but *not* the green one might expect from blue and yellow!

Subtractive mixtures of a color with itself

Using dyes, you can verify that a color mixed with itself subtractively may change its hue, saturation, and lightness. Rather than increasing the concentration of a dye, it is easier to decrease it—to dilute a concentrated dye. Food coloring is a convenient transparent dye. Notice that the highly concentrated yellow dye looks red in its bottle. Compare this color to the dilute yellow obtained by putting a few drops from the bottle into a glass of water.

PONDER

What do you think the transmittance curve of the dilute yellow dye looks like?

Also try diluting a spoonful of grape juice with a lot of water.
To see what happens when two different colors are mixed subtractively, you can mix two colored dyes together, or, more amusingly, use food coloring and gelatin desserts. First try to guess the resultant color, then add the dye and mix up the gelatin. Try to draw transmittance curves that explain your results.

9.7
DEPENDENCE OF SUBTRACTIVE COLOR ON THE LIGHT SOURCE

The color of the light reflected from an object usually depends on the color of the illuminating light—for example, gold looks more orange if the light shining on it is itself golden yellow. Thus, Monet repeatedly painted the same scene at different times of the day—the colors of the west facade of the Rouen Cathedral kept changing as the lighting conditions changed. You can see a more extreme example of this effect under the "golden white" sodium lamps one often finds on highways; some objects lose their color because there is very little green and red light emitted by this source (Plate 9.4). Colored lights

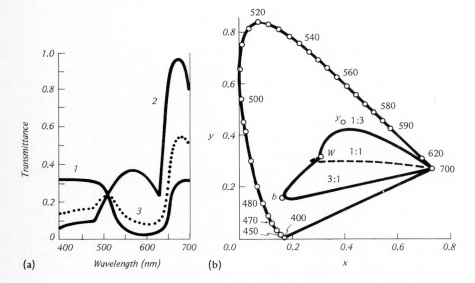

(a) Wavelength (nm)   (b)

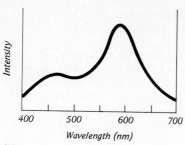

(a)

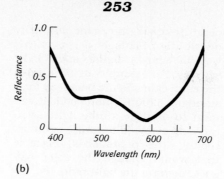

(b)

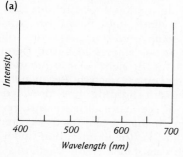

(c)

FIGURE 9.18

(a) Intensity-distribution curve of light from a Cool White fluorescent tube. (b) Reflectance curve of a magenta object. (c) The intensity-distribution curve of the light from that object under Cool White fluorescent illumination is given by the product (at each wavelength) of curves (a) and (b). Under this illumination, the object loses all hue.

the late-afternoon light is reddish. All of these lights may look white to you, since your eye adapts to the illumination (see Sec. 10.6B). For the same reason, makeup mirrors often come with a variety of lights so that you can adjust the color of your makeup under the illumination in which you expect to be seen. (You may not appreciate the fact that your lipstick, which appeared red when you put it on under incadescent illumination, looks purple by skylight. These effects were often quite dramatic before the advent of nonincadescent light sources and artificial, organic dyes. The first green silk dress that didn't look off-color by artificial illumination created a sensation in the last century.) Figure 9.20 gives the inten-

are, of course, quite common in the theater. There, the designer must pay careful attention to the colors exhibited by the sets and costumes as the stage lighting changes.

How can you know what color an object will acquire under nonwhite illumination? You know that the object's reflectance curve tells you what fraction of the incident light is reflected at each wavelength. Hence, if you know the intensity-distribution curve of the incident light, you need only multiply it by the reflectance curve to get the intensity-distribution curve of the reflected light (Fig. 9.18). That is, the subtractive properties of an object (its reflectance or transmittance curve) tell how that object will affect the light illuminating it. The object may look quite different depending on the nature of the illumination.

Even if two illuminating lights look the same, an object may still appear different colors in them. If a white light consists of two narrow bands of complementary colors, say cyan and red, objects illuminated by it can have only one or the other, or a mixture, of these colors: a "yellow" object may appear red or black, depending on how broad its reflectance curve is. However, the same yellow object will appear yellow

when illuminated by a broad-band white, such as daylight, even though both whites look the same when reflected from a white screen. This can have significant effects if you are trying to match the colors, say, of two pieces of cloth (Fig. 9.19). For this reason, when buying clothes, you are often advised to take the material to a window and look at it by daylight, as well as by the artificial illumination in the store. Even that may be insufficient. Light from the northern sky may be bluish, direct sunlight has its maximum in the green, while

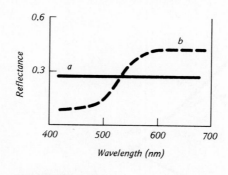

FIGURE 9.19

The reflectance curves of two pieces of cloth. In sunlight, a looks gray, b looks brown. Under incandescent illumination, which lacks the short-wave end of the spectrum, both have the same hue.

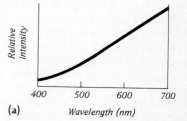

(a)

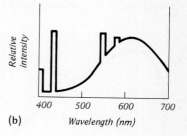

(b)

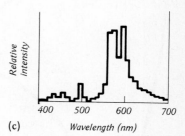

(c)

FIGURE 9.20

Intensity-distribution curves of white-light sources: (a) 100-W incandescent bulb, (b) Delux Warm White fluorescent tube, (c) 400-W high-pressure sodium high-intensity discharge lamp.

sity distributions of several artificial white-light sources.

If the source produces light because it is *hot,* for example, an incandescent bulb or the sun, its color (and consequently the color objects appear when illuminated by it) depends on its *temperature.* The intensity-distribution curves of most hot, glowing bodies are just the black-body spectra we saw in Figure 1.19. The hotter the source, the more the relative intensity at the shorter wavelengths. A cool body radiates almost exclusively in the infrared. As it is heated, it begins to glow red. Further heating may make it yellow, white, or even blue (Fig. 9.21).

Broad-spectrum sources are often classified by their **color temperature,** the temperature of a blackbody source whose color matches that of the source in question. Two sources of the same color temperature give off light that looks the same. Of course, colored objects may appear differently colored under two such sources because, although the sources look the same, their intensity distributions may be different. But almost all *incandescent* sources of the same color tem-

perature have the same intensity distribution. This is a convenient way of standardizing such light sources. For example, in TV studios the color cameras are balanced for a standard 3200°K photoflood. If you want to dim the studio lights, you can't do it by "dimming"—that is, cooling—the photofloods, because that would change the color balance, making the light more red. Although that may not be apparent to the actors' eyes because they adapt, it would be to the home viewers, who can compare the color to that given off by their room lights. Consequently the studio lights are dimmed by the use of a diaphragm in front of them. For outdoor televising, the color cameras must be balanced differently. Similarly, color film is also balanced for a particular light source (see Sec. 11.5).

FIGURE 9.21

The location of the color of the light from incandescent sources at various temperatures. The temperature is in **degrees Kelvin (°K),** where °K = °C + 273. Shown are three standard white sources: *A* = tungsten filament (2854°K), *B* = noon sunlight (4870°K), and *C* = tungsten filament filtered to approximate "daylight" (6770°K). A candle would be about 1800°K, while a photographic flash would be 4300°K.

*A. Which white light source is best?

Since there is a variety of light sources, all of which appear white, which source should you use? This is a problem for which there is no unique answer. When you choose a light source, you should consider several issues. First, of course, a white light should *look white.* Further, a white light must also have good *color rendering* properties—it must make objects look the right color. But the right color is a matter of what you are used to. Because incandescent lighting has provided most of the artificial light until recently, it has been used as one standard for the illumination under which colors look right. Daylight is a more common illuminant but, as we've seen, its intensity distribution varies considerably throughout the day. Various standards based on daylight are also used. The choice here is a matter of taste, not science, and the standards change as our environment changes.

There are also other considerations that do not depend on color, but on economics and convenience. You want a light of high **efficacy**—one that gives you a large amount of light for a small electric bill. This is one reason for the popularity of fluorescent tubes, which have an efficacy three to four times that of incandescent bulbs. Table 9.1 gives the efficacy of several light sources, in *lumens per watt.* The *lumen* is a measure of the total amount of light given off by the light source, weighted by the response of the human eye. Roughly speaking, the more lumens (or the greater the *luminous flux*), the brighter the source will seem to a human observer, independent of the color of the source. The *watt* is a measure of the energy used each second, in this case to produce the light.

You may also be concerned with the size of the bulb (e.g., a small, localized source or a broad, extended source), whether you can direct the light or not, the cost and convenience of maintenance, safety (some sources require high voltages), or a host of other considera-

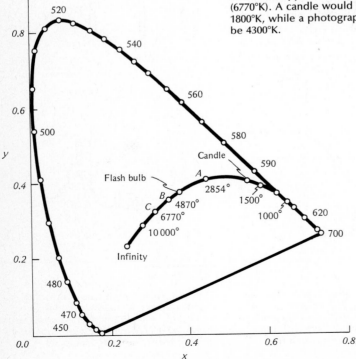

(a)

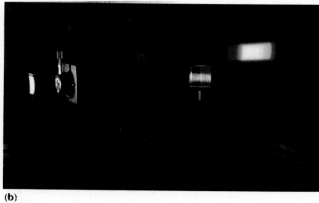

(b)

PLATE 2.1

(a) Spectrum of white light, obtained by sending light from a narrow slit through a prism, as in Figure 2.65a. (b) View of an extended white light source seen through a prism, as in Figure 2.65b.

PLATE 2.1

(a) Spectrum of white light, obtained by sending light from a narrow slit through a prism, as in Figure 2.65a. (b) View of an extended white light source seen through a prism, as in Figure 2.65b.

PLATE 2.2

A photograph of the many reflections of a single light beam falling on a diamond can serve as the diamond's "fingerprint." Rocking the diamond will bring several of these reflections in turn to your eyes—the flash.

PLATE 2.3

Photograph of primary and secondary rainbows.

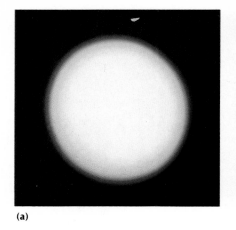
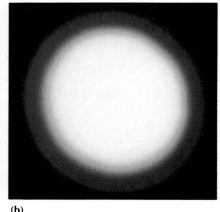

(a)   **(b)**

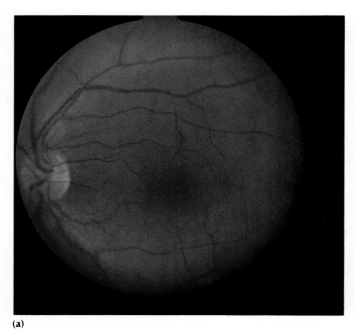
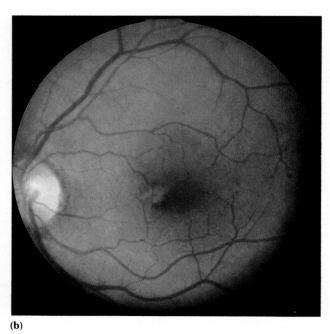

(a)   **(b)**

PLATE 5.3

"Redeye" resulting from the camera flash being retroreflected by the child's choroid. The blood in the choroid is responsible for the red color.

PLATE 7.1

"Arcturus II," Victor Vasarely. The four X's result entirely from the processing of your visual system. Each concentric square making up the blue region, for instance, is of uniform reflectance. Because the corners of each square are directly bounded by darker areas on two sides (and not just one, as are the sides of the square), lateral inhibition makes the corners appear lighter than the rest of the square. All these lighter corners are aligned in an X shape, so you see a light X in the blue (and in the three other colors).

PLATE 7.2

Georges Seurat's "La Poseuse en Profil." By exaggerating the edge enhancement due to lateral inhibition, Seurat makes the difference between the actual light intensity of each region appear greater.

PLATE 8.1

An anaglyph of an optical set-up (a laser shines light through a rectangular mesh, producing a pattern on a screen—Sec. 12.3A). To see this in depth, view with a red filter over your right eye, and a green filter over your left eye. Colored plastic filters usually work well.

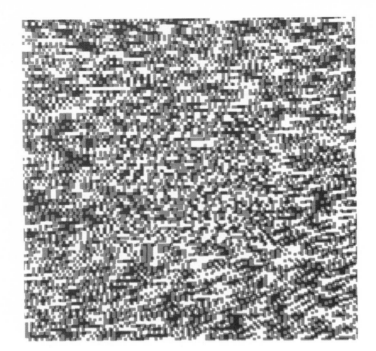

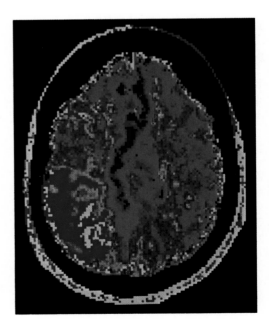

PLATE 9.1

The Munsell color tree.

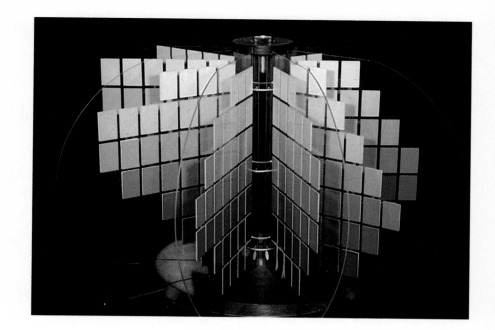

PLATE 9.2

Chromaticity diagram (reproduced as well as can be done using four-color printing—see Fig. 9.23).

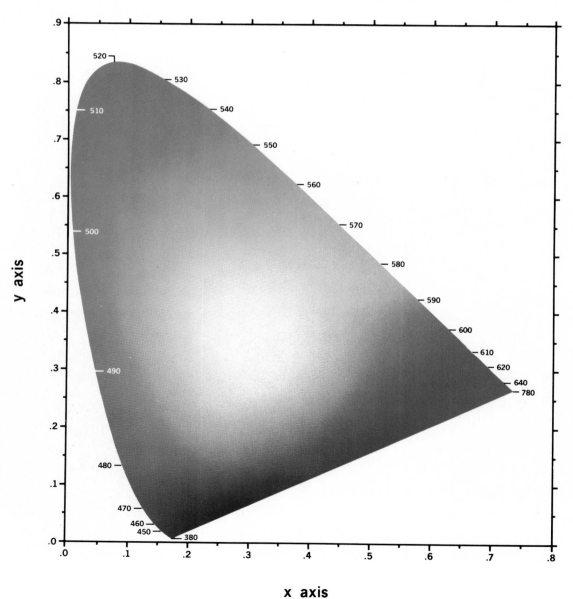

PLATE 9.3

Close-up of a color TV screen, showing the additive primary phosphors.

PLATE 9.4

Objects illuminated by **(a)** sunlight, **(b)** an Agro-Lite (plant light), **(c)** an incandescent light, and **(d)** a "golden white" sodium lamp.

PLATE 9.5

Enlarged colored halftone.

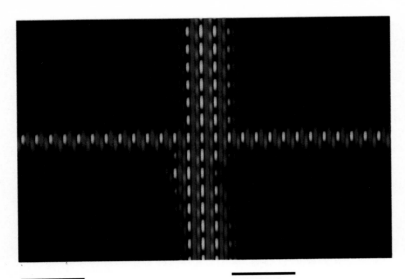

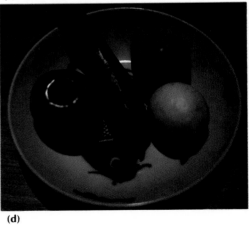

(a)

(b)

(c)

(d)

PLATE 9.6

Morris Louis, "Aleph Series V."

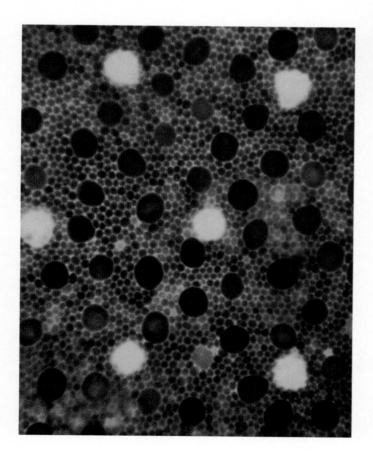

(a)

(b)

PLATE 9.7

Surface reflections considerably desaturate the colors: **(a)** Painting with surface reflections. **(b)** The same painting without surface reflections.

PLATE 10.1

Microscope view of the photoreceptors in a human retina (from an eye bank). The S cones have absorbed a special fluorescent dye that glows yellow, while the other cones have not.

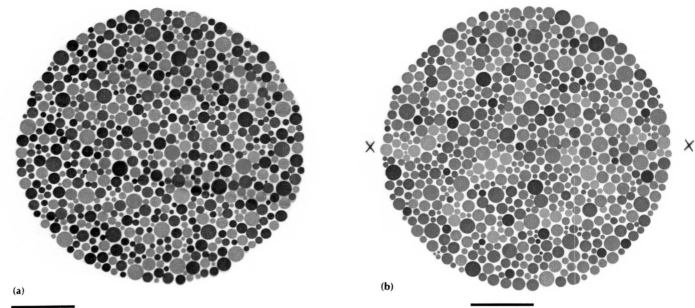

(a)

(b)

PLATE 10.2

Two plates from the Ishihara "Test for Colour-Blindness." The dots in the pattern and in the background are metamers for some observers but not for others. **(a)** Normals see 26. Protanopes and some protanomalous observers see only the 6. Deuteranopes and some deuteranomalous observers see only the 2. **(b)** Normals cannot trace the serpentine path between the two X's, but many color deficients can.

PLATE 10.3

Simultaneous color contrast. The four gray squares are objectively equal (as you can verify with a mask that reveals only the gray squares), but appear to have slight coloration of the complement of their respective surrounds.

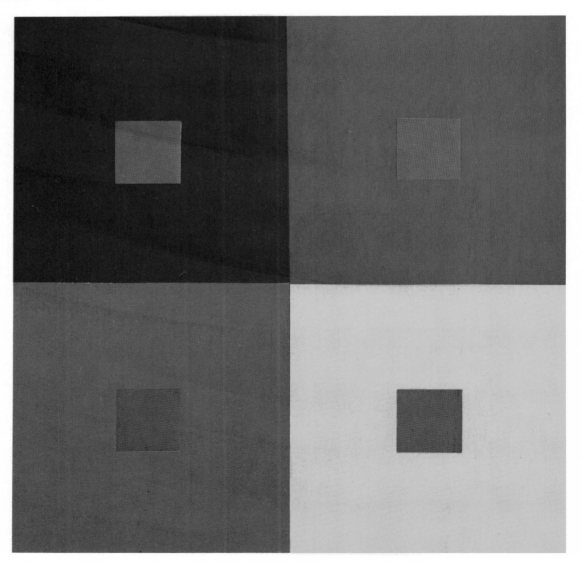

PLATE 10.4

Richard Anuszkiewicz, "All Things Do Live in the Three." The red background is physically the same throughout the painting, but appears to have a color similar to the dots as a result of assimilation.

PLATE 10.5

Colored afterimages. Adapt to the pattern by staring at the center of the plate for at least 30 seconds. Then, look at a white piece of paper.

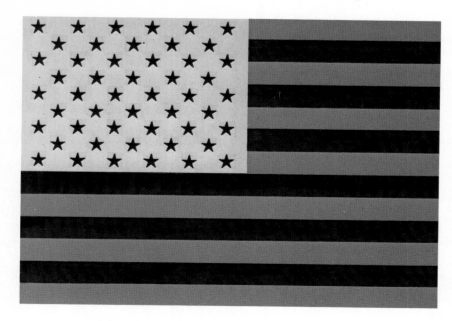

PLATE 10.6 Pattern for illustrating chromatic latency.

PLATE 10.7

McCulloch effect. Adapt alternately ten seconds on the green pattern, then ten seconds on the red for about 5 to 10 minutes. Sweep your eyes throughout the patterns while adapting to ensure a standard afterimage is not formed. Then view the black and white pattern. What colors do you see? Where do you see them? Rotate the black and white pattern by 90° and view it again.

PLATE 11.1

(a) An Autochrome photograph from the turn of the century. **(b)** Close-up showing how the colors are generated.

(a)

PLATE 11.2

Manipulated Polaroid SX-70 photograph.

(b)

PLATE 11.3 Infrared color film gives unusual colors to familiar objects.

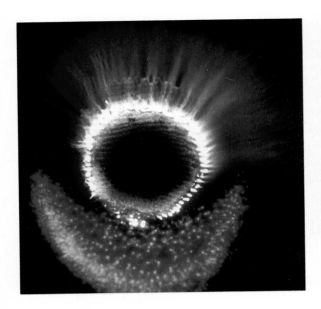

PLATE 11.4

Color Kirlian photograph of a fingertip.

PLATE 12.1

Interference colors in a film of oil floating on water.

PLATE 12.2

The colors of a peacock feather change as it is viewed from different directions (cf Fig. 12.17).

PLATE 12.3

Crawling diffraction gratings: **(a)** ground beetle, **(b)** mutillid wasp. Note the wide spread of the spectrum in the beetle. In the wasp, the spectrum is narrower, but several orders are visible (cf Fig. 12.19).

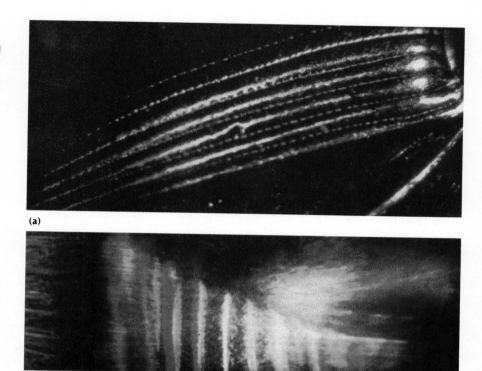

(a)

(b)

PLATE 12.4

Small portion of an opal's surface, illuminated from different directions.

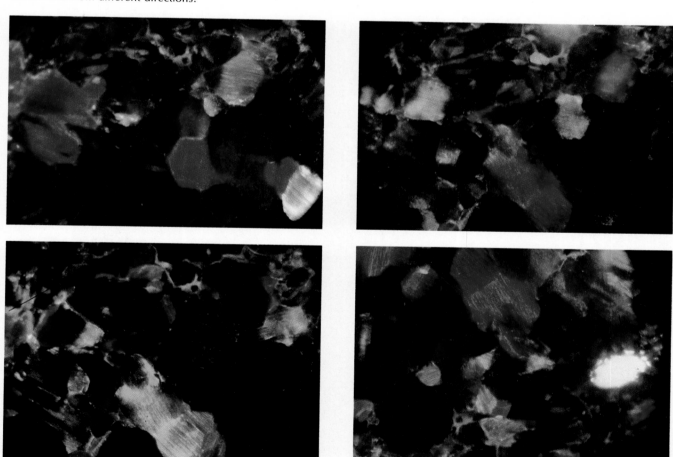

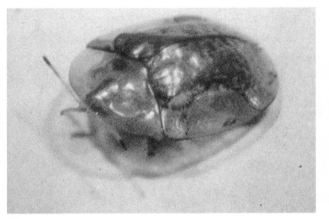
(a)

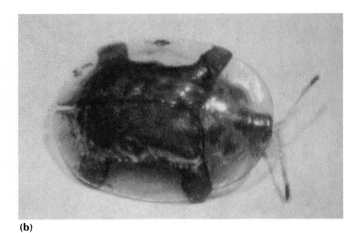
(b)

PLATE 12.5

Colors of the tortoise beetle. Since the spacing of the layers on the wing cases is a fraction of the wavelength of light, there is little change in color with angle of observation. The color changes, due to variation in moisture content, from **(a)** pale gold to **(b)** reddish copper (cf Fig. 12.30).

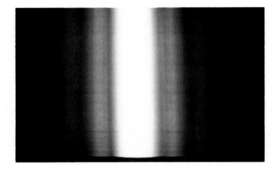

PLATE 12.6

Fraunhofer diffraction of white light by a single slit.

PLATE 12.7

A corona around the sun. (The lamp blocks the sun so the picture isn't overexposed.)

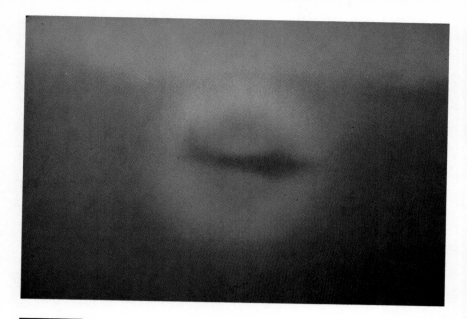

PLATE 12.8

A glory around the shadow (on the clouds) of the photographer's airplane.

PLATE 13.1

Birefringent crystals seen in a polarization microscope.

PLATE 13.2

A piece of transparent plastic exhibits stress birefringence when viewed between a crossed polarizer and analyzer. The colors show the patterns of stress in the plastic. Plastic in the shape of a Gothic cathedral (at Reims) shows the stress patterns in such a structure. The same stress occurs in regions linked by a common color.

PLATE 13.3

The polarization interference fringe pattern resulting from a slab of calcite cut perpendicular to its optic axis and viewed through a crossed polarizer and analyzer, in the manner of Figure 13.17b. The colors of this interference pattern can be reversed to their complements by orienting the analyzer parallel to the polarizer.

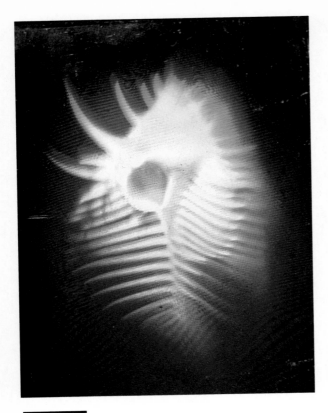

PLATE 14.1

Photograph of white-light transmission hologram of a spiny murex shell. Notice the color blurring, especially near the edges of the image.

PLATE 14.2

Photograph of true-color reflection hologram made using a set-up similar to that of Figure 14.32.

(a)

(b)

(c)

(d)

PLATE 15.1

Various spectra, as seen through a simple grating spectroscope. Only one first-order spectrum is shown. (a) White light from a carbon arc. (b) Sodium spectrum. (c) Mercury spectrum. (d) White light from the sun, showing dark absorption lines due to cooler gases, which surround the sun. The presence of sodium in these gases is indicated by the two dark lines in the yellow, corresponding to the two bright sodium emission lines seen in part (b).

TABLE 9.1 Efficacy of light sources

Source	Efficacy (lumens per watt)
Gas mantle	1–2
Incandescent (Fig. 1.22):	
40 watt	11.4
60 watt	14.5
60 watt (long-life)	13.0
100 watt (Fig. 9.20a)	17.5
High-intensity carbon arc (Fig. 1.22)	18.5
Fluorescent (Figs. 1.24 and 9.20b)	50–80
High-intensity discharge:	
Mercury	50–55
Metal halide (mercury plus metallic salts)	80–90
High-pressure sodium (Fig. 9.20c)	100–140
Maximum theoretically possible:	
Broad-band white	220
Monochromatic 555 nm (maximum photopic sensitivity)	680

tions that depend on your personal taste—you may actually *like* an electric light that looks like a flickering candle.

*9.8
WATER COLORS AND PRINTER'S INKS

Color mixture rules are of particular importance to artists and printers, so let's see how these rules can be applied to practical situations. Water colors are very much like filters. By and large, they reflect very little light themselves. Rather, light is transmitted through the wash of water color, with some wavelengths absorbed more than others, as in a filter. The light then is reflected by the paper and is transmitted back through the wash to the observer's eye. To determine the color of the reflected light, assuming incident broad-band white light, you multiply the transmittance curve of the wash by the reflectance curve of the paper, and then by the wash's transmittance curve again. Clearly, then, the color of the wash will depend on the color of the paper it's on. Further, if you put on too many different colored washes, almost all wavelengths will be absorbed (particularly so because the light passes through each wash twice), so the color becomes dirty.

Printer's inks are similar to water colors, in that they reflect very little. (Such inks are called process colors.) The incident light passes through the ink, reflects from the paper, and passes through the ink again—each step producing color by a *subtractive* process. A small amount of light, however, is reflected directly from the ink—color is produced here, also, by a *subtractive* process, determined by the reflectance curve of the ink (which may be different from the transmittance curve). The light from these two different mechanisms combines *additively* in your eyes (Fig. 9.22).

The simple subtractive rules (Sec. 9.6A) work pretty well for these inks, for the simple reason that only inks that obey these rules are used! The printer wants inks he can rely on to mix according to simple rules, so he can get as many colors as possible out of a small number of inks. Every time he uses another color, he must prepare another plate and print it in register with all other plates. By overlapping the subtractive primaries (yellow, cyan, and magenta—which he may call yellow, blue, and red), he can produce the subtractive mixtures shown in Figure 9.15. However, even his carefully chosen inks are not ideal; when all three are overlapped in the thicknesses he uses, they do not absorb all the wavelengths—the "black" that results has a noticeable color tinge. Therefore printers use a fourth printing plate with black ink **(four-color printing).** This improves the detail visible in the darker parts of the picture. It also allows deeper blacks and enables the printer to achieve better color balance with the other three inks. (Even more colors may be used for high-quality art.)

The main limitation in color printing is that the colors reproduced on paper cannot be too saturated or monochromatic. The narrower the reflectance curve, the less total light is reflected, so saturated colors would be dark. Color television and photographic slides needn't suffer as much from this limitation, because the intensity of their illumination can be increased

FIGURE 9.22

Color from printer's ink. Ray *a* reflects from the ink. Ray *b* passes through the ink, is reflected by the paper, and passes through the ink again. Each reflection or transmission produces color by a subtractive process, but rays *a* and *b* combine additively (partitive mixing).

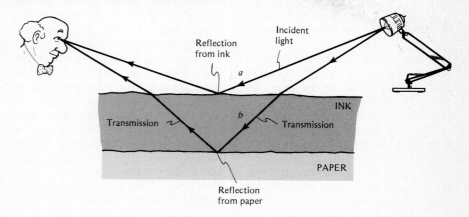

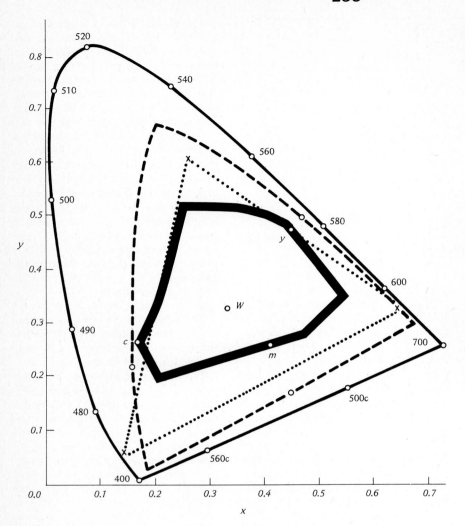

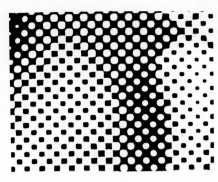

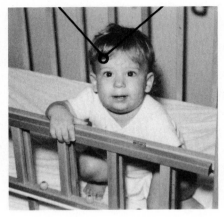

FIGURE 9.24

Close-up of a halftone picture.

FIGURE 9.23

Media colors. Inside the heavy boundary are the colors that can be printed using the subtractive primaries. The colors of the full-strength inks are indicated: y = yellow, m = magenta, c = cyan. Inside the dotted triangle are the colors available in color TV, with the colors of the three phosphors marked by x's. Inside the dashed curve are the colors available in photographic slides, with the colors of the three dyes marked by small circles.

to compensate. A printed color, however, is automatically compared to the white border of the page, so, in order not to appear dark, it should reflect an appreciable fraction of the illuminating light. The region on the chromaticity diagram that can be reproduced by a four-color print is near the white center—for the most part unsaturated colors (Fig. 9.23).

A. Halftones

The printer wants not only the six colors shown in Figure 9.15, and the darker colors that can be made by adding black, he also wants lighter, less saturated colors. He can't get these by diluting the inks, because that would require a separate plate for each different lightness. Instead, he achieves the lighter color by printing less ink in a given area, usually by the **halftone** technique. The black or colored ink is put on in a fine array of dots (Fig. 9.24). A region meant to look black will have large black dots that touch each other and almost completely cover the paper. A region meant to look gray will have smaller black dots—the smaller the dots, the lighter the gray. The colored inks are also put down in dots, the

smaller dots producing the lighter, less saturated colors by virtue of the partitive mixing of the light from the colored dots and from the white paper between them (Plate 9.5; also see the first TRY IT).

To mix colors, the printer uses a separate set of dots for each color. Each array of dots is rotated (say, by 15°) from the preceding one, to avoid the creation of large-scale, symmetrical clustering of dots, called a **moiré*** pattern, which would result if the two sets of dots

---

*A French word that derives from an earlier word for mohair, the lustrous fabric made from angora goat hair. The word *moiré* is usually translated as "watered," because the moiré pattern often looks like ripples on the surface of water. Silk exhibiting this pattern is called watered silk.

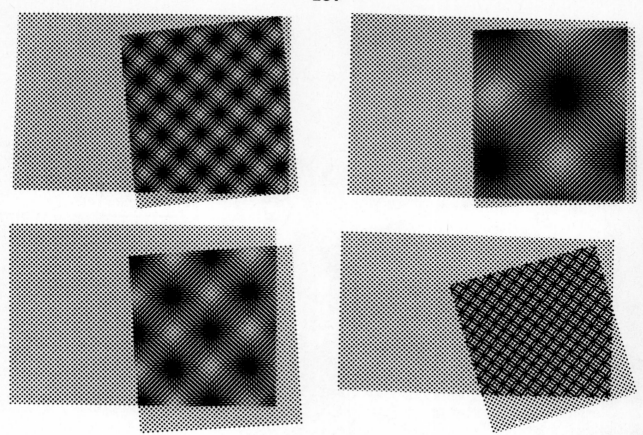

FIGURE 9.25

Halftone arrays at different angles produce different moiré patterns.

were almost, but not exactly, aligned (Fig. 9.25). (If you are unfamiliar with moiré patterns, see the second TRY IT.) Where dots of different colors overlap, the result is a subtractive mixture of the two colors. Where dots of different colors are next to each other, the mixture is additive (partitive).

There is no necessity that the halftone pattern consist of a rectangular array of dots. You can make irregular halftone dot arrays, halftone lines (straight, circular, or any other pattern), or any other fine mesh pattern that suits your artistic purposes (Fig. 9.26).

FIGURE 9.26

Dramatic effects of a circular halftone screen.

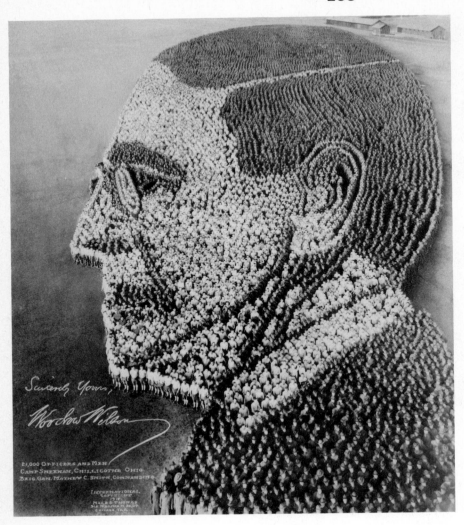

FIGURE 9.27

Arthur Mole's portrait of Woodrow Wilson. Notice that the soldiers appear nonuniformly distributed throughout the face, but the portrait from this particular camera position is undeformed—this is an example of slant anamorphic artwork (see Sec. 3.3C). (This figure itself was printed with a halftone array of 133 lines per inch.)

Nor is it necessary that the mesh be very fine. The pop artist Roy Lichtenstein frequently displays the dots prominently. Indeed, the halftone "dots" needn't be dots at all. For instance, Salvador Dali has used small figures, or even printed words, as the halftone "dots" in a larger picture. Computer-drawn pictures, often seen on tee shirts, are halftone pictures constructed of typewriter symbols. A halftone portrait of Woodrow Wilson was made by photographing an arrangement of 21,000 soldiers who served as the "dots" (Fig. 9.27).

First TRY IT

FOR SECTION 9.8A
Halftones

With a magnifying glass, examine newspaper pictures and the black and white pictures and color plates in this book. Notice how the grays are made. Compare the fineness of the halftone dot array in the color plates here with those in the Sunday comics, a magazine, a poster, an art book, etc. Also notice that the dots of different colors are arrayed at different angles. Usually the yellow is 15° from the cyan, the magenta another 15° from the yellow, and the black is 30° from the magenta, in a four-color print. (If your magnifier is not sufficiently strong, you may not be able to resolve the colors of the individual dots, but you should still notice the rosette pattern formed by the rotated arrays.) In two-color prints (duotones), the two colors are about 30° apart. Sometimes, in overprints, a halftone image in one color is printed over a solid image in a second color.

Second TRY IT

FOR SECTION 9.8A
Moiré patterns

When two periodic patterns (such as an array of straight lines, a mesh, or a series of concentric circles) are overlapped with imperfect alignment, a large-scale moiré pattern appears. You can see such a pattern quite easily. A fine mesh can be found in a sheer curtain, a handkerchief, or any finely woven fabric. Simply look at a double layer of such fabric against the light, and notice the large-scale wavy pattern produced where the lines of the two layers of mesh overlap. Two layers of a chain link fence (often seen on highway overpasses) will form a moiré pattern, as will two pieces of graph paper or window screen (the finer the mesh, the better). If you have some fabric, graph paper, or window screen, notice what happens to the moiré pattern as one layer is slid over the other. Also try rotating one layer.

Another technique is to hold a comb at arm's length a few centimeters in front of a mirror and look at the comb and through it at its reflection. Notice whether and how the moiré pattern moves as you slowly move the comb to the side or gently tip one end away from the mirror. Compare this to what you see when you hold two combs next to each other, about a centimeter apart, and slide the front one parallel to the rear one, or tip the front one.

You can also make moiré patterns without straight lines. For example, on semitransparent paper, trace or photocopy Figure 9.28. Place the semitransparent copy on top of the original figure, first with the centers of the two patterns coinciding, then slightly displaced.

Because large-scale patterns result from fine periodic arrays, moiré patterns provide a magnification effect; one can use them, for example, to determine the spacing of the fine array.

FIGURE 9.28

Concentric circles for making moiré patterns.

*9.9
PIGMENTS, PAINTS, AND PAINTINGS

No matter what color you see—purple or orange, pink or brown or green—it comes from mixing the three basic colors called the primary colors: blue, yellow and red.
Yellow + blue makes green.
Red + yellow makes orange.
Blue + red makes purple.
And all the colors smooshed together make—black.

from an ad for Health-tex® Stantogs

You probably learned rather simple rules of color mixing in kindergarten. These rules, summarized above, stood you in good stead during your Crayola stage and still serve as a good starting point for you when you mix paints. Nevertheless, they cannot tell the whole story. The very fact that the Crayola people put more than blue, yellow, and red in their crayon boxes suggests that you can't make all colors by mixing just these three. Further if you "smoosh" all the colors together, you don't get black—rather you get what one art text describes as a "dark brown muddle." Let's look at these artist's rules more carefully and then examine the optical complications that come about when the artist puts brush to can-

vas. We'll see that, when the chips are down, the artist must experiment and be familiar with her paints and cannot rely on simple rules.

A. Simple rules

What are the artist's primary colors? The Roman Pliny defined the primary colors as those "used by both sexes," a rather less than useful definition. Artists will tell you that blue, yellow, and red are the primary hues because "you can't mix these from any other hues." In fact, a mixture of appropriate magenta and yellow pigments yields a perfectly good red. However, magenta pigment was hard to come by. It was rare and expensive compared to red made from abundant iron oxides. Further, magenta is a fairly rare color in nature—a bit of a shrinking violet. For these reasons, combined with the fact that red and blue are among the psychological primaries (Sec. 10.4), the **artist's primaries** are *blue, yellow, and red.* (There is a certain amount of semantic difficulty here. We've used the word red for the long-wavelength end of the visible spectrum and magenta for a mixture of those wavelengths with short wave-

lengths. The psychological primary red, we'll see, actually *has* some short wavelengths mixed in—what we would call red-magenta. Further, we've already seen that what a printer refers to as red, other people call magenta. These ambiguities of hue names account for some of the differences in color lore among the various fields concerned with color.)

The artist arranges his colors with his three primaries equally spaced around a circle (the **color circle**). Midway between any two of these colors he places their mixture, which he calls the complement of the third primary. Thus, the artist says that blue and yellow give green, which he calls the complement of red. Similarly, orange, from red and yellow, he calls the complement of blue, and the color he obtains from blue and red he calls violet, the complement of yellow. The green, orange, and violet obtained by these mixtures he calls **secondary colors.** The artist's complementary color pairs are not complementary in the sense we've been using the term—mixed together they do not give an **achromatic color** (that is, white, gray, or black). The artist often prefers these mixtures to pure achromatic gray, which he finds lifeless. His mixtures give "richer grays," grays that are "warm, cool, soft, or clear"—that is, gray with some hint of hue. Since achromatic colors are rare in nature, the artist's mixtures are usually more suited to his purpose.

The artist's color-mixing rules, just cited, are far less definite than the additive rules for mixing lights. Not only do the complications of subtractive mixing make these simple rules somewhat idealistic, but a myriad of further complications, having to do with the physical structure of his paints and paintings, make the artist's rules only a starting point.

B. Complications

Paint consists of solid matter, ground into fine particles, in some transparent medium. The particles

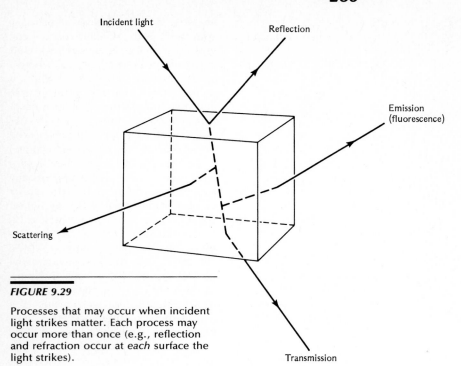

FIGURE 9.29

Processes that may occur when incident light strikes matter. Each process may occur more than once (e.g., reflection and refraction occur at *each* surface the light strikes).

may have color of their own *(pigment)* or may consist of a transparent base to which dye has been added *(lake).* The transparent *medium* or *vehicle* may be water, oil, egg yolk (the hardening agent in tempera—try leaving your breakfast plates unwashed for a few days), polyvinyl acetate, or acrylic resins. The mixture is painted on some **support** (paper, canvas, wood, etc.) or **ground** (previous layer of coloring). A number of processes may occur when light strikes matter, as illustrated in Figure 9.29. Each of these processes may be **selective,** occurring more strongly for some wavelengths than for others. Thus, the color of the paint will depend on the selective reflection, absorption, and transmission of the vehicle (particularly its surface), the pig-

ment or lake, and the support (Fig. 9.30). It also depends on how the properties of one of these, say the

FIGURE 9.30

Some of the processes contributing to the color of paint. Incident light *(I)* strikes the paint, some of it *(1)* is reflected at the surface of the vehicle, some *(2, 3, 4)* continues to a pigment particle. There it may be selectively reflected or transmitted. It may go on to strike the support *(2)*, another pigment particle *(3)*, or go directly to the viewer *(4)*. Each process may impart a different intensity distribution to the light. The successive reflections *(A and B)* and the transmission followed by reflection *(C and D)* constitute subtractive mixing. The different rays headed toward the viewer *(1, 2, 3, and 4)* combine additively, if they are close to each other.

pigment, compare to those of another, say the vehicle, since reflection at a surface, we know, depends on the *change* in index of refraction there. Additionally, the color will depend on the size and concentration of the pigment particles.

The different colors produced by the various processes involved may mix subtractively or additively (mainly partitively), as shown in the figure. The presence of additive mixing is verified by the difficulty in producing an achromatic black by mixing paints. If the mixing were solely subtractive, you'd need only apply enough different absorbing pigments and eventually all the light would be absorbed—you'd have an ideal black. In fact, mixing all the pigments together gives a dark brown or purple (Plate 9.6). Because there is some additive mixing, you always get *some* light reflected to your eye—usually more from the yellow and red pigments, which tend to be brighter because they are more broad band. The result is a weak, unsaturated color from the yellow-red side of the chromaticity diagram—that is, a dark brown or purple.

Additional complications arise from the many ways in which the paint can be applied: in layer upon layer, sometimes smoothly mixed, sometimes nonuniformly, and in a host of other techniques (the TRY IT shows some of the variety). The large range of possibilities means that no simple rules can suffice—the artist must rely on experience and trial and error, as well as on an understanding of the principles. To give an idea of the possibilities, we'll discuss a number of examples that illustrate some of the principles involved.

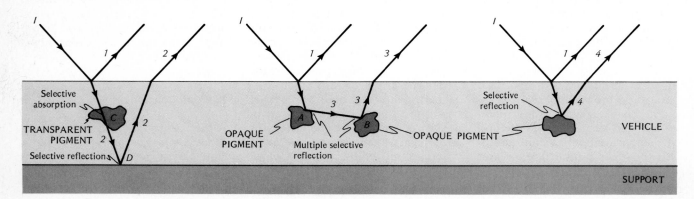

TRY IT

The range of colors that can be obtained from a set of pigments depends not only on the pigments, but also on the way they are mixed. We list here a variety of ways of mixing pigment colors; many other ways exist. Try these and any others you can think of, and examine the results. Carefully compare the differences and try to apply some of the ideas of Section 9.9 to explain them.

You will need three oil colors: white, Prussian blue, and Mars yellow (or any other equally transparent blue and yellow). You'll also need some mastic varnish, brushes, and some white cardboard.

1. Mixing on the Palette. Squeeze a small amount of each color on a palette, piece of glass, or other smooth surface. With a brush or palette knife, mix together equal amounts of the three paints until a smooth, homogeneous paint results. Paint a small area of white cardboard with this mixture.

2. Glazing (Three Layers). Paint an area of the cardboard with the white paint and let it dry thoroughly (this may take several days). When the white is dry, dip your brush first in the varnish and then in the blue, getting enough of the dilute mixture so the glaze can spread smoothly. Paint over the white, and let this layer dry. Then apply a glaze of varnish and yellow. Each colored layer should be somewhat uneven in thickness.

3. Glazing (Two Layers). Make a white ground and let it dry, as in method 2. Then make a glaze consisting of varnish and a palette mixture of blue and yellow. Paint this glaze over the white.

4. Mixing on the Brush. Using a broad brush (about 2 cm wide or more), lift up a large quantity of white and, on the same brush, pick up a similar portion of the blue and then the yellow. You'll then have three dollops of paint, one on top of the other, on your brush. With the paint side down, paint one broad stroke using the entire width of the brush.

5. Pointillist Painting. With a fine brush, make a random pattern of small dots. Each dot should be of only one color, and the three differently colored dots should be thoroughly intermingled but not overlapping. You can speed the process (at the expense of some overlap) by using your thumb to flick diluted paint off the bristles of a toothbrush.

6. Color Wheel Mixing. Out of white cardboard, cut a circle about 15 cm in diameter. Divide the inner half of the circle into six equal wedges, and paint the wedges alternately with the three colors. For comparison, mix the three colors together on your palette (as in method 1), and paint the outer half of the circle with the mixture. Spin the circle, as in the first TRY IT for Section 9.5C.

C. Example: dependence on pigment size

When light strikes a particle of pigment, some light is reflected and the rest travels into the pigment. Usually the reflection is only weakly selective—the reflectance curve of the pigment does not strongly favor one color, so the reflected light is only weakly saturated. The light entering the pigment, however, is selectively absorbed, becoming more saturated the deeper it goes, until either it is completely absorbed or it escapes out the other side of the pigment. Let's consider the light

that gets to the viewer's eye when pigment particles of different size are used. If the pigment occurs in large particles (Fig. 9.31a), weakly saturated light is reflected from its first surface. Any light entering the pigment is absorbed before it can traverse the large particle. The reflected light forms an additive mixture with the unsaturated light reflected from the vehicle surface, so the result is rather unsaturated. If the pigment particles are smaller (Fig. 9.31b), however, light is transmitted *through* them. This light is saturated and remains so when reflected by other similar particles. After traversing several particles, however, the light is eventually absorbed. The light reaching the viewer is then an additive mixture of the unsaturated reflection from the vehicle, the weakly saturated reflection from the first pigment particle, and the saturated reflections from successive particles. This light is then both more saturated and lighter than that in the large-parti-

---

FIGURE 9.31

Dependence of color on the size of the pigment particles. The light reflected from the vehicle is unsaturated (*u*), and that from the pigment is only weakly saturated (*w*). Light passing through the pigment is selectively absorbed. After traversing enough pigment, it is all absorbed (*a*), but if it escapes the pigment, the transmitted beam is saturated (*s*). All reflected light combines in an additive mixture here. **(a)** Large pigment particles result in a darker, unsaturated color. **(b)** Smaller pigment particles result in a lighter, more saturated color. **(c)** A fine powder gives a still lighter but unsaturated color.

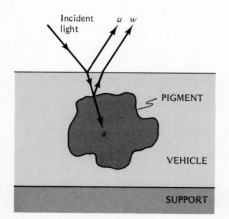

(a)

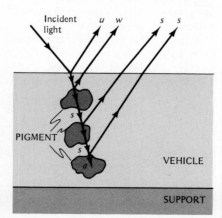

(b)

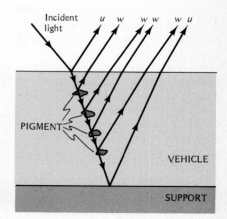

(c)

cle case. Finally, if the pigment is ground into a fine powder (Fig. 9.31c), each particle is too thin to absorb very much, so the transmitted light is only weakly saturated, if at all. The viewer now gets weakly saturated light reflected from many particles, additively mixed with the unsaturated light reflected from the vehicle and, perhaps, from the support. As in the large-particle case, the light is rather unsaturated, but much lighter here because of the extra reflections.

Thus, the same pigment in different sizes results in different colors. Since clustering of small particles into larger clumps may occur during mixing, the resultant color depends not only on the pigment but on the *method* of mixing.

The red squirrel fish *Holocentrus ascensionis* uses this technique to change its color for purposes of camouflage. Its red pigment granules can move together to form clusters in a matter of seconds, changing the fish color from red (as in Fig. 9.31b) to pinkish gray (as in Fig. 9.31a).

D. Example: dependence on relative index of refraction

We have seen that the amount of reflection and refraction at a surface depends on the *ratio* of the indexes of refraction on the two sides of the surface. This dependence affects the color of paint—the same pigment can look rather different in different vehicles. If the index of refraction of the pigment differs greatly from that of the vehicle, there will be a lot of reflection from the pigment, whereas if the two indexes are almost the same, the pigment will be almost invisible. Typically, the index of refraction of pigment is greater than that of oil, which, in turn, is greater than that of water. This means that the same pigment in water will reflect more and therefore transmit less than in oil—it will appear more opaque in water. Further, there will be additional differences as the paint dries. The index of refraction of oil increases during drying, so the pig-

ment becomes increasingly *transparent* as oil paint dries. (This is a common source of arguments with the house painter—your walls don't look the color you chose, and the painter tries to reassure you that they will when the paint is dry.) As water paint dries, however, the water evaporates and is replaced by air, with an even lower index of refraction—water paint increases in *opacity* as it dries.

A variation of this effect provides one reason for people to cry over spilt milk. When the spilt milk (or better still, one of the substances in the first TRY IT) dries, it leaves a deposit of small white particles whose index of refraction is close to that of water. Because of the many layers and rough surface of the particles, most of the incident light at each wavelength is eventually reflected to your eye and the dried deposit looks white. When you try to wash it up, you surround the particles with water, and they immediately become almost transparent, deluding you into thinking that the milk has been washed up. Later, when the water evaporates, the insidious deposit reappears. A similar effect occurs with ground glass. Because its surface has been roughened, the glass scatters light before transmitting it and hence is translucent—you can't see an image through it. A layer of water, however, by diminishing these surface reflections, makes the ground glass quite transparent. Similarly, scratched film can be duplicated without the scratches showing if it is first immersed in a fluid of the same index of refraction as the film. The second TRY IT gives another familiar example of this effect.

A rather less familiar example occurs in the wing cases of the large beetle, *Dynastes hercules*. This beetle likes bananas and wants to be yellow for protective coloring. At night, however, it prefers to be black. It accomplishes this change of color by using the principle we've been discussing. Its wing cases have a spongy layer covering a black cuticle. During the day, the spongy layer contains hollow air spaces, which we may think of as its pig-

ment. Because of the difference between the indexes of refraction of the air and of the material surrounding it, these hollow spaces reflect the incident light, primarily at the long-wavelength end of the spectrum, and the beetle appears yellow. At night, however, the humidity rises and the hollow spaces fill with water, which has an index of refraction close to that of the surrounding material. Light is no longer reflected from the hollow spaces, but rather passes on to the black cuticle where it is absorbed—the beetle now appears black. Thus, by choosing the proper relations between the indexes of refraction of its "pigment" and its "vehicle," the beetle paints itself the appropriate color to avoid its predators.

First TRY IT

FOR SECTION 9.9D
The reappearing spot

You can easily make a spot that disappears when wet and reappears upon drying. Stir more salt than will dissolve into half a small glass of warm water, and let the excess settle. Alternatively, mix enough water into some corn starch or baking soda to make a fairly liquid paste. In either case, pour a few drops of the liquid onto a dark plate, and set the plate aside. After the water has evaporated, you should see a white deposit left on the plate. The deposit will become invisible if you put a few drops of water on it, but will reappear when the water evaporates. The corn starch is particularly pernicious, reappearing after even a thorough rinsing.

Second TRY IT

FOR SECTION 9.9D
Seeing through the newspaper

You can read the print on the opposite side of a piece of newspaper with the help of a little water. Light is normally reflected from the rough surface of newspaper, except where there is absorbing, black ink. When the paper becomes wet, however, the air around the particles that make up the paper is replaced by water with its higher index of refraction, and thus reflections are significantly decreased. This allows light to pass through the wet newspaper, except at those places where there is black ink on the opposite side. You then need only have skill in mirror reading to read that print. The effect is even greater if you use oil instead of water to wet the newspaper.

PONDER

What do you expect to happen when the water dries? When the oil dries? Why? Try it!

E. Example: mixing in black pigment (shades)

Black pigment consists of opaque particles that, ideally, absorb all incident light. You might expect, therefore, that when you mix black with some color (that is, make a **shade** of the color) all you would be doing is decreasing the lightness of that color, without affecting its saturation or hue. In fact, however, you usually decrease the saturation, and sometimes you change the hue.

The black decreases the saturation for two reasons. First, because it diminishes the amount of colored light reflected from the pigment, it increases the importance of the unsaturated reflections from the surface of the vehicle. Since a higher fraction of the light then reaching the viewer's eye is unsaturated, the color appears less saturated. A second reason is that black pigment is never ideal—it always reflects some light, which often has some hue. The broad-band reflection from black pigment decreases the saturation of the colored paint with which it is mixed, and any hue reflected from the black pigment affects the hue of the mixture.

Nature rarely offers a true black (see the TRY IT). Look closely at any object you would call black and notice the subtle colors in it. Some paintings by Ad Reinhardt appear totally black on first glance, but careful viewing reveals regions of different subtle hues. Chinese ink, made by burning nuts, occasionally (by chance) comes close to the ideal

black. When such ink is discovered, it is treasured—passed down from generation to generation, to be used only when the blackest black is required.

An even more surprising result occurs when very fine black pigment is mixed into white paint. At first, when the black is quite dilute, the paint takes on a bluish cast. This phenomenon, noted by Leonardo da Vinci, has to do with the size of the black pigment particles. When light is scattered from small particles, the shorter wavelength light is scattered most (Sec. 13.2). Scattering from the fine black pigment thus results in a bluish hue—mixing black with white can result in blue!

TRY IT

FOR SECTION 9.9E
How black is black paint?

To see how black your paint is, compare it to a black body (Sec. 1.4B)—a body that absorbs all radiation that falls on it. You can make a reasonably good black body by cutting a small hole in an otherwise light-tight box whose insides have been painted, or covered with, black. A 2-cm hole in a $\frac{1}{3}$-m box should work well. Then any light entering the box is very unlikely to return through the hole, so the hole will look quite black (at normal temperatures). On the outside of the box, next to the hole, use your black paint to make a spot of roughly the same size and shape as the hole. Compare the blackness of the two black regions. Is the spot you made with the paint as dark and colorless as the hole?

F. Example: mixing in white pigment (tints)

White pigment consists of opaque particles that, ideally, reflect all light. Actual white pigment reflects strongly and nonselectively. What happens when we mix white into some colored paint, making a **tint** of that color? When the white pigment is added, two changes occur— the colored pigment is diluted, and highly reflecting opaque particles are mixed throughout.

The dilution of the colored pigment particles means that they are more widely separated, so the selective reflection is decreased in any area. It is replaced by nonselective reflection from the white pigment particles. You might therefore expect that the mixture would be of lower saturation and increased lightness. The effect on saturation, however, depends on the nature of the colored pigment.

If the colored pigment is highly transmitting (has **low hiding power**), then by itself it would selectively transmit light, which would be reflected by the support to the viewer's eye. The opaque white pigment, by blocking this reflection in addition to diluting the colored pigment, further decreases the saturation (Fig. 9.32a).

If, however, the colored pigment is more absorbing (has **high hiding power**), then, in the absence of white, much of the selectively transmitted light would be absorbed by subsequent layers of colored pigment before it could be reflected by the support. By replacing some of this colored pigment with highly reflecting white pigment, we allow a part of this selectively transmitted light to be reflected to the viewer's eye (Fig. 9.32b). The presence of the white pigment here increases the amount of reflected colored light— that is, it *increases* the saturation. Eventually, of course, as more and more white pigment replaces the colored pigment, the saturation decreases (Fig. 9.32c).

G. Example: dependence on surface reflections

Surface reflections are nonselective. Since the light from them mixes additively with the colored light from the pigment, surface reflections desaturate the colors. Even a weak surface reflection will desaturate the darker colors significantly, because so little light comes from these colors in the first place. The amount of surface reflection depends on the surface: whether it is rough or smooth, glossy or dull, has detectable brush strokes, etc.

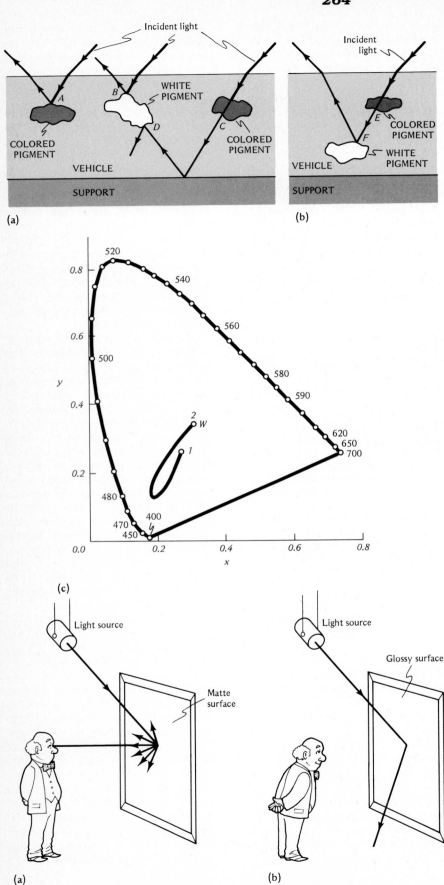

(a)

(b)

(c)

(a)

(b)

FIGURE 9.32

(a) White pigment mixed with colored pigment of *low* hiding power. The colored reflection, *A*, is weak but selective. The white reflection, *B*, is strong and nonselective, decreasing the saturation. Further, the selective transmission of the colored pigment, *C*, which is reflected by the support, is blocked by the opaque white pigment, *D*, further decreasing the saturation. **(b)** White pigment mixed with colored pigment of *high* hiding power. Here the colored pigment transmits less light than in **(a)**. Had the selectively transmitted light, *E*, struck another colored pigment particle, it would have been absorbed. Instead, it strikes the highly reflecting white pigment particle and is reflected, *F*, thus increasing the saturation. **(c)** The path in the chromaticity diagram of the color of a mixture of high hiding power blue paint, *1*, with white paint, *2*, as more and more white is added. First the saturation increases and then it decreases. (Of course, since the lightness increases, this really should not be drawn on one chromaticity diagram.)

If the surface is dull or **matte** (also spelled **mat**),* there will be *diffuse reflection* from it (Sec. 2.4B). Since light hitting it is scattered in all directions, the viewer gets light nonselectively reflected from it no matter where he stands relative to the light source (Fig. 9.33a). This reflection may be as large as 50% of the incident light and greatly desaturates the colors.

If the surface is smooth and glossy, however, most of the reflection is specular, obeying the law of reflection (Sec. 2.4). Now by positioning himself with a little care,

*Old French *mat*, mated at chess, confounded, exhausted, dull. Ultimately from the Persian for checkmate: *shah mat*, the king is dead.

FIGURE 9.33

Surface reflections. **(a)** Diffuse reflections from a matte surface are seen by the viewer no matter where he stands (likewise, he sees the light scattered by the pigment). **(b)** Properly located, the viewer does not see the specular reflections from a glossy surface (but does see the light scattered by the pigment).

the viewer can almost completely avoid the surface reflection, and see the properly saturated colors of the picture (Fig. 9.33b and Plate 9.7). Reflections from floors made of mosaic tiles that have been walked on are diffuse, and the colors tend to be dull and unsaturated. The mosaics are brought to life by spilling water on them. The water eliminates the diffuse reflections (as in Sec. 9.9D), replacing them by the specular reflections from the top surface of the water, and thus the colors of the wet mosaics appear much more saturated and alive. A similar effect occurs with pebbles on the beach. The TRY IT shows you some of the effects produced by reflections from textured surfaces.

TRY IT

FOR SECTION 9.9G
Textured surfaces

By varying the texture of the surface, you can get the same color to appear different. Select any water color and paint a small region of paper with it, in the normal way. Now cover another region with as much water as possible, and drip concentrated color into the water until the saturation is about the same as your first painting. Carefully let this dry completely (probably several hours). As a final technique, paint a region of paper in the normal way, let it dry completely, and wash the paint off under a faucet, lightly rubbing off the color with your brush. Keep repeating the painting and washing until the color is about as dark as your first painting. The resulting three paintings will have very different textures, and the reflections will depend on these textures. For example, in the third method, the reflections will be different in the ridges and valleys of the paper. Examine these differences in reflections and compare the way the texture affects the color in these three paintings.

H. Conclusions

What has reasoning to do with the art of painting?
                                        William Blake

It is clear that the color mixing rules that are of use to the painter are far more complicated than those for the additive mixing of lights or even for subtractive mixing by filters. Nevertheless, the light that allows you to see a painting obeys the laws of physics, and an understanding of the basic processes (as well as the way your eye perceives color—Chapter 10) can allow you to know, reasonably well, what effects are possible when paints are mixed. The trouble is there are so many different variables that are important; it is not sufficient to know only the chromaticity of the colors to be mixed. The intensity-distribution curves, the size of the pigment particles, their chemical properties, their relation to the vehicle, the method of mixing, the method of application, and a myriad of other considerations all may be relevant to the final effect—sometimes in subtle ways, and sometimes in a more dramatic manner. The skill of the artist depends on his ability to master these effects. Generally, the artist gains this mastery by arduous trial and error. An understanding of the principles discussed here may assist in systematizing the results of this experimentation. Ultimately the artist must know his medium in order to gain control over it, to be able to understand what effects are possible and how to achieve them. In the next chapter we'll see that he must know something about visual processes as well.

SUMMARY

The colors you see are usually not **monochromatic,** but rather are made of a **distribution** of visible wavelengths, characterized by an **intensity-distribution curve.** Colored lights appear the same to you, no matter what their intensity-distribution curves, provided that three of their properties are the same: their **hue** (main color or dominant wavelength), **saturation** (purity), and **brightness** (apparent overall intensity). For surface colors, the third property is **lightness** (whiteness, grayness, or blackness). All colors can be arranged according to these three properties in a **color tree.** Two colors that look alike but have different intensity-distribution curves are **metamers.**

Overlapping spots of light combine their colors by **additive mixture.** Any color can be matched by an additive mixture of *three* colors, if we allow **negative amounts.** Colors that additively combine to produce white are **complementary colors.** The results of additively mixing colors depend only on the colors' **chromaticities**—the rules are represented by **straight lines** on a **chromaticity diagram;** for example, complementary colors lie on opposite ends of straight lines that pass through the point representing white. The greatest variety of colors can be obtained by an additive mixture of **blue, green, and red,** the **additive primaries.** In addition to **simple addition,** produced by overlapping illuminations, additive mixtures can be obtained by **partitive mixing** (small, closely spaced spots of different colors), and by rapid, successive presentation of different colors at the same place.

The **subtractive process** occurs when certain wavelengths are removed more than others from a beam of light, say when it passes through a filter or when it reflects from a painted surface. The results obtained when white light passes through two filters successively are found by *multiplying* together the **transmittance curves** of the two filters. (For reflection, the **reflectance curve** of the surface plays the same role.) Subtractive mixtures can result in surprising colors. Colors formed by the subtractive process depend on the intensity-distribution curve of the **illuminating light** as well. Light sources are often characterized by their **color temperature**—the temperature of a matching incandescent light. The quality of a white light source depends on how **white** it looks, its ability to **render color** in a desirable fashion, and its **efficacy** (how efficiently electric power is converted into light).

Water colors and printer's inks are fairly transparent, but their mixtures involve both subtractive processes (similar to filters) and additive (partitive) processes. Printers make good colors by **four-color printing,** a process that uses inks of the three **subtractive primaries** *(yellow, magenta, and cyan)* plus black. Different colors are mixed by printing them in **halftones**—arrays of fine dots in each color, the size of the dots proportional to the desired amount of that color. Mixing colors with pigments and paints has many complications due to the heterogeneous nature of paints. The color of a paint depends on such details as the **pigment size,** the relative **index of refraction** of pigments and vehicles, the amount and nature of the **absorption, reflection,** and **transmission** of the various pigments added, whether these processes are **selective** or **nonselective,** even on the nature of the **surface reflections.** Simple artist's rules are only rough guidelines.

PROBLEMS

P1 Professor Farbmesser wishes to show her class an example of a monochromatic color and an example of a nonspectral color. Describe how she may set up a demonstration for each case.

P2 Draw intensity-distribution curves that can represent each of the following lights: (a) a saturated blue light, (b) an unsaturated green light, (c) pink light.

P3 Draw a color tree and locate the following colors: (a) black, (b) light pink, (c) saturated blue, (d) gray-green, (e) dark orange, (f) cream white.

P4 What is the hue of the color represented by the intensity-distribution curve shown in the figure? Is it saturated or unsaturated?

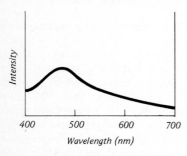

P5 Draw intensity-distribution curves for two complementary colors, labeling them with the names of their hues.

P6 (a) Using Figure 9.10, estimate what mixture of 460-nm blue, 530-nm green, and 650-nm red will match a monochromatic orange light. (b) Why is a negative amount of one of these colors required?

P7 (a) Using crayons, colored pencils, inks, or paints, make a partitive mixture of two colors. (b) Sketch approximate intensity-distribution curves for each of the colors and for the mixture. (c) Locate each of the colors and the mixture on a chromaticity diagram.

P8 A certain light has the intensity-distribution curve shown in the figure. If this light shines on a white screen, what will be the hue and saturation of the illuminated regions?

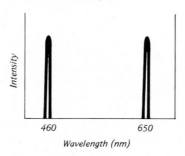

P9 (a) What are complementary colors? (b) Give names and wavelengths of two pairs of complementary colors.

P10 Use the chromaticity diagram of Figure 9.13 for the following. (a) Find the position of a spectral red of 700 nm, then find the wavelength of its complementary spectral color. (b) Repeat for a spectral blue of 450 nm. (c) Repeat with an orange of 600 nm. (d) Show that there is no spectral complement to green.

P11 When white light is additively mixed with colored light, which of the following properties of the colored light does not change (or changes very little): (a) saturation, (b) purity, (c) chroma, (d) hue?

P12 A slide projector projects a beautiful picture on a screen. A clever but misguided photographer wants to use his camera to photograph the beautiful picture from the screen, but there is not enough light. So, he illuminates the screen with a (white) floodlight. (a) Does this procedure mix the colors of the picture and floodlight by addition, subtraction, or neither? (b) When the floodlight is on, is the light intensity reflected from the screen and into the camera larger than, smaller than, or the same as without the floodlight? (c) One of the regions in the projected picture is saturated red. On a chromaticity diagram, locate a point corresponding to this red. (d) On the same diagram, draw a curve to show how the point that represents the color of the (projected) red picture moves as the photographer turns on more and more intense white floodlighting. Mark a point corresponding to equal amounts of the projected red and the floodlight. (e) What color light would be needed to illuminate the (projected) red region of the picture to change its color to white?

P13 (a) What is the complementary color to green? (b) Is it a spectral or a nonspectral color? (c) True or false: an unsaturated version of the color *complementary* to green can be obtained by an additive mixture of yellow and cyan.

P14 During a slide show, someone turns on the (white) roomlights. In the projected images of the slides, what changes are there in the following properties? (a) Brightness, (b) saturation, (c) image size, (d) hue.

P15 Prof. Fauxcouleur, the famous colorist, has found the following color match by mixing lights:

vivid yellow ≡ 10 parts red
        + 10 parts green − 1 part blue.

Which of the following (possibly more than one) are true? (a) To get vivid yellow, you project green and red lights on a pale yellow screen (since yellow is negative blue). (b) When you additively mix 1 part blue with vivid yellow, you can match the result with an additive mixture of 10 parts red and 10 parts green. (c) Prof. Fauxcouleur could never have made this match since it requires a negative amount of a primary. (d) This mixture is neither

additive nor subtractive but partitive.
(e) None of the above.

P16 (a) Roughly, what is the wavelength of monochromatic light of the following wavelengths: red, green, and blue? (b) Sketch a large chromaticity diagram and indicate, roughly, where the colors of part (a) are to be found on it. (c) Indicate the position of white light with a *W*. (d) Use your chromaticity diagram and your results from parts (b) and (c) to locate the complementary color of blue. What is its hue? (e) Mark on the diagram the color that results from an additive mixture of equal amounts of red and green. What is its hue? (f) Briefly describe two ways of forming additive color mixtures.

P17 Look at a colored candy wrapper or magazine photograph in sunlight and answer the following. (a) What color light is falling on each of the different areas? (b) What color reaches the eye? (c) What happened to the other colors?

P18 Take three pieces of colored cellophane, blue, red, and green, and hold them up to a white light, one at a time. (a) What color light falls on each filter? (b) What color enters the eye after passing through each filter? (c) What happened to the other colors?

P19 Hold each piece of colored cellophane you used in P18 close to your eye, and look at the colored regions you used P17. (a) Describe each color you see as you look through each filter at each of the regions, in turn. (b) Explain these colors.

P20 (a) Think of the color of leaves in spring or summer. What range(s) of wavelengths do you think that the light involved in photosynthesis has? Explain. (b) If you were designing a light that put its energy where chlorophyll absorbs most, what color would it be? (c) What color would the leaves appear under this light?

P21 Shown in the figure is the transmittance curve of a filter. (a) Draw a transmittance curve for a filter of the complementary color. (b) What color do you get when *light* from these two filters is superimposed (additive mixture)? (c) What color is transmitted through both filters when they are placed back to back?

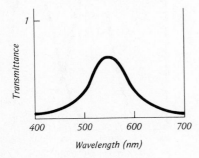

P22 Either magenta or yellow light is incident on three pigments, blue, cyan and yellow. Complete the table showing what the resultant color of each region is, when the region is illuminated by each of the lights.

	Magenta	Yellow
Blue	—	—
Cyan	Blue	—
Yellow	—	Yellow

P23 State which of the following processes involve additive color mixtures and which involve subtractive mixtures (some may involve both) and *why*. (a) You look at the world through yellow (or rose-colored) glasses. (b) A Christmas tree with many small green and red bulbs seen from a distance looks yellowish (actually, an orange yellow, because the red bulbs are brighter than the green ones). (c) Daylight falls into a room through a stained glass window that contains predominantly red and blue stained glass. A piece of "white" paper in the room looks magenta. (d) An orange, when illuminated with blue light, appears black.

P24 (a) What is the additive mixture of blue and green? (b) What is the subtractive mixture of (ideal) blue and green? (c) What is the complementary color to yellow?

P25 What would the subtractive mixtures summarized in Figure 9.15 look like if the filters passed only monochromatic light?

P26 White light passes first through a cyan filter, then a magenta filter. What color results?

P27 Just before the Battle of Midway in World War II, the U. S. submarine *Trigger* ran aground on a reef. The officers were wearing red goggles (in order to dark adapt—Sec. 5.3D). On their maps, the reefs were marked in red. Why did they run aground?

P28 TV color cameras, when shooting indoors, are balanced for 3200°K light. If by mistake, 2500°K lights are used, what will be the effect on the pictures seen by home viewers?

P29 A yellow object has the reflectance curve shown in the figure. What will be the color of the light reflected from

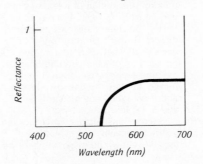

it if it is illuminated by a Cool White fluorescent tube (see Fig. 9.18a)?

P30 From this list of light sources— (i) sunlight on a sunny day, (ii) monochromatic red light from a neon tube, (iii) light from the blue sky, (iv) light from an incandescent 60-watt bulb—select the one that has the (a) lowest saturation, (b) lowest color temperature, (c) greatest brightness.

P31 Before modern color film, which uses three primaries, there were experiments with two-primary systems. Suppose we use orange (600 nm) and cyan (490 nm). Draw a chromaticity diagram, and show the approximate position of these primaries, as well as the locations of all the colors that can be obtained by an *additive* mixture of the two. How well would you expect a film based on additive mixtures of these two colors to reproduce: (a) skin color (all races), (b) sky color, (c) grass, (d) purple people-eaters?

P32 An incandescent bulb is covered with a blue dye so that the color temperature of its light is 5000°K. Which of the following is true? (a) The dominant wavelength of the light from the bulb is 5000 nm. (b) The bulb's filament heats up to 5000°K. (c) The hue and saturation of the bulb's light is the same as it would be if the bulb were not covered with blue dye, and its filament were heated to 5000°K. (d) The color temperature can't be 5000°K, because the filament would melt at that temperature.

P33 "Yes," I answered you last night.
"No," this morning, sir, I say:
Colors seen by candle-light
Will not look the same by day.
          Elizabeth Barrett Browning,
               "The Lady's 'Yes' "

(a) Explain the last two lines. (b) Give an example of a color for which those lines are true. (c) Give an example of a color that may look the same under both conditions.

P34 Two equally bright lights, *A* and *B*, have chromaticities $x = 0.4$, $y = 0.2$ and $x = 0.5$, $y = 0.4$, respectively. They are produced by letting white light shine through two filters, which are also called *A* and *B*. Assume the filters are ideal (i.e., transmit 100% in some wavelength range and 0% everywhere else). (a) Draw a chromaticity diagram and locate *A*, *B*, and the white point *W*. (b) Use your diagram to estimate the dominant wavelength, hue name, and saturation of *A*. (c) Repeat for *B*. (d) Now let equal amounts of lights *A* and *B* illuminate a white object. Find the chromaticity of the object in this light, showing your construction on your diagram, labeling the resultant point *A + B*, and estimate the dominant wavelength, hue name, and saturation

of that color. (e) What is the complementary color of your answer to part (d)? Show on your diagram how you obtained your result.
(f) Now, white light is sent through both filters in succession. What is the resulting color? (Use "simple" laws.) Does it make a difference whether the white light passes first through *A* then through *B*, compared to first *B*, then through *A*?

P35 Why are more than four different printer's inks used in high-quality art reproduction?

P36 The next time you buy postage stamps, get a sheet of the larger, multicolored kind often used for commemorative issues. Notice the various color squares printed on the border. (You may also find these in unobtrusive places on colorful packages, such as cereal boxes.) These color squares represent the pure printer's inks used for this particular stamp. (a) Why are they (usually) not the four standard inks mentioned in Section 9.8?
(b) What colors are these inks, and what colors will they produce better than the standard set? (Compare with colors present in the scene depicted in the stamp.) (c) What is the least number of inks that must be used (in addition to black) if the stamp shows only an American flag?

P37 With a pencil or pen, draw a simple figure using the halftone technique.

P38 Mixing oil paints results in color mixtures that are neither simply additive nor simply subtractive. Give some examples and discuss the mechanisms.

HARDER PROBLEMS

PH1 (a) Draw a large chromaticity diagram, and mark the white point with a *W*. Mark each of the spectral colors red, orange, yellow, green, cyan, blue, and violet with its initial. (b) Locate the spectral complement of orange and label it *Oc* (for orange complement). (c) If you start with this color *Oc* and gradually *additively* mix in more and more yellow, the color will gradually change from the original *Oc* to yellow (when you've added an overwhelming amount of yellow). Draw a line or curve indicating the path that the color mixture follows on the chromaticity diagram as it changes from *Oc* to yellow. Mark with an *E* (for equal) the location corresponding to an additive mixture of equal amounts of *Oc* and yellow. (d) Mark with an *A* a color with the same hue as *E*, but much more saturated. (e) Mark with a *Z* a color having approximately the same saturation as *E*, but with a slightly different hue (sufficiently different that it is unambiguous on your diagram).
(f) Two *pigments*, *P* and *Q*, have the

following properties: *P* is an unsaturated yellow, *Q* is an unsaturated red. Mark each of their locations on the chromaticity diagram with a dot, and label them. When a little *Q* is mixed with *P*, the mixture at first begins to become more orange and simultaneously becomes more saturated. With more *Q*, the mixture becomes a quite saturated orange. With still more *Q*, the mixture gradually becomes more like the color of the pigment *Q*. Draw a line or curve indicating the path the color mixture of these pigments follows on the chromaticity diagram as more and more *Q* is added to *P*.

PH2 On a blank page, draw a large, neat chromaticity diagram. (a) Label the position of red, green and blue with the letters *R*, *G*, and *B*. Label the white or illuminant point by *W*. (b) Well-used laundry, washed with plain soap, tends to acquire a yellowish color. Mark spectral yellow on your chromaticity diagram with a *Y*, and a reasonable guess at the laundry's unsaturated yellow with an *L*. (c) The modern wash day miracle whiteners contain a fluorescent substance that converts the ultraviolet component of sunlight to a colored light. The laundry's color will then be the additive mixture of this fluorescent color and the laundry's own yellowish color. Why is this mixture additive? (d) What should be the hue at which the miracle whitener fluoresces, in order that the laundry turns out white? (e) Suppose the manufacturer has made a mistake, and your "whitener" fluoresces cyan. Mark the position of cyan on your diagram and draw a line or curve corresponding to the possible colors of the laundry for various amounts of this "whitener." Mark one possible color on this line or curve with an *X* and describe its hue and saturation (high or low) in words.

PH3 Suppose we have two different beams of light, *A* and *B*, that appear to be identical white lights. We don't know which is which, but we do know that one of them, say *I*, is made up of an additive mixture of two monochromatic sources with wavelengths 490 nm and 640 nm. The other, say *II*, consists of a more or less uniform distribution of intensities of all wavelengths between 400 nm and 700 nm. If you shine these two beams through a filter that absorbs all wavelengths between 500 nm and 600 nm (and no others), you find beam *A* remains white, while beam *B* looks purple. Identify the unknown beams *A* and *B* with the known distributions *I* and *II*. Explain your answer.

PH4 (a) Sketch transmittance curves for real yellow, magenta, and cyan filters.
(b) The yellow and magenta filters are

overlaid to form another filter. Draw its transmission curve. (c) White light is incident on this combination filter. What color does the light appear after passing through? (d) Suppose that the incident light were cyan, rather than white. What color will be transmitted through the yellow filter? Through the magenta filter? Through the combination filter?

PH5 The common material for green eyeshades (favored by old-time poker players) appears red when doubled over, so you look through twice the normal thickness. This effect is known as dichroism (two colors). Sketch a transmittance curve that may be correct for one thickness of this material, and draw the path on the chromaticity diagram of the color of the light transmitted by thicker and thicker layers of this material (assume that the incident light is broad-band white and that its intensity is increased as more material is put in the way, in order to keep the transmitted intensity the same).

PH6 (a) Water absorbs light weakly at the red end of the visible. This gives a drop of water a little bit of color. What color? (b) Water near a sandy beach on a bright, cloudless day often ranges in color from green near the shore to blue farther out. Explain. (Hint: what color is the sand? The sky?)

PH7 The person in charge of lighting in a certain theater has two lights available, a red light and a green light (with spectra shown in the figure). The beams from these lights both shine (superimposed) on the backdrop. Region *I* of the backdrop is painted a certain English vermillion, region *II* is painted a certain emerald green, and region *III* is painted a certain cobalt blue. The reflectances of these paints are shown in the figure. Sketch roughly the spectra of the light reflected from each region and indicate what color it looks like.

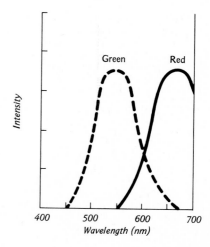

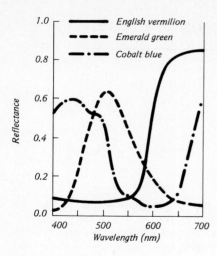

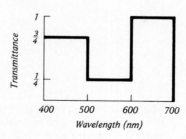

curve of the pair. (c) Ten layers of the plastic are stacked one behind the other. Plot the transmittance curve of the stack. (d) If broad-band white light is incident on the ten-layer stack, what will be the color of the transmitted light?

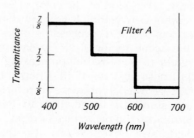

PM2 Two filters have the transmittance curves shown in the figure. (a) What colors result when broad-band white light passes through each filter? (b) Broad-band white light is incident on each filter. Plot the intensity-distribution curve for light that is an *additive* mixture of the light passing through the filters. What color is this mixture? (c) Starting with broad-band white light, use the two filters to make a *subtractive* mixture. Plot the intensity-distribution curve for the resultant light. What is the color of that light? (d) Repeat part (c) for the case where the incident white light is not broad band, but rather a mixture of complementary monochromatic 480-nm and 580-nm lights.

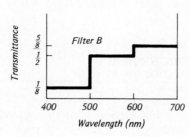

PM3 Repeat PM2 for two filters with the transmittance curves shown in the figure.

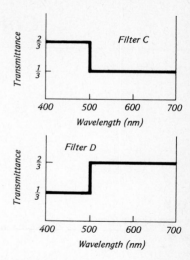

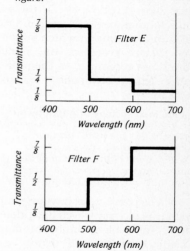

PM4 Repeat PM2 for two filters with the transmittance curves shown in the figure.

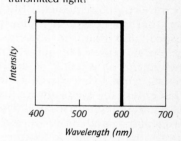

PM5 Light, with the intensity-distribution curve shown in the figure, passes first through filter *A* (PM2) and then through filter *F* (PM4). (a) Plot the intensity-distribution curve of the transmitted light. (b) What color is the transmitted light?

PH8 Draw a large, neat chromaticity diagram. (a) Label the positions of red, green, blue, and white with the letters *R*, *G*, *B*, and *W*. (b) A light bulb is connected to a slowly increasing line voltage. At first, the voltage is so low that the filament is not hot enough to glow. When it begins to glow, its hue is first a dark red, then (as it becomes hotter), orange, yellowish, and finally white. If it were heated even hotter, it would acquire a bluish hue. On your chromaticity diagram, draw a curve that represents the filament's color at these various stages. Label the various stages by their colors. (c) When the bulb is not up to full voltage, its light looks orangish. This light is reflected by a frosted glass window, behind which a green light shines. Is the resulting color of the frosted glass an additive or a subtractive mixture of the orange and green? Justify your answer. (d) What should be the hue of the light behind the frosted glass if you want the glass to appear white?

PH9 Take paints of two colors that an artist would call complementary. (a) Paint five small squares as follows: one with each of the pure paints, one with an equal mixture of the two paints, and one with each of the mixtures formed by taking three parts of one paint and one of the other. (b) Carefully examine the colors of the squares and describe, as well as you can, all the processes that seem to have occurred. (c) Sketch the locations of the five colors on a chromaticity diagram.

MATHEMATICAL PROBLEMS

PM1 A transparent piece of plastic has the transmittance curve shown in the figure. (a) If broad-band white light is incident on this plastic, what color will the transmitted light be? (b) Two layers of this plastic are placed behind one another. Plot the transmittance

PM6 Show that the additive mixture of equal amounts of the complements of two colored lights is complementary to the additive mixture of an equal amount of the two lights themselves.

Color Perception Mechanisms

CHAPTER 10

10.1
INTRODUCTION

Color phenomena, such as we studied in the last chapter, do not result from the properties of light alone. They depend intimately on the way your photoreceptors respond to light and the way in which your visual system processes the signals from those receptors. As the philosopher Arthur Schopenhauer wrote in 1816:

Colors, their reciprocal relations and the uniformity of their appearance, all reside in the eye itself, and are nothing else than specific modifications of the activity of the retina. External causes act only as stimuli, to cause this activity.

To a certain extent, your visual system acts as a wavelength detector; if you look at monochromatic 575-nm light, you will identify it as yellow and easily distinguish it from 600-nm light, which you'll recognize as orange. That is, as long as the light is confined to points along the horseshoe curve of the chromaticity diagram, color phenomena are wavelength phenomena. But the fact that the curve *is* horseshoe shaped, and that the area inside the curve represents actual color phenomena, is a result of the processing within your visual system. What enters the eye is always just a collection of light waves of different wavelengths and intensities. Yet rather than seeing a mixture of 575-nm light and 474-nm light as yellow and blue, you see it as white, a color not in the spectrum. A mixture of 410-nm and 690-nm light appears as a new saturated color, also not in the spectrum—purple. A broad-band stimulus of long-wavelength light appears yellow, much like the monochromatic 575-nm light. These effects, indeed the whole of the region inside the horseshoe, as well as your ability to distinguish monochromatic lights, can be understood if your color perception mechanisms are understood.

But there are more color phenomena than those described by the chromaticity diagram. You seldom experience individual isolated colored regions. Look at the colors around you now—more than likely, each colored region is surrounded by others of different sizes, colors, and proximity. Such spatial attributes of a colored pattern affect your perception of the color of each component region. It is your visual system's *spatial* processing of color that accounts for this—closely analogous to the spatial processing of lightness discussed in Chapter 7. Our knowledge of lateral inhibition will stand us in good stead here. Similarly, the principles of receptive fields and channels are also quite general and will find applicability, here more colorfully than before.

10.2
TRICHROMACY OF COLOR VISION

The first step in your perception of color is the response of your rods and cones to the incident light. We've mentioned that these cells contain photochemicals (pigments) that resonate (recall Fig. 1.12). A plot of the response of rhodopsin, the rods' photochemical, to light of different frequencies is shown in Figure 10.1.

Suppose you had only one system of receptors, each having the same photochemical (for instance the one shown in Fig. 10.1). A large amount of light at one wavelength, λ_1, could then produce the same response as a small amount at λ_2. As your brain can be aware only of the *response* of the photoreceptor, it would not be able to distinguish between the lights of two different wavelengths.

Scotopic vision (Sec. 5.3D) relies on the rods alone. Because there is only one type of rod, you cannot distinguish colors in dim light. Hence under scotopic conditions, the world appears black, white, and gray—not colored (see the TRY IT).

Under bright light (photopic conditions), when only the cones are operating, you *can* distinguish colors. Therefore, there must be more than one type of cone. How many are required—two, three, a million? The insight of the physicist Thomas Young led him in 1801 to postulate *three* types of receptors. Young made his hypothesis of **trichromacy** based on the fact that there are three independent attributes to color: hue, saturation, and lightness. He knew that any system (including the visual system) could have three independent *output* signals only if it has at least three independent *input* signals. He guessed that these input signals arose from three different types of cone.

Later, Hermann von Helmholtz promoted Young's theory by drawing three hypothetical response curves, representing the response of Young's proposed receptors (Fig.

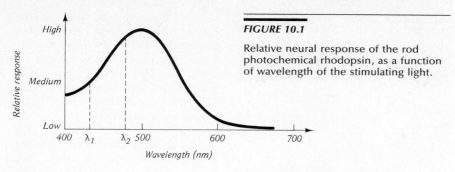

FIGURE 10.1

Relative neural response of the rod photochemical rhodopsin, as a function of wavelength of the stimulating light.

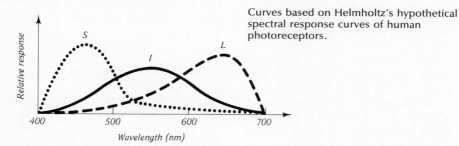

FIGURE 10.2

Curves based on Helmholtz's hypothetical spectral response curves of human photoreceptors.

10.2). Each curve has the general shape of a resonance curve, the difference between them being their resonant frequencies. One curve describes a receptor with best response to *short*-wavelength light (we'll call cones having this response S cones), one corresponds to *intermediate* wavelengths (I), and one to *long* wavelengths (L).

Spectral light (indeed *any* visible light) can excite all three receptor types, and so there are *three* signals associated with each color seen. For example, a spectral light of short wavelength will excite S cones most, I cones somewhat, and L cones least, as can be seen from the curves of Figure 10.2. This triplet of cone signals is (somehow) interpreted by the brain as "blue." No other spectral light can produce this relationship of all three cone responses, and thus no other spectral light looks like the blue we've chosen as an example. The same holds true for every *spectral* light; each causes a unique set of three cone responses.

Because the cone signals depend on the shape of these absorption curves, it is important that we determine the shape as closely as possible. What evidence will help us do that? Well, the evidence is plentiful

but often indirect. Like Sherlock Holmes, we can piece it together, hoping that the pieces will fall into place and give a consistent picture of what the response curves are.

A. Overlap of response curves

The first deduction we can make is that the curves must *overlap* throughout the region of visible wavelengths (or equivalently, frequencies). One reason is that resonance curves are generally broad and cover a range of frequencies, as we saw for paints and pigments in the last chapter. Second, if there were a region of the spectrum where cones of only one type responded, you would be unable to discriminate colors based on wavelength alone, for the reasons we saw above. But you know that spectral light of each wavelength appears different from that of other wavelengths, so at least two cone response curves must overlap at any wavelength of visible light.

B. Spectral complementaries

Recall that all wavelengths contribute equally to broad-band white

light. Such light will excite all three cone types approximately equally because there is short-, intermediate-, and long-wavelength light present. But we saw that an additive mixture of two complementary spectral lights can also yield the sensation of white. Thus the two complementary lights must produce equal photoreceptor responses, the same as broad-band white.

We can use this idea and the fact that the complements of green spectral lights are the nonspectral purples to determine where the S and I curves cross, and also where the I and L curves cross. Green spectral light has intermediate wavelength. Such light, with wavelength between the two crossover points, excites the I system most. Its complement has to excite *both* the S and the L system more than the I system to make the final excitations equal in all three systems. But *no* single wavelength light can do that! (Why?) Thus, the limits of the wavelength region with purple complements, that is, without monochromatic complements, must correspond to the two crossover points. These limits occur at about 490 nm and 565 nm (Fig. 9.9). Similarly you can argue that any spectral light having wavelength less than the S and I crossover point must have its spectral complement at a wavelength greater than the I and L crossover point, and vice versa. (These arguments hold rigorously if the curves of Fig. 10.2 are adjusted to have equal areas under each of them, ensuring that they have equal responses to broad-band white.) The revised curves of Figure 10.3 were drawn to incorporate the crossover points from the spectral complements.

The spectral complementary clue for people having one type of color deficiency ("color blindness," Sec. 10.5) also helps determine the crossover points. Color-deficient people who have only two of the three cone types can match white light with two spectral lights throughout the spectrum; there is no "gap" in their curves of spectral complementaries. Indeed, for these people, a certain *single* wavelength

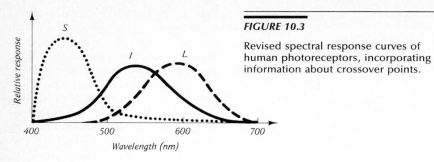

FIGURE 10.3

Revised spectral response curves of human photoreceptors, incorporating information about crossover points.

light (called a **neutral point**) appears achromatic (gray or white). Since this wavelength excites the two cones equally, just as broadband white light does, it must be at the crossover point of their two cone response curves. For instance, those color-deficient people that have only S and I cones see broadband white and 495-nm light as identical. Thus the crossover point for their S and I cones occurs there, at about the same place as the prediction based on the spectral complementary data from normal observers.

C. Hue discrimination

Suppose you are looking at two spots of spectral light that differ only in wavelength. If the wavelength difference is not too small, you are pretty good at detecting the difference in hues. But if the wavelength difference is sufficiently small, you will not be able to distinguish the two hues. The *difference in wavelength* that is *just noticeable*, $\Delta\lambda$, is a measure of your ability to discriminate hues. This just noticeable difference depends on the wavelength you start with, λ; you may be able to distinguish two yellows with wavelengths only one nanometer apart, but two reds may have to differ by three nanometers to appear different. Figure 10.4a shows the results of such **hue discrimination** measurements. The dips in the curve at about 430 nm, 490 nm, 590 nm, and 640 nm show that as the light's wavelength gets

close to these values, your hue discrimination improves. This implies that near these wavelengths the cone response curves must be steep, so that a small change in wavelength will result in a large (and thus noticeable) change in the cone responses. We've again revised the cone response curves (Fig. 10.4b) to take these hue discrimination results into account. Hue discrimination results for color deficients who have only two cone types also support the shape of the curves shown in Figure 10.4b.

D. Microspectrophotometry

We can also get a clue from **microspectrophotometry**,* that is, the physical measurement of the amount of light of each wavelength absorbed by each cone type. (This assumes that the signal produced by a cone is directly related to the amount of light absorbed by its pigment.) The absorption due to a single cone can be determined, one wavelength at a time, by carefully measuring the amount of light passing through it. Since there is also absorption in the part of the retina that supports the cone, this is separately measured (at a point between cones) and subtracted. For instance, if the cone and its supporting retina passed considerably less of the short wavelength light than the supporting retina alone did, the cone must have absorbed a lot of that short wavelength light.

Although many cones were measured in this way, only *three* types of absorption spectra were observed. This satisfied anyone who was still doubting Thomas; Thomas Young was correct about the number of types of cones. Further, the shapes of the curves agree very well with evidence from overlap, complements, and hue discrimination. The clues to our "mystery" fall into place! We have the cone responses (Fig. 10.5).

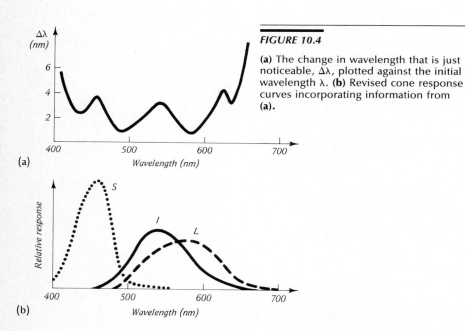

FIGURE 10.4

(a) The change in wavelength that is just noticeable, $\Delta\lambda$, plotted against the initial wavelength λ. (b) Revised cone response curves incorporating information from (a).

---

*Greek *mikros*, small, *phos*, light, *metria*, measure, plus Latin, *spectrum*.

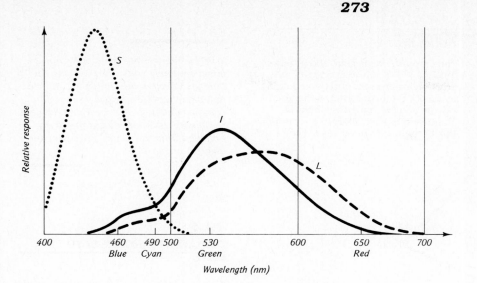

FIGURE 10.5

Spectral absorption of light by the three cone types, determined by microspectrophotometry. (We have modified these curves by including the short-wavelength absorption of the eyelens and macula, in order to make them comparable to the three previous figures.)

It is even possible now, by a new staining technique, to see the different cone types in the retina (Plate 10.1).

TRY IT

FOR SECTION 10.2
Scotopic vision and color

It's easy to demonstrate that you cannot distinguish hues under scotopic conditions. Have a friend arrange pieces of colored paper or unfamiliar colorful objects on a table and darken the room, while you sit with your eyes closed for about seven minutes. When you open your eyes, you will be properly dark adapted for the scotopic lighting conditions. You will not have previously seen the objects and so you cannot use past knowledge and memory to judge their hues. Notice that you cannot properly judge the hues of the papers and objects.

Because the peripheral parts of your retina contain chiefly rods, the TRY IT for Section 5.3D is closely related to this one.

10.3
COLOR MIXING AND MATCHING

The law of proportion according to which the several colors are formed, even if a man knew he would be foolish in telling, for he could not give any necessary reason, nor any tolerable or probable explanation of them.

Plato, "Timaeus"

In Section 9.4B we saw how different mixtures of lights could be metamers, that is, could appear identical to your eye although they might be of rather different spectral composition. One "tolerable and probable" explanation for this fact is that such mixtures produce identical triplets of cone signals. Now that we have the cone response curves (Fig. 10.5), we want to understand how they produce the color matching results.

For instance, suppose you want to match a 490-nm spectral cyan light, using as "primaries" 460-nm (blue), 530-nm (green), and 650-nm (red) lights. We saw in Section 9.4C that such a match can be made if you use a small negative amount of the red primary. What response does the cyan produce? From Figure 10.5 we see that it excites the S and I cones about equally, and the L cones about half as much. Now consider the response to the blue

and green primaries. Again using Figure 10.5 we see that the blue excites the S cones a lot. We must therefore use enough green to excite the I cones the same amount. When we do that, the L cones will have a bit too much excitation. We can compensate for this by using a small *negative* amount of red, thus *decreasing* mainly the L cone response. Because the cone stimulation is then the same, the mixture looks like cyan—the match is verified:

460 nm (blue) + 530 nm (green)
      − a little 650 nm (red)
      ≡ 490 nm (cyan)

This matching information is contained in Figure 9.10. You can similarly verify that other results in that figure agree with the cone response analysis.

PONDER

Verify this for matching a 630-nm light using these three primaries.

Thus, by considering the cone responses we can understand the matching of *spectral* colors. But what about the colors *inside* the horseshoe of the chromaticity diagram, for example, an additive mixture of 415-nm violet and 515-nm green? An equal mixture excites the S and I cones about equally, and the L cones slightly less. So the mixture appears like spectral cyan (S and I cones excited equally, and L cones about half as much) with some red mixed in (excess excitation of the L cones)—an unsaturated cyan, in agreement with the chromaticity diagram (Fig. 10.6).

PONDER

Verify from the cone response curves (Fig. 10.5) that 415-nm and 565-nm lights are complementary.

In sum, the information contained in the chromaticity diagram, both on and within the horseshoe, is consistent with that of the response curves of the three human cone types.

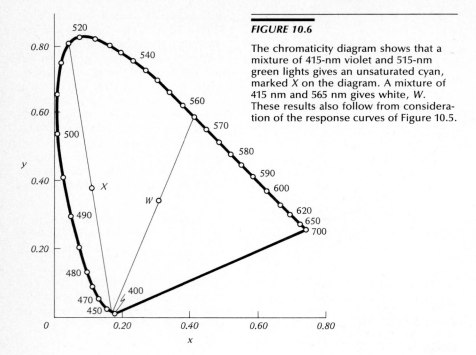

FIGURE 10.6

The chromaticity diagram shows that a mixture of 415-nm violet and 515-nm green lights gives an unsaturated cyan, marked *X* on the diagram. A mixture of 415 nm and 565 nm gives white, *W*. These results also follow from consideration of the response curves of Figure 10.5.

PONDER

What differences are there between complementary colors for bees and for humans? (Draw a bee line in Fig. 10.8 to determine a bee's spectral complementary to 480 nm. Compare your result with the complement of 480 nm obtained from the human chromaticity diagram of Fig. 10.6.)

10.4
OPPONENT PROCESSING

Most of the color phenomena we've been discussing up to now—such as detection, discrimination, matching—can be explained in terms of the response of the three cone types. But we have yet to understand the *appearances* of color mixtures. For example, an additive mixture of red and green does not appear reddish green, but instead yellow. A similar surprise is a subtractive mixture of cyan and yellow, which gives green, not a yellowish cyan. We must have a different set of primaries, other than the additive or subtractive primaries, in order to tell each other what colors *look like*. For this purpose, in fact,

Similar consistency is found in honey bees, which are trichromatic, but with their visible wavelengths ranging from ultraviolet to orange. Bees can be trained to respond whenever they experience a difference between two colors presented. In this way, color matching results have been obtained for the honey bee. These results agree quite well with the bee photoreceptor response curves obtained by recording nerve pulses from the bee's cones directly (Fig. 10.7). The bee chromaticity diagram (Fig. 10.8) shows the curve corresponding to the spectral colors to which the bee is sensitive, as well as the point we call "bee white"—broad-band light extending from 300 nm to 600 nm.

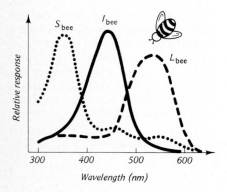

FIGURE 10.7

Spectral response curves of the three types of photoreceptor in the honeybee, as determined by directly measuring the neural response produced when light of various wavelengths is shined on the bee eye. The curves are somewhat idealized.

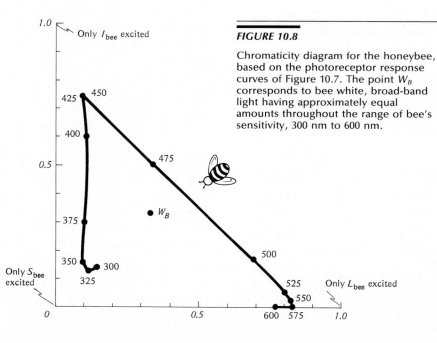

FIGURE 10.8

Chromaticity diagram for the honeybee, based on the photoreceptor response curves of Figure 10.7. The point W_B corresponds to bee white, broad-band light having approximately equal amounts throughout the range of bee's sensitivity, 300 nm to 600 nm.

you need *four **psychological primaries**: blue, green, yellow, and red.* All hues can be *verbally described* as combinations of these (e.g., orange looks yellowish red, cyan look bluish green, purple looks reddish blue, etc.). How can your brain process the output of the *three* cone types to obtain the psychological description in terms of *four* primaries?

To understand the appearance of colors, we must understand the neural signals that get to the brain. Conversely, to determine what signals get to the brain, we can study the appearance of colors. Only the last 30 to 40 years have brought the reconciliation of the trichromatic theory of Young and Helmholtz (for which we have already seen considerable evidence) with the apparently competing **opponent colors** theory of Ewald Hering.

A. Color naming

What's in a name?

One of the simplest approaches for studying the appearance of colors is to ask an observer to *name* the color of a spot of spectral light by rating its similarity to each of the four psychological primaries on a scale from 0 to 10. Figure 10.9 shows typical results for a normal color observer.

Note that there are some spectral **unique hues** that are described by just *one* psychological primary: blue (475 nm), green (500 nm) and yellow (580 nm). For example, unique yellow is what most people would call a pure yellow, without any red, green, or blue mixed in. But there is no spectral unique red—even at

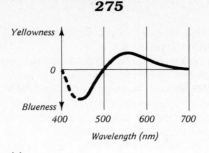

(a)

(b)

the longest wavelengths, the light appears slightly yellowish.

The most important point, however, is that you never experience a "reddish green" nor a "yellowish blue," though all other combinations of the psychological primaries are experienced. That is, there seems to be some opposition between yellow and blue, and between red and green—they are **opponents.** This opponency within these pairs extends to their presence in nonspectral colors; there is *no* color that simultaneously appears red and green, or simultaneously yellow and blue. You can't even imagine such a color!

B. Hue cancellation

The amount of subjective **chromatic response** for each spectral light can be determined more quan-

---

FIGURE 10.9

Typical color naming results for spectral lights. Subjects rated the apparent content of the four psychological primaries in each spectral light. For example, a 650-nm light appears primarily reddish and somewhat yellowish, but neither greenish nor bluish.

FIGURE 10.10

(a) Yellow versus blue **(y − b)** and **(b)** red versus green **(r − g)** chromatic responses for spectral lights determined using the hue cancellation technique. The two curves were separately determined. The values of the yellow chromatic response (plotted above the axis) are related to the amount of unique blue (475 nm) needed to cancel the yellow in all those wavelengths perceived as having yellow (above 500 nm). The blue chromatic response (plotted below the axis, to emphasize its opponency to yellow) is related to the amount of unique yellow (580 nm) needed to cancel the blue in all wavelengths perceived as having blue (below 500 nm). Likewise, the red and green responses are plotted above and below the axis. The red response was canceled using unique green (500 nm). A 700-nm light was used to cancel the green chromatic response. (The green chromatic response curve does not depend significantly on the choice of "red" wavelength used for cancellation.)

titatively by the technique of **hue cancellation.** To find the blue chromatic response for a certain wavelength light, say 430 nm (a deep blue), you add enough unique yellow to it so a color results that appears neither bluish nor yellowish. You find that the intensity of the 580-nm unique yellow must be *large* in order to cancel the blue in this light, so this blue is said to produce a *large* blue chromatic response. On the other hand, 490-nm light (which appears only *slightly* blue) requires only a *small* amount of the 580-nm yellow to cancel the blue there. The results of such hue cancellation are shown in Figure 10.10.

You're familiar with this type of cancellation technique, used in a simple balance (like that held by the statue of Justice). To find the

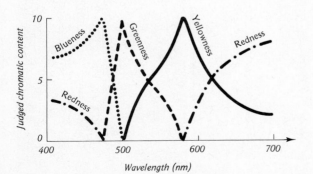

weight on one pan, you add a known weight on the other until the scale is level. Similarly in hue cancellation, to find how much blue is in a certain spectral light, you add yellow until the blue is canceled. Of course, it is as if there are *two* pan balances in the visual system; after one opponent pair has been balanced by this technique, the other pair may not be balanced.

PONDER

What hue name(s), chosen from the four psychological primaries, would describe the resulting color after you canceled the blue in the 430-nm light?

You can use Figure 10.10 to describe the appearance of colors. Thus, the curve of Figure 10.10a shows that a 550-nm light appears fairly yellowish, and a 430-nm light quite bluish, whereas a 500-nm light (unique green) appears neither.

C. Neural connections

How can the signals originating in the three cone types get mixed, or processed, to yield signals corresponding to the subjective responses described by Figure 10.10? As we saw in Chapter 7, signals to the brain are often comparisons between different receptor responses. Figure 10.11 shows hypothetical connections, two of which involve *differences* between signals from the different types of cone. These correspond to **chromatic channels** and explain the observed chromatic responses. Similar connections are directly observed by electrophysiological measurements on goldfish and monkeys.

Consider, for instance, the **yellow minus blue** chromatic channel **(y - b),** which compares the long-wavelength part of a light with the short-wavelength part. When light excites either the L or I cones, there is *excitation* in the cells of the **y - b** channel. If, however, light excites the S cones, there is *inhibition* of the **y - b** cells. If the net result is excitation, the light appears yellow-

ish. A net inhibition makes the light appear bluish.

Likewise, for the **red minus green** chromatic channel **(r - g)**, a net excitation leads to the sensation of red, a net inhibition leads to the sensation of green. (Note that an excitation of the S cones leads to a red sensation, hence the short-wavelength end of the spectrum, "violet," appears to have some red in it as shown by Fig. 10.9.)

We've stressed that there are three attributes of any color and that there are three cones. Unless there is a third channel, your brain could be aware of only the two attributes of color represented by "red-greenness" and "yellow-blueness." This third opponent channel—the **white minus black** channel **(w - bk)**—relays *lightness* information. The **w - bk** channel has a positive response whenever *any* of the three cone types is stimulated. The opponent nature of this channel is slightly different from that of the chromatic channels, though. Inhibition in this channel does not result from excitation of any cone type. There is inhibition in all three channels due to adaptation and lateral effects. The spatial processing discussed in Chapter 7 involves such effects in **w - bk** channels. (Color television, interestingly enough, also employs two chromatic channels and one gray-scale channel. The latter is necessary so the signal can be used by black and white sets. Reception of the two chromatic channels is controlled by knobs sometimes labeled "color" and "tint.")

So, we find a reconciliation between the once apparently competing ideas of trichromacy and opponency; the initial photoreceptor stage is trichromatic, but the signals from that stage are subsequently processed through three opponent channels (two chromatic channels and one achromatic). Each chromatic channel accounts for two hue responses. Thus there are four fundamental hue responses and hence four psychological primaries.

D. Chromaticity diagram

We saw in Section 10.3 that the relative stimulations of the L, I, and S cones determine the location of any color on the chromaticity diagram. But the relative stimulations of the cones also determine the responses in the opponent channels. It is possible, then, to relate the activity in the chromatic opponent channels **(r - g** and **y - b)** to the chromaticity diagram. (The activity in the **w - bk** channel, related to the lightness, is not contained in the diagram.)

The two lines on the chromaticity diagram of Figure 10.12 divide it into regions corresponding to the activities in the chromatic chan-

FIGURE 10.11

Highly schematic hypothetical connections between the cones and the cells of the two chromatic channels, **r − g** and **y − b,** and of the achromatic channel, **w − bk.** Excitation is denoted by + and inhibition by −.

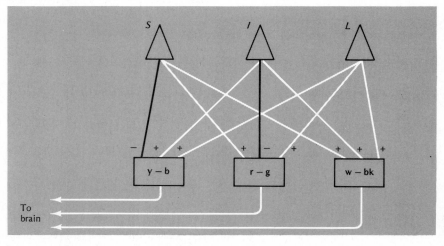

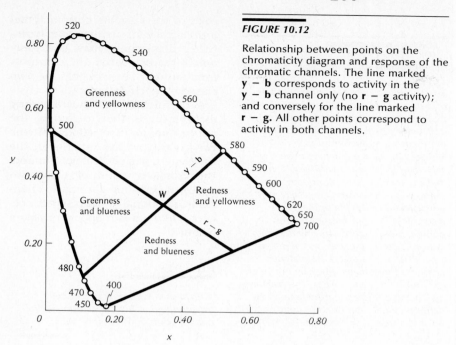

FIGURE 10.12

Relationship between points on the chromaticity diagram and response of the chromatic channels. The line marked **y − b** corresponds to activity in the **y − b** channel only (no **r − g** activity); and conversely for the line marked **r − g.** All other points correspond to activity in both channels.

nels. The neutral point, *W*, is where there is as much blue response as yellow and as much red as green. Thus, the lines must cross there. From Figure 10.10, you see that spectral light of 580 nm or 475 nm does not lead to activity of cells in the *r - g* channel. Since any color lying on the line connecting these two points looks like a mixture of these lights, *any* color falling along this line appears neither reddish nor greenish—this line denotes *no* response in the *r - g* channel. Colors above this line have an inhibitory signal in the channel; such colors appear greenish. Colors on the other side of this line appear reddish.

The line that passes through the *W* point and 500 nm (unique green) corresponds to no activity in the *y - b* channel. Colors falling along this line appear neither yellowish nor bluish (though they may appear reddish or greenish). Colors to the right of this line yield excitation in the *y - b* channel and thus appear yellowish, whereas colors to the left of this line yield inhibition and appear bluish. Notice that this line does not strike the horseshoe at the lower right. This reflects the fact that the unique red is not spectral,

but the complement of 500 nm, that is 500c. (Notice from Fig. 9.9 that the complement of unique blue, 475 nm, is *not* unique yellow, 580 nm, as opponent processing would predict, but 575 nm instead. We may take this discrepancy as a measure of the error accompanying these measurements.)

Thus, any point on the chromaticity diagram can be identified by the amount of activity in the two chromatic channels and vice versa. These two channels carry all the chromaticity information.

*10.5
COLOR DEFICIENCY

A person is called *color deficient* when his color matches are not the same as those for most of us. (The term "color blind" has fallen out of favor primarily because most color-deficient people do see color, only somewhat differently from most of us.)

While color matching is the most reliable and revealing test for color deficiency, it is simpler and faster to have the subject look at colored plates similar to those of Plate 10.2,

which appear different to normals and color deficients. For example, normals can see two digits in Plate 10.2a, but some color deficients can see only one.

Little is known directly about the way in which the visual system of a color deficient differs from that of a normal color observer, and you should consider the explanations presented here as simply informed guesses. For instance, some color deficients lack the use of one or more cone type. This might result from any of a number of causes: the nonfunctional cones may simply be missing; they may be present but lack connections to subsequent nerves; perhaps the cones are connected to subsequent nerves meant for *another* cone type; maybe the cones are present but are filled with the "wrong" photopigment, either a modified one, or one used in other cones; or any combination of these (and yet other!) reasons. There may even be problems in the neural processing of the opponent channels. All this makes it difficult to be definite about conclusions concerning color deficiency.

One thing is clear, though: different people require a different number of colors to match any light. Thus we can group those requiring, one, two, or three such colors as **monochromats, dichromats,** and **trichromats,** respectively.

A. Monochromacy

By suitably adjusting the intensity of only *one* spectral light, a monochromat can match a light of any color. These unfortunate people can be considered truly color *blind,* in that they cannot distinguish any wavelength from any other—like watching a color TV program on a black and white set. (This relatively rare color deficiency is often linked with other visual problems, such as inability to see fine detail.)

A standard photocopier is monochromatic, as you can see from Figure 10.13 and the TRY IT.

*Greek *mono,* one; *di,* two; *tri,* three; plus *chroma,* color.

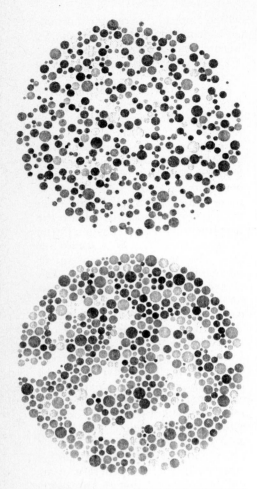

TRY IT

FOR SECTION 10.5A
Color blindness of a photocopier

You can investigate the color blindness of a standard photocopier using colored pens, colored photographs, or the color deficiency plates of Plate 10.2. First photocopy a multicolored picture. To which colors is the machine most sensitive? Least sensitive? If you can make two colors that photocopy the same, you can construct a pattern of dots (similar to the plate) that is discernible by you, but not by the photocopier. Next photocopy the color deficiency testing plates, as we did for Figure 10.13. The plates are matched in lightness for humans, but may not be so for your photocopier. Thus, the machine may be able to distinguish some of the patterns. What kind of color deficiency is most like that of the machine?

B. Dichromacy

Dichromats require only *two* colors to match any color. Some of these people may have only two of the three *cone* types fully functioning. There are three main classes of such dichromats, corresponding to

each of the possible nonfunctional (or relatively insensitive) cone types. Other dichromats may have only one of the two *chromatic channels* functioning. There would be two classes of such dichromats.

A useful test for understanding dichromacy is that of finding the *neutral points* (Sec. 10.2B). If only two cone types are functional, the wavelength of a neutral point marks the crossover point of these two cones; if only one chromatic channel is functional, a neutral point occurs where its two opponent colors balance.

PONDER

A dichromat with only two cones functioning has a second (nonspectral) neutral point in the purple. Why?

FIGURE 10.14

Rayleigh test. **(a)** *L* and *I* cone response curves for a normal observer, as in Figure 10.5. (The *S* cones do not respond at the wavelengths used in this test.) **(b)** A protanomalous observer may have the *L* curve shifted to shorter wavelength, *L'*.

FIGURE 10.13

Photocopy of Plate 10.2 made using a Savin photocopier. The photocopier cannot "see" the numerals, but it can "see" the serpentine path.

There are two main types of monochromats, depending on which single photoreceptor responds to light. **Cone monochromats** have functional cones, but of only one type (or they lack the two chromatic channels). They can see under photopic conditions. On the other hand, **rod monochromats** lack all cone function and respond using their rods alone. They have great difficulty in seeing under photopic conditions. Lacking foveal cones, their acuity is usually quite poor, and they must avert their eyes slightly when they want to look at something.

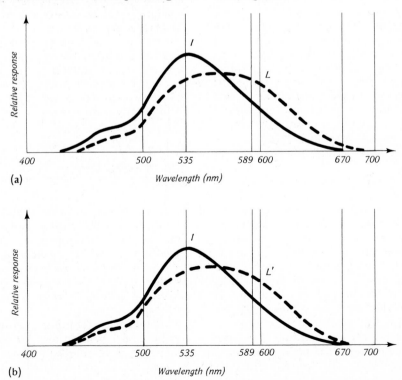

The responses of the L and I cones can be compared by the **Rayleigh test.** The subject adjusts the relative intensities of particular green and red lights until the sum appears identical to a certain yellow. (In practice, 670-nm and 535-nm lights are used to match a 589-nm light.) If both L and I cones respond normally, a certain, "normal" ratio of the two lights' intensities is required for the match (Fig. 10.14a). If, however, one type of cone is not functioning, say the L cones, then the subject can make a match using a variety of relative intensities. He need only keep the total excitation in the I cones the same, since the S cones do not respond appreciably in this wavelength range. In fact, given many chances, the subject will not show consistency in his settings.

An analogous test for the S and I cones would reveal information about the relative response of these cones. But, because the composition and color of the macula lutea (Sec. 5.2B) varies considerably with age and from person to person, such a matching test at short wavelengths is less reliable.

A further test is that of hue discrimination. As we saw in the Section 10.2C, hue discrimination is good ($\Delta\lambda$ small) at wavelengths where the cone response curves are steep, and poor ($\Delta\lambda$ large) where they are fairly flat. If a cone type, and hence the steep regions of its response curve, is missing, there are fewer wavelength regions where hue discrimination is good. Their location tells indirectly which cone types are present.

With these tests in mind, check to see that the results in Table 10.1 are consistent if **protanopes** lack the L cones, **deuteranopes** lack I cones, and **tritanopes** lack S cones.

What would happen if a subject had the photoreceptors and pigments of a normal person, but lacked one of the chromatic channels, say the y - b channel? Such a person would respond using his **r - g** channel, and would consequently have *two* neutral points: about 580 nm (i.e., near unique yellow) and about 470 nm (i.e., near unique blue). This behavior is similar to that of a few observers called **tetartanopes.**

PONDER

*Suppose a subject lacks the **r - g** channel instead. Where would his neutral points be? Would his response differ from a deuteranope's?*

C. Trichromacy

Trichromats (including normal color observers) require *three* colors to match an arbitrary color. As such, trichromats must have three cone pigments. Their differences in color vision may arise from pigments with slightly different response curves than those of the normal pigments. Alternatively, the connections between one type of cone and the subsequent nerve cells may be defective.

The Rayleigh test is particularly useful for describing **anomalous trichromats.** Suppose the L cones of a trichromat were shifted to wavelengths shorter than normal (Fig. 10.14b). The shifted cones then have their sensitivity at the 535-nm green increased by a small percentage, compared to normal cones; however, their sensitivity at the 670-nm red is reduced by a large percentage. To compensate, such an observer requires *extra red* in his Rayleigh match with the 589-nm yellow. (This kind of color deficient would also have reduced sensitivity at long wavelengths, due to the shifted L cone curve.)

Anomalous trichromats are usually classified as **protanomalous** (excessive red in the Rayleigh

---

TABLE 10.1 *The symptoms of the principal types of color deficiency*

	Number of colors needed for match	Approximate wavelength of neutral point(s) (nm)	Approximate wavelength of peak sensitivity (nm)	Reduced sensitivity at long wavelengths?	Rayleigh test excess green	Rayleigh test excess red	Percentage of males affected*
Monochromacy							
Rod monochromacy	1	all	505	yes	inconsistent		.003
Cone monochromacy	1	all	three types	some may be	inconsistent		very small
Dichromacy							
Protanopia	2	495 and red-purple	540	yes	inconsistent		1.0
Deuteranopia	2	500 and purple	560	no	inconsistent		1.1
Tritanopia	2	570 and blue-purple	555	no			very small
Tetartanopia	2	580 and 470	555 (?)	no			very small
Trichromacy							
Protanomaly	3	none	540	yes		yes	1.0
Deuteranomaly	3	none	560	no	yes		4.9
Tritanomaly	3	none	560	no			very small
Neuteranomaly	3	none	555	no			very small
Normal	3	none	555	no			91.

*Greatest percentages of occurrence are reported for males.

match), **deuteranomalous** (excessive green in the Rayleigh match), and **tritanomalous** (normal Rayleigh match but yellow-blue problems), but there is a continuous range of variation. Indeed, there is not even a sharp division between anomalous trichromats and dichromats.

Some of the problems associated with these trichromats may arise in the neural processing beyond the photoreceptors. Indeed, a fourth type of anomalous trichromacy, **neuteranomaly,** is believed to involve normal cone responses but relative ineffectiveness of the **r - g** channels compared to that of the **y - b** channel. One symptom of this deficiency involves the neuteranomalous observer's wavelength settings for unique blue and yellow, which are the same as for normal observers on the average, but show great variation; because his **r - g** chromatic response curve is reduced, it is indistinguishable from zero for a *range* of wavelengths (Fig. 10.15).

In short, it seems that anything and everything can go wrong in a color vision system. One, two, or all three cone responses may be lacking; a chromatic channel may be nonfunctioning or may not respond efficiently; the photoreceptor pigments present may not have the normal response curves, or they may be combined differently in the chromatic channels; or there may be a combination of these defects. Further, each set of symptoms exhibited in Table 10.1 may be achieved by more than one such physiological defect.

Are there any "cures" or techniques that can help color deficients, such as there are for people with focusing problems? In some cases, yes. One technique is to have the color deficient wear a red filter over *one* eye (in the form of an eyeglass lens or a contact lens). Suppose a color deficient, who had a neutral point in the cyan region of the spectrum, wears such a lens. Although to him a broad-band gray and a certain cyan appear *identical* in the uncovered eye, they appear *different* in the covered eye. Thus, the red filter over one eye has converted physical information of color difference into binocular information, which the subject can detect. With practice, the deficient comes to associate such binocular information with different colors—permitting a greater range of distinguishable colors.

10.6
SPATIAL PROCESSING OF COLOR

In Chapter 7 we saw several fundamental effects of lightness perception: simultaneous lightness contrast, edge enhancement, lightness constancy, etc. In this section, we'll see that these spatial effects have counterparts "in living color" that come about because of similar neural processing in the chromatic channels.

A. Chromatic lateral inhibition

Even though the four small squares in Plate 10.3 are objectively gray (use a mask to check this), they appear different colors because of their surrounding regions. Notice that the yellow region makes the gray appear somewhat blue, the green region makes the gray appear red, and so on. Such **simultaneous color contrast** suggests that there is lateral inhibition (Sec. 7.4). That is, there must be a *spatial* opponency ("center minus surround") of the chromatic channels, in addition to their chromatic opponency ("red minus green," for example). Indeed, some cells of the chromatic channels are opponent both in color and in space, and hence are called **double-opponent cells.**

A double-opponent cell of the **r - g** channel (Fig. 10.16) has the cones of the center connected in the usual **r - g** fashion. The surround is connected in the opposite way, **g - r** (e.g., stimulation with red here leads to inhibition in the pooling cell). Thus, the stimulus that leads to the greatest excitation of the pooling cell has long-wavelength light striking the center and intermediate-wavelength light striking the surround. Likewise, a double-opponent cell of the **y - b** channel would have the familiar **y - b** response from the center and a **b - y** response in the surround.

Consider such a cell of the **r - g** channel whose center is viewing an area just inside the gray region surrounded by the green in Plate 10.3 (Fig. 10.17). The gray stimulates both the *L* and *I* cones in the center, leading to no net excitation there. The green in the surround, however, primarily excites the *I* cones there, leading to excitation of the **g - r** in the surround, and hence to a net *excitation* in the pooling cell. This excitation is interpreted as red because it stimulates the pooling cell in the same way that a central red stimulation would. The gray square therefore appears slightly reddish. If the surround had been red, instead of green, there would have been net *inhibition* in the pooling cell and the gray square would appear slightly greenish, in accord with what you see in Plate 10.3. A similar process occurs for the double-opponent cells of the **y - b** channel.

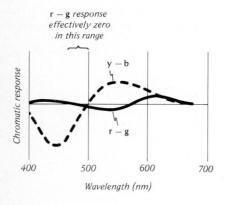

r − g response
effectively zero
in this range

FIGURE 10.15

Hypothetical chromatic response curves of a neuteranomalous observer showing reduced response in the **r − g** channel.

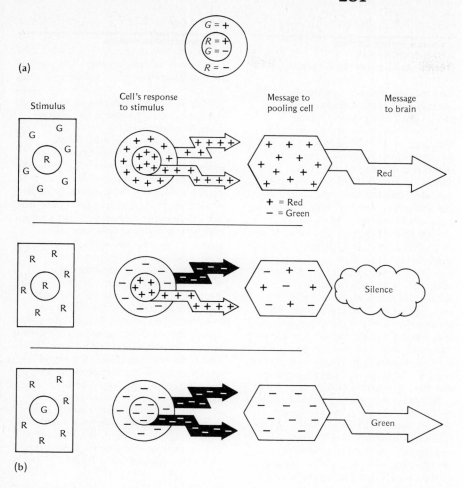

(a)

Stimulus   Cell's response to stimulus   Message to pooling cell   Message to brain

+ = Red
− = Green

(b)

FIGURE 10.16

(a) Receptive field of double-opponent cell of the **r − g** channel. Long-wavelength light in the center excites the pooling cell, while intermediate-wavelength light there inhibits the cell, as in Figure 10.10b. In the surround, however, intermediate-wavelength light excites the pooling cell, and long-wavelength light inhibits it (Figure 10.10b with redness and greenness interchanged). (b) Examples of this cell's response.

PONDER

The gray square surrounded by light yellow appears somewhat dark. Why?

In general, any large colored region tends to make its neighboring regions appear more like the color opponent of the original region. For example, a green region when sur-

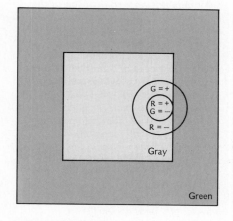

FIGURE 10.17

An **r − g** double-opponent cell explanation of simultaneous color contrast. The green in the receptive field's surround leads to excitation of the cell, and thus to some perception of redness in the objectively gray region.

rounded by yellow appears more bluish, that is, more like cyan. The TRY IT shows how you can further verify that the appearance of a color is influenced by colors nearby.

Although the basic phenomena of simultaneous color contrast were fairly well known to artists for centuries, it was only in 1839 that the director of the Gobelins tapestry works, Eugène Chevreul, formulated the relevant principles somewhat scientifically. He had been receiving complaints about the quality of the colors in the tapestries. He found no deficiencies when he thoroughly checked the dyes, so Chevreul considered the *perception* of color. He discovered that he could make colors appear more saturated and bright by changing the *designs* of the tapestries. For instance, a yellow surrounded by dark blue appears more saturated and lighter than if it were surrounded instead by orange. The brilliance of the Impressionists' color was achieved using these ideas, by the placement of large dabs of complementary colors adjacent to one another. The painter J. M. W. Turner may have used this principle in reverse; it is said that he made last-minute modifications in the colors in his paintings to diminish the effectiveness of the colors in his rivals' paintings, hung nearby. More recently, color contrast has been used by many artists—for example, the saturation of the green region in Plate 5.2 results, in part, from the red surrounding it.

The effects of edge enhancement that we saw in Chapter 7 have counterparts in color as well. For example, in "Side Show (La Parade)," Georges Seurat paints one side of an edge extra bluish and the other side extra yellowish, thereby making the two regions each appear more saturated (a chromatic analogue of the Craik-O'Brien effect). Although most of the Impressionists disdained intellectual approaches toward color, Seurat had an obsession with color theories and principles and was greatly influenced by Chevreul among others.

TRY IT

FOR SECTION 10.6A
Simultaneous color contrast

In the middle of a piece of colored construction paper, cut a small hole (about 2 mm across) to form a mask. Hold the mask at arm's length, close one eye, and look through the hole at colored objects. Compare the colors you see with and without the mask. Notice that the mask tends to make the region in the hole appear more like the color complementary to that of the mask.

B. Color constancy

Objects tend to retain the same perceived color even though the coloration of the overall illumination may change. This **color constancy** certainly is both useful and a biological necessity. An edible berry may be distinguishable from a poisonous one only by its color, so it is essential that this color persist, no matter what the illumination (blue noon sky, red twilight, gray clouds, etc.)—there must be color constancy if the berry eater is to survive.

Like lightness constancy (Sec. 7.4A), color constancy involves lateral inhibition. For example, an overall excess in red illumination is "ignored" by the double-opponent cell of the r - g channel (Fig. 10.16), because the increased stimulation of the receptive field's center and inhibition in its surround cancel each other. To look red, the berry must reflect *more* red light than the average, so that the center of a double-opponent cell viewing the berry can be excited, but not the surround. Hence colors depend on the *relative* amount of colored lights. The first TRY IT shows how you can demonstrate these effects.

Color constancy isn't perfect, however, and may be affected by your overall state of adaptation, even achromatic adaptation, as the second TRY IT shows.

First TRY IT

FOR SECTION 10.6B
Colored shadows and Hering papers

*You can demonstrate simultaneous color contrast with **colored shadows.** Place an object, such as a beer can, some distance in front of a white screen. Use two lamps or slide projectors as point sources to form two nonoverlapping shadows of the can on the screen. Now hold a red filter in front of one of the lamps, so that the lamp illuminates the can and screen with red light. Notice the color of the shadow regions on the screen. One of them receives light from the red lamp only and appears red as you might expect. The other shadow region receives only white light but appears cyan. (You may have to wait a few moments.) This is because the regions surrounding that shadow receive red (and white) light, and by simultaneous color contrast, make the shadow appear cyan; the shadow is less red than the average illumination. What would be the color of the shadow if you replaced the red filter by one of another color? Try it!*

This phenomenon was first observed by Count Rumford in the eighteenth century without lamps or slide projectors. He used a candle as one (slightly yellowish) source and the (bluish) noon sky as another. To balance the intensities of the two lights, he brought the candle close to the screen. What color were the shadows he produced? Try it!

*Another way to demonstrate that a color depends on a comparison of the light coming from different regions involves a sequence of achromatic regions ranging from black through gray up to white (Fig. 7.4b). (Hering originally used small pieces of achromatic papers of various lightness, which are now called **Hering papers.**) View that figure under strong red light; be sure to eliminate all other illumination. Notice that (after a few moments) the intermediate gray region in the middle of the scale still appears gray. The upper region, reflecting more long-wavelength light than the average, appears red. The lower region, reflecting less long-wavelength light, appears cyan, or green. Your perception of color is determined by the relative amount of light at different wavelengths, not the overall amount.*

Second TRY IT

FOR SECTION 10.6B
Dependence of color on your state of adaptation

That color depends on you, as well as on the light coming to your eyes, is easily demonstrated by taking advantage of the fact that you have two eyes. Find a picture with lots of different colors, or arrange a variegated scene. By alternately closing one eye and then the other, compare how these colors look to each of your eyes. Most people will notice no difference between the eyes (although there are very rare people with color deficiency in only one eye—if you are such a person, rush to your nearest psychophysicist and volunteer as a subject). Now close one eye, and stare at a bright white light (not the sun!) with the other for half a minute. Move your eye around, so no one spot on the center of your retina receives more light than its neighbors. Having adapted one eye in this manner, repeat the original experiment; closing first one eye, and then the other, look at the colored scene. Compare the colors as viewed by your two eyes now. Note that the colors change even though you have adapted to achromatic (white) light. What would you expect to happen if you adapt by staring at a bright colored light? Try it!

C. Spatial assimilation

Even though the red background in Plate 10.4 is physically uniform throughout, the red in the central square appears slightly yellowish, while the red in the outer portions appears slightly bluish. Simultaneous color contrast would make the red in the central square appear bluish (in contrast to the yellow dots there) and the red in the outer portions appear yellowish (in contrast to the blue dots there). The effect shown in the plate, in which the color of a region **assimilates** that of neighboring regions (rather than taking on the opponent hue), is called the **von Bezold spreading effect.** If two colors are in adjacent large areas, they will affect each other via contrast. However, if they are small, intermingled areas, they produce assimilation, as in partitive color mixing (Sec. 9.5B).

Although the mechanism for assimilation is not fully understood, one idea is plausible—the receptive fields of the double-opponent cells of the chromatic channels come in a variety of sizes. The large receptive fields contain the color information while the small ones primarily relay information about fine detail. Cells with large receptive fields do not see the dots as separate, so the red area in the center appears somewhat yellowish. You can still see the dots as separate because the cells having small receptive fields relay the spatial information.

Because contrast and assimilation depend on the size of the patterns involved, the colors of a given region will change as you change your distance from it. When you stand very close to a painting like the one shown in Plate 7.2 you can see each dot and its color as individual or there may be contrast effects due to nearby dots that affect the color. From an intermediate distance, however, you still see a dot as individual but its color is affected by that of neighboring dots via assimilation—the colors tend to mix partitively. If you view the painting from a greater distance, you see the partitive mixture, but you can't resolve the individual dots. It is unfortunate that some museums hang pointillist paintings in rooms so small that you can't move far enough from the painting for some of these changes to occur.

10.7
TEMPORAL PROCESSING

We can extend the parallel between achromatic (Chapter 7) and chromatic processing by discussing the temporal response of the chromatic system. Many of the temporal phenomena of the earlier chapter have analogies in color. In addition, because the temporal properties of the response to different colors can

themselves differ, we'll find temporal phenomena in color that have no counterpart in achromatic vision.

A. Standard negative afterimages

Adapt (Sec. 7.5) to Plate 10.5 to form a negative afterimage. Note particularly the colors in the afterimage. Where you adapted to yellow, you now see blue (and vice versa), and similarly for cyan and red. Notice here that the afterimage of a color is its *complement*. As in black and white, the color afterimages do not transfer between your eyes. (Try it!)

The explanation for such color afterimages is a simple extension of that used for achromatic afterimages. Consider the cones in the area of your retina responding to the yellow region during adaptation. The yellow light stimulates (and thus desensitizes) your L and I cones much more than it does your S cones. Therefore, when you subsequently look at the white piece of paper, your S cones respond vigorously whereas your desensitized L and I cones do not. The resulting excess of S cone response makes the afterimage appear blue in that region.

As you continue to observe the afterimage carefully, it fades and its *color* changes slightly. This is because your different cones (and chromatic mechanisms) recover from adaptation at different rates.

Negative afterimages can be used to explain the phenomenon of **Bidwell's disk** (Fig. 10.18). When the black and white sectored disk is rotated *counterclockwise*, the red bulb appears properly *red*, whereas rotating the disk *clockwise* makes the bulb appear *cyan*. While rotating clockwise, first the notch reveals the red bulb, and then the *white* region covers the bulb. The brief exposure of the red bulb desensitizes your L cones primarily; when you then view the white re-

FIGURE 10.18

Bidwell's disk. As the black and white sectored disk rotates, its notch intermittently reveals a bright red light bulb, whose apparent color depends on the direction of rotation.

gion, the standard cyan afterimage results. At the proper rotation speed, the afterimage can last much longer than the presentation of the bulb, and so your impression is that the bulb is cyan. When the disk is rotated counterclockwise, the notch revealing the red bulb is immediately followed by the *black* sector. By the time the white region comes into view, your L cones have recovered, so you don't see a negative afterimage. The bulb thus appears red.

B. Positive afterimages

When producing a *positive* afterimage in the first TRY IT for Section 7.7A, you undoubtedly used a colored stimulus, for example, the blue sky. That the positive afterimage retained the original colors (at least initially) demonstrates that your chromatic channels have persistence. Our discussion of TV and movie presentation and flicker is just as valid for your chromatic response as it is for lightness, as you know if you watch color TV.

Notice that the color of the afterimage *changes* as you wait. As in the negative afterimage case, your different cone types recover at different rates, leading to a changing set of signals as time goes on.

*C. Other temporal effects

Because the different chromatic mechanisms respond at different rates, it is possible to elicit color responses by presenting black and white stimuli with the proper temporal behavior. An example is **Benham's disk** (Fig. 10.19), which was invented by a toymaker in the nineteenth century who noticed colors appearing in concentric rings when the disk was rotated at the proper rate (he mounted his on a toy top—see the first TRY IT). Consider what happens as the disk turns and white light, reflected by the disk, strikes your retina. Figure 10.20 shows the sequence of white and black produced at various positions while the disk is rotating. A receptive field pointed at any one of these

FIGURE 10.19

Modern version of Benham's disk. When rotated at 33⅓ rpm, colors appear in the concentric bands: red on the inner set, blue on the outer set.

places, therefore, can have a different sequence of white and black stimulation on its center, compared to that on its surround. Because the persistence exhibited by a receptive field's center may differ from that of its surround, such unusual intermittent stimulation can lead to an imbalance in chromatic response between the center and surround. This imbalance is interpreted as color. (For a related effect, see the second TRY IT.)

Like your cones, the different color sensors in a television camera also have different temporal response properties, as can be verified when the TV camera sweeps across a very bright spot, such as a stadium light; there is often a colored streak across your color TV tube, because the different sensors each respond, and decay, at different rates.

We've discussed how eye movements can translate spatial variation of a stimulus into temporal variation in signals (Sec. 7.6). These variations induced by eye

movements can produce subtle colors even when the stimulus is black and white, as in Figure 10.21. Such **Fechner's colors** can also be experienced as your eye scans any black and white pattern having fine detail such as Figure 7.17.

First TRY IT

FOR SECTION 10.7C
Benham's disk

You can easily see colors from Benham's disk. Photocopy Figure 10.19, and use a black felt tip pen to get good blacks, if required. (Just like the visual system, photocopiers process edges, often at the expense of large, uniform regions. Since the photocopier does not have a brain to fill in the solid regions, you, armed with your black pen, must do it.) Place the copy on a phonograph turntable and set it rotating at 33 rpm, well illuminated by incandescent light or daylight. Notice the colors in the bands produced by the rotating disk. If you can change the speed of your turntable, try 45 rpm and note any change in colors.

Second TRY IT

FOR SECTION 10.7C
Latency and color

Under bright, white light, hold this book on your lap and look at Plate 10.6. Shake the book back and forth as quickly as you can, as you would shake a pan when making popcorn. The red square appears locked to the cyan, and the two move together, as you would expect.

Dim the room lights until you can just barely recognize the colors, and then shake the book as before. Note now how the cyan area appears "rubbery"—the red square seems to slosh back and forth within it.

Under bright light the cones and channels respond rapidly. Thus, they can "keep up" with the moving squares of the plate. Under dim light, however, there can be significant differences between the latencies of your response to the red compared to the cyan. (Recall the Pulfrich phenomenon, Sec. 8.5C.) As you shake the plate, then, one region is processed more slowly and appears to lag behind the other.

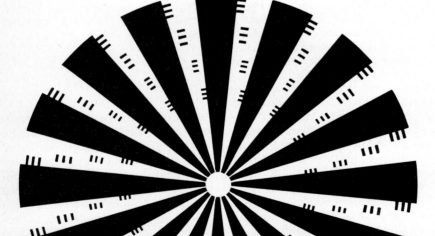

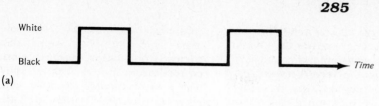

(a)

(b)

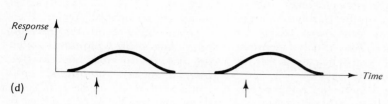

(c)

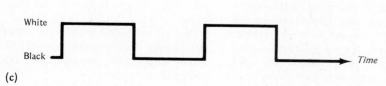

(d)

FIGURE 10.20

One effect that may contribute to the Benham's disk illusion. **(a)** Stimulation and **(b)** response versus time for the *center* of a receptive field centered at the point shown in the inset. **(c)** Stimulation and **(d)** response for the *surround* of the same receptive field. The receptive field is of the **r − g** channel. For the sake of simplicity we take it with *L* cones in its center and *I* cones in the surround. Suppose the *L* cones respond faster than the *I* cones. Then, the intermittent stimulation would lead to the responses shown. Because at certain times the signals from the center and the surround are not balanced (arrows), your **r − g** channel responds and you see color at that point on the disk.

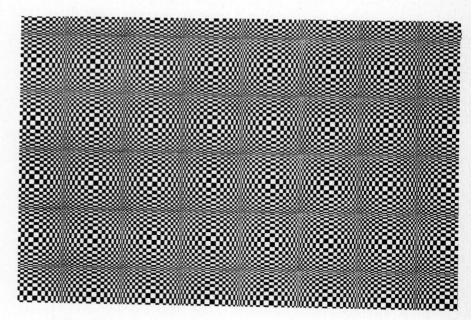

FIGURE 10.21

Fechner's colors. As your eye scans across the black and white pattern, the spatial pattern is translated into a temporal pattern of stimulation. Because the different cones and chromatic channels respond at different rates, their signals are not always balanced. Consequently, you may see unsaturated colors in the pattern.

*10.8
CONTINGENT AFTEREFFECTS AND MEMORY

A curious example of an aftereffect is the **McCulloch effect** (Plate 10.7). After prolonged adaptation, alternating between the vertical red bars and the horizontal green ones, vertical white bars will appear slightly greenish while horizontal white bars appear reddish. Because you adapt to as much red as green, there cannot be a standard *color* afterimage—after all, what color would such an afterimage be? There cannot be a standard *pattern* afterimage because you move your eyes while adapting. Also, the aftereffect can last much longer than a standard afterimage. With adaptation of about 10 minutes, you should be able to experience the aftereffect for hours or even weeks! (Try it! You needn't be concerned about the prolonged aftereffect. If it bothers you, you need only take the antidote—rotate the book 90° and adapt again.) Check to see that the effect transfers.

How can we explain the McCulloch effect? Presumably it involves cells in the visual cortex that respond to *both* color (in the usual opponent combinations) *and* form. Here is one possible explanation: Adaptation to the plates selectively desensitizes cells of the *r - g* system that also respond well to vertical lines. Later, when you look at the black and white vertical bars, those cells respond less than the *g - r* cells that also respond well to the vertical bars. This imbalance in response after adaptation is then interpreted as green. Naturally, the analogous process occurs with the horizontal bars, only here it is the *g - r* cells that are desensitized.

A similar aftereffect links color to *motion*. You adapt alternately to an *upward*-moving grating of red and black horizontal bars, and a *down*-*ward*-moving grating of green and black horizontal bars. Then, an upward-moving, black and white grating appears slightly greenish whereas a downward moving grat-ing appears slightly reddish. This suggests that there are cells sensitive to both motion and color.

Contingent aftereffects such as these are common; any two after-effects we've mentioned before can be linked together as in the examples here.

Knowledge and memory can also affect color perception. For example, if you're looking at a well-lit lemon, you'll have no problem seeing that it is yellow. If the lights are dimmed, you'll continue to see yellow even under scotopic conditions, where the cones cannot give color signals. The reason you still see yellow is poorly understood, but it most probably involves higher processing in the brain that links lemons and yellowness based on past experience. Such **memory color** is thought to play a role in color constancy.

A related experiment further supports this view. A piece of gray paper, when cut into the shape of a leaf, will appear slightly greenish even when viewed under photopic conditions. Try it!

SUMMARY

That all colors have three attributes demands that there be three cone types in the normal color observer. Based on information of **spectral complementaries, hue discrimination** and **microspectrophotometry,** the rather broad and overlapping **spectral response curves** of the S, I, and L cone types are found. Every color produces a triplet of responses in these cones, and colors that produce such identical response triplets are metamers even though their physical characteristics may differ. The response curves are consistent with such color matching data as summarized, for instance, in the chromaticity diagram.

The signals from the initial (cone) stage are processed through **opponent channels** to form signals coding **red minus green (r - g),** and **yellow minus blue (y - b), chromatic responses.** The evidence for chromatic opponency comes from **color naming, hue cancellation,** and electrophysiological measurements. Because your perception of color is related to the activity in the opponent channels, you need **four psychological primaries** to describe the appearance of any hue; blue, green, yellow, and red. A third, **achromatic** opponent channel, **white minus black (w - bk),** carries lightness information.

The three broad classes of **color deficiency** are **monochromacy, dichromacy,** and **anomalous trichromacy,** reflecting that either one, two, or three colors are needed to match an arbitrary color. Such deficiency may result from **missing pigments, shifted pigments, missing chromatic channels, ineffective chromatic channels,** or a combination of these problems. Color deficiency can be revealed by matching data, a **Rayleigh test, hue discrimination, overall sensitivity,** or through patterned **color plates.**

Lateral inhibition within a chromatic channel can result from **double-opponent cells,** having a general center-surround structure. Such cells can account for **simultaneous color contrast, chromatic edge enhancement,** and **color constancy. Spatial assimilation** of color (such as the **von Bezold spreading effect**) occurs when small areas of one color affect another to make it appear more *like* that of the small areas. **Negative afterimages** in color are the **complement** of the color used for adaptation and can explain the phenomenon of **Bidwell's disk. Positive afterimages** illustrate persistence of the chromatic response. Because there are different response times for different colors, temporally modulated black and white patterns can produce colors, as in **Benham's disk** and **Fechner's colors.** Different aftereffects can be produced simultaneously to form **contingent aftereffects,** such as the **McCulloch effect. Knowledge** and **memory** can influence your perception of color.

PROBLEMS

P1 Suppose a Venusian is trichromatic and has the response curves in the figure. (a) What wavelength ranges can a Venusian detect? (b) What wavelengths can a Venusian detect but a normal human cannot? (c) What wavelengths can a normal human detect but a Venusian cannot? (d) Where is a Venusian's overall sensitivity probably greatest? (e) Suppose Venusian white contains broad-band light between 200 nm and 500 nm. The Venusian spectral complements are rather peculiar. One color will have several complements, while another has none. Approximately for what ranges of wavelengths does the Venusian have spectral complementaries? (f) What is the Venusian spectral complement of 425 nm? Of 450 nm?

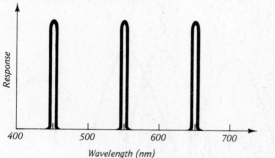

P2 Suppose that, like Earthlings, Martians have three different color-sensitive pigments in their retinas. The sensitivities of the Martian pigments are shown in the figure. (a) What wavelengths can a Martian detect but a human cannot? (b) What wavelengths can a normal human detect, but a Martian cannot? (c) Does a Martian have any spectral complementaries?

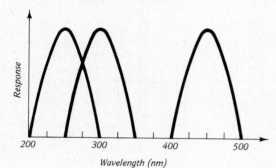

P3 A Martian (with color-sensitive pigments as described in P2) is watching a show about street lights, on Earth color TV. One of the lights discussed is the sodium light that emits only 589 nm. The Martian is impressed by the effectiveness of this light in illuminating streets. That night he rushes out to a country road that is illuminated by sodium lights. To his chagrin, everything appears dark—it seems to him that the lights are not even turned on! Explain why the Martian could see objects illuminated by sodium light in the TV program, but not on the country road.

P4 How does the stimulation of your L, I, and S cones differ for two lights that differ in: (a) brightness, (b) hue, and (c) saturation?

P5 (a) What is an advantage of having significant overlap of the cone response curves? (b) What is a drawback of such significant overlap?

P6 (a) Use the bee's chromaticity diagram (Fig. 10.8) to determine the bee's spectral complement of 500 nm. (b) Use Figure 10.7 to check your answer by analyzing the bee's cone responses to the two lights. Describe how each cone is stimulated by 500 nm and by its complement.

P7 What (if it exists) is the spectral complement of 650-nm light for (a) a human, (b) a bee?

P8 Why are *two* light beams required to determine the cone absorption curves by microspectrophotometry?

P9 (a) What is meant by a psychological primary color? (b) Name all of them. (c) Which are opponent to which? (d) What is a unique hue? (e) Which unique hue does not lie in the spectrum?

P10 (a) How is hue cancellation used to determine the chromatic response in the chromatic channels? (b) Suppose only the red chromatic response due to a 650-nm light is to be canceled. What wavelength should be used? (c) What hue name would describe the color after the red is canceled?

P11 (a) How many colors does a normal person need to match an arbitrary color? (b) How many does a protanope need? (c) A tritanomalous observer? (d) A cone monochromat? (e) A neuteranomalous observer?

P12 (a) What is a neutral point? (b) What kinds of color deficients have them?

P13 A person who is color deficient and described as deuteranomalous might be missing: (a) rods, (b) cones, (c) one visual pigment, (d) two visual pigments, (e) no visual pigment, just one modified. (Choose one.)

P14 Describe *two* different physiological defects that might cause a person to be deuteranopic (neutral points at about 500 nm and in the purple). Specifically state what is missing, nonfunctional, or ineffective.

P15 How does placing a colored filter over *one* eye help some color deficients? Why won't placing identical filters over *both* eyes help?

P16 Suppose you have two strips of uniformly colored paper, one a dark blue and one a light blue. If you place them edge to edge, neither will seem to be uniform any longer. (a) Describe and explain the apparent nonuniformity. (b) Repeat for a yellow and a blue strip of paper.

P17 (a) Why does the color produced in the last part of the TRY IT for Section 9.4B look desaturated? (b) Does it look more desaturated than it did before you rotated the paper? Why? (Think about edge information.)

P18 By analyzing the response of a double-opponent cell of the **y − b** channel, explain why the gray square that is surrounded by blue in Plate 10.3 looks yellowish.

P19 If a gray hat is brought into a room illuminated by blue light, it may still look gray. Explain this by considering what happens to a double-opponent cell of the **y** − **b** channel.

P20 Small yellow dots on a blue background make this background look desaturated. On the other hand, a blue region surrounded by a large yellow region looks very saturated. (a) What are these two phenomena called? (b) What accounts for the appearance of the blue background when the small dots are present? (c) What accounts for the appearance of the blue region when it is surrounded by yellow?

P21 Mention and explain two instances where the human eye may see complementary colors that are not "really" there.

P22 You view a blue book for a minute, then shift your gaze to a white piece of paper. (a) What color do you then see? (b) What is this effect called? (c) What accounts for it?

P23 Suppose the red bulb in Figure 10.18 were replaced by a yellow bulb. (a) What color would you see when the disk rotates counterclockwise? (b) What color would you see when it rotates clockwise? (c) Explain why you see these colors.

P24 What are Fechner's colors and what causes them? Would you experience them under retinal stabilization?

P25 A subject adapts alternately to a *rightward*-moving grating of blue and black vertical bars, and a *leftward*-moving grating of yellow and black vertical bars. After adaptation, she is shown a vertical, *leftward*-moving black and white grating. What color does it appear to her?

P26 How may memory color play a role in color constancy?

HARDER PROBLEMS

PH1 From the bee's photoreceptor response curves (Fig. 10.7), deduce and draw a bee's hue discrimination curve as best you can.

PH2 Use the Martian photoreceptor response curves (shown in P2) to answer the following: (a) One type of dichromat is a color-deficient person who is missing the *I* cone. Does a Martian dichromat of this type have a neutral point? (b) Devise a color plate analogous to Plate 10.2 that would reveal such deficiency in a Martian. What colors would you use for the pattern? What colors would you use for the background? (c) Devise another simple test that could tell you quickly if a Martian were deficient in this way. (One such test involves having the Martian look at a certain light.)

PH3 Use the Martian sensitivity curves of P2 for the following. (a) Construct a chromaticity diagram for the Martian. The horizontal axis should denote the relative excitation in the Martian *L* photoreceptor, while the vertical axis should denote relative excitation of the *I* photoreceptor. Mark on the chromaticity diagram the points corresponding to 450 nm, 550 nm, 650 nm, as well as to broad-band white (*W*). (b) What point (if any) corresponds to 500-nm light? (c) Use a *Y* to mark the point corresponding to broad-band yellow. (d) Show that, for Martians, this yellow is the complement of 450-nm blue. (e) Is *Y* complementary to 400-nm blue? Why or why not?

PH4 The relative absorptions of the three retinal cones of the cichlid fish *Cichlasoma longimanus* are shown in the figure. Note the large number of crossover points of the curves, which we've labeled *A* through *G*. Because of all these crossover points, the fish's spectral complementaries are strange. (a) For these fish, wavelengths above 550 nm (*G*) have spectral complements lying only between *D* and *E*. Explain why these complements lie in this, and no other, region. (b) Are there spectral complements, and if so, where, to wavelengths lying between *C* and *D*? Explain. (c) Repeat (b) for wavelengths lying between *E* and *F*. (d) Repeat (b) for wavelengths lying between *A* and *B*. Compare your result with your previous results and comment.

PH5 Using a cone response analysis similar to that of Section 10.3, state the approximate hue and saturation of a mixture of equal amounts of 470-nm and 600-nm light. Use a chromaticity diagram to check your answer.

PH6 What two different wavelengths would you use to make a unique red—one with neither blue, yellow, nor green chromatic response? Explain your answer using Figure 10.10.

PH7 Suppose in some spectral region, ranging from some point *A* to another point *B* on the horseshoe curve of the chromaticity diagram, only *two* photoreceptors respond. (a) Show that any spectral light in the region can then be matched by a suitable additive mixture of *A* and *B* only. (b) What is the shape of the horseshoe curve between *A* and *B*? Identify this region in Figure 10.6 and give approximate wavelengths at the limits of the region, *A* and *B*. (c) Show that your result is consistent with Figure 10.5.

PH8 (a) One way to distinguish between the three kinds of cone monochromat is to have the monochromat match a given broad-band stimulus with each spectral light, one at a time, and measure the amount of each spectral light needed for the match. How would this allow you to determine which kind of monochromat the subject was? (b) Another technique would be to measure absolute sensitivity to light of various wavelengths. Again, how would this

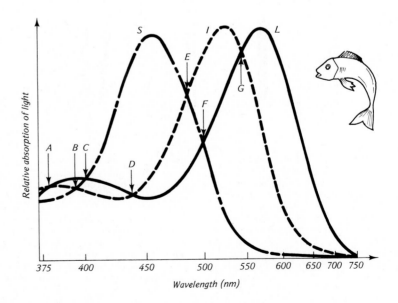

Relative absorption of light

Wavelength (nm)

information allow you to determine which kind of monochromat the subject was?

PH9 Draw the hue discrimination results you'd expect from a deuteranope.

PH10 (a) Suppose a certain color deficient has his L cone response curve shifted to longer wavelengths. Would he require extra red or extra green to form a Rayleigh match? (b) What kind of color deficient might this describe? (c) What other symptoms might he have?

MATHEMATICAL PROBLEMS

PM1 Use the hypothetical photoreceptor response curve of the figure to answer the following. (a) About how much light at 650 nm will produce the same photoreceptor response as 100 units of 550-nm light? (b) About how much light at 450 nm must be added to 50 units of 600-nm light to produce the same response as 150 units of 500-nm light? (c) About how much light at 625 nm must be added to 100 units of 500-nm light to match 100 units of 450-nm light? Interpret this result.

PM2 Use the photoreceptor response curves of the figure to analyze the following matching experiment. A 500-nm test light is to be matched using three "primaries" of 450 nm, 550 nm, and 650 nm. (a) About how much light at 450 nm will produce the same S photoreceptor response as 100 units of 500-nm light? (b) About how much light at 550 nm will produce the same I photoreceptor response as 100 units of 500-nm light? (c) About how much light at 650 nm will produce the same L photoreceptor response as 100 units of 500-nm light? (d) Notice that the 550-nm light you used in (b) also produces a response in the L photoreceptor. How much 650-nm light would you need to produce the same response in the L photoreceptor as that produced in it by the 550-nm light of (b)? (e) Use your results from above to determine how much light at 450, 550, and 650 nm (when presented together) will match the 100 units of 500-nm light.

PM3 (a) Show that, if only two receptors respond to light in some region of the visible spectrum, the horseshoe curve on a chromaticity diagram must be a straight line in that region. (b) Using Figure 10.5, find where the corresponding straight-line region on the chromaticity diagram lies.

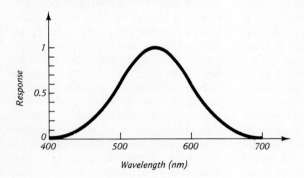

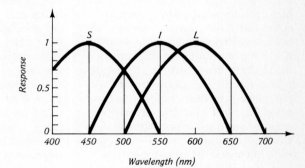

Color Photography

11.1
INTRODUCTION

It may seem hard to believe today, but there was once a time without photographically reproduced color—without sunset scenes on billboards, travel posters in your office, even without plastic, imitation wood-grain furniture. But the demand for colored photographs was so powerful that before natural color film was available, it was popular to hand-color portraits and to tint movie film. A Nobel Prize was even awarded for the invention of one system of color photography (Sec. 12.3D)! Yet, after the introduction of black and white photography, the world had to wait over half a century for a comparably faithful and feasible system of color photography.

The reason was not the lack of understanding of the principles of color vision and reproduction. In 1861, James Clerk Maxwell projected full-color photographic images; and as early as 1869, Ducos du Hauron described the principles of virtually every system of color photography known today. But the *technology* needed to make these principles practical was lacking. Du Hauron did not tell how to achieve some of the crucial steps—in effect he waved a magic wand of then unavailable technology. The development of this technology has occupied inventors ever since then and continues even today.

Color film provides a marvelous example of how technology builds on itself. A given process may be quite complex and originally difficult to achieve, yet once mastered it becomes a building block in an even more elaborate process. Thus, we already know how black and white film works; we may now use that as a starting place for color film. The early color film consisted of three black and white film records, suitably combined—a technique dependent on our knowledge of human color perception. Later the introduction of couplers allowed dyes to be produced directly in the emulsion. The development process became an elaboration of the original black and white development. These technological building blocks were piled layer upon layer on top of each other (quite literally—modern instant color film may have over a dozen layers), and the resulting film may itself serve as a building block in an artist's creation.

11.2
PRINCIPLES

Color photographs are meant to be viewed by humans. Therefore, they do not need to record the exact intensity distribution of the light that strikes the film, but only those characteristics that are important for human color perception, that is, the light's chromaticity. We have seen that human color vision makes it possible to match almost any color with only three colors. If three emulsions, sensitive in three suitably chosen spectral regions, are used to record the same scene, it should be possible to use these three "records" to reproduce most colors acceptably for the human eye. The images on these three emulsions, whether latent or developed, are called the **red, green,** and **blue records.** When they are developed into separate negatives or positives, they are called **separation negatives** or **positives,** respectively. It would be nice if the three records could be made directly of colored dyes, but only silver compounds are sufficiently sensitive to light. Hence, the three records (when developed) almost always consist of (black) silver instead. For example, the red record is a black and white recording of the amount of red light that hit the color film.

Modern films carry these three emulsions as a sandwich on a common base, called a **tripack** (Fig. 11.1). Typically, the top emulsion has only the basic blue sensitivity of all silver halide film. The next emulsion contains sensitizers (Sec. 4.7C) that make it sensitive to green light; the bottom emulsion is sensitized for red. Because the green- and red-sensitive emulsions retain their sensitivity to blue light, there is a yellow filter just before them, which prevents the blue light from reaching them.

After the blue, green, and red records have been obtained by exposing the film, the emulsions must be developed, and the black silver images must somehow be changed to color and reunited into one color picture. This process can proceed either by additive or by subtractive color mixture.

The **additive system** is conceptually simplest and was historically the first to be practicable. For projection (such as slides or movies) we need only make black and white positives of the three records and use the appropriate colored lights

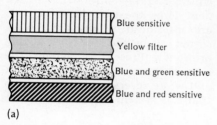

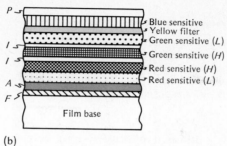

Blue sensitive
Yellow filter
Green sensitive (L)
Green sensitive (H)
Red sensitive (H)
Red sensitive (L)

Film base

FIGURE 11.1

Cross section of a tripack color film.
(a) The principle of the three emulsion layers plus the yellow filter. **(b)** The many layers of an actual film. In addition to the emulsions there is a protective layer P, interlayers I to separate the emulsions, an anti-halation layer A, and a foundation layer F. Also, the green- and red-sensitive emulsions often consist of two separate layers each, one of high sensitivity H, and one of low sensitivity L, to increase the latitude of the color negative film. The whole film is less than 0.2 mm thick, and one emulsion may be as thin as 2 μm.

to project them in register on the same screen. Many commercial systems of the early years of this century used this system, as does modern projection TV (Sec. 9.5A). Alternatively, the other methods of additive mixing (Secs. 9.5B and 9.5C) can be, and have been, used. A disadvantage of all additive systems that derive the three primary colored lights by filtering white light is that at least two-thirds of the light is thrown away by the filters.

The **subtractive system** of putting the three color records back together is also simple in principle; we need only superimpose the records as they were at the time of exposure (or leave them in the sandwich tripack, if we can process them simultaneously). The technological problem here was how to convert the black and white records into the appropriately **colored records.** The *colored records* would then serve as *filters* that, placed behind one another, would mix the colors subtractively. When the technological problem was solved, this system became the foundation of most modern color film.

11.3
ADDITIVE COLOR FILM

The first successful types of color film were photographic **mosaics.** Recall that a mosaic achieves many of its colors by partitive mixture, similar to color TV screens (Sec. 9.5B).

To create a mosaic color film, it is sufficient to cover an ordinary black and white film with a mosaic of tiny red, green, and blue *filters.* The filters perform a dual role. During the exposure, they assure that the film behind them is exposed to just one region of the spectrum. During projection, when light strikes the developed film, they filter that light, assuring that only the proper color emerges from each part of the film.

After exposure, the emulsion behind a red filter, say, carries one picture element of the red record. Similarly, a blue record lies behind the blue filters, and a green record behind the green filters. These three records are intimately intertwined, and it is neither possible nor desirable to separate them. The film is developed like ordinary black and white *reversal* film because, except for the filters, that's all it is.

When viewed in transmitted light, the colors from the individual filters of the mosaic mix partitively to reproduce the original colors. For example, a yellow region in the object exposes the film behind the red and green filters, but not behind the blue filter. After reversal development, the emulsion is clear (transparent) behind the red and green filters, and black (opaque) behind the blue filter. When white light is shined through the film during pro-

jection, only the regions behind the red and green filters let the light through, and this filtered light combines additively to yellow, as it should. Figure 11.2 shows how some of the other colors are produced.

PONDER

Can you see the colors in a developed mosaic color film from the "wrong" side, that is, if the light in Figure 11.2 came from the bottom and the eye were at the top? What about exposing the film from the "wrong" side—does that work?

Different brands of mosaic color film differed mainly in the arrangement of the mosaic color filters. The earliest and most famous was the Autochrome film of the Lumière company, introduced in 1907. Here the filters consisted of colored starch

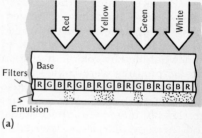

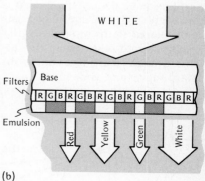

FIGURE 11.2

Additive color reversal film. **(a)** The latent image produced by exposure to the colors indicated. Exposed crystals are denoted by black dots, to symbolize the silver nuclei. **(b)** The film after black and white reversal development, with white light shining at it. After passing through the film and partitive mixing with light from adjacent color dots, the light reproduces the original colors.

grains that were mixed and pressed on the film base, with the gaps filled by a paste of lampblack. The emulsion was then poured on top of this. The exposing light went first through the *base*, then through the colored grains, and then to the emulsion. The pastel colors of these mosaic films led to some remarkably beautiful pictures (Plate 11.1).

PONDER

Why would leaving transparent gaps between the colored grains desaturate the colors? (For a similar reason some color TV screens use a much touted black matrix between the colored phosphors.)

Autochrome film fit standard cameras and could be developed by standard black and white chemistry. However, since the colored grains were randomly arranged, there tended to be accidental single-color clusters, which resulted in somewhat blotchy colors; and when mosaic pictures were copied, there was a loss of color purity because grains of the same color were not necessarily aligned in the original and the copy. Therefore, later additive films (e.g., Dufaycolor) used a more regular grid of filters, consisting of colored strips or squares. Recently, Polaroid resurrected the additive system with line filters in instant movie and slide films.

PONDER

If an Autochrome is copied on another Autochrome film, the colors in the copy are improved (though some detail is lost) if the original is jiggled slightly during the copying exposure. Why?

A different additive system, developed in the 1930s, used intermeshed pictures arranged so cleverly that the three records could be separately projected through different filters. Figure 11.3 shows the optics during projection. The film carries many long, narrow, cylindrical lenses, called lenticules (see Sec. 8.5B). Behind each lenticule are three narrow strips (each about 10 μm wide) of the three colored records. During projection, light passes through the film, lenticules, colored filter, projection lens, and onto a screen. The lenticules send the light from each strip of the red record through the red filter. The projection lens then sends it to the screen to form the red image. Similarly, light from each strip of the green record is sent through the green filter, and likewise for the blue. Thus on the screen the additive mixture gives us the full-color image. During exposure, the optics of Figure 11.3 are reversed—light from the subject passes through the lens, colored filters, and lenticules, finally exposing the film. The lenticules ensure that red light strikes the proper thin strip, while the

green and blue light strike different, appropriate thin strips on the film. Thus the three color records are intermeshed, but are always sorted out by the lenticules during projection. Processing, as with the mosaic films, involves only black and white reversal development, so that this system was well suited for

FIGURE 11.3

Projection of a lenticular film. In this cross-sectional view the emulsion behind each lenticule is divided into three regions. The bottom region of such a triplet carries a picture element of the red record *(r)*, the middle carries an element of the green record *(g)*, and the top carries an element of the blue record *(b)*. All the bottom regions send light rays only toward the red filter, because each lenticule directs these rays to this filter. Similarly the middle regions send light rays only toward the green filter, and the top regions send light rays only toward the blue filter. The projection lens focuses rays from *one* lenticule, which have passed through *different* filters, onto *one* screen location, giving an additive mixture there. For example, all three regions behind lenticule 1 are transparent, hence all three primary colors are sent to the screen at the image of lenticule 1, giving white at that position. Only the top and bottom regions of lenticule 2 are transparent, hence only red and blue rays hit the screen at the image of lenticule 2, forming a magenta spot there. Thus, this arrangement makes the light from each of the three intermeshed records pass through its appropriate filter and then mixes them additively on the screen.

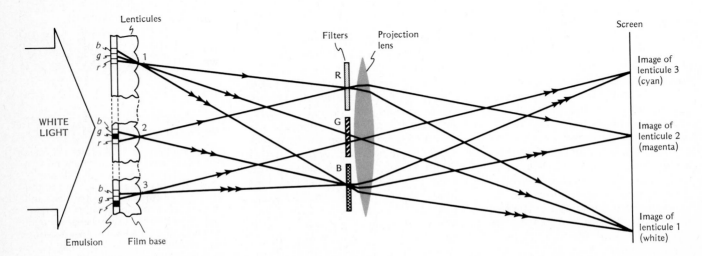

amateur movie films. However, copying was also possible, either by contact or by replacing the screen in Figure 11.3 by another lenticular film.

11.4 SUBTRACTIVE COLOR FILM

Consider for a moment the red, green, and blue records as separate black and white positive transparencies (i.e., not intermeshed as in the systems just discussed). For an *additive* system these only have to

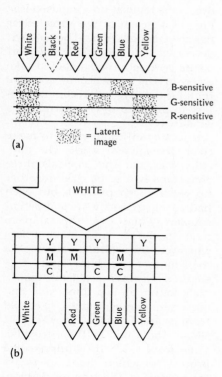

(a)

(b)

FIGURE 11.4

Principle of subtractive color film. **(a)** The latent image produced in each emulsion layer of the tripack film by the colors indicated. (We indicate black in the region where there is no incident light.) **(b)** After reversal development, the black silver image in each layer is replaced by a dye of color complementary to the color to which the layer was sensitive. Light passing through all three layers gives back the original colors by subtractive mixing.

be *tinted*—turned into black and *color*—which can be easily done by shining colored light through them. The developed silver stays black, and this is why development is usually no problem in the additive system. But suppose we have du Hauron's magic wand, so the developed silver can somehow be replaced by a dye of color complementary to that of the record. This would allow the three records to be *toned*—turned into *complementary color* and white (clear)—and they could then be put back together *subtractively*.

We can understand subtractive systems by considering Figure 11.4. A white part of the original becomes a transparent region in each of the three layers, hence the superimposed layers project as white in this region. A black area of the original becomes the complementary color in each layer; filters of all three complementary colors (yellow, cyan, magenta—the subtractive primaries) superimposed transmit nothing, hence this region projects as black. So far, so good. What about a colored region, say red? Light from it exposes the red layer, which then becomes transparent after reversal development; but the red light does not expose the blue and green layers, hence yellow and magenta are created in those layers. On projection, red light emerges from this subtractive combination of yellow and magenta, as it should. Similarly, you can check that all other colors are properly reproduced.

*A. Dye destruction

Now we come to some practical methods of waving the magic wand over the emulsions to turn them each into a color image. As mentioned, for reasons of sensitivity virtually all systems use silver halide. The **dye destruction** process gets color in the film by placing a *complementary* dye throughout each silver halide emulsion layer during manufacture. As the film is being developed, the dye is chemically removed only in the vicinity of ex-

posed silver halide, thus forming a *positive*. The important advantage of this process is that the dyes are in the emulsion to start with—they do not have to be put there later by chemistry or little people with tweezers. Unfortunately the dyes absorb light during exposure, so the film's sensitivity is reduced. However, the method has been successfully applied to color print paper (e.g., Cibachrome), where high sensitivity is not essential.

*B. Technicolor and dye transfer

Although the idea of dye destruction hails back to du Hauron and has been experimented with since the turn of the century, its successful application is rather recent. A more complicated process was necessary to produce the first subtractive full-color film, **Technicolor**,* whose dramatically successful results advanced color movie film to its modern quality. The Technicolor camera is an example of a color separation camera, making three separate records without intermeshing them. It uses a **beam splitter** (e.g., a half-silvered mirror—Sec. 2.3E) that transmits the green image to one film, and reflects the complementary magenta (blue plus red) image to a blue-sensitive front film and a red-sensitive back film (Fig. 11.5). The three films are first developed normally. Then, a photographic copy is made of each film, and its emulsion's gelatin is chemically hardened near the exposed silver. The unhardened gelatin is washed away, leaving so-called **gelatin matrixes,** which carry the picture of each color record in varying thickness of gelatin. For example, the more red light in a given area, the thicker the gelatin in the corresponding matrix. When placed in dyes of the complementary color, the matrixes absorb more dye where the gelatin is thick—and

*Invented in 1932 by Herbert T. Kalmus and named after his university, MIT.

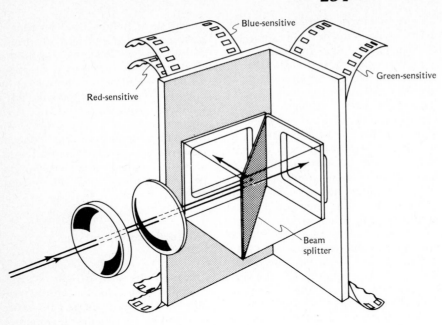

FIGURE 11.5

The principle of the original Technicolor camera. Note the backward telephoto lens (shown schematically) to give enough room between lens and film for the beam splitter, while keeping the image size and perspective normal. Later Technicolor movies were shot in regular cameras on color negative tripack, from which the matrixes were printed using filtered light sources.

presto finito you have your complementary color and clear records. If you now sandwich these records together properly and shine light through them, the colors will mix subtractively and you get one full-color film, as in Figure 11.4.

Technicolor of course wanted more than one copy, so the gelatin matrixes were used (mechanically, like the plates in a three-color printing press) to squeeze the complementary dyes, one color at a time, onto a film base. Since each color was printed separately, the color balance could be controlled in the final copy. As in four-color printing, the Technicolor process uses a fourth, black and white image for good contrast and dark blacks. That is, the film base onto which the three matrixes are pressed is in fact not blank, but is black and white film carrying a somewhat

faint ordinary photographic image (as well as the sound track).

The immediate success of Technicolor was due in large part to labor-intensive quality control, which also explains it eventual decline. Only in China are true Technicolor films still being made.

The same principle can be applied to get prints of single frames on paper—the **dye transfer** process. Here, the major differences are that larger gelatin matrixes are used and are printed directly on special paper.

C. Couplers

In the 1910s two chemists, Rudolf Fischer and Hans Siegrist, originated the process of color coupling, which allows you to turn silver images into color. A **color coupler** is a transparent, colorless substance that combines with some particular developer to form a dye while the developer reduces silver halide to silver. It's easiest to think of the action of the coupler in two steps. First, the developer reduces the silver halide and is itself changed, used up, or *oxidized*, as the chemist calls it. Next, this oxidized developer combines with the coupler and forms a dye. The whole process

is called **color development.** The color of the dye depends only on the coupler used, not on the silver; but the *amount* of color formed depends on the amount of oxidized developer present at any place in the emulsion, and hence depends on the amount of reduced silver, which in turn depends on the exposure there. So by this two-stage process, in which the silver is used only as an intermediary, we are able to start out with a silver halide emulsion, which has good light sensitivity, and end up with a colored, dye image. The silver and unexposed silver halide are then removed by a process called **bleach-fixing,** leaving a pure dye image.

Since we want the most colored dye in the regions least exposed (see Fig. 11.4), each of the three silver halide emulsions should be *reversal* developed. The *second* development of the reversal process should take place in the presence of couplers that form a dye color complementary to the color of each emulsion's sensitivity.

It took another 20 years to make color development work. The difficulty was that the tripack's three emulsions had to be close together (to get a sharp picture), but each had to be treated with a different coupler. When wet, the couplers tend to diffuse and make colors wherever they please, outside of their proper emulsion layer. One solution to this problem was finally found by two musicians and amateur photographers, Leopold Mannes and Leopold Godowsky, Jr. ("Man and God" to their friends). Their work led to the **Kodachrome** process, in which the couplers are introduced at the time of development, one at a time. (After several years' work at Kodak these inventors turned back to music, with Mannes directing the college of music in New York founded by his father.)

D. Kodachrome development

Figure 11.6 shows a schematic representation, by now familiar, of the

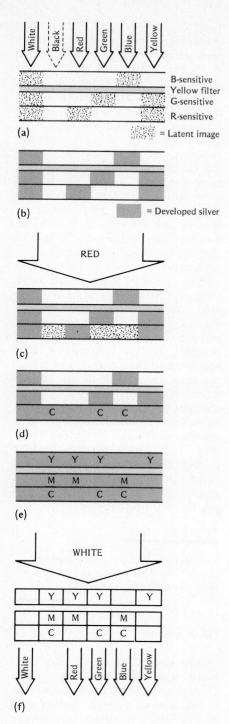

(a)

= Latent image

(b)

= Developed silver

RED

(c)

(d)

(e)

WHITE

(f)

FIGURE 11.6

Kodachrome film at various stages of exposure and development (see text).

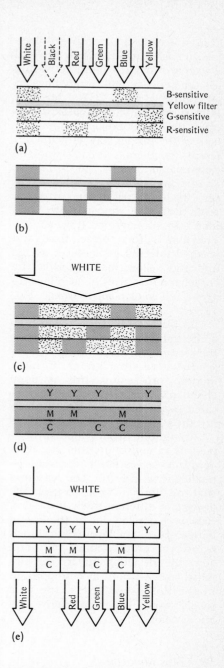

(a)

(b)

WHITE

(c)

(d)

WHITE

(e)

FIGURE 11.7

Various stages of exposure and development of color reversal film containing incorporated couplers; for example, Agfachrome or Ektachrome (see text).

You should check that you agree on the layers that carry the latent image for each of the colors. Now we send the film in for development. The first thing they do is develop the latent image as in ordinary black and white film, giving the silver image (Fig. 11.6b). Now comes the second exposure in the reversal process, not to white but to *red* light, which gives a latent image of the remaining silver halides in the red layer only (Fig. 11.6c). The film is then developed in a developer with cyan coupler. It produces a cyan dye and, of course, developed silver, where the latent image was—that is, *only* in the bottom layer (Fig. 11.6d). So we already see that it is possible to make a dye image in one emulsion layer only. Then there are two more "second" exposures and color developments, first with blue light and yellow color development (affecting only the top layer), and then with green light and magenta color development. Figure 11.6e shows the structure at this stage, containing silver everywhere and dye in selected places. If you were to look through the film at this stage, it would look completely black. Finally the silver (as well as the yellow filter) is made soluble by a bleach-fix and washed away, leaving a transparency that reproduces the colors correctly (Fig. 11.6f).

*E. Home development

It is clearly advantageous if each emulsion of a tripack already contains a colorless coupler so that the proper colors can be produced at once in all three emulsions by one color development. (Such *transparent* couplers won't reduce the film's sensitivity.) Various methods were found to prevent the couplers from wandering between emulsions, for example, attaching them to long, heavy organic compounds that like to hang on to the emulsion's gelatin, or by enclosing them in globules with a skin that is impenetrable to water but can be penetrated by the alkaline developer.

All films suitable for home development today incorporate the cou-

three emulsion layers and the yellow filter. In Figure 11.6a, light of selected colors falls on the film and forms a latent image as indicated.

plers. The latent image formation (Fig. 11.7a) and the first development (Fig. 11.7b) is similar to that of Kodachrome. At this point the exposed silver halide crystals have, as usual, been converted to silver grains, leaving the unexposed crystals in each emulsion. Now the film is exposed to white light, activating *all* the remaining silver halides in all emulsions (Fig. 11.7c). The film is then treated with a **color developer,** that is, a developer that interacts in the proper way with the different couplers, already in the different emulsions, to form the appropriate dyes. Like all developers, the color developer turns the remaining (now exposed) silver halides into silver grains (Fig. 11.7d). The film is then bleach-fixed and washed to remove the silver grains and the yellow filter, leaving the complementary dyes in each layer to give a full color positive image (Fig. 11.7e).

The purpose of the exposure to white light in step (c) was to start the conversion of the unexposed silver halide to silver grains. In more modern processing, this is done chemically, by chemicals incorporated in the color developer of step (d), thus avoiding that exposure.

Throughout the development process, it is important that the temperature be carefully controlled—usually to within $\frac{1}{4}$°C. The reason is that the different dyes will generally develop at different rates. If the process were carried out at the wrong temperature, one layer would finish before the others. Since you can't stop the development in the finished layer without also stopping it in the unfinished layers, the final result would be improperly balanced color.

*F. Color negatives

Historically, color films were first made for *reversal* development, to yield positive transparencies or movies. However, both additive and subtractive color films can also be developed into *negatives*. There is then only one development, which for the subtractive films should be a color development (i.e., using a developer that reacts with the couplers to give complementary colors). In either case the result will be a negative in colors complementary to the original. Figure 11.8 shows the

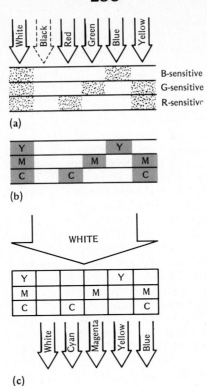

(a)

(b)

(c)

FIGURE 11.8

Exposure and development of a subtractive color negative film (ideal dyes, no masking): **(a)** after exposure, **(b)** after color development, **(c)** after bleach-fix, showing colors complementary to those of the exposing light. The letters indicate the colors of the dyes.

steps in the formation of a subtractive color negative.

PONDER

Would the same complementary colors result if the additive film of Figure 11.2 were developed into a negative?

As in black and white photography, color negatives are useful for making prints on paper. For color prints, the paper has to carry three emulsions with couplers, like the film, and has to be developed as a color negative. Since the complement of a complementary color is the original color, this double negative process gives a positive that is correct in intensity and color. At

least this would be so if the colors at each stage were ideal. Since practical colors are far from ideal, color shifts can occur and will be compounded by each copying.

For example, suppose one of the three emulsions of a color film is a little too contrasty. If you copy such a film by photographing it using the same type of film, then the already too contrasty record is made even *more* contrasty, and the associated color shifts are exaggerated. (In copying positive transparencies, these shifts are not too objectionable, but for best results a special copying film should be used.)

Because there are two steps in a negative-positive system, any color shifts are more likely to be exaggerated. On the other hand, such a system has all the advantages of control enjoyed by black and white printing, and in addition the color balance can be controlled by printing through suitable filters. For example, if a positive film balanced for daylight is used under ordinary incandescent illumination, the transparency will have an orange tinge (Sec. 11.5). In a negative-positive system, the orange tinge in the final print can be avoided by inserting a colored filter in the light beam during the printing process.

PONDER

The filter should be faintly orange. Why does that remove the orange tinge, rather than making it worse?

*G. Masking

Filters can change the color *balance* of a print, but they cannot change an incorrect *contrast* of one of the emulsion layers. (A filter cuts all intensities of a given color by the same factor, say $\frac{1}{2}$; but the contrast, which is measured by the *ratio* of intensities, is *unchanged*. If it was 10:1 before filtering, it is $5:\frac{1}{2}$ after filtering, which is the same ratio as 10:1.) Contrast can, however, be changed by a technique called **masking.** A mask is a faint black and white transparent copy of an original, mounted in register with

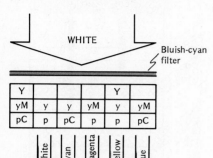

(a)

(b)

FIGURE 11.9

Integrally masked subtractive color negative film. The cyan dye is not ideal, absorbing too much green and blue, so it looks an unsaturated cyan—we call it pinkish cyan, pC. To compensate for this unwanted green and blue absorption in the dye, the coupler in this red-sensitive layer also absorbs a little green and blue; that is, it has a pinkish color, p. Similarly, to compensate for the blue absorption of the yellowish magenta dye, yM, the coupler in the green-sensitive layer has a yellowish color, y. The unused couplers form the color mask. The yellow dye, Y, has no unwanted absorptions, and its coupler is colorless (so it forms no mask). **(a)** Exposure. **(b)** Finished negative in an enlarger with an unsaturated bluish cyan filter, yielding the same complementary colors as the ideal dyes of Figure 11.8. The filter is chosen to be the complement of the orange produced by the y and p, which are now present everywhere. (The orange can be seen in color negatives, even between frames.)

it. If the copy is a negative it decreases the contrast; if it is a positive, it increases the contrast. (The TRY IT gives examples.) If the mask is made by exposing the black and white film with colored light, it will correct contrast deficiencies of the original related to that color. This can be useful if there are differences in contrast between the three emul-

sions of a color negative film. The black and white mask is mounted in register with the color negative in the enlarger, to make the positive print.

Since masking is an additional complication, it is advantageous to have it done automatically. If, moreover, the mask itself is colored, it can compensate not only for incorrect contrast but also for other imperfections in the color film's dyes. Such automatic masking is incorporated in modern color negative films such as Kodacolor (Fig. 11.9). For example, cyan dyes should absorb only red, but actually they also absorb fairly large amounts of green and blue. To balance this extra absorption, there must also be some absorption of green and blue in those places in the film that have *no* cyan dye after development. A pink mask in such places will do just that. The resulting *overall* shift to the pink is corrected by a filter in printing. This pink mask is made automatically by using a coupler in the red-sensitive layer that is pink before it reacts, and changes to cyan when it reacts with the oxidized developer—the pink remains wherever there is no cyan.

TRY IT

FOR SECTION 11.4G
Masking and contrast

Masking can increase or decrease the contrast of a photograph, depending on the type of mask. To see how a mask can decrease the contrast, you need a (preferably large-size) negative and a contact print of it. For increasing the contrast, you need a transparency (again, a large size is best).

Place the negative in exact register with the contact print. The light areas of the print will be covered by the dark areas of the negative and vice versa, and if the print paper has a gamma of 1 (Sec. 4.7G), the result will be a uniform gray—the contrast will be reduced to zero. However, this exact cancellation seldom happens, so you may see a positive or negative image, but of considerably reduced contrast; masking with a negative decreases the contrast.

Try shifting the negative a little so the registration is not perfect. Along sharp edges some white regions of the print will then show through transparent regions of the negative. You see a (low-contrast) picture as before, but with enhanced edges (not unlike the result of the visual system's processing, Sec. 7.4A). This technique is called **bas-relief**.

To see the result of positive masking, place the transparency in contact with a white paper and illuminate strongly from above. Because the light must pass through it twice, the transparency is its own mask. Note that this positive mask increases the contrast. Try lifting the transparency away from the paper somewhat, so that its shadow becomes fuzzy; now the contrast of large areas is still increased, but the contrast of small detail is not changed.

*H. Instant color photography

We saw in Section 4.7E how black and white prints can be obtained "instantly." Can a similar film and development be designed to yield color prints? The film should be a tripack negative, and between frames it can carry developer pods, whose contents will be spread over the negative as in the black and white process. The three resulting silver images must then be converted to the appropriate colors. The colors, in turn, must be transferred away from the developed silver to a receiving layer (like the support paper of Sec. 4.7E) to make the final print, as in the dye transfer process. To do this with a coupler-formed dye image would be difficult, since one of the basic properties of couplers is that they should not diffuse (Sec. 11.4C). So Polaroid had to develop yet another way to generate subtractive color images, the so-called method of dye developers. Not even our old friend du Hauron thought of this one!

A **dye developer** is a chemical combination of dye and developer molecules—it appears colored and acts as a developer. Moreover, when the developer is oxidized (i.e., when it reduces exposed silver halide to silver), it becomes immobile and thus the color associated with it does not transfer. Color transfers only in places where the silver halide was *not* exposed. There it transfers to the receiving layer, yielding a color and white positive image on the print—just what is required for the subtractive process.

In order to obtain a subtractive mixture of three different colors, three differently colored dye developers are used. Since each dye developer must act on only one of the emulsions, each emulsion carries near it a layer of appropriate dye developer. The trick is to place the dye developer just below the emulsion it develops, because the developer will then also act as the proper color filter for the emulsion layers beneath it when the picture is exposed. For example, we saw that ordinary color film needs a yellow filter layer below the blue-sensitive emulsion. If a layer of yellow dye developer is used instead, it will still act as a yellow filter during exposure. Later, when the pod is broken and a reagent is spread over the exposed negative, this yellow dye developer becomes mobile in the liquid environment. Where there was exposure in the blue-sensitive layer, the diffusion of this yellow dye developer is stopped; but where there was no exposure, this developer moves to the image-receiving layer, forming a yellow and white positive image as required for proper subtractive mixture. Similarly the two other sets of emulsion and dye developer layers transfer the magenta and cyan components of the image.

All popular instant color films work by one or another variation of this basic process. Let's consider what happens in detail in one particular film, Polaroid's Time-Zero Supercolor. Figure 11.10a shows a cross section of this film as it is being exposed. We see the usual sequence of blue-, green-, and red-sensitive emulsions. Underneath each there is a dye developer layer of the complementary color and a spacer. Consider the action of one particular color of exposing light, say red. It passes through the transparent layers on top, as well as through the top two emulsions and their dye developer layers; these two emulsions are not sensitive to red

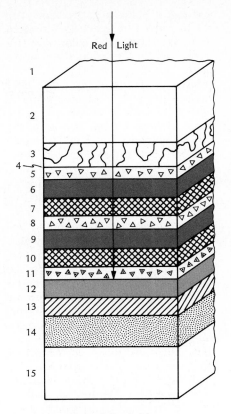

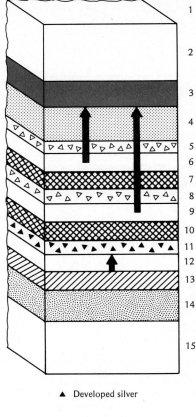

△ Exposed silver halide

△ Unexposed silver halide

(a)

▲ Developed silver

(b)

FIGURE 11.10

Cross section of an instant color film (Polaroid's SX-70 Time Zero Supercolor). (1) Nonreflecting surface (see Sec. 12.2B), (2) clear plastic, (3) image-receiving layer, (4) space for reagent, (5) blue-sensitive emulsion, (6) yellow dye developer, (7) spacer, (8) green-sensitive emulsion, (9) magenta dye developer, (10) spacer, (11) red-sensitive emulsion, (12) cyan dye developer, (13) timing layer, (14) neutralizing layer, (15) plastic base. **(a)** During exposure. **(b)** Diffusion of the dyes during development.

light, and the yellow and magenta dyes both transmit red. The red light then exposes the red sensitive emulsion at the bottom. Any light that gets through this emulsion is swallowed up by the cyan dye developer underneath, which, as we know, absorbs red. (Here we see a third function of the dye developers—each serves as an antihalation backing!) Note that the dye developers are dry, undissolved, and hence (before the reagent is spread) cannot reach the silver halide, which remains undeveloped.

Development starts when the film is squeezed between rollers on its way out of the camera, breaking the pod and spreading its reagent contents between the emulsions and the image-receiving layer (Fig. 11.10b). The reagent "turns on" a number of chemical reactions. It quickly penetrates all the emulsions down to the timing layer, but takes a while to penetrate the timing layer. When it does, it is neutralized by the acid layer underneath. The reactions are thereby "turned off," and the development is complete. What are these reactions? For one thing, the developers are activated by the reagent and become able to move about. In particular, they move into the nearby emulsion layer. Consider again the area originally exposed to red light. There, the cyan dye developer meets exposed silver halide. The two react and the developer thereby becomes immobile, that is, it does not move beyond the red-sensitive layer. The magenta and yellow dye developers, on the other hand, find only unexposed silver halide as they diffuse upward through this area exposed to red light only. Therefore, they do not react, and continue their way to the image-receiving layer. There these two dyes combine to form red. By this time the reagent has reached the acid, which renders the dyes less mobile. Thus the image is fixed in the image-receiving layer. The other colors are generated in a similar way. (Eventually—e.g., by diffusing downward—color developers reach the "wrong" emulsion layers. The point of the spacers is to prevent this until the emulsion has been already developed by the "right" developer.)

The Time-Zero film we have described stays together to make the finished print and is to be viewed from the top. It must therefore be transparent at the top both during exposure and at the end of development. However, while it is developing outside the camera, the emulsions must be protected from further exposure to light. For that reason the reagent also contains an opaque substance, which serves as the door to the darkroom of the developing emulsions. The originally opaque substance becomes clear when the acidity increases toward the end of development, opening the darkroom door to reveal the developed picture. The transferred colors then become visible against a highly reflective white pigment, which is also contained in the reagent. (For fun and games with this film, see the TRY IT.)

The Kodak PR-10 system uses a variation on this technique. Somewhat different chemicals make an initially immobile dye become mobile near the unexposed silver halide during the development process. The resultant instant color prints are viewed from the bottom, because the image-receiving layer is at the bottom of the PR-10 film.

PONDER

Which of these two films (Time-Zero, PR-10) needs a camera with one left-right reversing mirror to take the picture? Which needs no (or two) mirror(s)? Check your reasoning by comparing the shapes of Polaroid and Kodak instant color cameras!

TRY IT

FOR SECTION 11.4H
Manipulating instant photographs

Because you can hold a Polaroid SX-70 photograph while it develops, you can manipulate its emulsions and thereby produce some unusual effects not possible with other photographic formats.

Expose a Polaroid photograph of a scene that has good contrast and a few brightly colored regions. As the photograph develops, place it face up on a hard surface. Use a dull point, such as the tip of a fork tine or the end of a pen cap, to "draw" on the print. Use moderate pressure. Try outlining the regions in your print as they appear (Plate 11.2). If you're careful, you may be able to identify the colors and sequence of the dye layers by drawing on a gray region.

*11.5
FALSE COLOR, INTENDED AND UNINTENDED

We have stressed that, for natural color reproduction with subtractive color film, the emulsions must be sensitive in the red, green, and blue, respectively, and the color couplers must produce their complementaries, cyan, magenta, and yellow. Of course, if we do not insist on natural color reproduction, then the emulsions may be sensitive in *any* spectral regions we please, and the dyes formed by the couplers would not necessarily have anything to do with the complementaries (if any) of these emulsion sensitivities. Any such change from natural color reproduction is called *false color.*

Why would anyone want pictures in false colors? The reason is that these colors can be used to convey specialized information in a readily apparent fashion. Such information may be the amount of reflection of some nonvisible radiation, for example, infrared. While both healthy and unhealthy green foliage may look alike in the visible, only the healthy foliage reflects strongly in the infrared. To locate diseased foliage, you should use film that is sensitive in the infrared, but also in the visible so you can recognize the scene. **Infrared color film** is made for this purpose. Its emulsions are sensitized to green, red, and infrared, and they are exposed through a camera lens with a yellow filter to suppress the blue entirely. The couplers form the usual colors: yellow, magenta, and cyan. Thus, green reproduces as blue, red as green, and infrared as red. So our healthy green leaves come out reddish purple, whereas the unhealthy green plants, which don't reflect infrared, come out blue. A red rose that reflects infrared becomes yellow on this film—and so on. By showing a region of diseased foliage, or for that matter a camouflaged enemy installation, in a different color than it does healthy green foliage, this film makes such a region stand out noticeably. The unusual coloring that this film gives to familiar subjects makes it useful not only for enemy detection and forestry, but also for illustrations in bizarre colors (Plate 11.3).

PONDER

In false color film, the couplers usually produce the standard colors (yellow, magenta, cyan). Why?

Color in photographs can also be used to indicate features of a picture that have nothing to do with color. For example, to make subtle changes in time easily visible, a double exposure can be made through complementary color filters. Objects that did not move between exposures will be shown in their natural colors, whereas those that moved between exposures will acquire the shades of the filters (see the first TRY IT).

Less drastic false color is occasionally achieved unintentionally by amateurs using the "wrong" light source. The sensitivities of the three emulsions in ordinary color film are chosen so that each color record is properly exposed when the picture is taken in daylight. As we know, incandescent ("tungsten") light sources have considerably less blue than daylight. If they are used for illumination, the blue record will be underexposed, resulting in a yellowish transparency—unintentional false color. To avoid this, the sensitivities of the other two emulsions can be effectively decreased by taking the picture through a filter that removes some of the red and green, that is, a light blue filter. However, this leads to an overall decrease of sensitivity. Preferably, a special tungsten film should be used, which has a more sensitive blue emulsion. Each color reversal film is balanced for a certain color temperature. The two most common types are daylight (5000 to 6000°K) and photoflood (3400°K). If the light source has a spectrum with one or several narrower peaks, such as a mercury lamp, its color temperature does not suffice to describe the often unpredictable results it will give on color film (see Plate 9.4). In fact, the sensitivity of color film can vary considerably with wavelength, particularly in regions where the sensitivity of one emulsion ends and another begins. To check this yourself, see the second TRY IT.

The "false" color due to a mismatch between the film and the color temperature of the light source (as opposed to effects due to nonstandard processing) is due as much to human color perception as it is to the film and light. The world *is* less bluish in incandescent light than in daylight, but normally we don't notice, because our eyes adapt to the incandescent light (see Sec. 10.6B). A photograph taken with daylight film under incandescent illumination, however, *will* have a lower reflectivity in the blue, and will therefore appear yellowish, no matter what the illumination under which it is viewed.

First TRY IT

FOR SECTION 11.5
Motion as color

While photographic film is being exposed, it "adds" all the light that strikes it. Hence a multiple exposure through different color filters will show additive color mixtures, and will introduce unusual colors in those regions where motion has occurred between exposures. Although any set of filters will do, the effect is most vivid for filters in the additive primary colors. You can make filters from three pieces of strongly colored plastic or gelatin (or you can purchase them in photographic, hobby, scientific, or theater supply stores). Hold the filters in front of your light meter to find out by how much they decrease the light. Ideally they should decrease it by a factor of 3, or about 1½ stops; if so, make the three exposures of a triple exposure at the normal setting (i.e., that indicated when your exposure meter is not looking though a filter). On most automatic cameras you can override the double exposure prevention by cocking the shutter while pressing the rewind button. The camera should be held rigidly throughout the three exposures.

Suitable subjects are anything moving slowly—drifting clouds, crawling snails, etc.—even a building being erected or growing plants, if you have patience. Choose subjects whose principal color is white, such as a melting snowman or slowly swimming swans.

Second TRY IT

FOR SECTION 11.5
Spectral response of color film

You can get an idea of a film's spectral response simply by photographing a spectrum. Use as bright a spectrum as you can get by one of the methods described in the TRY IT for Section 9.6A. Take a color picture of this spectrum, starting at normal exposure (as measured by your exposure meter) and then decreasing the exposure by factors of 2, for two or three more pictures.

Your developed pictures may show weak or missing spectral regions, particularly in the underexposed frames. This happens when the response of one emulsion layer decreases with wavelength before the next layer's response increases sufficiently, and

would happen in your vision if the spectral response curves of your different cone types didn't overlap (see Sec. 10.2A). Does your film respond well to spectral yellow? Is only the red emulsion exposed by wavelengths above spectral yellow—that is, are spectral red and orange distinguishable in your photograph? If not, why don't you notice this in photographs of ordinary red and orange objects?

*11.6
KIRLIAN PHOTOGRAPHY

Before leaving color film, let's digress a bit to discuss a process of some notoriety—a way of obtaining a picture that shows something different from that shown by an ordinary photograph, a picture that is sometimes thought to reveal strange and wonderful properties.

The process is based on the action of strong electric fields near an object in the vicinity of film. A strong electic field means there are strong forces on the charges in the object. If the forces are large enough, they will pull electrons right out of the object and accelerate them until they collide with air molecules. The collisions excite the molecules into oscillation, making them emit highly energetic light waves. The collisions, as well as these light waves, also knock further electrons out of the molecules, escalating the effect. This whole phenomenon is called an **electrical corona** because it looks like a crown of sparks. Much of the light is in the UV and blue. With your naked eye you can see the blue streamers of the corona, and they are a favorite background for horror shows. Film, which is also affected by UV, will respond to the large amount of UV present.

Corona discharges can easily be seen (and felt) on a dry day. Simply rub your feet on a rug (building up electrical charge on your body) and touch a metal object, such as a doorknob (discharging your body through a corona between your hand and the doorknob). During a thunderstorm, pointed objects sometimes emit coronas known as St. Elmo's fire.

Kirlian photography uses such discharges to make pictures, by placing film in the corona so that it will be hit by the UV and blue light given off. The electric field is set up between a metal plate and the object to be photographed, the two being separated by a glass plate and the film (Fig. 11.11). The field is turned on (usually a rapidly changing AC field) and the corona's light exposes the film, much like a contact print. However, unlike a contact print, the Kirlian picture shows "radiations" coming from the edges of the object, that is the light from the corona.

It has been suggested that Kirlian photography of living objects somehow captures the "aura" of the object and carries some mystical significance. Indeed, Kirlian color photographs can look full of colors that were not apparent during the process, and these colors can vary significantly from one photograph to the next. Also the shape of the discharge, the length, density, and curvature of the streamers, as well as the appearance of weaker, fuzzier corona images, vary considerably depending on slight changes in conductivity, water content, and other physical properties of the object and surrounding air.

We can understand the false colors of a Kirlian photograph made with color reversal film if we re-

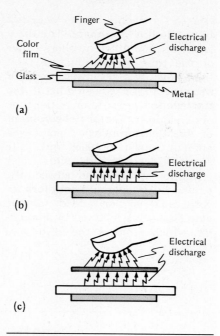

FIGURE 11.12

UV light from a discharge falls on color film: **(a)** from the front, **(b)** from the back, **(c)** from both front and back.

member how such film is made. All of the film's emulsions can be exposed by the UV, depending on where the discharge takes place. This in turn depends on where and how tightly the contacts are made. In Figure 11.12, you see three cases: where the UV hits the film from the front, from the back, and simultaneously from both. If the UV comes from the front, it will excite the blue-sensitive layer only, because the yellow filter prevents it from reaching the other layers—the developed picture will be blue. Conversely, if the UV comes from the back, it can expose only the red- and green-sensitive emulsions. For weak coronas most of the UV will then be absorbed by the red-sensitive emulsion. Hence at low levels of UV from the back, the developed reversal film will be red. At higher UV intensities, the green emulsion is also exposed, so the developed film will show yellow. If UV light comes from both front and back, at low intensities only the red- and blue-sensitive layers are exposed, giving purple, while at higher intensities

all emulsions are exposed, giving white (Plate 11.4).

Anything that can affect the electric field enough to change the location of the corona will change the appearance of the photograph. Water, which affects the electrical conductivity of the film's emulsion, can have a significant effect. Because it tends to lower the electric field, moisture diminishes the number of streamers in any given place—it causes streamers to curve around the damp region and shifts the corona to the back side of the film. This probably is the most important variable in determining the form of the corona from a given object. To the extent that the moisture of your finger or of a leaf tells you something about the life of the object, Kirlian photography can also reveal it. Another way of saying this is that your "aura" is primarily sweat.

SUMMARY

Color photographs are taken by making three photographic black and white **records** of the scene, in three wavelength ranges corresponding to the colors *blue*, *green*, and *red*. The records may be physically separate **(separation negatives),** or sandwiched together in three emulsion layers **(tripack).** To recreate a full-color image from the three records, an **additive system** (largely obsolete: **color mosaics, lenticular film**) or a **subtractive system** can be used.

Additive color mosaic films have a pattern of tiny blue, green, and red filter patches in front of an emulsion that is reversal developed. On projection with white light through these same filters, the original colors are reconstructed by partitive mixing. Lenticular films are embossed with a pattern of cylindrical lenses. They are exposed and projected through lenses with three filter strips in the lens plane. Dis-

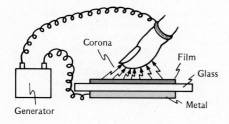

FIGURE 11.11

Arrangement used to take Kirlian photographs. A high-voltage, high-frequency generator is connected to a metal plate and to the object, O. Between them is a glass plate and the film. The corona exposes the film.

advantages of additive systems are *loss of* projection *light* and *difficulty in copying.*

Subtractive systems convert the red record, say, into a cyan and white (clear), positive transparency. The three complementary color transparencies formed this way are layered together. This can be done by making a matrix from each separation negative, dying the matrixes, and transferring the dye from the three matrix images onto one film (**Technicolor**) or onto a paper base (**dye transfer system** for color prints). Modern films use a tripack, which stays together throughout development. The three layers acquire different colors either at the time of manufacture (and then have dyes selectively removed by action of the exposing light, **dye destruction** process) or at the time of development (**color development** or **dye development**). Color development consists of black and white reversal development, with color formation by a reaction between the oxidized developer and **couplers** incorporated in, or added to, the emulsion. It is followed by **bleach-fix,** which removes the silver and leaves a pure dye image. Dye development in **instant color films** uses three different dye developers, located in a layer behind each emulsion. Diffusion of dye developer, controlled by the emulsion's latent image, creates a color image in a receiving layer.

False color can be used to represent IR radiation or other phenomena (e.g., motion), and may even result unintentionally from using the wrong film for the light source. **Kirlian photography** produces false color from the UV radiation of a corona discharge.

PROBLEMS

P1 (a) Why are three (rather than, say two or ten) color records used in color photography? (b) What are the three colors that are recorded? (c) Are these colors (broadly speaking) the same for additive and for subtractive color films?

P2 A photographic pioneer from the turn of the century has just produced his first mosaic color film and finds that his pictures come out too blue. (a) To get a better emulsion the next time should he make the emulsion more or less sensitive in the blue?
(b) Alternatively, should he use more or fewer blue starch grains in his mosaic filter?

P3 Region *A* in a color film has been exposed to a bright yellow light, and region *B* has been exposed to a dimmer yellow light. By means of a diagram, explain how the bright yellow and the dim yellow are both correctly reproduced in the developed transparency, if the film is: (a) an additive mosaic film, (b) a subtractive film.

P4 A mosaic color film is used to photograph a scene containing a red object. Suppose the film is developed so it becomes a *negative* rather than the usual positive. (a) When viewed (so that the partitive color mixture takes place as usual), in what color does the negative reproduce the red object? (b) Comment on any problems one would encounter in copying this negative onto another mosaic negative film to obtain a positive color transparency.

P5 Cameras designed for lenticular film needed another special feature, in addition to the filters on the lens; the circular, variable "iris" stop, used to control the f-number, was replaced by a variable slit. (a) Why was this necessary? (b) How was the slit oriented, assuming the lenticules ran vertically?

P6 In an *ideal* dye-destruction film no silver would be necessary, and the dyes in the three layers would be removed by the bleaching action of light. (Red light would remove cyan dye, green light would remove magenta dye, and blue light would remove yellow dye.) (a) Make a diagram of such a film before exposure, showing the layers of dyes. (b) Consider three regions of this film that are exposed with red, yellow, and blue light, respectively. Show which dyes would be removed. (c) Processing of the film would consist of making the dyes fast, so that they no longer bleach. In a third diagram show how the final film would reproduce the original colors.

P7 A Technicolor consultant wants a scene that has already been shot to look as if it had been taken at night. That is, it should be dark, and have an overall blue cast. She does this by making new gelatin matrixes that have different thicknesses than the ones that would give the scene a daytime look. How should each of the new gelatin matrixes differ from the corresponding old one?

P8 After a modern color film has been exposed, but before it has been developed, each emulsion layer carries a: (a) color filter, (b) antihalation layer, (c) colored image, (d) guarantee by Kodak, (e) latent image. (Choose one.)

P9 The parts of the red-sensitive emulsion layer in Kodachrome (and other reversal color films) that were *not* exposed to light acquire (via developer and coupler) a dye in the final transparency. The color of this dye is: (a) blue, (b) cyan, (c) green, (d) yellow, (e) red, (f) magenta. (Choose one.)

P10 A modern color slide film (e.g., Kodachrome) is exposed to white light in one region, green light in another, and yellow light in a third. Make a sketch of the cross sections of such a film: (a) After exposure, showing the location of the silver nuclei of the latent image. (b) After color development, showing the color in the three layers. (c) During projection, when light passes through the slide, showing the colors that are transmitted by all three layers.

P11 In a certain transparency made by the subtractive process, there are three regions of the same hue but differing in saturation and lightness (e.g., pink, maroon, and red). (a) Draw a diagram showing what color is present in each emulsion. (Indicate greater amounts of dye by double and triple letters.) (b) Describe how the image in each emulsion differs in the three regions of the same hue.

P12 Most subtractive reversal color films carry a yellow filter, often made of very finely dispersed, metallic silver. (a) Why is this filter needed during exposure, and why does it have to be removed during development? (b) What development step removes the silver of the yellow filter? (c) If the manufacturer forgets to put this filter into the film (but makes no other change in the film structure and its development), how do the following colors reproduce photographically: blue, green, red, cyan, magenta, and yellow? (d) Draw a chromaticity diagram and indicate by arrows how these colors would change.

P13 Figure 10.8 gives a bee chromaticity diagram. (a) Design an additive color film that would reproduce color faithfully as judged by a bee. (b) How would the bee transparencies look to a human?

P14 You want to expose an outdoor color film under illumination whose color temperature is 8000°K. What color filter should you use if your film is (a) a positive film, (b) a negative film?

P15 Two Kirlian photographs are made of a man's thumb: the first while the man does nothing but press his thumb on the film, the second while he kisses his wife. The first picture is a rather dark blue, whereas the second is rich in purples and reds. Give a *plausible* explanation of what happened.

HARDER PROBLEMS

PH1 Pantachrome uses an obsolete three-color film combining both the additive and the subtractive systems. It consists of two films, sandwiched together with the emulsions facing each other. The front film is orthochromatic (sensitive only to blue and green) and carries a lenticular screen, so the green and blue records are intermeshed in adjacent strips, similar to ordinary lenticular films. The back film is sensitive only to red (which passes through the front film). The camera lens carries two color filters that allow both the front and back films to receive their proper color of light. (a) What goes wrong if the two camera lens filters are green and blue? What colors should they actually be? Why do the correct filters give the same result as green and blue filters, as far as the front film is concerned? (b) Make a diagram, like Figure 11.3, showing how various colors are recorded.

PH2 Lenticular film can be copied by replacing the screen of Figure 11.3 by another lenticular film. (a) Which way should the copy's lenticules face? (b) Why is the copy a mirror image? (c) How could you project such a copy so it looks correct?

PH3 An (obsolete) two-color film has a red-sensitive emulsion on one side of the base, and a cyan-sensitive emulsion on the other side of the base (cf Chapter 9, Problem P31). Both emulsions are reversal developed to yield black and white positives. By floating the film on a toner with one or the other side down, the two emulsions can be toned in two different colors. (a) Why can this film not reproduce all colors correctly? (b) What should be the toning colors, in order to reproduce as many colors as possible correctly? (c) Make a series of diagrams, similar to Figure 11.7, that shows which parts of the film are exposed, where the toning dyes are after development and toning, and what colors are reproduced when the exposing lights were white, black, red, yellow, green, and blue.

PH4 Suppose that the color couplers in a subtractive color film are not diffusion proof, but can wander freely between the two bottom layers of the film, but that the top layer contains only its own coupler. Describe the effect on the final transparency by telling how the following colors are reproduced: red, yellow, green, cyan, blue, magenta, white, black.

PH5 The image in an exposed silver halide emulsion that has just finished one development (but has not yet been washed or fixed) is present in four ways: (1) the developed silver, (2) the undeveloped silver, (3) the fresh developer in the emulsion, (4) the used-up developer in the emulsion. One or another of the images (1) to (4) is used in the next step toward the final image. Identify which image is used for (a) color development to a positive, (b) color development to a negative, (c) dye development to a positive, and (d) black and white development to a negative.

PH6 Suppose a film manufacturer has couplers available that will form any desired color (one color per coupler). He wants to design a tripack, false-color transparency film that will reproduce each color in the original by that color's complement, but that reproduces the brightnesses correctly. (That is, the film should, for example, show the sky as a bright yellow, and burgundy wine as dark green.) Draw transmission curves for three dyes that will work.

PH7 Figure 10.8 gives a bee chromaticity diagram. (a) Specify the sequence and color sensitivities of the emulsions, and the colors that the couplers should produce, for a bee's subtractive color film, (b) What might be some special requirements for a bee's slide projector?

Special effects in the movies

You see a panoramic view of New York City, in which half the familiar buildings have been smashed to pieces. The camera pans in on this scene of terrible destruction toward the Empire State Building, whose tower is broken off. At the entrance of that building, surrounded by flames, the Lieutenant and the Sergeant are discussing the destruction. Suddenly, around the corner amidst falling debris, comes an immense dinosaur, crushing cars, knocking over buildings, scattering people. Its tail, swinging wildly, severs the Lieutenant's head, which flies across the screen. The Sergeant, with a burst of her ray gun, vaporizes the dinosaur.

As you munch your popcorn and watch this commonplace scene, you might ask how it was filmed. After all, dinosaurs are hard to come by, harder still to direct. New York would undoubtedly charge a fair amount to allow you to tie up Fifth Avenue, let alone destroy Manhattan, and ray guns capable of vaporizing dinosaurs are not particularly plentiful. Further, actors with severable heads charge an arm and a leg. Yet clever use of many of the principles we've been discussing has enabled the makers of the movie to capture this edifying scene, almost within budget.

1. THE COMPONENTS

The movie has a number of visual components. First, of course, there are the *actors.* They usually talk, so sound is important. This creates a serious problem, because outdoor sound is rarely good—it is much better to film the talking scenes on a sound stage, where everything is under control.

The *scenery* is another component. Background shots of New York (albeit somewhat modified), with moving traffic and people, must be blended into the action. Controlling the lighting and weather is a problem here.

Because the camera uses primarily the *ambiguous* depth cues to convey depth, a small, close object can be made to look like a large, distant one—the viewer has no way of discerning the difference. Thus, **scale models** are used as an important component in films: models of dinosaurs, broken buildings, or whatever is needed. A separate, larger model may be used for the dinosaur's decapitating tail. (The proper scaling of moving objects is not easy. To ensure that the apparent speed of the miniature in the film is not unreasonably fast, the speed of either the miniature or the camera recording its motion must be properly chosen. Moreover, some things don't scale. For example, in "The Incredible Shrinking Man," the scenery was built on a giant scale to make the actors look small. At one point they needed giant water drops, but the surface tension that holds a small water drop together won't hold a giant one together. The solution was to make each drop out of a water-filled condom. While the 100 gross of condoms in the expense statement raised some questions about the nature of the cast party, the trick worked quite well, producing tear-shaped drops that landed with a splat.)

Because the result of the filming is *two dimensional,* another component filmed can be **two-dimensional pictures.** Painted or photographed backdrops are used to allow indoor filming of outdoor shots. More impressive are the **glass shots**—paintings or photographs on a glass plate that is placed between the camera and the scene being filmed. The painting may be transparent, to superimpose clouds on an otherwise blue sky, or opaque, to block out unwanted details. An opaque glass painting would obscure the actual top of the Empire State Building, replacing it by a truncated building. Clearly, the painting must match the actual scene (visible through the transparent bottom part of the glass) in all the ambiguous depth cues: perspective, shading, color, etc.

Through-the-lens viewing allows the photographer to check these details, as well as alignment, lighting, and so on. Because the glass painting is close, while the rest of the scene is distant, the camera must have great depth of field. Such scenes are usually shot with short-focal-length lenses at between f/11 and f/22. Very small openings may require long exposures for each movie frame, up to several minutes per frame for scenes using only models. To maintain the proper alignment, the camera cannot move, except for pivoting about the center of the lens.

PONDER

Why can the camera pivot about this point, and no other, and still maintain the alignment? (It may help to draw a picture of two objects at different distances from a camera lens and examine what happens to the central ray when the camera pivots.)

The ability to replace one part of a scene with another, by means of glass paintings, allows buildings to be torn down and reconstructed, composite scenes to be constructed, etc. During World War II, publicity photos were taken through glass paintings at the Boeing Aircraft plant. The paintings depicted rearranged instrument panels of the planes so as not to reveal military secrets.

2. TIME EFFECTS

A great many special effects take advantage of the fact that any movie is made of a *succession of still frames.* For example, *stop action* photography allows the replacement, between one frame and the next, of a real actor by a dummy with a removable head. It also allows for *animation*—the positions of the dummy and the model dinosaur tail can be changed slightly between successive frames to give the appearance of motion. Similarly, an animated series of drawings can simulate the destruction in the background. The film of the crumbling of a small model building can be *slowed down* to stimulate the demise of a larger building, while a mock auto accident can be *speeded up* to heighten the excitement. Gravity can be defied by *reversing the film direction.*

Yet other effects can be achieved through *multiple exposure* of each frame, Thus, the scene can be changed by a *dissolve;* in a sequence of frames, one gradually decreases the exposure to one scene *(fade-out),* and then reexposes the same frames to another scene with gradually increasing exposure *(fade-in).* Alternatively, in a *wipe,* one exposes a smaller part of each successive frame to the old scene, and a correspondingly larger part to the new scene. One actor can appear in different parts of the same frame if each part is separately exposed. If the exposure *time* of each frame is divided into two parts, one with the dinosaur present and one without, the latter taking an increasingly larger fraction of the exposure time with each successive frame, the dinosaur will fade before the viewer's eyes—vaporized.

3. MIRRORS

The dinosaur may instead be vaporized with the help of a *half-silvered mirror* (or an angled piece of glass), as in the TRY IT for Section 2.4. Such a mirror can also be used to superimpose titles over a scene (how?), to provide highlights (e.g., flames can safely "surround" actors), and to make dissolves. A dramatic wipe can be made by smashing an ordinary mirror, thus revealing the scene behind the mirror.

A mirror can be used in a *part* of a scene. In a *Schüfftan shot,* the actors are photographed by means of a mirror as they stand on a sound studio model of steps and a doorway, say. The silver on the mirror is removed in all places other than the part reflecting the actors and their immediate surroundings. Behind the mirror, visible through the now transparent glass, is a large photograph of the Empire State Buidling. The camera thus sees the actors walking down the steps of the actual building, without the necessity of outdoor filming or of reconstructing the entire building indoors.

In all mirror shots, front-silvered mirrors are used to eliminate unwanted reflections. The camera is focused on the virtual image, of course, so unavoidable dust on the mirror is not in focus. The lighting between the two scenes (reflected and transmitted) must be carefully balanced, and the left-right reversal must be taken into account (say by a second mirror).

Mirrors can also be used to create kaleidoscopic effects or distortions (by the use of bent, fun-house mirrors). Reflections from a mirror in a tray of water can simulate the reflections from a rippling lake surface.

4. MATTES

If one part of a frame is to be exposed to one scene and the rest to another, some way of covering each part of the film must be found. A device that covers one part of the film while exposing the rest is called a *matte*. Early trick shots didn't use mattes. For example, the film might be exposed to a building with a black doorway, and then reexposed to an actor standing in front of a black background. The result, if the actor was properly positioned, showed him standing in the doorway.

Modern films require more flexibility than a black surround allows. One simple procedure is to cover part of the first scene with a black matte of the required shape. The matte can be cut out of black cardboard or painted on glass. It can be placed in front of the camera or inside it just in front of the film. It should be slightly out of focus to avoid a sharp *matte line* at the edge of its image on the film. The second scene is then covered with a *counter-matte:* the negative of the matte, black where the matte was transparent and vice versa. The matte line is harder to notice if it lies along lines in the background, such as the horizon, or in regions of darkness.

This procedure is adequate when the same matte is to be used for a long sequence of frames. However, if an actor is to appear to walk through a ring of fire, the matte must move as the actor moves—there must be a moving or *traveling matte.* Further it is undesirable to expose the original film a second time, as this creates the risk of ruining expensive footage. One method of solving both these problems is to film the actor walking, develop the negatives, and carefully trace them to make a hand-painted matte for each frame. The negatives are then printed, in contact with these mattes, onto unexposed film. *This* film is subsequently reexposed, through counter-mattes, to negatives of the ring of fire.

Such a tedious, expensive, and not very accurate method has been replaced by a number of *self-matting* techniques. One such technique, suitable when animated figures (or titles) are to be superimposed over a live background, is called *aerial-image printing.* The animated picture is drawn, with opaque inks, onto transparent sheets, called *cels.* The film of the background is projected from the rear to the location of the cel, which is in turn photographed from the front. As the animation is opaque, it produces its own matte—the backgound is visible only where there is no animation. The art work is visible, as well, if the cel was also front illuminated when photographed.

The *blue-backing process* can be used for self-matting when both foreground and background are live and moving. It depends on film processing after the pictures are taken. The actor is filmed in front of a deep-blue background. The resultant color negative is used to produce the traveling matte, as shown in Figure FO.5. By printing the original color negative in contact with this matte, one gets a color transparency of the actor alone, with the background parts unexposed. Color negatives of the desired background scenery can then be printed in contact with the counter-matte to fill in the background on the unexposed color transparency, without affecting the image of the actor.

PONDER

In the blue-backing process, what happens to a foreground object, such as a glass of gin, that is partially transparent and partially visible?

The *sodium process* avoids the high-contrast printings, and thus can better handle transparent foreground items. It also doesn't require the avoidance of background colors in the foreground. Here the background is a screen illuminated only with sodium light, an almost monochromatic light of 589-nm wavelength. The actors are illuminated by incandescent lights from which any 589-nm component has been removed by a special filter. The camera used here is rather elaborate, containing a half-silvered mirror (or other *beam splitter*), which allows it to take two pictures simultaneously. The beam splitter, equipped with appropriate filters, reflects 589-nm light to one side and

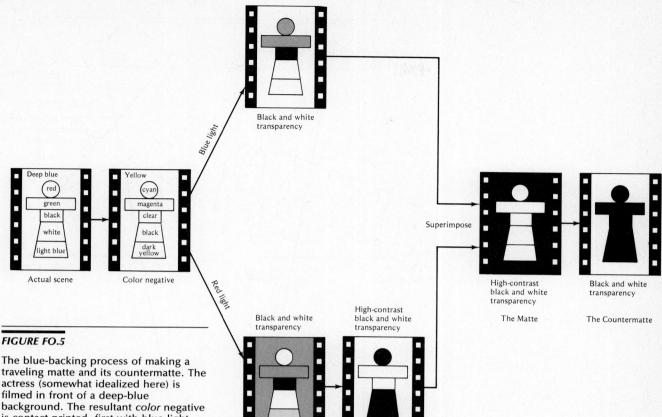

Black and white
transparency

Deep blue | Yellow

red | cyan
green | magenta
black | clear
white | black
light blue | dark yellow

Actual scene | Color negative

Blue light

Red light

Black and white
transparency

High-contrast
black and white
transparency

High-contrast
black and white
transparency

The Matte

Black and white
transparency

The Countermatte

Superimpose

FIGURE FO.5

The blue-backing process of making a
traveling matte and its countermatte. The
actress (somewhat idealized here) is
filmed in front of a deep-blue
background. The resultant *color* negative
is contact printed, first with blue light
and then with red light, to make two
black and white transparencies. (At every
step, the printing is done by the
ordinary negative process.) The one
made with red light is then contact
printed to make a new *high-contrast*
black and white transparency. This, in
turn, is superimposed on the black and
white transparency made with blue light,
and the pair is contact printed to make
another high-contrast black and white
transparency: the traveling matte—the
deep-blue parts of the original scene are
black here, the rest is transparent. A
countermatte is then produced by make
a black and white negative of the
traveling matte.

transmits the rest. The transmitted light
exposes the color film. Since most
natural yellows are broad band,
elimination of a narrow region around
589 nm doesn't affect their appearance,
so the actors' yellow hard hats will still
be properly photographed. The reflected
589-nm light exposes black and white
film to produce the matte—black
wherever the background sodium light is
visible, clear elsewhere.

With all of these matting techniques,
which insert a foreground into a
separately filmed background,
considerable effort must be taken to
ensure that the lighting is properly
matched between the two parts, the
perspective blends smoothly, the relative
sizes and parallax are appropriate, etc.

5. PROJECTIONS

A foreground shot in a sound studio can
be combined with a background filmed
elsewhere by a technique that avoids the

necessity of separate mattes. The actors
in the foreground are filmed in front of a
ground-glass screen, on which the
background film is shown by **rear
projection.** The actors themselves then
block any unwanted background, as they
would in real life.

There are a number of technical
difficulties associated with this apparently
simple scheme. Since now the
background is also a succession of still
frames, it is necessary to synchronize the
rear projector with the camera. Further,
to ensure that in the final film the
background is evenly illuminated, both
the projector and the camera must be at
distances from the screen that are large
compared with the field of view. This
means the camera must have a long-
focal-length lens, which implies a small
depth of field. This, in turn, requires the
actors to be close to the screen, so both
will be in focus. The trouble here is that
the light that illuminates the actors then
tends to splash onto the screen,
producing flare and washout of the
background projection. Additionally, if a

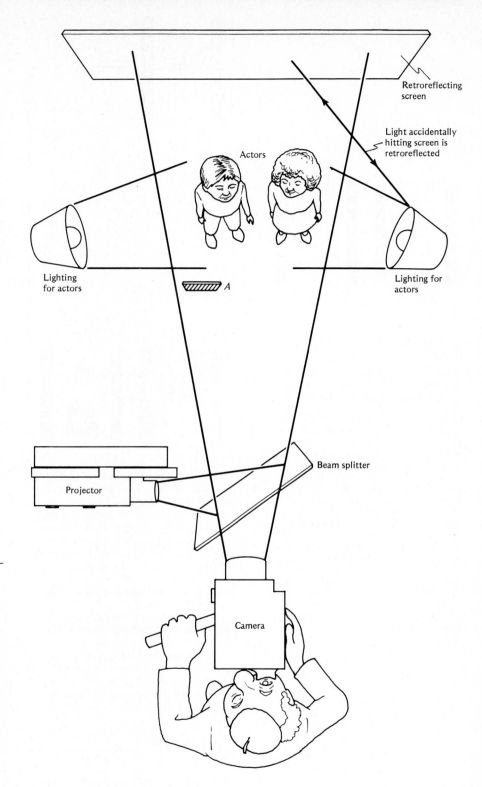

FIGURE FO.6

Front projection. The actors are in front of a retroreflecting screen, on which is projected the background scene. Note that the projection is virtually from the same location as the camera (due to the beam splitter), so any shadows cast by the actors in the projection are not seen by the camera. Projections on the actors themselves are washed out by the lighting. Because of the retroreflecting properties of the screen, any light accidentally striking it won't get to the camera. An auxiliary screen placed at *A*, say, allows the actors to move behind an element of the projected scenery.

close-up is desired, the background image must be changed. That is, the parallax that the camera sees is between the actors and the nearby screen, while the desired parallax is what would occur between the actors and a distant scene. Hence any motion of the camera or zooming of the lens must be compensated by an appropriate change in the background projection.

An alternative approach uses *front projection* (Fig. FO.6), a technique in which an actor can appear to walk *behind* an element of the background. Because of the careful alignment of the camera and projector, however, it is difficult to move the camera in this method.

6. THE OPTICAL PRINTER

Many of the complicated processes described above can be greatly simplified, and other tricks can be performed, with the help of a device called an *optical printer*. Basically, this is just a projector and a camera placed head to head, with their shutters synchronized. This allows film in the projector to be imaged directly on the raw film in the camera, frame by frame. Two rolls of film, in contact, may be run through either the camera or the projector or both. This *bipack* process allows film to be projected and/or exposed through a traveling matte. Some optical printers have a beam splitter in front of the camera, so that several projections can be made onto the same film simultaneously.

Fades, dissolves, wipes, and split-screen projections are easily accomplished on the optical printer, without endangering the original film. A fade-out, for example, merely requires a gradual decrease of the projector illumination. Further, the film may be sped up by recording every other frame,

slowed down by recording each frame twice, reversed in direction, or completely stopped. Superpositions of two images are easily done, especially if the optical printer has more than one projector. By using different lenses, the size of an image can be changed. Lighting and color changes can be accomplished by means of the insertion of filters. This enables, for example, a scene shot under ample lighting in the daytime to appear as a dimly lit night scene (*day-for-night* photography). Because of the space between the projector and the camera, optical devices can conveniently be inserted. These can be filters, lenses, prisms, or anamorphic or kaleidoscopic devices. They enable the image to be moved, rocked, rotated (as when the hero is knocked out and the room spins), inverted, distorted, discolored (e.g., a gradual desaturation of the colors, changing to black and white), or otherwise modified. The possibilities are unlimited.

7. FINAL REMARKS

We've only touched on some of the special effects tricks available. There are many others, such as computer-controlled camera and model positions, and new techniques are being developed every day. (We have, of course, completely ignored nonoptical tricks, such as the use of stunt men, exploding devices, and the like.) Clearly, the adage that the camera doesn't lie is itself a falsehood. By means of these tricks, cameras record entire worlds that do not, and could not, exist. Further, even with only realistic elements, special effects allow great artistic freedom and control to the filmmaker. They can both save and cost a lot of money. In any event, special effects surely beat using a live dinosaur.

Wave Optics

12.1
INTRODUCTION

We have known since Chapter 1 that light is a wave, and its wavelength has played an important role in succeeding chapters. But we have not yet explained *how* we know it is a wave or how we can be so sure, say, that spectral yellow light has wavelength near 580 nm. In fact, by confining our attention to *geometrical optics* we have assiduously avoided situations where light shows itself to be a wave. Now, on the contrary, we want to consider **wave optics** (also called **physical optics**)—phenomena for which the wave nature of light is essential.

Geometrical optics works when the relevant dimensions of the objects that interact with light (lenses, mirrors, stops, etc.) are much larger than light's wavelength (about $\frac{1}{2000}$ mm). To see wave phenomena, then, we must let light interact with tiny objects. Fortunately, the sizes needed are often not as small as the wavelength itself, but may be as big as a fraction of a millimeter. Indeed, we'll see that using just your hands you can "squeeze" a light beam to a sufficiently narrow width so it reveals its wave nature!

Of course, only a small part of an incident, wide light beam will strike a tiny object, and so you might expect the wave effects to be rather faint. But suppose many identical small objects are placed in a beam; their effects can add to give you a new beam of intensity as great as, or even greater than, one propagating according to geometrical optics.

Thus, wave optics is not confined to small or faint corrections of geometrical optics. In fact, all of optics (including geometrical optics) is really a particular kind of wave behavior. An understanding of wave optics allows you to control light with greater versatility; to modify images; to make nonreflective surfaces, narrow-band filters, extremely accurate calibrators of lengths and angles, and many other useful devices. Nature, too, exploits the wave properties of light to produce some of the purest and most striking colors around you—in birds, insects, rocks, and across the sky.

12.2
INTERFERENCE

What happens when two waves, which may previously have been separate, come to the same place? If the two waves are of the same type (e.g., two water waves, but *not* a water wave and a sound wave), their displacements will *add together*. For example, suppose we have two waves traveling on the same string. At any moment, each wave attempts to move a given point of the string up or down. But, of course, the string doesn't split into two pieces, one for each wave; the two waves must combine into *one* disturbance. If the two waves pull in the same direction, the displacement where both are present will be larger than the individual displacements; but if they pull in opposite directions, the resultant

displacement will be less. In Figure 12.1a two humps are traveling on a string and will meet head on. A little later they *both* pull point P up, so the string has the shape shown in Figure 12.1b, where point P has moved up further than if either wave alone had been present. (To obtain this shape you first pretend that each wave is propagating by itself, disregarding the other wave, and then you add at each point the heights of these two individual waves.) At yet later times (Fig. 12.1c) the two humps become separate again and propagate, each in its original direction, as if the other

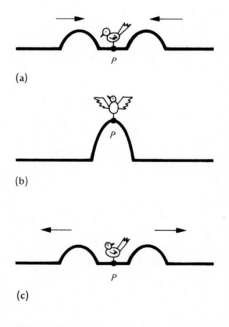

(a)

(b)

(c)

FIGURE 12.1

Two equal humps traveling on a string add together when they coincide, to make a hump twice as large.

wave had never been there. In Figure 12.2a is a similar situation, but with one "up" and one "down" hump meeting head on. Here the point P will be pulled up by one wave and down by the other, with the result that a little later (Fig. 12.2b) P and the string near P will not be displaced at all—the two waves will have canceled each other! Note, however, that this situation is not the same as a totally undisturbed string—it lasts for only an instant. The reason is that, although the string has no *displacement*, it still has a vertical *velocity* (downward left of P, upward right). Therefore, at the next instant (Fig. 12.2c) the two humps build up again and separate, each continuing on as if the other hump had never been present.

Thus we see that the size of the combination of two waves depends not only on the *sizes* of the original waves, but also on their **phase relations.** If the crests of two waves coincide, that is, if they are *in phase*, they will *reinforce* each other—the net motion of the string will be large. But if the two waves simultaneously try to move the string in opposite directions, with the crest of one coinciding with the valley of the other, that is, if they are *out of phase*, they will tend to *cancel* each other.

This adding together of the

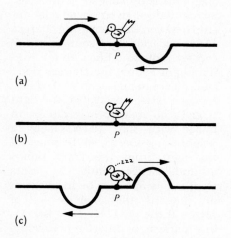

(a)

(b)

(c)

waves, each with its appropriate sign, is called **superposition.** By flipping pages 311–431 you can see this process for waves of several crests and valleys. The waves here are sometimes in phase (their sum is then twice as big as the two original waves), and sometimes out of phase (their sum is then zero). At times in between they add to an intermediate value. Superposition is valid quite generally—it holds for waves of other types, for example small water waves. In particular, it holds for light waves.

When two arbitrary waves are superposed, the reinforcement and cancellation generally occurs in a complicated, rapidly moving pattern. However, if the two waves have the *same wavelength*, the locations of reinforcement and cancellation may be *fixed in space* for an appreciable time, making it possible to observe the superposition effects. We use the general term **interference***** to describe the combining of two waves of the same frequency in the same region of space. We call the interference **constructive** if the waves are in phase (amplitudes add) and **destructive** if they are out of phase (amplitudes subtract). If the two waves can be made to have the same phase at all times, there will be definite regions of either constructive or destructive interference. If the waves are traveling in nearly the same direction, these regions will be large compared to a wavelength. In the case of light, they can then be observed as light and dark bands, even with the naked eye, in spite of the smallness of the light's wavelength.

A. Interference from two point sources

Let's consider what happens when waves of the same wavelength, from two monochromatic, point sources, are allowed to interfere. We need to

*****Latin, to strike one against the other, originally of a horse that strikes the fetlock of one leg with the hoof of the other.

know the phase difference between the two waves at any particular point in order to tell whether the interference is constructive or destructive at that point. The simplest arrangement is to place the two sources *on top of each other,* that is, at a separation d that is much less than a wavelength of the radiation they emit ($d \ll \lambda$).***** If the two sources themselves are *in phase* (that is, the emitted waves are in phase at the instant of emission, Fig. 12.3a), then the two waves will interfere constructively *everywhere,* because to reach any point, the two waves travel equal distances and therefore never get out of step. So the waves from such a pair of sources spread out in the same pattern as the wave from a single source, but with greater intensity.

If the two sources are *out of phase* (but still on top of each other), however, then their waves will also be out of phase everywhere and interfere destructively (Fig. 12.3b). An example of this is a

*****The symbol $\ll$ means "is much less than."

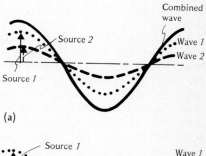

(a)

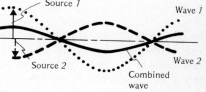

(b)

(a) Two sources close to each other and oscillating in phase emit two waves that are everywhere in phase, and thus interfere constructively. (b) Two close, out-of-phase sources emit waves that interfere destructively everywhere.

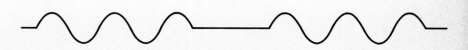

small, unmounted loudspeaker that is trying to emit a low-frequency sound wave (a low note). A loudspeaker makes its sound by compressing or expanding the air—this disturbance, which travels, is the sound wave. But as the front of the speaker pushes air out, the back of it sucks air in; a moment later the situation is reversed. Thus the front and the back of the speaker constitute two sources of sound that are out of phase. For low frequencies, the emitted wavelength is much longer than the size of the speaker, so these two sources interfere destructively almost everywhere, and you hear nothing. For wavelengths comparable to the speaker size or smaller, the sources can no longer be considered on top of each other (after all, a wave has to travel *some* distance from the back of the speaker around to the front), interference ceases to be destructive everywhere, and some sound is heard. That's why small, unbaffled speakers sound very "tinny"—lacking low frequencies. A baffle (usually a piece of wood on which the speaker is mounted) is a device that, merely by separating the front source from the back source, improves both the volume of the sound and the low frequency response.

Another simple situation occurs when the two sources are *separated by half a wavelength* ($d = \lambda/2$), as in Figure 12.4c. Suppose the sources are in phase, and consider first what happens on the x-axis (the line passing through the sources). Because the sources are half a wavelength apart, by the time a crest from S_1 reaches S_2, S_2 will be emitting a trough. Therefore, the two waves will be *out of phase* to the right of S_2 (and similarly to the left of S_1). Thus the interference on the x-axis is always *destructive* (except on the segment between the two sources). Now consider the y-axis. Any point there is equidistant from the two sources, so it takes crests the same time to travel from either source to that point. Hence, the waves arrive there *in phase*, and always interfere *constructively*. In directions intermediate between

the x and y directions the phase difference has an intermediate value. Thus the intensity is highest on the y-axis and drops off to zero for directions closer to the x-axis. This means that the radiation from the pair of sources is "beamed" mainly north and south, that is, along the y-axis, with little inten-

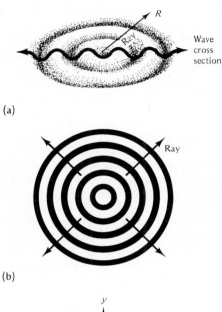

(a)

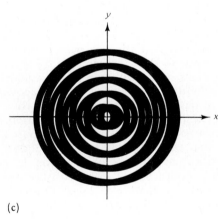

(b)

(c)

FIGURE 12.4

(a) A water wave spreading out from a source. **(b)** Schematic representation of a wave radiating in all directions from a source. Troughs are represented by black lines and crests by the white spaces in between. **(c)** Interference of *two* sources that radiate in all directions. The sources are half a wavelength apart and oscillating in phase. The waves are in phase along the y-direction (crests coincide with crests, troughs with troughs) but in the x-direction they are out of phase (crests coincide with troughs).

sity toward the east and west (x-axis).

Suppose instead that the two sources are kept at the same separation ($d = \lambda/2$), but are now *out of phase.* Since any point on the y-axis is still equidistant from the sources, the waves will now arrive there out of phase—they will interfere destructively there. The constructive interference now occurs along the x-axis instead. (Why?)

Electromagnetic radiation will be emitted in this pattern of interfering waves if the two sources are, for example, transmitting radio antennas driven by the same amplifier to maintain their phase relationship. Indeed, beaming of radio signals is achieved by this principle, but with more elaborate antenna arrangements to produce a narrower beam. (The reason for beaming of radio signals is to enable intercontinental or interplanetary transmission without wasting the electromagnetic energy on places for which the signal is not intended.)

B. Coherence

Interference of the *light* from two separate sources is *not* commonly observed. For example, we don't see it in the light of two distant automobile headlights, even though the beams reaching us are nearly parallel. This is because we cannot easily maintain a given phase relationship between the oscillating charges of two separate light sources, as we can for radio antennas. In an ordinary light source the charges don't all oscillate in step with each other. As each charge oscillates, the light it gives off may or may not be in phase with that given off by a neighboring charge, so the phase of the light from the lamp jumps around. Hence, while the lights from *two* lamps may be in phase at one instant, they may be out of phase at the next (10^{-8} sec later)—any pattern of interference would shift around so quickly that you could not see it.

Thus, most ordinary light sources change their phase so rapidly and randomly that we cannot observe interference between light from two such sources. We call such sources **incoherent.*** On the other hand, sources that stay in step with each other are called **coherent.** *Interference can only be seen with coherent beams.* Hence, to observe the wave nature of light we need coherent sources. Where do we find them?

The trick is to use *one* ordinary, monochromatic, point light source and somehow split its light into two beams. Then, whenever the phase of the source changes, *both* beams change their phase in step and thus maintain their phase relationship. If the paths followed by the two beams are not too different, we can recombine them and see their interference.

C. Thin films

We already know that one way to split a beam in two is to use a half-silvered mirror or other partially reflecting surface. This method is called **amplitude splitting,** because the reflected and transmitted beams look just like the incident beam except that they have smaller amplitudes.

A very thin layer of transparent material (called a **film**) can provide two coherent beams by reflecting light from its first and second surfaces, as in Figure 12.5. Each surface reflects a small portion of the incident light. (The additional beams from multiple reflections have even smaller intensity and are negligible.) These two coherent beams then *appear* to come from two sources, the two *virtual images* of the real light source, as shown in the figure.

Now we have our choice as to how these beams will interfere, because we can make the thickness of the film anything we like. Suppose we make the thickness so that the distance between the virtual sources is half a wavelength. We then have created for light just the situation that we discussed in part A of this

*Latin *in*, not, plus *cohaerens*, hanging together.

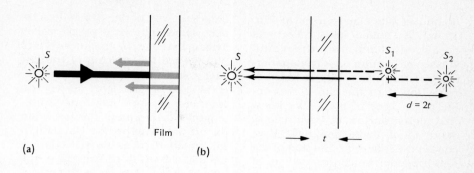

(a)

(b)

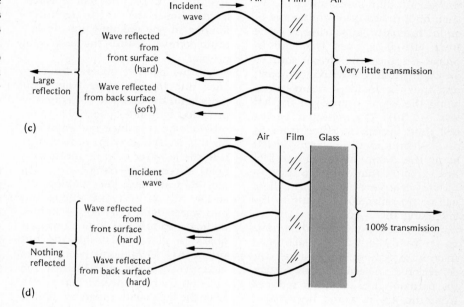

(c)

(d)

FIGURE 12.5

(a) Light from a source S is partially reflected from the first and second surface of a thin film. The beam reflected from the first surface (hard reflection) suffers a phase change. (b) The reflected beams *seem* to come from S_1 and S_2, the virtual images of S. (c) If only one of the surfaces gives a hard reflection, S_1 and S_2 are out of phase. If the distance between S_1 and S_2 is half a wavelength, then the two reflected beams interfere constructively. (d) If glass with a higher index of refraction is behind the film, both reflections are hard and the interference is destructive. Consequently there is no net reflected beam, and 100% of the light is transmitted.

section. Since the first reflection is a hard reflection (Sec. 2.3) the first reflected beam suffers an extra 180° phase change (Fig. 12.5c). The second reflection is soft, however, so it introduces no extra phase change.

In terms of the virtual images this means that the source S_1 is *out of phase* with S_2. Since we've made these out-of-phase sources half a wavelength apart, the reflected beams interfere *constructively* along the *x*-axis. This means that such a film will reflect considerably more light than a single surface does. Of course, the two virtual, coherent sources can be seen only from the front of the film (the side of the actual source), so there is no constructive interference behind it. In fact, the energy of the more strongly reflected beam has to come from somewhere, and the only supply is the incident beam. Hence, as usual, if there is more reflection, there must be less transmission.

It is not hard to arrange the converse situation, where the two virtual source images, S_1 and S_2, are

in phase (like the sources in Fig. 12.4c). To make *both* reflections hard we need only place the second surface of the film on a transparent support that has a higher index of refraction than the film, for example on a piece of glass. We know that a glass surface by itself reflects about 4% of the incident light. If we choose the index of refraction of the film properly, we can see to it that equal amounts of the reflection come from the first (air-film) and the second (film-glass) surfaces. We then obtain two *coherent* beams of *equal* intensity, appearing to come from virtual sources that are in phase. Suppose the virtual sources are again half a wavelength apart, and that the incident beam comes along the *x*-axis from S on the left (Fig. 12.5d). The reflected beams will than appear to come from S_1 and S_2 and move toward the left along the *x*-axis. As we saw in Figure 12.4c, in that direction they interfere destructively. Hence there will be *no* reflected intensity to the left along the *x*-axis. Since the intensity rises slowly in directions away from the *x*-axis (reaching a maximum only on the *y*-axis), the reflection of the coated glass surface for paraxial rays is much reduced from the 4% it would be without the film.

PONDER

The proper thickness of this antireflection film is λ'/4, where λ' is the wavelength of light in the film. Why?

This interference by amplitude splitting on a thin film is one of the most common examples of interference. Your camera's lens is coated with a thin film to reduce reflections. There is not much absorption in these thin films, so the light must go someplace. Thus, because it is *less reflective*, the coated surface is *more transparent* than the uncoated surface. In fact, unwanted reflections prevented the use of multielement camera lenses until the technique of antireflection coating was perfected in the 1930s.

If you look carefully at your cam-

era's lens, you notice that it does reflect some light, of a purple color. This is because the thickness of the film (hence the spacing of the virtual coherent sources) can be exactly right to give *total* destructive interference for only *one* wavelength. To *minimize* the reflection for *all* wavelengths, the wavelength for which there is no reflection is usually chosen in the middle of the visible range, that is, in the green. For other wavelengths the source spacing is not quite half a wavelength, hence the destructive interference in the red and blue is not quite complete—thus the lens looks purple.

Another place to observe interference due to reflections from thin films is on the oil slick so frequently provided by modern technology on otherwise untroubled bodies of water (" . . . the dirty grey-green surface of the water with its mother-of-pearl sheen of gasoline . . . ," as Maj Sjöwall and Per Wahlöo describe it). As everyone knows, oil and water don't mix, so the lighter oil floats on top. On chicken soup it floats in globules; it spreads out into a thin film if some detergent is also present (as is more likely in the oil slick than in the chicken soup). Like the lens coating, such a film is nonreflecting or hardly reflecting for *some* wavelengths and more highly reflecting for others, depending on the film's thickness. That's why you see colors that vary with the thickness (Fig. 12.6, Plate 12.1). Since the oil's index of refraction is higher than that of water, one reflection is soft and one is hard, so when the film is much thinner than a wavelength, the reflected beams interfere destructively (the out-of-phase sources are very close, as in Fig. 12.3b), and the film seems to disappear. On the other hand, where the film is too thick, the different colors that all interfere constructively overlap and give you white. Somewhere between these cases, for film thickness of the order of a wavelength of light, there is an optimal film thickness for the most brilliant colors (see the first TRY IT). You can also observe these

effects with soap bubbles, as in the second TRY IT. The most colorful soap bubbles are not more than a few wavelengths thick. (Antireflection lens coatings, however, are only fractions of a wavelength thick, hence they do not show strong colors. Window glass, on the other hand, is much too thick a "film" to show interference colors.)

Further examples of interference due to thin film reflection are found in the oxide layer that coats many smooth metal surfaces, particularly when heated. Look for the delicate colors in the tarnish of your copper pots before you repolish them! The third TRY IT suggests further interference experiments you can perform.

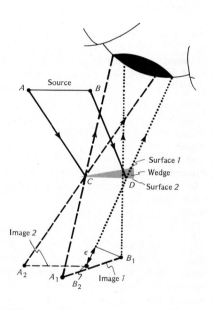

FIGURE 12.6

A broad source illuminating a wedge-shaped film. Surface 1 forms image 1, surface 2 forms image 2 of the source. The two effective sources are out of phase, because one reflection is soft and the other hard. The coherent sources at points A_1 and A_2 send rays of equal path length to the eye, hence the eye sees a dark fringe at C (destructive interference). The coherent points B_1 and B_2 send rays to the eye that differ in path length by ε. If ε = λ/2, a bright fringe will be seen at D in the color corresponding to λ.

First TRY IT

FOR SECTION 12.2C
Oil on troubled waters

Notwithstanding the ease with which modern industry does it, it is remarkably difficult to make your own oil slick. Many oils you find around the home, for example salad oil, are too heavy and won't spread—they tend to form globules. A light penetrating oil such as "3 in 1" can be used, and lighter fluid works quite well.

Fill a wide, dark dish with warm water and position yourself so you can see the reflection of a broad light source in the water's surface. Then put one drop of the oil on the surface, letting it spread out to form a thin film. It may take several minutes for the film to become thin enough, or it may happen rapidly, depending on the oil you use and on the water temperature. Notice the colors as the film gets thin. Also notice that the colors disappear when the film gets thin enough.

Second TRY IT

FOR SECTION 12.2C
Interference in soap films

You can make soap films from household detergent, but commercial soap bubble solutions make films that last somewhat longer. With a horizontal wire loop, your fingers, or the thumb hole of a pair of scissors, pick up a film from a soap solution. Experiment with lighting so you can see the colors well. Have a dark background below the soap film and an extended source, such as a light-colored wall illuminated by a shielded lamp in an otherwise dark room. Tilt the wire loop toward the vertical so the soap solution collects at the bottom edge and the film gets thin on the top. Note the sequence of colors that appear at the top and wander toward the bottom. When your film becomes very thin, a well-defined colorless region appears at the top that is much less reflective than the rest of the film.

Insert a straw into a dark-colored cup containing some detergent solution and blow some soap bubbles (a shot of lime juice in a glass of beer works in a pinch). Note how the colors depend on the tilt of the soap film. Blow up one bubble till it bulges out over the rim of the cup. The top of the bubble becomes thinnest first (provided there are no air currents). Why do you see colored rings about this thinnest place?

Third TRY IT

FOR SECTION 12.2C
Newton's rings and mirror

Newton, who did not believe that light is a wave, nonetheless made one of the first observations of thin-film interference. He used the wedge of air formed between the bottom curved surface of a lens and the top surface of a flat glass plate on which it was placed. Use a glass plate with a dark background, illuminated from above. Place your eye so you can see a good reflection from the plate, then put a lens on its surface. Press down gently and rock the lens back and forth. You will see a small disk of decreased reflection where the lens touches the surface, surrounded by colored rings, called **Newton's rings.** Unless your lens is only slightly curved (long focal length), the rings may be too small to see with the naked eye.

A larger pattern can usually be seen with two flat pieces of glass, such as plate glass or microscope slides. Such glass is always bent a little, but when you place the two pieces together, they usually touch somewhere near the edge. Turn the pieces all possible ways to find a combination of surfaces that touch somewhere in the middle. Make sure that the glass is clean and slide one piece around on the other until you see colors in the reflected light, usually in oblong ring shapes. Press harder near the center of the pattern until a dark spot appears at the center. (Why is it dark?) The dark spot is surrounded by a light (reflective) ring, where the thickness of the air film between the two glass surfaces is near $\lambda/4$; the beam reflected from the bottom travels an extra path length of $\lambda/2$ and suffers an additional change in phase due to the hard reflection, hence interferes constructively with that reflected from the top. Do the colors you see agree with the idea that the shorter wavelengths (blue) interfere destructively at smaller film thickness than the longer wavelengths? Do the colors change as you look at the pattern from various angles?

The second ring has a green border. Here the film thickness is about $2\lambda_{red}$ as well as about $3\lambda_{blue}$, that is, about 1300 nm, so that these colors are both removed by destructive interference. How thick is the air film at the magenta part of the third ring? How many rings can you see? Now gently release the pressure on the top glass and watch the rings move. Why do they move toward the center? (Recall that each ring occurs where the air film has a certain thickness, e.g., 1300 nm for the green border of the second ring.) Why do the rings disappear as you lift the top glass even a little bit? When the rings have just disappeared, about how far have you lifted the glass?

To observe another interference pattern also mentioned by Newton, you need a mirror, about the size of a bathroom mirror or larger, that is good and dusty. (If your house is spick-and-span, spill some diluted milk on a mirror and let it dry, or press your kid sister's play dough against it to dull its surface.) In a dark room, illuminate the mirror with a flashlight or a candle, held about halfway between your eye and the mirror so it just obscures its own mirror image. In the light scattered by the dust you will see colored rings similar to Newton's rings.

It is tempting to think that this pattern is due to thin film interference between beams reflected from the mirror and from its front glass surface, respectively. However, not only is mirror glass much too thick to give visible colors from this type of interference, but also, if this were the explanation, you could see the pattern better without the dust! Rather, the two beams that interfere are (1) a beam that is first scattered by the dust and then reflected by the rear, silvered surface of the mirror, and (2) a beam that is first reflected by the rear of the mirror and then scattered by the same dust. Since these two beams have nearly the same path length to your eye, interference of white light can and does occur. According to this explanation, should the central spot be white or dark? Is it?

If particles on the mirror are larger and diffract light, rather than scatter it, you see a corona pattern (see Sec. 12.5B) superimposed on the interference pattern. The corona is quite prominent and colorful if you breathe on a clean mirror and use a candle as a light source, in the manner described above. By varying the distance of the candle from your eye you can find positions where the interference pattern is also visible. Of course, as described in Section 12.5B, you can also see a corona by breathing on your window and then looking through the condensation at a street light.

D. Young's fringes

Another way of making two beams from a single monochromatic point source is to use two neighboring beams. Such beams can be selected by, say, two parallel, narrow slits. As a wavefront from the source reaches the slits, part of it will pass through each slit. This method is called **wavefront splitting.** If the slits are equidistant from the source, each wavefront reaches both slits at the same time. Since a wavefront connects crests, say, of the wave, the light on the two slits must then be in phase.

In Figure 12.7a two neighboring, coherent beams from a point source Q are separated by two slits, S_1 and S_2, in an opaque plate. We'll see later (Sec. 12.5A) that if the slits are narrow enough, the beams do not go straight through, but spill out in all directions (Fig. 12.7b). The two slits then act as coherent sources that are in phase if Q is equidistant from them, as shown.

At large distances from these *coherent* sources there are many places of constructive and of destructive interference. If the waves are light waves, it is also easy to *observe* these regions of interference. You simply place a screen far from the sources (Fig. 12.8). (Alternatively, you can interpose a lens between the sources and the screen, placing the screen closer, at the focal plane of the lens—Parallel Rays Rule, Sec. 3.4C.) Wherever the interference is constructive the screen will be bright; where it's destructive the screen will be dark. For example, if the two beams must travel the same distance to reach a given point on the screen, they will arrive there in phase and make a bright spot. If one beam must travel half a wavelength more than the other to reach a given point, it will arrive half a wave out of phase, and that point will be dark. These regions of alternating bright and dark intensity are called **interference fringes,** and the fringe that is equidistant from the sources is called the **central fringe** (Fig. 12.9). (The TRY IT after Sec. 12.2F tells how to see these fringes.)

FIGURE 12.7

(a) Two slits, S_1 and S_2, are illuminated by beams *1* and *2* from the same light source Q. The slits act as coherent sources, radiating into the space above the opaque plate. **(b)** Photograph of water waves spreading out from two slits in a barrier.

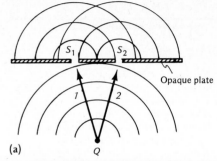

(a)

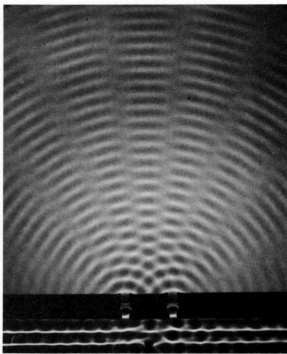

(b)

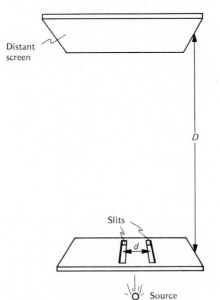

(a)

FIGURE 12.8

Two ways to observe distant interference: **(a)** screen at a distance D that is large compared to the slit separation, d, **(b)** screen at the focal plane of a lens, L.

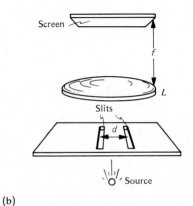

(b)

FIGURE 12.9

Young's fringes—interference fringes seen on either screen of Figure 12.8.

Such fringes were first seen by Thomas Young in an arrangement similar to this. The constructive and destructive interference he saw showed conclusively that light had wave properties.

A variation of Young's experiment uses only one slit, S, and its mirror image, S′ (Fig. 12.10a). In the region to the right of this so-called **Lloyd's mirror** each point receives a direct beam from the slit source S and a reflected beam via the mirror. The reflected beam seems to come from a second slit source S′ that is coherent with the first. These beams interfere and produce a screen pattern similar to Young's fringes; however, the bright and dark bands are interchanged. This is because the reflected wave suffers an additional 180° phase change upon reflection, so the slit and its virtual image are 180° out of phase.

This experiment showed for the first time that there is such a phase shift for hard reflections of *light*, just like for the rope waves of Figure 2.14.

Occasionally you are an involuntary witness to this experiment, done with TV or FM radio waves to which your receiver is tuned (Fig. 12.10b). An airplane passing overhead takes the place of Lloyd's mirror. As the plane moves, the path length of the reflected beam changes, so that this beam arrives at your antenna sometimes in phase and sometimes out of phase with the direct beam, and the fringes drift over your antenna. Therefore, the signal you receive becomes alternately stronger and weaker.

E. Spacing between the fringes

The fringes seen by Young not only demonstrated the wave nature of light, they also allowed its tiny wavelength to be measured! To see how this is possible, we must find out how the spacing between the bright fringes depends on that wavelength. Consider the *difference in path length* from the slits, S_1

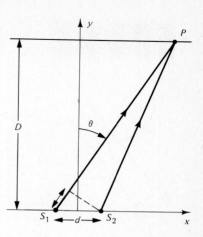

FIGURE 12.11

Path lengths from two sources, S_1 and S_2, to a distant point P in a direction θ from the y-axis. The difference in the two path lengths is ϵ.

and S_2, to some point P on the screen, *very far from the slits* (Fig. 12.11). We have drawn a perpendicular to the path S_1P from S_2, intersecting S_1P at R. Now the distances $\overline{S_2P}$ and $\overline{RP}$ are equal, so the extra path length that the beam from S_1 has to cover is $\overline{S_1R}$, which we have denoted by ϵ (the Greek letter epsilon). Whenever this extra path length is an integral multiple of the wavelength, the two waves arrive at P *in phase*, and add *constructively*.

For example, for $\epsilon = \lambda$, a crest from S_2 arrives at P at the same time as the crest that left S_1 one cycle earlier. Hence, if P is located so that the path difference is $\epsilon = \lambda$ (or 2λ, or 3λ, or . . .), P is a point of *constructive* interference, and hence at a bright spot.

Between these places, where P is located so that the path difference is $\epsilon = \frac{1}{2}\lambda$ (or $\frac{3}{2}\lambda$, or $\frac{5}{2}\lambda$, or . . .), the two waves arrive at P *out of phase*— a crest from S_1 arrives at the same time as a trough from S_2. At such points, then, there is *destructive* interference—the two waves cancel, so there is a dark spot.

If you look at different points P along a line that is parallel to the x-axis, as shown in the figure (but, in practice, located much farther from the sources than in the figure), the interference you see at P will vary

FIGURE 12.10

(a) A slit and its virtual image in Lloyd's mirror form a coherent pair of sources and cause interference on the screen. **(b)** Reflection from an airplane causes interference of TV waves.

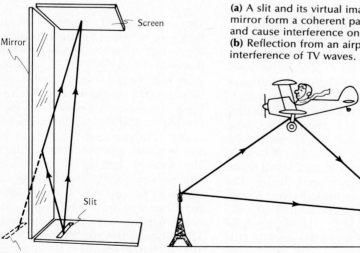

Mirror

Screen

Slit

Virtual image of slit

(a)

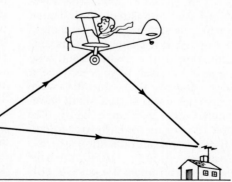

(b)

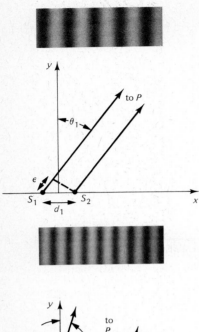

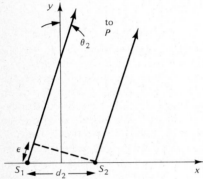

FIGURE 12.12

As the distance between the sources increases from d_1 to d_2, the angle θ decreases from θ_1 to θ_2 in order to keep the same path difference ϵ.

periodically. At the point directly above the sources (on the y-axis), the path lengths from the sources are equal ($\epsilon = 0$), so there is a bright spot. For points farther to the right, ϵ increases, first reaching $\epsilon = \frac{1}{2}\lambda$ (dark spot), later reaching $\epsilon = \lambda$ (bright), then dark at $\epsilon = \frac{3}{2}\lambda$, and so on, alternating bright and dark fringes. You'll find a similar fringe pattern for points to the left.

The *spacing* between the fringes is determined by the wavelength λ of the light, the distance d between the sources, and the distance D from the sources to the screen. First, how does it depend on D? Consider $\epsilon = \lambda$, corresponding to the first bright fringe near the central fringe. From Figure 12.11 we see that ϵ is determined by the *direction* of the two beams. That is, the first bright fringe is in a definite *direction* θ, no matter how far the screen. The farther the screen, the farther from the y-axis the first bright fringe will be. So the *fringe spacing is proportional to the screen distance* D.

Further, Figure 12.12 shows that with a larger distance d_2 between the sources, a smaller angle θ_2 gives the same path difference ϵ—the condition $\epsilon = \lambda$ now occurs at a *smaller* angle. So the larger d is, the closer the fringes are together; the *fringe spacing is roughly inversely proportional to* d.

Finally, if you use a larger wavelength λ, you will need a larger path difference ϵ before you reach the first bright fringe (where ϵ must equal λ). Hence a larger λ means the first bright fringe occurs at a larger angle θ, for a fixed d. So the *fringe spacing is roughly proportional to the wavelength*.

These three proportionalities can be expressed by one formula (valid for fringes near the central fringe):

$$\text{fringe spacing} = \lambda \frac{D}{d}$$

(Appendix K gives a mathematical derivation.) For a typical optical set-up we may have $d \simeq 1$ mm or less, $D \simeq 1$ m, so $D/d \simeq 1000$—the wavelength is translated into fringe spacing by an "amplification factor" D/d of about a thousand. This amplification is the main reason why fringes are useful—you can see a large display that exhibits wave properties, even with a small wavelength, such as that of light. The TRY IT uses moiré patterns to demonstrate this amplification. (In the arrangement of Fig. 12.8b, the lens' focal length f takes the place of D.)

TRY IT

FOR SECTION 12.2E
Moiré model of two-source interference

Two-source interference can be illustrated with the second TRY IT for Section 9.8A. Each of two sets of concentric circles (Fig. 9.28 and its copy) can be thought of as a snapshot of the wave crests from a point source. The dark circles represent troughs, and the light circles represent crests.

Superimpose one set of circles upon the other, shifting one set to the side so that the centers don't coincide (as in Fig. 12.4c). Notice the moiré pattern that results, that is, the regions that are on the whole darker and those that are on the whole lighter. The lighter regions are white (where white is on white) crisscrossed with black (where black is on black), hence they correspond to constructive interference. The darker regions are nearly uniformly dark, where white is on black or black is on white, corresponding to destructive interference. Note how the spacing of these regions of constructive or destructive interference changes as you vary the separation of the sources.

Imagine a screen placed at the edge of your moiré pattern. With a ruler, measure the screen distance D, the source separation d, and the fringe spacing. Check whether our formula for the fringe spacing (valid for fringe spacing small compared to D) is satisfied.

F. White-light fringes

You don't need monochromatic light to observe Young's fringes. White light also works (as you can discover doing the TRY IT), at least near the central fringe. Each color (wavelength) contained in the white light interferes only with itself, and the white light fringe pattern is the additive mixture of the fringes in the various spectral colors. Since the central fringes of all wavelengths coincide, this fringe is white in the middle. Because blue light has a shorter wavelength, the blue fringes are more narrowly spaced than the red fringes. Hence the region of destructive interference is reached first for blue light. At that point the central fringe looks yellow—white minus blue

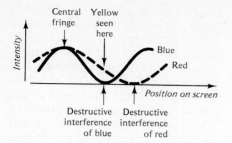

FIGURE 12.13

White-light Young's fringes. The resultant colors are produced by additive mixture of the fringes in various colors. The intensity versus screen position is shown for red and blue light only; the other colors have bright and dark spots in between those for blue and red.

(Fig. 12.13). Farther out, the other wavelengths reach their regions of destructive interference. At the point where the red interferes destructively with itself, the next bright blue fringe is almost at its brightest. Thus a few fringes are seen in pretty colors on either side of the central fringe, but farther away the maxima and minima in the different colors get totally out of step, and the fringes "wash out."

TRY IT

Make two thin slits, less than a millimeter apart, in an opaque material. For example, slit aluminum foil with a razor blade or scratch the soft paint on a recently painted piece of glass with a sharp knife. Make parallel slits of various separations, as well as a single slit, well separated from the rest, for comparison. (If you cut your pair of slits in the shape of a narrow wedge, you will get the effect of different spacings by looking through different parts of this wedge.) Pick a small concentrated light source: a bare bulb with a long straight filament (or a slit mask in front of an ordinary bulb) in a dark room, or a distant streetlight. Orient the slits parallel to the longest dimension of the light source. Place your eye close to the slits and focus on the

light source. Your eye's lens and retina take the place of the lens and screen of Figure 12.8b. You will see a combination of Young's fringes and the single-slit diffraction pattern (Sec. 12.5A). The larger pattern is the diffraction pattern. You can identify it by comparing the pattern with what you see by looking through the single slit. Young's double-slit fringes are the fine lines in the central part of the diffraction pattern. The narrower, more uniform, and closely spaced your slits are, and the smaller and more distant your light source is, the better will be the fringes you observe. Experiment until you see good fringes.

Using your narrowest slits, note the sequence of colors in the fringes. Estimate how many fringes you can see with white light, then try a color filter anywhere in the light path and see whether more fringes are visible. Verify that the fringe spacing changes with slit separation as advertised. If you have several color filters, switch them back and forth to see how the fringes move as the dominant wavelength changes.

G. Interference of many coherent sources

If we have *many* monochromatic, coherent, in-phase sources, we can arrange them in many different ways. A regular, equally spaced arrangement can send a lot of light in certain directions, whereas Young's double slit sent only a little. For simplicity, let's put all the sources on the same line with equal distances between them and in phase—like a chorus line.

In such an arrangement (Fig. 12.14) we get *constructive* interference from all the sources not only in the forward direction ($\theta = 0$), but also in the same directions that would give constructive interference from any *two* nearest neighbors—if light from each source is in phase with that from its nearest neighbor, then the light from *all* the sources is in phase. For example, at an angle for which $\epsilon = \lambda$, some crest, say the first, emitted by a given source is in phase with the second crest emitted by its nearest neighbor, and with the third crest emitted by its next nearest neighbor, and so on. Thus, on a distant screen

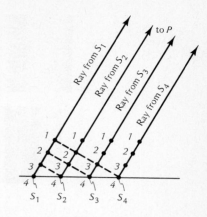

FIGURE 12.14

Constructive interference from four in-phase sources. Crest 1 from source S_1, crest 2 from source S_2, crest 3 from source S_3, and crest 4 from source S_4 are all equidistant from P. Hence they all arrive simultaneously at P and interfere constructively there.

bright fringes occur at the *same* spacing, $\lambda(D/d)$, as for Young's fringes, but these are *quite* bright, since *all* of the beams interfere constructively.

The story is different, however, for *destructive* interference. In addition to sources canceling in nearest neighbor pairs, a source can cancel with its neighbor two doors down, or three doors down, and so on. Thus we get many more places of complete destructive interference than in the two-source case. For example, if there are 100 sources, then the smallest angle of cancellation occurs when source *1* is out of phase with source *51*, source *2* is out of phase with source *52*, and so forth up to source *50*, which is out of phase with source *100*. Thus, in this case the first dark fringe does not occur at half the distance to the first bright fringe, but instead at $\frac{1}{50}$ of that, since the sources responsible for interference are 50 times farther apart. So at $\frac{1}{100}$ of the distance between the central fringe and the first bright fringe the intensity has already dropped to zero. This means that the central fringe, and the other bright fringes as well, must be much *narrower* than Young's fringes. As we move farther

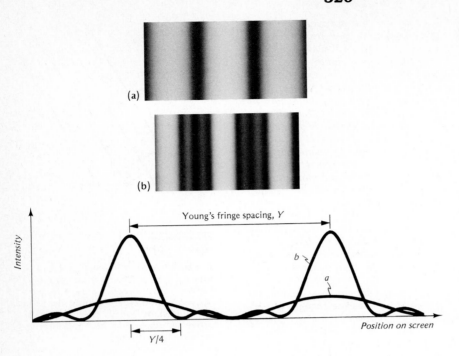

FIGURE 12.15

Plots and photographs of intensity distribution on a screen from: **(a)** two narrow slits (Young's fringes), and **(b)** four narrow slits with the same spacing between slits.

away from the central fringe, we encounter many more dark fringes, due to the many possible ways the sources can group together to interfere destructively. Between these dark fringes there is not much intensity because the sources are never *all* in phase until we reach the next bright fringe. The upshot is that the more sources, the sharper and brighter the fringes, with negligible intensity between them (Fig. 12.15).

12.3
APPLICATIONS OF INTERFERENCE

Interference definitely shows the wave properties of light; but if we really want to sell interference as something other than a physics demonstration, we had better show some other, useful things it can do. In fact, we already know that it can

overcome some of the limitations of geometrical optics, for example, by making surfaces nonreflective, and that it is sensitive to very small dimensions, such as the spacing between the slits that produce Young's fringes. Thus interference invites numerous technological applications. In Nature it has been a growth industry for millennia, as many of the most striking natural colors are produced by interference.

A. Gratings

A **grating** (also called a **diffraction grating**) allows us to realize the array of many coherent light sources discussed in Section 12.2G. One such device, a *transmission grating*, is an opaque plate with many closely and regularly spaced slits in it (Fig. 12.16a). The spacing of the slits is called the **grating constant.** When parallel coherent light falls on a grating, this light is broken up into many coherent beams by wavefront splitting. Another way to realize such an array is to use many regularly spaced *reflecting* surfaces, instead of the many slits. Such an arrangement is called a **reflection grating** (Fig. 12.16b).

The key property of a grating is

that it sends a lot of light (i.e., a large fraction of the incident beam) into extremely well-defined directions: 0, θ_1, θ_2, . . . These directions depend on the wavelength λ of the light and the grating constant *d*. Because of the sharpness of the bright fringes, the colors produced when white light is used will not overlap as much as the colors produced by two slits (Fig. 12.13). For a grating there is a central (also called zeroth-*order*), white fringe, as always. The next bright fringe on either side (called first order) has the smallest λ, blue, on the inside, followed by wavelengths running through the spectrum to the largest λ, red, on the outside. Similarly the next bright fringe (second order) spreads light into a spectrum, and so on; as the angle θ increases, several fringes or spectra are seen.

Viewed from a particular direction, then, a grating with its fixed grating constant will transmit (or reflect) a narrow range of wavelengths to the observer—the gratings will look colored. As you shift your position and change your angle θ with respect to the grating, the wavelength you receive will also change—the grating changes color. Such colors that change with the angle of view are called **iridescent.***

Many natural structures are sufficiently regular to form good gratings for radiations of suitable wavelength. For example, the "eye" of a peacock feather has iridescent blue-green and violet colors due to a regular array of very fine melanin rods that reflect light (Fig. 12.17, Plate 12.2). Similarly, hummingbirds have stacks of thin platelets in the barbules of their feathers, giving them their iridescent colors. If you rotate an iridescent feather slowly you can observe its color change. Virginia Woolf described this subtle change: ". . . after a flight through the sunshine the wings of a bird fold themselves quietly and the blue of its plumage changes from bright steel to soft purple." The common pigment colors in birds are blacks, browns, yel-

*Latin *iris*, rainbow.

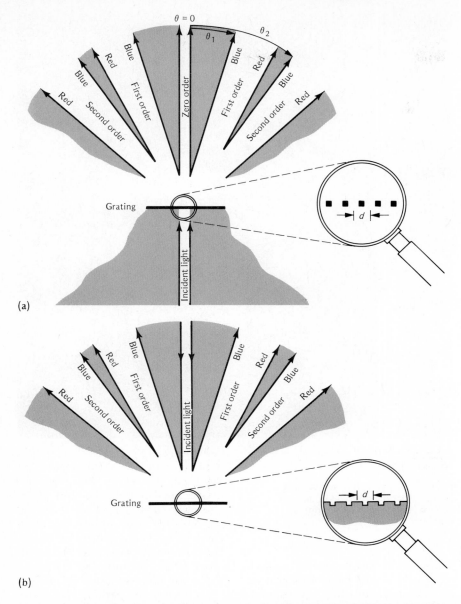

(a)

(b)

FIGURE 12.16

(a) A transmission grating. **(b)** A reflection grating.

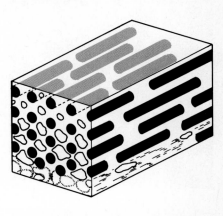

FIGURE 12.17

Regularly spaced melanin rods in a peacock feather form a reflection grating and produce the colors shown in Plate 12.2.

lows, reds, and oranges, whereas most blues and cyans are iridescent. The rare green colors, such as those of parakeets, are iridescent cyan mixing with pigment yellow.

Reflection gratings are also found in the scales of butterfly wings, such as that of the South American *Morpho* (Fig. 12.18). African artists make designs with such iridescent wings that rival what pop art does with technology. The scales of snakes of the *uropeltid* family have ridges spaced by about 250 nm, which produce iridescent colors. Some beetles (Fig. 12.19, Plate 12.3) are literally crawling reflection gratings. Electron microscopy reveals the remarkably constant spacing of the gratings, as is necessary for such sharp spectra, and confirms that the narrower pattern comes from the wider spaced grating.

FIGURE 12.18

Electron micrograph of a cross section of a scale of a *Morpho* butterfly. Each vane, shaped like a Christmas tree, consists of ridges (branches of the tree) perpendicular to the plane of the picture. The ridges are spaced by about 220 nm. Blue light of wavelength 440 nm coming from above is strongly reflected by this elaborate grating.

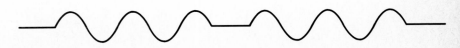

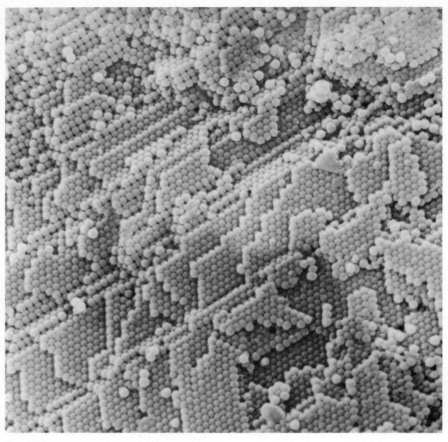

FIGURE 12.20

Electron micrograph of the silicate spheres that make up an opal. Each sphere is about 150 to 300 nm in diameter.

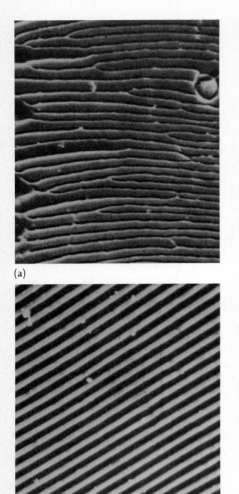

(a)

(b)

FIGURE 12.19

Scanning electron micrographs of the gratings formed by ridges in (a) the ground beetle's wing case and (b) the mutillid wasp's underbelly. Views of the insects themselves are shown in Plate 12.3.

PONDER

Will exposure to bright light make iridescent colors fade as fast as colors formed by organic dyes, which are bleached by light?

You frequently find structures that are periodic not only in one, but in two or three directions. A common example is a window screen. Its openings form a *two-dimensional grating*—you can consider it as both a horizontal grating and a vertical grating. Correspondingly, the interference pattern of such a grating is spread both in the horizontal and vertical directions (Plate 8.1). The TRY IT tells you how you can see this and other patterns.

If you stack many two-dimensional transmission gratings parallel to each other you get a *three-dimensional grating*. The beautiful play of colors in an opal is produced by interference from a three-dimensional repetitive structure of minute silicate spheres (Fig. 12.20). When illuminated they reflect light and thus act as a three-dimensional array of coherent point sources. As you rotate the opal, you change the effective separation between the spheres and thus the colors reflected to you (Plate 12.4). The size of the area of the opal that sends one color to your eye is determined by the size over which the array maintains the same regularity. Large areas are rare, and opals with large areas of uniform color are therefore more valuable.

An even finer periodic structure enables crystals to act as three-dimensional gratings. Simple table salt crystals consist of sodium and chlorine atoms aranged into submicroscopic, regular cubes. As usual, these atoms reflect electromagnetic waves by reemitting waves from their oscillating charges, which therefore form a regular array of coherent sources. The grating constant of this reflection grating is too small to give salt iridescent colors in visible light, but x-rays work just fine. Although those of us who don't have x-ray eyes cannot enjoy these "colors," we can record the x-rays on photographic film, develop it, and enjoy the resulting geometrical

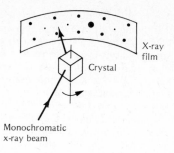

(b)

FIGURE 12.21

X-ray diffraction by a crystal. **(a)** Principle for photographing the pattern. The crystal is rotated during exposure, so that many ways of grouping the atoms into stacks of plane gratings contribute to the pattern. **(b)** The x-ray diffraction pattern of table salt, sodium chloride (taken by a slightly different technique).

pattern (Fig. 12.21). This so-called **x-ray diffraction** pattern reveals a great deal about the structure of the crystal, which is far too small (less than a nanometer) to be seen under any ordinary microscope. In particular, since the spots define the angles of constructive interference, (analogous to θ_1, θ_2, . . . of Fig. 12.16a), and since the wavelength λ of the x-rays used to make the photograph is known, one can compute how far apart neighboring atoms are in the crystal. An atom in a more complicated crystal has neighbors at a variety of distances, so the

x-ray pattern for such a crystal consists of a complicated arrangement of spots, from which the whole crystal structure can be deduced. For example, one can distinguish the modern, pearlescent lead white pigment from its older form by x-ray diffraction, and thus detect art forgeries.

X-ray diffraction uses the pattern obtained with a known wavelength to determine the crystal structure. Conversely, a grating of known structure can be used to measure wavelengths. But it is not easy to make gratings comparable in accuracy even to that of a lowly iridescent bluebottle fly—the art was not learned until the late nineteenth century. Early gratings were made by using a diamond to scratch grooves in glass plates. The rough scratches correspond to the opaque spaces. If the unscratched glass is transparent, we have a transmission grating. Alternatively the glass can first be silvered so that the unscratched places reflect, to make a reflection grating.

The key part of the engine that rules the parallel lines is a big screw that moves the ruling tool by exactly the same distance from one line to the next. To make a large grating of 10-cm size, consisting of some 100,000 lines, takes several days and nights, in temperature-controlled surroundings. Even a small periodic error in the slit spacing makes a single sharp fringe break up into several "ghosts." More recent ruling engines use interferometers (see Sec. 12.3B) to control the spacing of the grating lines.

Once you have a good grating you can replicate it by letting a layer of plastic solidify on the grating and then carefully lifting it off. Depending on the care taken, replica gratings can be of high or low quality. Cheap ones have become a common source of interference colors and are adequate for most purposes, certainly for decals on vans and tie clips. They find artistic application from belt buckles with the name of your favorite rock group to Roy Lichtenstein's pop art picture of a rainbow, which is actually an arc-shaped reflection grating.

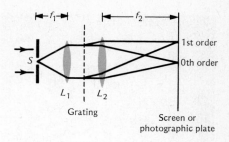

FIGURE 12.22

Principle of a grating spectroscope. Light of an unknown wavelength is shined on a narrow slit S. Lens L_1, separated by its focal length from S, makes this light parallel and directs it toward the grating. After emerging from the grating, the light is focused on the screen by another lens L_2. If the grating were not there you would see on the screen a fine line of light, the image of the slit S. With the grating in place, some of the light propagates undeviated (the zeroth order), while some of it is deviated by well-defined angles. In each order, therefore, the screen shows several colored images of the source, one for each wavelength contained in the original light.

Gratings have also been made by a photographic technique. The principle is to use an interference pattern, like Young's fringes, that can be relatively simply produced and that has very accurately spaced fringes. Such a pattern can be photographically transferred to etched lines in glass, giving a very accurate grating.

To measure wavelength, a grating is set up as a **spectroscope**, as in Figure 12.22. The screen pattern it produces consists of several line images of the source slit (Fig. 12.23), spaced at the distance given by the formula of Section 12.2E. The zeroth-order line of course occurs on the axis for all wavelengths. The distance to the first-order line, however, is proportional to λ. Thus the first order image of the source is spread out horizontally on the screen into a spectrum of the source (Plate 15.1). The screen carries a wavelength scale, obtained either from the formula (knowing the grating constant) or by calibration with light of a known wavelength, and thus enables you to measure an unknown wavelength.

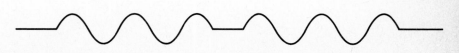

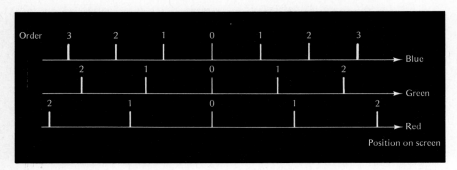

FIGURE 12.23

Dependence of fringe spacing on wavelength. The plot shows the intensity of the light produced by the grating for monochromatic sources of blue, green, and red light as a function of position on the screen.

PONDER

Is the sequence of colors seen in a grating spectrum the same as in a prism spectrum? Which color is deviated least, and which is deviated most, in the two cases?

TRY IT

FOR SECTION 12.3A
Diffraction gratings

A number of common items can be used as diffraction gratings. You can look at the light from a small bright light source reflected by the grooves in an LP record. Depending on the tilt of the record you will see the light source's component colors, provided the source is small or distant enough. How many different orders of interference can you see as you tilt the record? Verify that there are interference beams on either side of the zeroth order.

Objects with an array of holes that are periodic in two directions can make two-dimensional gratings. Examples are window screens and fine-meshed, thin fabrics. Use the same method of viewing as in the TRY IT for Sections 12.2D, E, and F. To see the two-dimensional pattern, however, you need a point source, such as a distant street light or candle flame. Observe what happens when you tilt your head, when you rotate the grating about the line of sight, and when you rotate the grating into the

line of sight. How can you tell which of two gratings have the finer mesh, whether the grating constant in the two directions is equal, and whether the two directions of periodicity are perpendicular? (Try distorting a piece of screening and see what happens to the pattern.)

You can purchase replica gratings cheaply—as "spectrum glasses" in novelty stores (coarse gratings) or from scientific supply houses. For a bit more you can purchase gratings made to be used as camera "filters" for special effects, such as rainbow (line grating) and crostar (square grating) filters. The scientific filters specify the grating constant by telling you that the grating has, say, 13,400 lines/inch. (So in this case d = 1/13,400 in = 2.54/13,400 cm ≃ 0.0002 cm = 2,000 nm.) Other types should be calibrated using a known wavelength, such as the strong blue line of mercury, which has λ = 436 nm. If you do not have a mercury lamp (sun lamp) you can see this line in a black light or as a brighter line against a continuous background in the light from a fluorescent tube. You can get precision

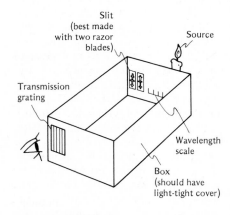

FIGURE 12.24

Design for a simple grating spectroscope.

spectra if you mount your grating so that it is illuminated by a narrow slit and stray light is excluded (Fig. 12.24). Interesting spectra to observe are those of a light bulb, of fluorescent lights, of black lights, and of sodium lights used for highway illumination. Also note the effect of colored filters or of a glass of colored liquid held in the light beam.

You can photograph the grating pattern by holding the grating closely in front of your camera lens, focusing and exposing as you would normally. Scenes with sharp highlights acquire new colors. For colored rings around each bright spot, such as Christmas tree lights, rotate the grating about the line of sight while taking a time exposure. It does not matter if you jiggle the grating a little while rotating it, but the camera should be firmly mounted on a tripod. Why do you get rings, and why are they centered around the bright spots (rather than the center of the picture)?

*B. Interferometers

Interferometers are devices that make accurate measurements by the use of interference. For example, if we send one of two coherent beams through a region in which the air temperature varies, then the corresponding small variation in light speed will change the arrival time of this beam. Even if this change is as small as 10^{-15} sec, it will be sufficient to alter the interference with the other beam. The pattern may change from constructive to destructive, say, and thus the temperature variation will become visible as an altered fringe pattern.

Interferometers usually split the amplitude of the incident light by means of a half-silvered mirror called a **beam splitter.** The two coherent beams may then travel in different directions and separate (Fig. 12.25). Fully reflecting mirrors or optical fibers bring the beams back into the same region. An additional beam splitter may be used to make part of each beam travel in nearly the same direction, so that interference fringes can be seen easily. Depending on the application, many different arrangements of mirrors have been designed following this general scheme.

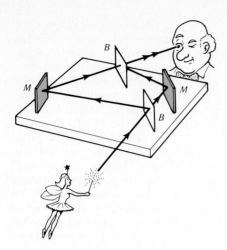

FIGURE 12.25

Principle of an interferometer, consisting of two partially silvered mirrors (beam splitters) *B* and two fully reflecting mirrors *M*. The two emerging coherent beams cross at a small angle, so that interference fringes can be seen.

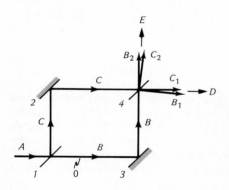

FIGURE 12.26

The Mach-Zehnder interferometer. Beam splitter *1* splits the original beam *A* into the coherent beams *B* and *C*. Mirrors *2* and *3* bring the beams back together, but crossing at 90°. Beam splitter *4* recombines the beams so that half (*B₁*, *C₁*) of each travels in the direction *D*, and the rest (*B₂*, *C₂*) in the direction *E*. If there is destructive interference at *D*, there will be constructive interference at *E*—the energy of the light is conserved, so it has to go somewhere.

For example, Figure 12.26 shows the ***Mach-Zehnder interferometer.*** If the mirrors are exactly at 45° you see only a uniformly bright field from *D*. However if mirror *2*, say, is rotated slightly, the path length of beam *B* depends on where it reflects from mirror *2*, and so does the interference. Therefore, if you now look from *D*, you will see fringes in the form of parallel lines perpendicular to the plane of the figure, located at mirror *2*.

With the beams so separated, we now have the opportunity to take anything that affects light, such as variations of air temperature due to a hot body or a shock wave from a speeding bullet, and place it at *0*. The air temperature variation is ordinarily invisible because it only affects the index of refraction, not the transparency of the air. But these variations do affect the speed of light, and hence the arrival time of beam *B*. Therefore, if we photograph the object from *D* through the interferometer, the resultant picture (e.g., Fig. 12.27) shows any visible parts of the object together with fringes distorted from parallelism by the air's varying index of refraction. Thus, effects like temperature variations and shock waves can not only be made visible but even quantitatively evaluated in

FIGURE 12.27

Photograph of a hot wire taken through a Mach-Zehnder interferometer. The wire runs perpendicular to the page, and the interference pattern shows the way the hot air forms around the wire.

terms of the amount of phase change, since this is related to the fringe displacement.

C. Multiple layers of thin films

The pretty colors and narrowly defined angles of the light from a grating stand in sharp contrast with the faint color and broad angle of interference obtained from a thin film, such as that of a lens coating. One difference is that a grating involves interference from *many* effective sources, whereas the film involves only *two*. The effect of many-source interference can be obtained from films as well, however, by stacking many films on top of each other. If the film layers have alternating high and low indexes of refraction, then at each interface there will be reflections that are alternately hard and soft. By choosing the film thickness appropriately (for a particular wavelength), you can make all the reflected beams be in phase (Fig. 12.28). You can consider these beams as coming from many, coherent, virtual images of the light source (Fig. 12.29). Except for small effects due to the fact that these virtual images become suc-

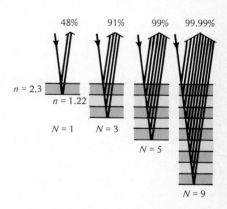

FIGURE 12.28

Multiple reflections from a stack of films with indexes of refraction alternating between 2.3 and 1.22. The thicknesses of the films are chosen so that all reflected beams are in phase. The resulting reflectance for normal incidence is shown for different numbers of films, *N*, in the stack.

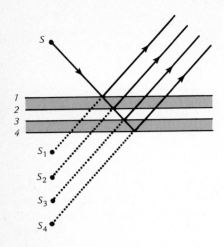

FIGURE 12.29

The effective coherent sources whose interference causes the selective reflections from film layers.

cessively fainter, the interference of the beams is similar to the many-source interference we discussed in Section 12.2G, for example in its wavelength selectivity.

Note in Figure 12.28 that not very many films are needed to get nearly all the incident light reflected, even though each interior interface by itself reflects only 9%. The high reflectivity of multiple-layer films of course occurs only near the particular wavelength for which the thickness of the layers was designed, but at this wavelength the film layers can be a better mirror than a metal surface (which cannot reflect much more than 90% of the incident light). Such multiple layers, sometimes on top of a metal surface, are used where high reflectivity at one wavelength is needed, for example in lasers (Sec. 15.4). Film layers acting as mirrors also occur in animals (which cannot easily grow metal surfaces) in reflection eyes (Secs. 3.3D and 6.4D). Similarly, the silvery scales of many fish are made reflective by such multiple layers. These mirrors can reflect a *range* of wavelengths by having a variety of layer thicknesses.

Other animals (as well as inanimate objects) have multiple layers of more uniform thickness, so that

only a narrower range of wavelengths is reflected. Such layers then appear in brilliant, iridescent colors. Like the colors caused by gratings, these colors vary with angle of view: the more obliquely the film layers are viewed, the more the color shifts toward the blue.

PONDER

Why toward the blue? Keeping the virtual images in Figure 12.29 fixed, redraw the picture for several angles of incidence and draw the path difference ε in each case—doe ε increase or decrease when the layers are viewed more obliquely?

An interesting example is the tortoise beetle (*Aspidomorpha tecta*, Fig. 12.30, Plate 12.5), which has multiple layers in its wing cases, making them iridescent. The beetle can vary the moisture content of

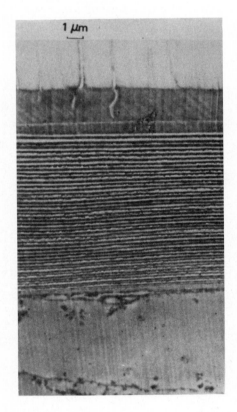

FIGURE 12.30

Electron micrograph of interference layers in the wing cases of the tortoise beetle (Plate 12.5). The spacing of the layers is about 200 nm.

these thin films, so as to change the layers' thicknesses, and hence the wavelength of the reflected light. Thereby it can change its color from pale gold to reddish copper. Among the iridescent inanimate objects, pearls and abalone shells (mother-of-pearl) derive their luster from layers of semitransparent nacre. What Sylvia Plath called the "rainbow angel's finger-nail interior" of the shell of the blue mussel is similarly formed. Some minerals, such as labradorite and mica, are deposited in many thin layers and therefore have a bright and shiny appearance.

It is not surprising that artists have tried to incorporate these subtle interference colors in their work. Larry Bell, for example, makes thin films of minerals on glass squares, which he then uses to form the faces of an iridescent glass cube. The great glassmaker Louis Comfort Tiffany incorporated thin films within the glass to produce his splendidly iridescent favrile glass.

D. Standing waves

Interference can also occur when two waves meet head-on. Consider the two equal, coherent, in-phase sources in Figure 12.31. *Between the sources, their waves meet head-on and are therefore sometimes in phase and sometimes out of phase*, as are the waves of the flip pages 311–431. However, there are some places where the two waves always interfere destructively—they always cancel at these places. These places are called **nodes*** and are half a wavelength apart. Midway between the nodes the interference is always constructive, so the amplitude of oscillation is largest at these points—the **antinodes.** At other points the nature of the interference varies. Such a combination of two waves traveling in opposite directions is called a **standing wave,** because it does *not* travel.

---

*Latin *nodus*, a knot, such as that by which the string of Figure 2.14 is tied to the wall.

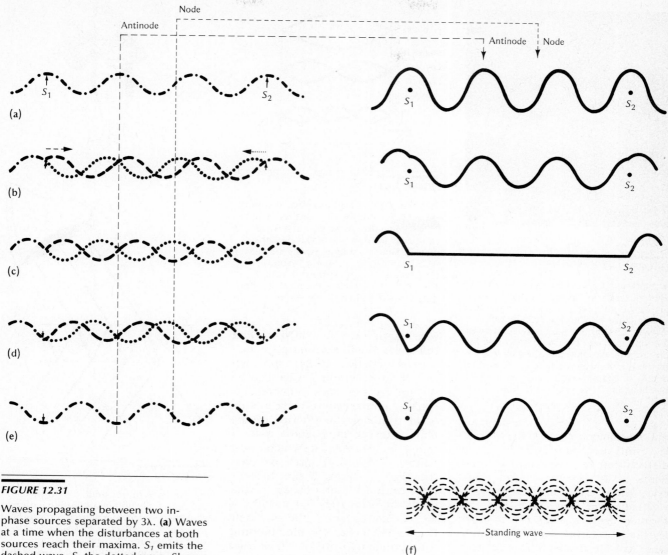

FIGURE 12.31

Waves propagating between two in-phase sources separated by 3λ. **(a)** Waves at a time when the disturbances at both sources reach their maxima. S_1 emits the dashed wave, S_2 the dotted wave. Shown are the individual waves and (to the right) their sum. **(b)** to **(e)** At later times, at intervals of an eighth of a period. **(f)** Between the sources the waves add to form a standing wave, as shown in the "multiple exposure."

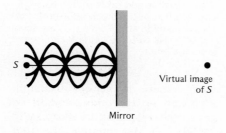

FIGURE 12.32

"Multiple exposure" of a standing wave of light caused by an incident beam that is reflected back on itself by a mirror.

The two sources need not be real. One can be a virtual image of the other—its reflection. The incident and reflected waves then meet head-on and interfere. If the reflection is soft, the two sources are in phase, so there is an antinode at the reflecting surface, which lies equidistant from the sources. As in Figure 2.14, a hard reflection makes a node at the reflecting surface (Fig. 12.32). You can easily set up such standing waves by shaking a rope that is tied at one end.

A photographic process using this phenomenon earned Gabriel Lippmann the Nobel Prize in 1908. A mirror surface is coated with a thick layer of fine-grain photo-graphic emulsion. When monochromatic light is reflected by the mirror, a standing wave is set up in the emulsion. Since there is no wave amplitude at the nodes, there is no exposure there (Fig. 12.33). The exposure occurs at the antinodes—in thin layers parallel to the mirror, separated by half a wavelength. The film is developed so that the exposed and unexposed layers are given different indexes of refraction. These layers then behave like multiple layers of thin film. Because of their spacing, these layers will preferentially reflect light of the same wavelength as that used for exposure. Such a **Lippmann plate** can take a photograph in color that

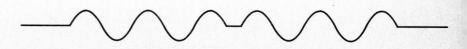

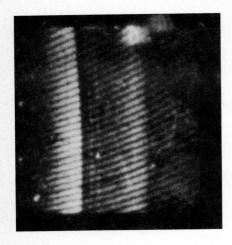

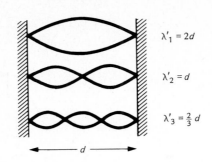

FIGURE 12.34

Standing waves of different wavelengths (λ'_1, λ'_2, λ'_3) set up by the hard reflections at two reflecting surfaces. Such waves can occur for only those wavelengths that fit. Similar standing waves are set up when a musical instrument is sounded; for example, on the strings of a violin. The particular wavelengths possible then determine the pitch of the sound.

FIGURE 12.33

A standing light wave causes regions of light and dark in a photographic emulsion, here reproduced from the first such picture obtained in 1890 by Otto Wiener. The distance $\lambda/2$ between nodes was enlarged sufficiently to be visible by placing the very thin emulsion almost parallel to the wavefront. Recording such details of a light wave photographically has become commonplace in holography (Fig. 14.9a)

not only looks natural to the human eye, but that even has the same distribution of wavelength components as the original light! However, these iridescent colors vary with angle of view, and—like the tortoise beetle's wing cases—with the moisture content of the emulsion (which makes it swell, and hence not so swell as a color photograph). Therefore the process remained a curiosity until the advent of holography (Chapter 14).

If a standing wave is set up by a source and one mirror, what happens between *two* mirrors? Suppose a single thin film is made reflective on both sides—for example, by coating them with thin layers of silver. Of course, since the film's first surface is now more reflective, it is more difficult for light to enter the film in the first place; but once it does, it bounces back and forth between the two reflective surfaces. These reflections inside the film interfere with each other and set up a standing wave. The hard reflections require that there be a node at *each* surface. For this to happen, the

wavelength must just "fit"—the film thickness must be a multiple of half a wavelength (Fig. 12.34)—in which case a large standing wave can be set up in the film; otherwise it can't. (Another way to look at this is to think of the interior of the film as an oscillator whose natural frequency is that at which light bounces back and forth between the film's surfaces. If the incident light contains that frequency, this oscillator resonates and absorbs a lot of energy from the incident beam.) The energy in the standing wave builds up until the small fraction of it leaking out on the far side (because the surfaces are not perfectly reflective) just equals the energy being fed in. Thus, at this particular wavelength (resonance frequency) the film is completely transparent.

Viewed from the far side, the multiple reflections (Fig. 12.35) appear as interfering beams from many coherent virtual sources. At just the right wavelengths *all* the virtual sources are *in phase*, so all the incident intensity is transmitted. Nothing is reflected, then, because the beam reflected from the first surface *(R)* is out of phase with *all* the beams reflected by multiple reflections. Hence, there are narrow ranges of wavelengths (characteristic of many-source interference)

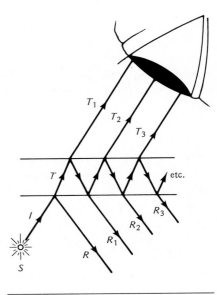

FIGURE 12.35

Interference of transmitted light by multiple reflection in a thin film with two reflective surfaces. A source *S* sends light *I* toward the reflective first surface. Most of it is reflected *(R)*, but some is transmitted into the film *(T)*. Most of that bounces back and forth between the film's surfaces, but it looses a constant fraction at each bounce. Half of these beams (T_1, T_2, . . .) go to the side opposite the source and get to the eye. The intensity seen will usually be quite low; however, if *all* of these beams are *in phase*, they add up to an intensity equal to that of the incident beam *I*. (*All* the reflected beams R_1, R_2, . . . are then *out of phase* with the beam *R* and cancel it completely.)

where there is perfect transmission and no reflection, and in between there is no transmission (Fig. 12.36). As you change the incident angle, the transmitted colors change, like iridescent colors.

Films with highly reflective surfaces therefore are very good, selective color filters. Called **inter-**

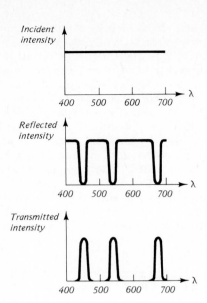

FIGURE 12.36

Incident, reflected, and transmitted intensity-distribution curves for a thin film with two reflecting surfaces.

ference filters, they can be constructed by mounting a thin film between partially silvered glass plates (Fig. 12.37). Such a filter can be designed, say, to transmit only the wavelength characteristic of sodium. It will then reflect all other wavelengths and can therefore be used, for example, as the beam splitter in the sodium traveling matte process of movie film (see the FOCUS ON Special Effects in the Movies). By combining several films and metal layers of suitable reflectivities, interference filters with almost any transmittance curve can be made. For example, **dichroic filters** transmit one region of wavelengths and reflect the rest, thus splitting a white beam into complementary colors. Such filters are used in color TV cameras to separate the image into its three primary color components. Since they absorb no light, no light intensity is lost (unlike ordinary subtractive color filters, which absorb some of the energy of the incident light, converting it to heat). This lack of absorption also makes dichroic filters suitable for floodlights, because they do not heat up.

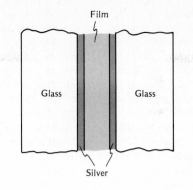

FIGURE 12.37

Construction of one type of interference filter.

Another application of interference by multiple reflection is a very sensitive measuring device called a **Fabry-Perot interferometer.** Here the film consists simply of the air between two highly reflecting plates with adjustable spacing. This device is usually operated with the plates many wavelengths apart, so that a pattern of very sharp fringes is seen (each fringe corresponding to a different incident angle from a broad light source). It can then be used to measure motions of one of the plates by a fraction of a wavelength, because the fringe pattern then shifts by a fraction of a fringe. Alternatively, the Fabry-Perot interferometer can be used to detect very small differences in wavelength, because the corresponding small differences in the very sharply defined fringe pattern can be detected.

With the help of two very general principles due to Babinet and Huygens, we can gain an understanding of a host of situations in which the wave nature of light shows itself. Both principles can aid us in determining the light distribution obtained on a screen when we shine coherent illumination through openings in an opaque plate. We'll shine parallel light through openings of arbitrary number and shape and observe the pattern of the interfering light (we call it the **screen pattern**) on a distant screen (Fig. 12.38). Since the plate is opaque, *immediately* behind it there is no light wave at all. Similarly, *just* behind any opening in the plate, the light is the same as it would be if there were no plate present—the wave in the opening is just the incident wave.

A. Babinet's principle

Babinet's principle compares the screen pattern produced by a plate

FIGURE 12.38

Set-up to which Babinet's and Huygens' principles can be applied. Lens 1 makes the illuminating light parallel. Lens 2 allows us to bring the screen close and see on it the screen pattern that would otherwise be on a distant screen, as in Figure 12.8.

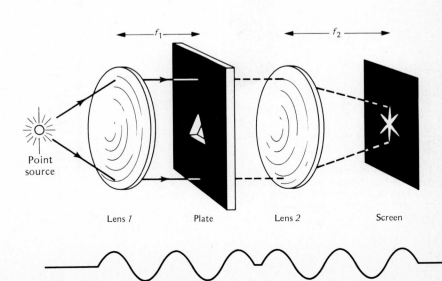

with that produced by the negative of the plate (in the sense of photography). That is, we have two plates, *1* and *2*, and wherever plate *1* has an opening (i.e., is transparent), plate *2* is opaque, and vice versa. Such plates are called *complementary*. (Plate *1* might be made by cutting holes in an originally opaque plate, and plate *2* would be made by freely suspending the cut-out parts in their original relative positions.) If we could combine the *openings* of the two plates, we would get a completely transparent plate.

In order to relate the screen pattern obtained with plate *1* to that obtained with plate *2*, we imagine a plate *3* that combines the openings, letting light pass where *either* plate would (Fig. 12.39). Since lens 2 fo-

cuses the parallel light, the screen pattern of plate *3* is simply the geometrical-optics image of the point source, that is, a point of light on the axis, and darkness everywhere else. We can think of this overall darkness (except for the on-axis point) as resulting from the *sum* of two waves, one from the openings of plate *1* and one from those of plate *2*. From this point of view, the darkness almost everywhere must be due to complete destructive interference of the waves from the two complementary plates. For the darkness to occur, then, the two waves must be out of phase and *have the same amplitude* at each point on the screen (except for the on-axis point). This somewhat surprising result is called **Babinet's principle:**

The screen patterns due to complementary plates are the same, except at the geometrical-optics image of the light source.

Let's check this out in a few situations where we already know the answer. Suppose plate *1* is completely opaque, without any holes.

Then the screen is of course completely dark. The complementary plate *2* is open everywhere and *its* screen pattern is just the image of the point source, which is *also* dark everywhere, except at this image—just as Babinet's principle predicts.

Next, suppose that plate *1* is a grating with many slits, having transparent and opaque strips of *equal* width. The complementary plate *(2)* looks just like plate *1*, except that it is shifted by the width of a slit (Fig. 12.40a). We already know that the screen pattern from

---

FIGURE 12.40

Complementary gratings and the intensity of actual screen patterns due to them. Of interest are not the details of any one pattern, but the relation between the patterns of complementary plates. (We often plot the intensity as a function of position on the *same* diagram, as is done here.) **(a)** A plate with open and closed spaces of equal width is identical with its complement. **(b)** A plate with thin slits gives the same screen pattern as a plate with very wide slits, except for the central fringe. The central fringe is much brighter in the latter case because more light gets through the larger openings.

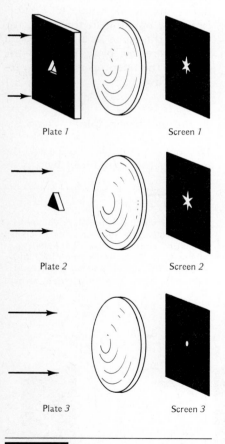

Plate *1*

Screen *1*

Plate *2*

Screen *2*

Plate *3*

Screen *3*

FIGURE 12.39

A plate *1*, its complement *2*, and the combined openings *3*. Babinet's principle states that the patterns on screen *1* and screen *2* are identical except at the center.

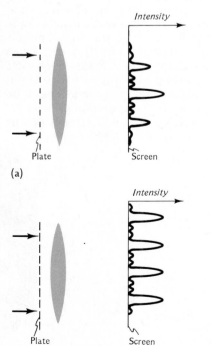

Intensity

Plate   Screen

(a)

Intensity

Plate   Screen

(b)

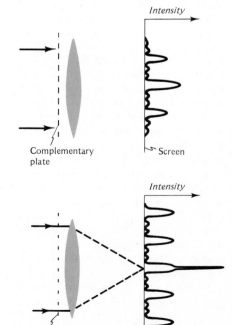

Intensity

Complementary plate   Screen

Intensity

Complementary plate   Screen

such a grating depends only on the angle at which the light leaves the grating, not on the exact position of the grating. So the screen pattern is the *same* for this grating as for its complement, again in agreement with Babinet's principle. But now we can learn something new. Since complementary plates give the same pattern even if they are *not* identical to each other, the transparent and opaque spaces need not be of equal width in plate *1*. Thus a grating with narrow slits and one with narrow opaque lines of the same spacing will give the *same* screen pattern (except for the central fringe), as in Figure 12.40b.

We can use Babinet's principle to find out what happens when light passes through a *small hole in a plate.* First, consider the complementary plate, which consists only of a *small opaque speck* and is open everywhere else. Most of the wave goes through this plate undisturbed, and forms an image of the source as if the speck weren't there. But if the speck is opaque, it must absorb *some* of the wave's energy. The only way the wave can give up this energy is by doing work shaking the speck's charges. The oscillating charges then emit electromagnetic waves in all directions. So the speck emits a "scattered" wave in all directions, which we can see on the screen *in addition* to the source image (Fig. 12.41a). Next,

consider the original plate, with a single small hole located where the speck was. By Babinet's principle, the screen pattern due to the plate with a tiny hole must be the same as that of the speck (except for the on-axis point). Therefore, a tiny hole must *also* emit a wave in all directions (here *without* the strong, undisturbed wave that forms the on-axis source image, Fig. 12.41b). In previous sections we have assumed that a small illuminated hole acts like a point source, emitting light waves in all directions. Now we have proved it! (Similarly, a narrow illuminated slit acts like a line source.)

B. Huygens' principle

We've just seen that the wave coming out of a little hole looks like that from a little source. Huygens' principle says that the waves coming through the holes in *any* plate look just like those from a suitable collection of coherent, little sources located where the holes are. Think of water waves—as a wave travels through the water, it moves the water up and down in one region, and this in turn pushes the water in neighboring regions up and down. It would seem to make little difference whether the motion of the water in each region is due to that in its neighboring regions, or whether each region of the water is moved up and down by external "sources," such as the feet of many children on a dock, paddling in unison. The waves propagating into previously still water are the same in the two cases. **Huygens' principle** states that there is in fact no difference for

the region into which the wave is propagating:

> **Any wavefront of a traveling wave can be replaced, as far as effects further along in the propagation direction are concerned, by a lot of sources located uniformly all over the wavefront, radiating in phase.**

Huygens' principle for light can be viewed in another way, by imagining each of the openings in a plate divided into many small pieces. Each small piece by itself acts like a small hole in an opaque screen. Above we saw that small holes emit waves in all directions toward the screen. So the wave traveling through the openings in a plate acts as if *each point* in the openings is emitting such waves (Fig. 12.42).

Huygens' principle can be used to find the screen pattern for a plate

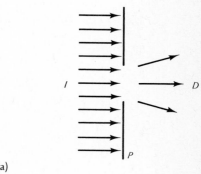

(a)

(b)

FIGURE 12.42

Huygens' principle. **(a)** Incident light *I* falls on a plate *P*, and some of it gets through, *D*. **(b)** The light *D* can be considered as coming from a large number of in-phase sources, S_1, S_2, . . . located all over the holes in the plate.

FIGURE 12.41

Babinet's principle relates **(a)** the scattered wave from a small opaque particle to **(b)** the wave that spreads out from a small hole in a plate. Wavefronts are shown.

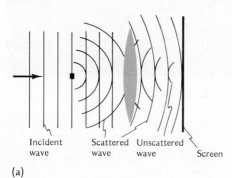

Incident wave   Scattered wave   Unscattered wave   Screen

(a)

Incident wave   Diffracted wave   Screen

(b)

with openings of any shape. We forget about the actual light that illuminates the plate. Instead, wherever there is an opening in the plate we imagine in-phase sources uniformly distributed, radiating toward the screen. The interference of their waves gives a screen pattern the same as that from the original plate. (Of course, the actual addition of the light from all these sources, each with its correct phase, may be a formidable task.)

Huygens' principle also implies a new concept of light *propagation*. So far we have considered propagation and interference as two separate things that light can do. We

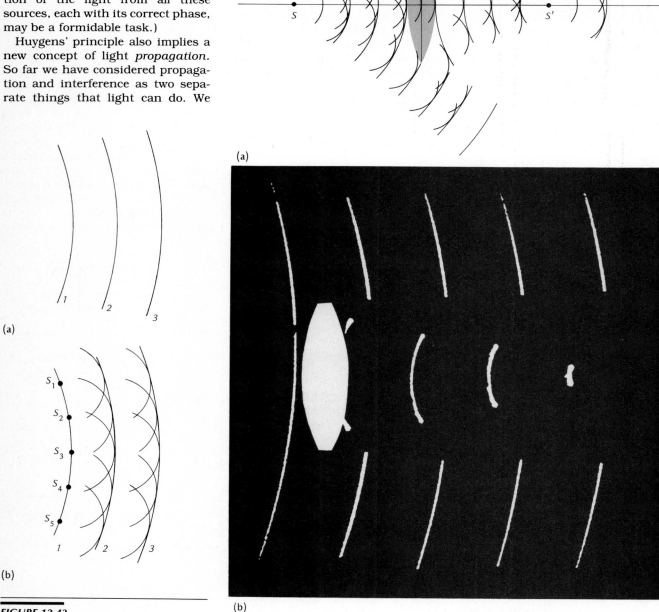

(a)

(a)

(b)

FIGURE 12.43

(a) Wavefronts propagating. (b) Huygens construction to find a later wavefront *2* from an earlier wavefront *1*. Each point on wavefront *1* emits a Huygens wavelet. Wavefront *2* is where these wavelets interfere constructively.

(b)

FIGURE 12.44

(a) Huygens construction for a wave from a point source *S* passing through a converging lens to an image *S'*. (b) A composite photograph of the wavefronts (see Sec. 14.4).

have propagated two or more coherent beams by geometrical optics, and interference only happened when they met. Now we see that interference can be thought to occur *all the time* while a beam propagates; every wavefront sends out a lot of new wavelets, and they interfere in the next instant to form the new wavefront. For example, consider the wave of Figure 12.43a. Notice that it has just passed through a completely transparent plate (provided by our imagination). Huygens says we would get the same result from a set of in-phase sources at the location of the plate (Fig. 12.43b). The interference of the light from these sources must be just such as to recreate the original wave. The *rays*, perpendicular to the wavefronts constructed for us by Huygens, will follow the law of geometrical optics wherever it is valid—geometrical optics is a special case of wave optics, valid as long as the holes and obstacles are very large compared to the wavelength of light.

No one in her right mind would attempt to solve geometrical-optics problems by adding a lot of Huygens wavelets, since the geometrical construction is so much simpler. Yet it is often useful to know that the wave viewpoint is always available. For example, take a point source S and form an image S' of it by means of a lens (Fig. 12.44). In terms of waves, the reason S' is bright (and everywhere else is dark) must be that there is constructive interference at S' (and destructive interference everywhere else). Hence the waves emitted in all directions from S must arrive *in phase* at S'. Now, the rays that get to S' via the edge of the lens have further to travel than those that travel via the center. But the central rays have to travel the largest distance through glass, where their speed is smaller than in air; thus the central thickness of the lens just compensates for the outer rays' greater path length, so that all the light arrives in phase at S. This new viewpoint will allow us to get some unfamiliar results from familiar lens situations (Sec. 12.5D).

12.5
DIFFRACTION

We have seen that if a beam of light is narrow enough, effects that are not described by geometrical optics become important. We call these **diffraction** effects. Thus, diffraction effects are interference effects within a single beam—light passing through a small hole spreads out. You can look at another diffraction effect right now. Take a sharp edge (of a knife blade or sheet of paper) and hold it between your eye and a small, bright light source. Focus on the edge and note that it is luminous near where it covers the light source—it seems to glow even when the knife completely covers the source. You are actually seeing the Huygens wavelet "emitted" *just* above the edge. (You see no wavelets produced by the bulb from places farther above the edge because of destructive interference with waves from the places between them and the knife.)

A. Diffraction from a slit or a hole

Most light beams are confined not by just one edge, but by at least two edges forming a slit. When light from a point source is transmitted by such a slit, there are two places we can put a screen and get a simple pattern. One location is very close behind the slit; since there is intensity only behind the slit, the screen pattern in this case is a good

geometrical-optics shadow of the slit. The other location is very far from the slit (or, equivalently, at the focus of a lens as in Fig. 12.8—see Fig. 12.45). The diffraction effects seen on the distant screen are called **Fraunhofer diffraction.** In the region of intermediate screen distances the diffraction pattern makes a transition between geometrical-optics shadow and Fraunhofer pattern as the distance increases. These transitional patterns are called **Fresnel diffraction** and are quite complicated—the luminous edge mentioned above is one example. Except in the first TRY IT, we won't discuss Fresnel diffraction further.

Let's apply Huygens' principle to the slit of Figure 12.45. Imagine the aperture divided into many parts, say 100, and illuminate the slit with monochromatic light of wavelength λ. Each part of the slit emits a Huygens wavelet. Thus we can consider the slit to be just like the 100 sources of Section 12.2G. (Here we are interested only in the *central* fringe. When we divide the slit into more and more parts, the spacing

---

FIGURE 12.45

Two set-ups for observing Fraunhofer diffraction, using a lens instead of a distant screen. Parallel light falls on an opening in a plate, say a slit of width *b*. **(a)** The pattern is seen on a screen at the lens's focal plane. **(b)** The pattern is directly projected on the eye's retina, which serves as the screen through the focal point of the eye's lens.

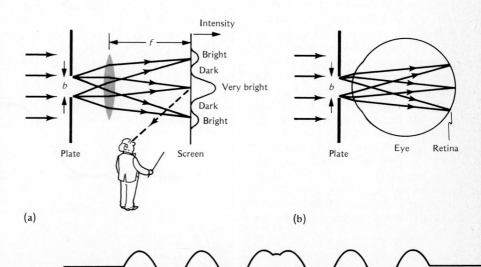

(a)

(b)

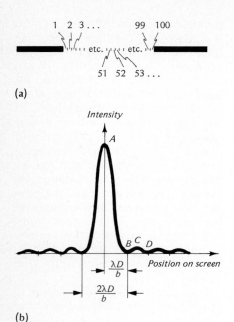

(a)

(b)

FIGURE 12.46

(a) A slit, considered as divided into 100 parts, is equivalent to 100 Huygens sources of light. **(b)** The intensity distribution from this slit. *A* is the central maximum, where waves from all the sources interfere constructively. *B* is the first zero, where waves from halves of the slit cancel one another. *C* is the next maximum, where waves from the first third cancel those from the next third, leaving the last third uncanceled. *D* is the next zero, where waves from successive quarters of the slit cancel each other.

between Huygens sources becomes smaller and smaller, so the higher-order fringes move away, out of the picture.) As before, the intensity first reaches zero near the central fringe when the contribution of source *1* is out of phase with that of source *51*, and so on (Fig. 12.46). That is, the zero occurs at the same place as it would for Young's fringes from sources separated by the distance b/2 between sources *1* and *51* (or between sources *2* and *52*, etc.)—at the first-intensity zero the waves from one half of the slit exactly cancel those from the other half. In Section 12.2E we saw that the distance between Young's fringes (or between zeros of the fringe pattern) was $\lambda(D/d)$, where *d* was the source separation. Hence,

since the source separation here is $d = b/2$, the distance between the two zeros on either side of the central maximum is $2\lambda(D/b)$, or:

$$\textbf{Distance from central maximum to first intensity zero} = \lambda \, \frac{\textbf{D}}{\textbf{b}}$$

As we move even further from the central fringe, the screen intensity increases again, reaching a maximum approximately where the path differences are such that sources *1* and *34* of the slit contribute out of phase, and so do sources *2* and *35*, etc., but sources *67* to *100* remain uncanceled. This first-order fringe is not as bright as the central fringe because the waves from the first third of the slit cancel those of the second third, leaving only the last third uncanceled. Next, another zero occurs when sources *1* and *26* cancel, as well as *2* and *27*, etc., up to *25* and *50*. Further, *51* and *76* cancel, etc., up to *75* and *100*. (There is cancellation by quarters—the first and second quarters cancel, and likewise the third and forth quarters.)

Thus we get a sequence of maxima and zeros on both sides of the

central maximum, but the intensity of the side maxima decreases with distance from the central one. This intensity distribution is shown in Figure 12.46b.

PONDER

What changes in this argument if we imagine the slit divided into 1000 parts, instead of 100 parts? In particular, is the distance to the first zero unchanged?

As in the case of the grating, the spacing of the fringes is proportional to the wavelength λ. If white light is used, the first- and higher-order fringes therefore appear in a sequence of colors (Plate 12.6) similar to those of Young's fringes (Fig. 12.13). Additionally, the fringe spacing varies inversely with the slit width *b*. If we make *b* very large, the higher-order fringes are imperceptibly close to the central one—

FIGURE 12.47

(a) Fraunhofer diffraction pattern from a wide slit. **(b)** Fraunhofer diffraction pattern from a narrower slit, $\frac{1}{4}$ as wide as in **(a)**.

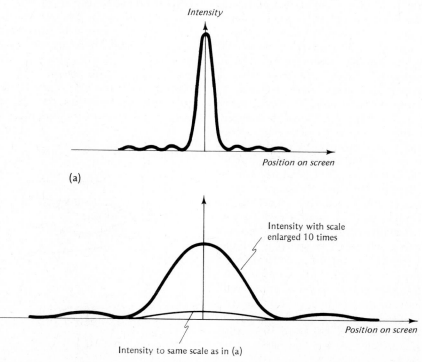

the screen pattern is essentially a point—the good geometrical-optics image of the point source we expect when the slit is very wide. As the slit is made narrower, the pattern expands in the direction crosswise to the slit (Fig. 12.47). You can see this **diffraction pattern** right now, by using your eyelids to form a slit. Look at a small, bright source of light (a light bulb or a candle flame). Squint, closing your eyelids until only a narrow slit remains open. (Use only one eye, and cover the other.) The streak of light crosswise to the edges of your eyelids is the diffraction pattern. As you tip your head, it tips with you.

Very thin slits with very wide central fringes approximate well the sources radiating in all directions that we considered in Section 12.2. If any of the arrangements discussed there (Young's fringes, grating pattern, etc.) are produced by somewhat *wider* slits, then the screen pattern will still have zeros where there is destructive interfer-ence *among* the slits. *In addition,* however, it will also show the single-slit diffraction zeros, where there is destructive interference *within* each slit. Thus a realistic Young's pattern (as you saw in the TRY IT for Sec. 12.2D, E, and F) is really a broad single slit pattern within which are the fine lines of Young's fringes.

If the beam is restricted in *two* directions, the pattern expands in two directions as well. If the aperture is a circular hole rather than a slit, the fringe pattern is a series of rings around the geometrical-optics image of the source (Fig. 12.48, and the second TRY IT). Its intensity depends on the distance from the center of the diffraction pattern in nearly the same way as the vertical slit's intensity depends on the horizontal position.

FIGURE 12.48

Photo of Fraunhofer diffraction pattern from a circular hole.

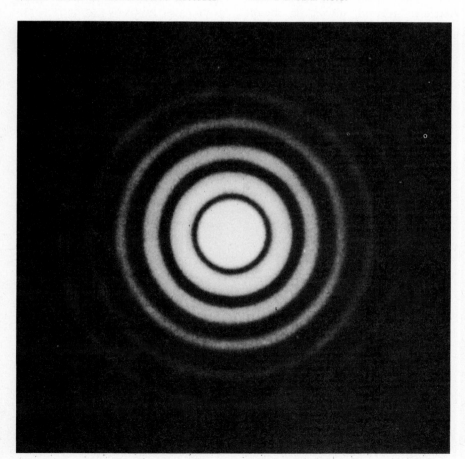

First TRY IT

FOR SECTION 12.5A
Fresnel diffraction

With a good point source of light you can see fringes in ordinary shadows. When the screen on which you display the shadows is relatively close to the object, you see Fresnel diffraction, and as the screen is moved farther away, the pattern changes into the Fraunhofer diffraction pattern.

You can easily see the Fresnel patterns using the coherent light from a laser beam as a source. To see it in ordinary light, mask the projection lens of a slide projector by aluminum foil with a pinhole (of diameter about $\frac{1}{2}$ mm). Shift the foil and change the focus until you get the brightest disk of light projected on a distant screen.

Use this light beam to cast shadows of simple objects that have sharp outlines, such as a razor or knife blade, a pin with a large round head, etc. Vary the distance of the object from the screen and observe the change in the pattern. The thin lines surrounding the main shadow of these objects are due to Fresnel diffraction. With a straight edge, such as a razor blade, observe that the intensity drops off gradually toward the dark side (there is some intensity inside the geometrical-optics shadow—this is the light from the luminous edge). Outside the geometrical-optics shadow there are fringes, which you can think of as interference between the light forming the geometrical-optics shadow and that from the luminous edge.

At the center of the shadow of the shaft of a pin you see constructive interference of the light from the two luminous edges of the shaft. Fresnel predicted, on the basis of theory, that an opaque sphere (pin head) should have a bright spot at the center of its shadow. (At Fresnel's thesis defense, Poisson opined that the result was clearly ridiculous, hence the dissertation must be wrong. This "ridiculous" phenomenon henceforth carries the name "Poisson bright spot.") The spot is not easy to see—but try it. Cast the pinhead's shadow on a piece of tissue paper and view the back of the paper with a magnifying glass, looking into the projector. Hold the pinhead so it is at the center of the brightest part of the beam from the projector. Have a friend move the tissue back and forth rapidly so as to blur the irregularities of the tissue's fibers.

Another way to see Fresnel diffraction patterns is by reexamining the floaters you saw in Section 5.2, but this time illuminated by a pinhole. (The purpose of the pinhole is to provide a point source, which will give coherent light over a region larger than the floater's size.) Make a very small pinhole in aluminum foil, for example, by crumpling and then flattening it, picking one of the smaller holes so produced. Hold the pinhole close to your eye, but focus your eye beyond it—don't look at the pinhole— and wait for floaters to drift past. You should see the Fresnel fringes around the outline of the floater, and maybe the Poisson bright spot. If you vary the pinhole distance from your eye, you can verify that the floaters must be close to the retina, because their sizes do not changes appreciably.

Perhaps the simplest way to see Fresnel diffraction is to form a narrow slit between two adjacent fingers. Hold the fingers about 5 cm from your eye, and look through the slit while focusing on a distant light source. The dark lines within the slit are the Fresnel pattern.

Second TRY IT

FOR SECTION 12.5A
Diffraction pattern of a hole

To see the fringe pattern of a circular hole you need a small pinhole in aluminum foil and a point light source. Place the aluminum foil on a piece of heavy paper and gently press the tip of a pin on it to make a tiny hole. For the light source, mask a light bulb with a larger pinhole or use a distant street light or the sun reflecting in a distant shiny object. Holding the pinhole very close to your eye, look through it at the light source and notice the pattern of diffraction. Try pinholes of various sizes and see how the size of the rings varies. Also try pinholes of other shapes, for example, made by cutting the foil with the tip of a sharp knife.

*B. Coronas and glories

A single hole that is small enough to show appreciable diffraction lets very little light pass through it. If we arrange *many* identical holes in a *random* way, there is no interference among them, *only diffraction* from the individual holes—the screen pattern is simply the single-hole diffraction pattern, made brighter by the many holes. More-over, Babinet's principle says that diffracted light from a random array of small identical holes is the same as that from a random array of small identical disks. Such arrays are frequently found in nature.

For example, if you breathe on a cool window pane, moisture con-denses on the pane in approxi-mately equal-size drops at random positions. By looking through the condensed moisture and focusing your eye on a distant street light, you can project the diffraction pat-tern on your retina. The pattern is that of Figure 12.48, except that the amount of light at the geometri-cal-optics image is much brighter than shown there (as is usual for patterns derived by Babinet's prin-ciple). It appears as a series of con-centric rings about the light, called a *corona.* The size of these rings depends on the size of the drops, and if the drops are not quite uni-form, the rings tend to spread out and overlap. You frequently see this pattern around oncoming head-lights when mist condenses on your windshield.

The atmosphere between your eye and a distant light source, say the moon, often contains water droplets or *tiny* ice crystals of more or less uniform size. You then see a corona around the moon (Plate 12.7). Usu-ally it is just a fuzzy area of light near the moon (whereas a halo, due to refraction, is a much larger cir-cle—Sec. 2.6C).

PONDER

Why does the corona have the ring shape? (Recall how the circular shape of halos and rainbows comes about.)

The reflected rays of geometrical optics that form the rainbow are also affected by diffraction if the water drop is small enough. For very small drops the diffraction is most important and gives rise to co-ronas in reflected light, similar in appearance to heiligenschein. Oc-casionally the drop size is just right so that refraction, reflection, and diffraction effects reinforce each other, and you see bright, colored rings around the shadow of your head when you are on a high moun-tain above the clouds. This marvel-ous phenomenon is called the **glory** (Plate 12.8). (Local superstition in the Harz Mountains associates seeing the glory, called "ghost of the Brocken," with impending death— you are seeing yourself as you will be in heaven.) Like the heiligen-schein and the rainbow, your co-rona and glory belong to you alone; you can never see it around your friend's head. In *The War Lovers,* John Hersey describes this effect as "a shining nimbus, a halo, like a ring around the moon" that is formed around the shadow of his airplane, and he marvels about "the way the sun singled us out from the six shadows" of the other airplanes nearby.

When the water droplets are somewhat larger, a rainbow is seen in reflected light, but diffraction modifies its intensity by sending light to angles slightly different from those given by geometrical op-tics. For example, one or another color will be brighter than the rest, depending on the drop size. Since the colors in a rainbow overlap to some extent, this can lead to new hues in the mixture. It is even pos-sible to estimate the drop size from the appearance of rainbow colors; the smaller the drops, the more the individual colors are spread out and overlap, giving a more pastel, or even white, rainbow. Also, just *in-side* the primary bow there may be other, so-called **supernumerary bows,** caused by higher-order dif-fraction (see Plate 2.3).

C. Resolving power

In the light of diffraction, you may wonder about even as basic a geo-metrical-optics result as the focus-ing by a lens. And rightly so, for we know that any lens has a finite aper-ture. Even if there is no plate with a circular hole in front of it, the beam is limited by the size of the lens itself. So no actual lens ever forms a point image of a distant point source, even if all aberrations are removed; this image, in the focal plane, is instead the Fraunhofer diffraction pattern corresponding to the aperture of the lens. Each point of an extended object becomes more or less fuzzy in the image, accordingly as the aperture is smaller or larger. For many purposes this is a small effect and can be ignored, as we did in our previous chapters. However, this diffraction makes it *impossible to see very fine detail* in the image, and it is useless to enlarge the image beyond a certain point—you will only see enlarged Fraunhofer patterns.

To describe the **resolving power** of a lens we use a criterion due to Rayleigh; two sources are distinguishable (can be *resolved*) if the central maxima of their images are at least as far apart as the maximum and the first zero of one source (Fig. 12.49). That maximum and zero are separated by a distance proportional to λ/b, where b is the diameter of the lens' aperture

(Sec. 12.5A). Hence the bigger the aperture, the finer the detail that can be resolved. With the appropriate numerical factor, this means that

$$\theta_{\text{Rayleigh}} = 70\,\frac{\lambda}{b} \quad \text{(in degrees)}$$

is the *minimum* **angular separation** *that can be resolved*. Thus, whenever a light beam is constricted by a lens or mirror of finite size, some fine detail about the image is irretrievably lost. Points on the object separated by an angle less than θ_{Rayleigh} will blur together in the image (Fig. 12.50). So it makes no sense to correct aberrations much beyond this limit, nor to use a film with much finer grain to record the image.

Very cleverly, eyes are optimally designed with this limitation taken into account. The pupil of the human eye under bright light conditions has an approximate diameter $b \simeq 3$ mm, and it is most sensitive near $\lambda = 550$ nm. This gives a resolution limit $\theta_{\text{Rayleigh}} \simeq \frac{1}{80}°$. An actual measurement of the human eye's resolution gives about $\frac{1}{60}°$—pretty close to the absolute limit. (An eagle has a larger pupil, $b \simeq 10$ mm, so its resolution is better, about $\frac{1}{260}°$.) Next, let's see whether the graininess of the retina is consistent with the Rayleigh criterion. In the fovea the distance between cones is about

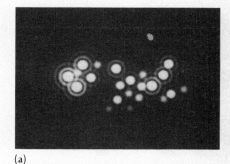

(a)

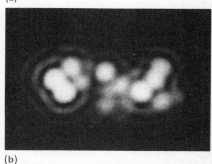

(b)

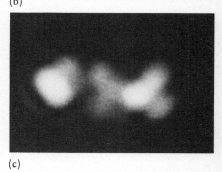

(c)

FIGURE 12.50

Photo of sources that are **(a)** resolved, **(b)** barely resolved, and **(c)** not resolved.

1.5 μm. From θ_{Rayleigh} we can find that the two diffraction images of two point sources must be at least 4 μm apart on the retina to be resolvable. In this distance there are three cones, which are just enough cones to see the two sources (two cones) with a darker spot between them (one cone) to be sure they are separate. It would do no good to have more cones, because they would see more detail of the diffraction pattern, but not of the sources. If there were fewer cones, they could not make use of all the detail of the object that the image on the retina contains. So the human eye is pretty well optimized.

FIGURE 12.49

The Rayleigh criterion for resolution. Shown is the intensity distribution for the focal-plane images of two distant, incoherent point sources. The individual sources' intensities as well as their sums are shown. **(a)** The two points are well resolved ($\theta > \theta_{\text{Rayleigh}}$), **(b)** they are just resolved ($\theta = \theta_{\text{Rayleigh}}$), **(c)** they are not resolved ($\theta < \theta_{\text{Rayleigh}}$).

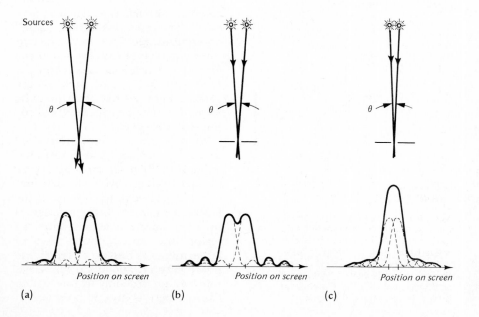

Sources

θ

Position on screen

(a)

θ

Position on screen

(b)

θ

Position on screen

(c)

95
874
2843

				distance equivalent
				$\frac{20}{800}$
				$\frac{20}{400}$
				$\frac{20}{200}$

638 ЕШЭ XOO — $\frac{20}{100}$

8745 ЭШ ОХО — $\frac{20}{70}$

63925 ШЕЭ XOX — $\frac{20}{50}$

42836 5 ШЕ Ш ОХО — $\frac{20}{40}$

37425 8 ЭШЭ x x o — $\frac{20}{30}$

9 3 7 8 2 6 ШШЕ x o o — $\frac{20}{25}$

ЕШШ o o x — $\frac{20}{20}$

DESIGN COURTESY J. G. ROSENBAUM, M.D., CLEVELAND, OHIO

FIGURE 12.51

A miniature eye chart. Read this chart from a distance of 14 inches to get a rough idea of your acuity for close vision. (The standard Snellen chart is read from a distance of 20 feet.)

Most people do not use their eyes primarily to resolve point sources. Other aspects of the eyes' acuity, besides resolving power, are important in connection with more common tasks. For example, **Snellen eye charts** (Fig. 12.51) measure how accurately you can read letters. The result is traditionally expressed as 20/X, where X is the distance (in feet) from which a normal eye can read the same size letters that you can read from 20 feet. A normal eye is considered one that can read letters subtending an angle of $\frac{1}{12}°$ containing lines of width equal to the average resolution limit of $\frac{1}{60}°$. Typical good eyes can actually read letters subtending $\frac{1}{24}°$, corresponding to 20/10 vision.

Other acuity tests involve objects closer in shape to some of the patterns for which the eye seems to have built-in detectors, such as lines. Whereas the resolution for two parallel lines is not much different from that for two points, two lines can be aligned to form one straight line without a break (Fig. 12.52) to an accuracy **(vernier acuity)** of only about 0.1 μm on the retina, much less than the size of a cone! (This ability is used, for example, in the split-image rangefinder, Fig. 4.5, and for precision measurements using a vernier.) Your eye does this by gathering information all *along* the lines (not just at the point of the break), for example by using its line detectors and by eye movements.

Insect eyes are designed entirely differently from vertebrate eyes, but they also take into account the limitations due to diffraction. An insect eye consists of a large number of facets, each of which acts as a lens and concentrates light from one direction onto a *single* photoreceptor. Each facet-photoreceptor unit, called an **ommatidium*** (Fig. 12.53a), points in a slightly different direction from its neighbor and accepts light from a small range of angles around the direction toward which it points. This range of an-

*Greek *omma*, eye, plus Latin *-idium*, a small structure or form.

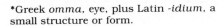
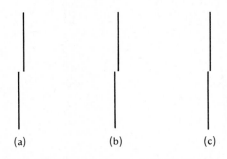

FIGURE 12.52

Measure your vernier acuity by viewing this figure and noting which pair of lines has a barely visible break between them. Move back until the break in **(a)** is visible, that in **(c)** is not, while that in **(b)** is just barely visible. With good visual acuity you can see this from about 10 m under these crude conditions.

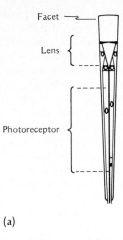

(a)

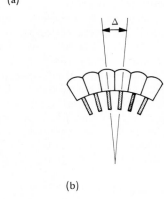

(b)

FIGURE 12.53

(a) Structure of an ommatidium. **(b)** Two adjacent ommatidia at the optimum difference in pointing angle, Δ.

gles, determined by both diffraction and by the finite size of the photoreceptor, should be as small as possible for good resolution. Such **compound eyes** are more efficient than our type of lens eye when the size of the head is small, because the entire surface of such an eye gathers light, rather than just the small fraction represented by the pupil of a lens eye.

What is the optimum facet size? If the facets are too large, not enough of them fit on the insect's limited head to give a detailed image (like coarse-grained film). If the facets are too small, the diffraction by their aperture allows light entering the ommatidium from a large range of angles to reach the photoreceptor, and again the receptors cannot respond to fine detail. Clearly there is an optimum, intermediate facet size for any given head size. The most efficient design occurs when the difference in pointing angle of adjacent ommatidia, Δ, is the same as the diffraction limit $\theta_{Rayleigh}$ of the facets (Fig. 12.53b). In this way no two ommatidia get appreciable light from the same direction, and light from every direction is received by some ommatidium. Sure enough, the central region in the eyes of locusts, dragonflies, and wasps have Δ nearly equal to the Rayleigh limit. These insects sit still in bright light while examining their prey. For most other insects, however, Δ is twice the diffraction limit or more. Part of the reason is that these insects move and hence cannot in any case see detail that would be blurred by the motion. Most insects trade angle of view for better resolution in a central, fovea-like region, where the facets are large (for a small angle of view) and point in nearly the same direction (for detailed sampling of small regions).

A typical value for the facet diameter (of a bee or a fly) is $b \simeq 30$ μm. At $\lambda \simeq 400$ nm, where the bee is most sensitive, its diffraction limit is about 0.9°, according to the formula for $\theta_{Rayleigh}$. Since your pupil size of 3 mm is about 100 times the bee's facet diameter, your resolution is about 100 times better than the insect's. If you were reading this book from a distance of 25 cm and had bee's eyes, you would barely be able to tell that there are separate lines of print. This gives you some idea how "coarse" the world looks to a bee. A fly does not make use of all the resolution possible with its facet size, for the angle between ommatidia directions is about 3°, so its world looks coarse indeed! Among the compound eyes of highest resolving power are those of the dragonfly with Δ ≈ 0.25°. At this large head size, however, compound eyes cease to be advantageous compared to lens eyes; the smallest hummingbirds have heads of comparable size, but a lens eye with ten times better resolution.

The problems in designing a pinhole camera are similar to those in the design of an insect's eye. Sup-

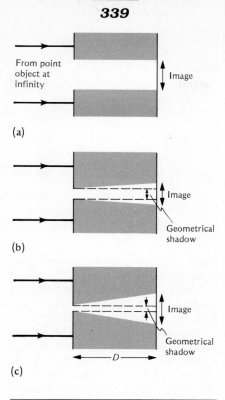

(a)

(b)

(c)

FIGURE 12.54

A pinhole camera with three different pinhole sizes. **(a)** Blurring of image due mainly to geometrical-optics shadow. **(b)** Minimal blurring when geometrical-optics and wave effects are equal. **(c)** Blurring of image due mainly to diffraction.

pose light from a distant point source enters a camera with a pinhole of diameter b (Fig. 12.54). The geometrical-optics image then is a "point" of size b. To sharpen the geometrical-optics image we should therefore decrease b. However, as b gets smaller, diffraction becomes important and spreads the "point" into a diffraction pattern of size

$\lambda(D/d)$, as we saw in Section 12.5A. The best you can do is choose b so that the geometrical-optics image of a point is the same size as the blur due to diffraction (see Fig. 2.10).

*D. Image reconstruction and spatial filtering

When Ernst Abbe was trying to understand the resolution of microscopes around the turn of this century, he found an entirely different way of thinking of image formation. This viewpoint is useful for many of the modern methods of improving the quality of optical images.

Consider again an object consisting of two points of light close together and an imaging lens (which might be the objective lens of a microscope). Suppose first that the two points emit *coherently*. The lens then seems to perform two different roles. If we think about the interference pattern produced by these two coherent point sources, we know the lens causes *Young's fringes* to fall on its *focal plane*. However, we also know that the lens projects a *geometrical-optics image* of the two point sources onto the *image plane*, which lies beyond the focal plane (Fig. 12.55). Abbe realized that, since the light that reaches the image plane must *first*

FIGURE 12.55

Two coherent point sources, S_1 and S_2, at a finite distance from a lens. If we put a screen at the focal plane F, we see a Young's fringe pattern. If we move the screen to the image plane I, we see a geometrical-optics image of the sources.

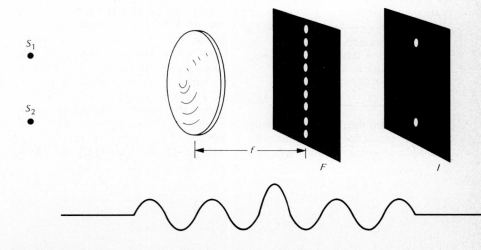

have passed through the focal plane, these two processes must be related.

Let's first see how, by looking at the *interference pattern* produced in the focal plane of a lens, we can tell something about the quality of the *image* produced in the image plane. Specifically, we can tell whether the resolution is good enough to separate the images of the two point sources. We know that the different-order Young's fringes are sent in different directions from the sources. Depending on the size of the lens aperture, it will transmit the central (zeroth-order) fringe, as well as, possibly, some of the higher-order fringes.

Now, no matter what the separation of the point sources, the central fringe always lies in the forward direction (along the axis, $\theta = 0$). Since it is the same for sources of *any* separation, the light in the central fringe cannot contain information about the source separation. Hence, if *only* the central fringe hits the lens, the resultant image must be the same no matter *what* the source separation. That is, the two points are not resolved by such a small lens—the image is just one blurred spot.

If it is to transmit any information about the source separation, that is, to resolve the two points, the lens must be at least large enough to catch the first-order interference beam, whose direction *does* depend on the separation (Sec. 12.2E)—the image resolution of the lens depends on its size. Thus, by considering the *interference pattern* produced by the two coherent point sources, Abbe was able to obtain a criterion for the *image resolution* produced by a lens of a given aperture:

If two points are to be resolved by a lens, the lens must be big enough to pass their first-order interference beams.

(This criterion leads to essentially the same result for large source-to-lens distances as the Rayleigh cri-

terion, which considered the finite-sized lens as producing *diffraction*.)

Abbe's argument assumed the sources were coherent. But suppose, as is usually the case, they are incoherent. Such sources are equivalent to coherent sources with a rapidly changing phase difference. Therefore, the Young's fringes produced by two noncoherent sources shift back and forth so quickly that you can't see any interference pattern. However, no matter how rapidly they shift, the fringe *spacing* always remains the same. Hence, if the aperture is too small, at no time will more than one fringe enter the lens. Since it is the *fringe spacing* that carries the information about the *source separation*, again the two points won't be separately seen in the image—Abbe's criterion is valid whether or not the sources are coherent.

Let's now turn to the question of how the light, which forms an interference pattern in the *focal plane* of the lens, goes on to produce an image at the *image plane*. Abbe's viewpoint is that a number of interference beams from the sources strike the lens (at least the zeroth- and first-order beams if the image is to be resolved) and continue on to produce the same number of Young's fringes in the focal plane. Each of these fringes can now be considered as a source of Huygens wavelets (Fig. 12.56). Since the Young's fringes all come from the same source (the pair of points), these wavelets must be coherent. Hence, the Huygens wave-

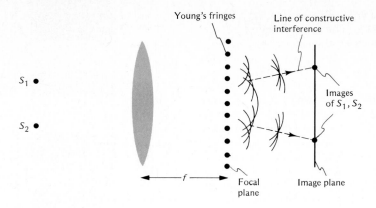

FIGURE 12.56

The Young's fringes in the *focal plane* serve as sources of Huygens wavelets. The interference pattern of these wavelets is the image in the *image plane*.

lets, originating at the Young's fringes, will interfere with each other. When they reach a screen at the image plane, their interference pattern will be the geometrical-optics image of the original two points.

In short, Abbe considered the formation of the geometrical-optics image of *any* object through *any* lens aperture as the result of two successive interference processes. The first, between light from points on the original object, produces the interference pattern in the focal plane (Fig. 12.57). Each point on *this* pattern, in turn, serves as a source for the second interference process, which results in the image. If the lens aperture is not infinitely large, it cuts off some of the interference pattern in the focal plane. There are then not enough wavelets in the second process for perfect reconstruction, and the image becomes fuzzy.

Just what features of the object correspond to the outer parts of the interference pattern, which are cut out by the lens aperture? Suppose the object is a *grating*. Its first-order beam gives a well-defined spot in the focal plane, and the location of this spot depends on the grating constant—the finer the grating, the farther from the axis is the interfer-

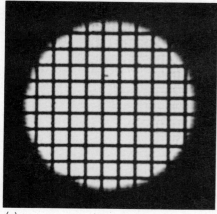

(a)

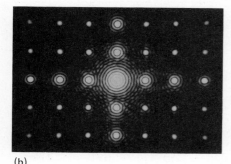

(b)

FIGURE 12.57

Photograph of **(a)** a grid and **(b)** its interference pattern in the focal plane.

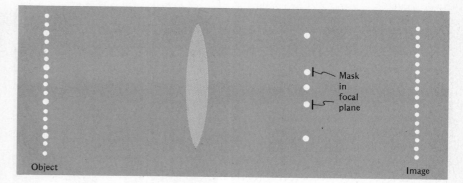

FIGURE 12.58

Set-up for spatial filtering. The object is a combination of a coarse grating and a fine grating. The spots in the focal plane corresponding to the coarse grating are blocked by a mask. Only the fine grating survives in the image plane.

ence spot. If the grating is too fine, the lens aperture will block the first-order interference beams, while if the grating is sufficiently coarse, the lens will pass the first-order beams. Hence, the smaller the lens aperture, the coarser the grating must be to have its first-order beam passed and thus for its slits to be resolved by the lens. Therefore, small lenses can't image fine gratings.

Even if the object is not a grating, it generally has various amounts of coarse and fine detail (called low and high spatial frequencies, respectively.) Therefore, it can be considered as composed of gratings. The finite aperture removes all the details finer than a certain size—the image does not contain the spatial frequencies higher than some value, determined by the size of the aperture. No amount of magnifica-

tion can then restore the spatial frequencies that have been lost. (It's like playing an old 78-rpm record, which lacks the high-frequency sound response, through your hi-fi system—it still sounds like a dull old record.)

Each spot in the focal plane of the lens corresponds to a particular spatial frequency in the object. When the Huygens wavelet from that spot interferes with the central, zeroth-order beam, it reconstructs the appropriate spatial frequency in the image. Now, the lens aperture does not have to be the *only* thing that modifies the interference pattern in the focal plane. If we want *to remove a particular spatial frequency* in the image (but not lower and higher ones), we can simply *block the corresponding spot in the focal plane* with a small opaque disk (Fig. 12.58). Of course, in order for these spots to be found, they must be stationary, hence *coherent light*—typically from a laser—must be used for the illumination. This process of **spatial filtering** is useful, for example, if some periodic structure is to be removed from a picture with minimal disturbance of the rest of the picture's detail. Thus, you can remove the halftone dots from a newspaper photograph or the scanning lines from a television picture by this method (Fig. 12.59). Other spatial filters can consist of a more complicated mask in the focal plane and can enhance or delete some shape present in the picture, no matter where it occurs.

Whether or not the illumination is coherent, if no object is present,

the only spot of light in the focal plane is the image of the (distant) light source. When the object is present, that image will always remain as the zeroth-order fringe. Hence, you can do some spatial filtering with *incoherent light* if all you want to do is block *that* fringe (the image of the source) in the focal plane. Since, in the absence of an object, the image plane will then be dark, this is like **dark-field microscopy** (Sec. 6.3A)—any object that diffracts light is able to send *some* light to the image plane, where it will stand out in good contrast against the dark background, even if that object would ordinarily appear quite faint (Figs. 12.60a and b).

If the zeroth-order beam is not blocked, but instead changed in phase, you get **phase-contrast microscopy.** A small piece of glass inserted in the focal plane on the axis will achieve this phase change, because light traveling in glass is slowed down. This changes the interference pattern between the zeroth-order beam and the higher-order beams, that is, it changes the reconstructed image from what it would be without the glass. Objects previously invisible may now become visible (Fig. 12.60c).

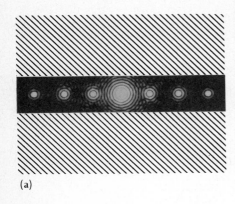

(a)

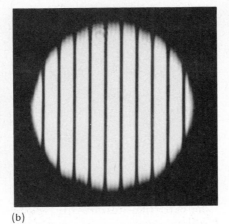

(b)

FIGURE 12.59

(a) Interference patterns in the focal plane, produced by the same object as in Figure 12.57, but with parts of the pattern blocked off as indicated. **(b)** The corresponding images in the image plane. Compare these to the original object in Figure 12.57.

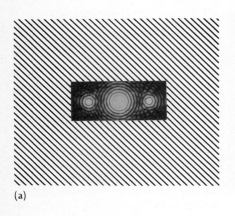

(a)

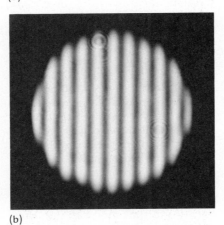

(b)

FIGURE 12.60

(a) Normal microscope view of two barely visible diatom shells (single-celled plants). **(b)** View of the same diatom shells in a dark-field microscope.
(c) View of the same diatom shells in a phase-contrast microscope.

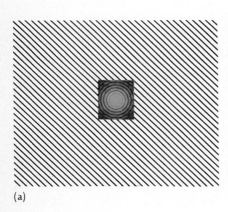

(a)

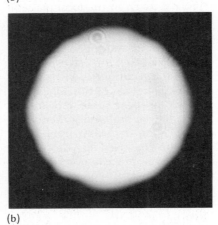

(b)

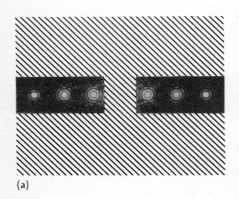

(a)

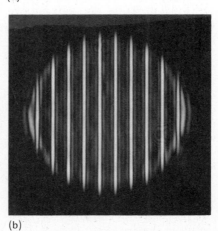

(b)

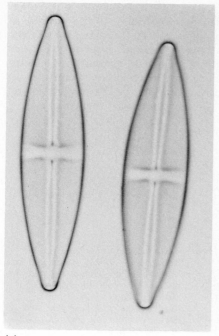

(a)

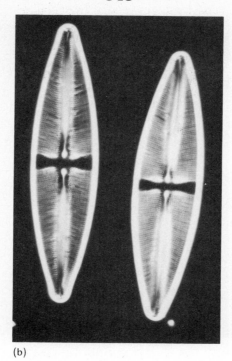

(b)

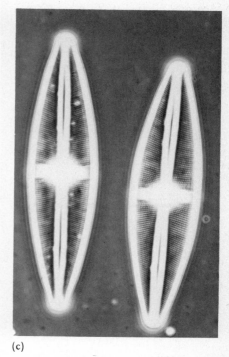

(c)

SUMMARY

When the crests and troughs of two waves keep in step or have a constant phase difference, the waves are **coherent** and will **interfere** in regions where both waves are present. The interference is **constructive** (increase in intensity) where crest meets crest and trough meets trough, that is, when the waves are **in phase**. It is **destructive** (decrease in intensity) when crest meets trough, that is, when the waves are **out of phase**.

To obtain two coherent light beams from a single point source, one can **split the amplitude** by means of a **beam splitter,** a partially reflecting, partially transmitting surface, or one can **split the wavefront** by means of closely spaced slits. The two resulting coherent beams seem to come from two coherent, effective sources. Their interference at any point depends on the **path difference,** ϵ, of the paths from each of the two effective sources to that point. If $\epsilon = \lambda, 2\lambda, 3\lambda, \ldots$, the beams interfere constructively; if $\epsilon = \lambda/2, 3\lambda/2, 5\lambda/2, \ldots$, they interfere destruc-

tively. On a screen illuminated by both beams you see bright and dark **interference fringes,** regions of alternating constructive and destructive interference.

Young's fringes are the fringes produced by *two* slits, that is, two coherent sources, illuminating a distant screen. *Many* coherent sources are obtained by a **diffraction grating,** an array of many, regularly spaced slits. The interference fringes of a grating are narrow lines emitted by the grating into sharply defined angles that depend on the slit spacing and the wavelength, hence the grating can be used in a **spectrometer.** The colors emitted by a grating are **iridescent**—they change with angle of view. Tiny gratings occur in nature and cause the iridescent colors of many birds, insects, and gems. The atoms in crystals form gratings suitable for **x-ray diffraction.** Other visible iridescent colors are due to **multiple reflections** from the top and bottom surfaces of thin films.

Interferometers are devices to make accurate measurements by the use of interference. In the **Mach-Zehnder interferometer,** the coherent beams circumscribe a

rectangle before recombination. Changes in the fringe pattern indicate very small disturbances of one of the beams.

Huygens' principle shows that interference occurs all the time when light propagates; a later wavefront is due to the interference of Huygens wavelets emitted by each point of an earlier wavefront. The Huygens construction implies **diffraction** of light, that is, the spreading of a light beam. **Fraunhofer diffraction** patterns are seen very far from an aperture (or at the focal plane of a lens). When parallel light is incident, geometrical optics predicts only a spot of light at the focus, but diffraction spreads the light out. Except at this focus, the Fraunhofer diffraction pattern is the same for two complementary plates, according to **Babinet's principle.** The Fraunhofer pattern of a hole has an intense central maximum, surrounded by less intense, higher-order maxima, with vanishing intensity in between. When water droplets in the atmosphere cause this diffraction, the result is a **corona** or, in combination with refraction and internal reflection, a **glory.**

Diffraction by the aperture of a lens or a mirror or by the eye's iris limits the **resolving power**—the ability of the lens or mirror to show separate images of closely spaced objects. The cones of your eyes are packed just closely enough to send separate signals for objects that are barely resolved by your eye's lens. Similarly, in an insect's eye, diffraction determines the range of angles from which light can pass through an **ommatidium**'s lens (**facet**) and reach its single photoreceptor. With too small an aperture, diffraction blurs the image.

The blurring due to a finite aperture is equivalent to the removal of higher spatial frequencies from the image. **Spatial filtering** removes only selected spatial frequencies and, with coherent illumination, can improve images by removing unwanted periodic structures.

PROBLEMS

P1 Light from two coherent, in-phase point sources interferes at some point P. Is the interference at P constructive or destructive if the distance from the first source to P differs from that from the second source to P by: (a) 3λ, (b) $\lambda/2$, (c) $5\lambda/2$, (d) 0?

P2 Repeat P1 for the two sources out of phase.

P3 Manufacturers of good camera lenses coat them with a thin film to cut down: (a) reflection, (b) deflection, (c) refraction, (d) distortion, (e) dispersion. (Choose one.)

P4 (a) Is a lens coating that is designed to transmit the blue region of the visible spectrum thicker or thinner than one designed for the central (green) region? (b) What color would be the light reflected from such a blue-transmitting lens (see Sec. 9.4B)?

P5 Two long, extremely narrow slits are situated close to each other in an otherwise opaque plate. Parallel monochromatic light is incident on the slits. A white screen is placed at a great distance beyond the slits. A pattern of light and dark fringes is observed on the screen. (a) Explain why some regions are light and others dark. (b) Sketch a graph illustrating the pattern of light and dark fringes that is obtained. (c) What happens to the pattern if one of the slits is covered up? (d) What pattern of fringes results if the incident light is white rather than monochromatic?

P6 The two slits of P5 are replaced by *many* narrow slits with the same space between adjacent slits that the original slits had. Sketch the pattern of light and dark fringes that would be obtained under the same circumstances, using the same position scale as in P5.

P7 Radio waves of wavelength 1 km are sent by the transmitter T shown in the figure and received by the receiver R, 5 km away. The receiver also picks up some reflected wave from an airplane overhead. A

snapshot of the reflected wave is shown. (a) Redraw the figure and draw the direct wave from T to R. (b) Describe what happens to the signal at R, compared to the case when the reflecting airplane is not there.

P8 Redraw Figure 12.31f for the case where the sources are separated by $2\frac{1}{2}\lambda$.

P9 The figure shows plots of intensity distributions on a screen obtained from interference between 2, 4, and 10 identical sources. (The vertical scale is not the same in each plot.) (a) Identify which number of sources goes with each pattern. (b) Which vertical scale has been shrunk the most?

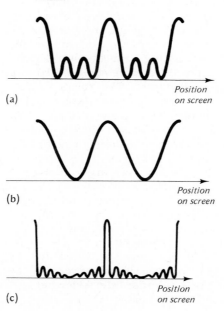

P10 Using the grating spectroscope of Figure 12.24, look at a variety of (at least three) light sources and describe the colors present. Identify the light source for each spectrum; note relative brightness if possible, and decide whether it is continuous, discrete, or both.

P11 Look through a diffraction grating at a small white light source. Slowly look more and more to the side until you see the first-order interference pattern. (a) As you move from looking straight at the source, which color comes first in the first-order pattern? (b) Continue until you come to the second-order pattern. Which color comes first there? Continue to higher orders, if possible. (c) Explain, with the use of a diagram, the sequence of colors in each order.

P12 The figure shows two unknown devices, each of which separates white light into its spectrum. The positions of two wavelengths are shown on each screen. For each case, describe an optical device that gives the result shown.

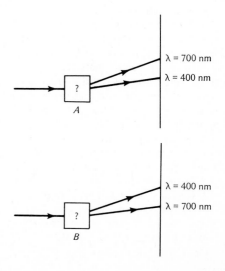

P13 What are Newton's rings and what causes them? Use a diagram to explain your answer.

P14 The colors of a soap bubble or of an oil film on water are produced by: (a) selective absorption and reflection, (b) diffraction, (c) interference, (d) refraction, (e) pollution. (Choose one.)

P15 (a) What are iridescent colors? (b) Will a color photograph of such colors display iridescence?

P16 Draw a Huygens construction analogous to Figure 12.44 for: (a) plane waves reflecting from a mirror, (b) plane waves refracting at a flat glass surface.

P17 Why does the image get fuzzier in the pinhole camera when the pinhole gets too small?

P18 Sketch a graph of the Fraunhofer diffraction pattern you would expect from an opaque wire in front of the lens of Figure 12.45.

P19 Compare and contrast a rainbow and a corona, particularly with respect to the mechanism of color production, the sequence of colors seen, and the (angular) size of the arc.

P20 A stack of films, such as that of Figure 12.28, is made of alternating layers with index of refraction 1.3 and 1.0 (air) and thicknesses so that the stack is highly reflective for light of wavelength 500 nm. How does the reflectivity of the film change if the air film is replaced by water, which has an index of refraction 1.3?

HARDER PROBLEMS

PH1 A traffic light that goes through a cycle of red and green is a periodic source for a wave of cars, emitting a car "crest" once every cycle. There is such a traffic light on each of two highways that merge down the road, one kilometer from each traffic light. Assume that all cars travel with the same speed. (a) What must be true about the traffic lights' timing in order that the waves of cars be coherent? (b) What must be true about the traffic lights' timing in order that they be in phase, considered as sources of car waves? (c) If the lights are in phase, it is found that lots of cars simultaneously get to the place where the highways merge, causing collisions and destruction. How does a physicist describe this interference of the waves of cars, destructive or constructive? (d) Offer some constructive suggestions on how to avoid the collisions, either by changing the timing of the traffic lights or the location of one of them. Humor the physicist and describe the changes in *his* terms.

PH2 The figure shows two coherent sources of waves (such as two stereo speakers hooked up to a monaural signal), 1 m apart. They can emit

waves whose wavelength λ is either 1 m or 2 m, and they can be hooked up either in phase or out of phase. For each case we want to know the intensity at the points x and y. Fill in the table below, using L (loud) for high intensity, Q (quiet) for little or no intensity.

	λ = 1 m		λ = 2 m	
In phase	$I_x=$	$I_y=$	$I_x=$	$I_y=$
Out of phase	$I_x=$	$I_y=$	$I_x=$	$I_y=$

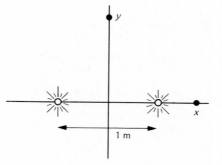

PH3 A "coated lens" is covered with a thin film of material with an index of refraction of 1.2 so that there is a "hard" reflection (180° phase change) at the two surfaces, air-coating and coating-glass. Yellow light is completely transmitted, that is, there is perfect cancellation of the reflected waves. The film thickness is found to be one-quarter the wavelength of yellow light. (a) What will be the color of the reflected light (see Sec. 9.4B)? (b) On the other hand, for a soap bubble, reflections cancel if the film thickness is one-half the wavelength of the light. Explain the difference.

PH4 Two in-phase coherent sources of light are on the x-axis, four wavelengths apart. (a) In what directions is the path difference ε as large as it can possibly be? What is ε there (in terms of λ)? Do the waves interfere constructively or destructively there? (b) How many other directions of this kind of interference are there? (Remember, light can go up, down, right, or left.) (c) Repeat the problem for the case that the sources are out of phase.

PH5 Repeat PH4 for the sources two and a half wavelengths apart.

PH6 In this problem, each section requires a sketch of an interference pattern. Draw intensity versus position graphs for each. Draw the sketch for each section directly below that of the preceeding one and to the same scale, so that if a maximum in intensity (a bright spot on the screen) in one section is meant to occur at

the same position as a maximum in another section, the two maxima will be lined up one below the other. Draw the interference pattern produced by: (a) two slits with spacing d, illuminated with monochromatic blue light; (b) a diffraction grating with grating constant d, illuminated with monochromatic blue light; (c) the same grating as in (b), illuminated with monochromatic yellow light; (d) a single slit of width b, illuminated with monochromatic blue light [here take b smaller than d of parts (a) through (c)]; (e) a single slit of smaller width than in part (d), illuminated with monochromatic blue light.

PH7 (a) What does it mean to say that two or several beams of light are *coherent*? (b) How can you test whether two given beams are coherent? (c) Are the zeroth- and first-order beams from a grating coherent with each other? Could a grating be used as a beam splitter?

PH8 The figure shows two filters with white light approaching them. The influence of each filter is shown. (Refer to Sec. 9.4B.) (a) Which filter is dichroic? Which is gelatin? (b) Describe what happens to the blue, green, and red components of the incident light in each case. (c) If the reflected and transmitted beams are both shined on a common point on a white screen, what will be the resulting color for each filter? Explain.

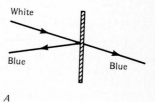

PH9 The figure shows a Michelson interferometer. M_1 and M_2 are front-surface mirrors, M_3 is a half-silvered mirror, and M_4 is a plain (unsilvered) glass plate of the same material and thickness as M_3. (a) Copy the figure and draw what happens to beam B from the source. Label coherent but separate beams B_1 and B_2, and the recombined beam B_3. (b) If M_1 and M_2 are equidistant from the silver of M_3, is the interference at B_3

constructive or destructive? (Assume that the half-silvered mirror surface causes a 180° phase shift for external reflections, but no phase shift for reflections inside the glass.) (c) M_2 can be moved in the direction of the double arrow. How far should it be moved to change the interference from constructive to destructive, or vice versa? Give your answer in terms of λ.

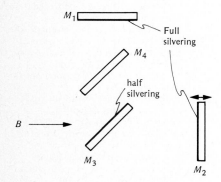

PH10 The figure shows a drawing of the interference fringes in a Mach-Zehnder interferometer when a wedge of transparent material is inserted perpendicular to the beams. Draw the outline of the wedge and identify its thickest and its thinnest edge.

PH11 A square grid of parallel and perpendicular lines is scratched in a layer of black paint that covers the surface of a flat pane of glass, so that the glass shows through at the scratched lines. The glass is used as the plate in a set-up such as that of Figure 12.38. The screen pattern is compared with that obtained when a piece of window screen is used as the plate. (a) In what respects are the two patterns similar? (b) In what way do they differ?

MATHEMATICAL PROBLEMS

PM1 A thin film of oil of thickness 200 nm is floating on water. The index of refraction of the oil is 1.5, that of water is 1.3. Light is reflecting from the film. (a) Is the reflection at the top surface hard or soft? Is the corresponding phase change 0°, 90°, or 180°? (b) Is the reflection at the bottom surface hard or soft? Is the phase change 0°, 90°, or 180°? (c) What is the extra *path length* for the beam reflected perpendicularly

from the bottom surface compared to that reflected from the top? (d) What is the total *phase* difference between the two beams if the wavelength in the oil is 400 nm? If it is 800 nm? (e) If the reflectance of the film is 0.1 at 800 nm, sketch the reflectance *versus* wavelength in the oil from 100 to 800 nm.

PM2 (a) What is the fringe spacing seen with light of wavelength 500 nm on a screen 1 m from a pair of slits that are 1 mm apart? (b) Could you actually see these fringes with your eye, viewing from 25 cm? (c) What would be the minimum fringe spacing that your eye could resolve at 25 cm?

PM3 Suppose the concentric rings in your moiré set (TRY IT for Sec. 12.2E) differ in radius by 2 mm. For a "screen" separated from two "sources" by 10 cm, you find the fringe spacing to be 5 mm. (a) What is the separation of the "sources"? (b) If you double this "source" separation, what is the new fringe spacing? (Use your moiré set to check your results.)

PM4 Suppose that the picture of Plate 15.1c was taken with a grating of 1093 lines/mm and a screen distance $D = 1.00$ m, and that the actual interference pattern was 1.9 times larger than it appears on the plate. Compute the difference in wavelength of the bright green and blue lines of the mercury spectrum.

PM5 A Lippmann plate is exposed with green light of wavelength 500 nm. (a) After the plate is developed, what will be the distance between layers of silver grain in the emulsion? (b) After drying, the thickness of the developed emulsion has shrunk to 90% of what it was when the exposure was made. What wavelength does the plate reflect now? (c) Can you compensate for this color shift by looking at the plate from an angle, rather than straight on?

PM6 In a Michelson interferometer (see PH9), when M_2 is moved by 1 mm, an observer looking at B_3 counts 5000 dark interference fringes crossing his field of view. What is the wavelength λ of the light used?

PM7 The figure shows parallel, monochromatic light incident on a plate that contains a pinhole. You want to get as much intensity at point P as possible by adjusting the size of the pinhole. Assume that the pinhole is large enough when ray 2 first interferes *destructively* with ray 1. (Note: unlike in the Fraunhofer diffraction shown in Figure 12.45, the rays that are brought to interference here are not parallel, and the point P is not infinitely distant.) Suppose the wavelength is represented by 1 cm in the diagram.

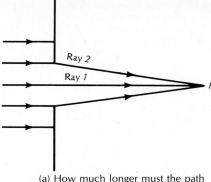

(a) How much longer must the path of ray 2 be than that of ray 1? (b) Using a ruler, find where ray 2 should actually go so that its length corresponds to your answer to (a). Hence find the optimum diameter of the pinhole (different from that shown). (c) Compare your result with the formula $d = 2\sqrt{\lambda f}$, where d is the optimal pinhole diameter, and f is the distance from the pinhole to the image at P. (This relation is sometimes used to define f as the "focal length" of the pinhole.)

PM8 Light of wavelength 500 nm passes through a narrow slit and is projected on a screen 1 m away. The first zero of intensity occurs at 1 cm from the central maximum. (a) What is the width of the slit? (b) For the same light and screen, what slit width would make the first zero occur 2 cm from the central maximum?

PM9 A 6° corona is seen around the moon. This means that the first zero of diffracted intensity is at 3° from the center (moon). Draw a diagram of a drop, an incident beam from the moon, and a beam diffracted by 3°. At a convenient distance D, draw a screen and measure the distance of the first zero from the central maximum. (Alternatively, you may use trigonometry.) Hence compute b, the approximate diameter of the drops that cause this corona, for 500-nm light.

PM10 (a) According to the Rayleigh criterion, what is the resolving power of an astronomical telescope with a 10-cm objective diameter lens for $\lambda = 500$ nm? (b) By comparing the resolving power of this telescope with that of your 20/20 eye, express the former in the usual Snellen ratio form, 20/X.

PM11 (a) Determine the Rayleigh angle of resolution for a camera lens of focal length 5 cm set at f/8 for $\lambda = 500$ nm. (b) Find the size of the blur due to diffraction on the film plane. (c) You have a coarse-grain film able to record details as fine as 50 lines/mm, and a fine-grain film able to record 200 lines/mm. Can you get appreciably more detail in a picture taken with this camera by using the fine-grain film than by using the coarse-grain film?

Scattering and Polarization

13.1
INTRODUCTION

So far we have considered what happens when light encounters material obstacles of a size much greater than the wavelength (geometrical optics) or of a size so small as to be comparable with the wavelength of light (wave optics). But you can also observe effects due to even smaller obstacles, much smaller than the wavelength of visible light. When light interacts with an isolated object that small, it shakes all the charges in the object, which then radiate in all directions. This phenomenon is called **scattering.**

To scatter with appreciable *intensity*, light must encounter *many* isolated small objects: for example, the molecules that constitute the air. Indeed, it is only because light does scatter in air that you can see the beams of Figures 1.3, 1.4, and 8.19b. Without scattering, the light's path through the air would be invisible and the camera could not record it.

This simple explanation, offered when these figures were introduced, now bears closer scrutiny. It would seem that when propagating in any dense medium, such as glass, light should be scattered by the many molecules that are present. Instead, as we know, it continues to propagate in a sharp beam, as in vacuum, only with a different *speed*. This is because there are many molecules present in glass, and whenever there is one molecule to scatter light, there will be another one for which the light path to your eye is half a wavelength longer. The scattered light from

these two molecules interferes destructively, as in Figure 12.4 (along the *x*-axis), and this is so for any sideways direction from which you look at the beam. Air, on the other hand, is a gas, so there is no guarantee that there will be another molecule half a wavelength beyond the first—sometimes there may be a few extra molecules around one point, sometimes a few less. You see the scattering from these points because of these fluctuations. (The TRY IT suggests ways to enhance the scattering in air.)

Scattering is *selective* in several ways: light of certain *wavelengths* is scattered more than light of other wavelengths, and light of one *polarization* (Sec. 1.3B) is scattered more than light of another polarization. Because our eyes are not very sensitive to the polarization of light, we have not yet discussed the phenomena associated with it. Scattering provides us with the opportunity to do so, even as it actually provides most of the polarized light around us. Much of this chapter will therefore deal with polarized light, produced by scattering as well as by other means.

TRY IT

FOR SECTION 13.1
Light beams

To see the path of light rays, you need tiny objects in the path that will scatter part of the light to your eye. To see the beam of your flashlight, put larger particles in the air, such as dust motes, smoke, or chalk dust. In their presence you should be able to trace the light's path, including reflections from mirrors and water surfaces, to test the law of reflection (Sec. 2.4) and Snell's law (Sec. 2.5).

13.2
RAYLEIGH SCATTERING

When white light scatters from some molecules, it scatters selectively because part of the light is absorbed at the resonant frequencies of the molecules—the scattered light is then colored. For many other molecules, however, the important resonant frequencies are significantly higher than visible frequencies. White light nevertheless becomes colored when it scatters from these molecules—the higher the frequency of the incident light, the more light will be scattered. This type of scattering is called **Rayleigh scattering** and occurs whenever the scattering particles are much smaller than the incident wavelength and have resonances at frequencies higher than those of visible light. Equivalently, we may write the rule for Rayleigh scattering:

The shorter the wavelength of the incident light, the more light is scattered.

This result was worked out in detail by the same Lord Rayleigh we met in Sections 10.5B and 12.5C. It says that blue light will be scattered more than red light. In fact, for incident broad-band white light, the intensity of scattered 400-nm light is almost ten times as great as that of 700-nm light.

One consequence of Rayleigh scattering is the sky's blue color. Light reaching your eyes from the sky is sunlight that has been scattered by the air molecules (Fig. 13.1) and is therefore predominantly blue. Since the direct rays from the sun have some of the blue part of the spectrum scattered out of them they should look slightly yellowish. When the sun is overhead, and if the sky is very clear, this is a small effect. However, if there are lots of tiny dust and smoke particles in the air, the effect is larger. It becomes even larger as the sun sets; the direct rays from it to your eyes must pass through more and more atmosphere, so these rays are depleted of more and more of the shorter wavelengths, and the sun looks redder and redder. (See the TRY IT.)

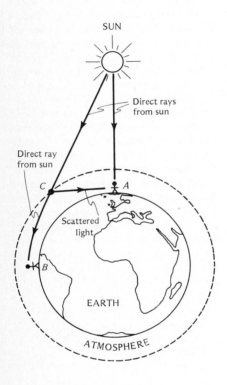

FIGURE 13.1

When A looks at point C in the sky, only scattered light from the sun reaches her eyes. As short-wavelength light is scattered most, the sky looks blue to her. Direct rays from the sun, from which the blue end of the spectrum has been removed by scattering, vary from white to yellow to red, depending on how much atmosphere they have traversed and how much dust is in the air—that is, on how much scattering the light has suffered. Thus to B, who sees the sun setting, the sun appears much redder than to A.

There are a number of other examples where blue coloring is due to Rayleigh scattering. We've already noted (Sec. 9.9E) that fine black pigment mixed into white paint gives it a bluish cast for just this reason. In fact, da Vinci noted this and attributed it to the same cause that makes the sky blue. The writer George MacDonald correctly identifies another blue with the blue of the sky when he writes, "Where did you get those eyes so blue?/Out of the sky as I came through." Your beautiful blue eyes are due to scattering from small, widely separated particles in your irises. Similarly, the moonstone owes its blue sheen to Rayleigh scattering.

You can see the same effect in fine smoke, say from a wood fire. The smoke looks bluish when illuminated from the side and viewed against a dark background, so only scattered light reaches your eyes. If instead the smoke is seen against a bright background, it looks red or brown (due to the *removal* of blue by scattering). If the smoke gets too thick, the particles become dense enough that *all* light is repeatedly scattered, and the light that emerges sideways is white or gray.

A similar effect can occur when the particles become larger. For example, the smoke rising from the end of a cigarette is bluish, but after you inhale and exhale it, the smoke looks gray or white. Here you have covered the smoke particles with moisture, making them much larger. They are then large enough to scatter light of all wavelengths equally—as in the reflections of geometrical optics—and thus they look white. This is also why clouds are white: The water droplets in clouds may be fifty times as large as the wavelength of visible light. With so many droplets, and thus so many surfaces to reflect the light, the clouds scatter almost all the light and look white even though the individual drops are nearly transparent. (Of course, very dense clouds will not transmit light—they either absorb it or reflect it upward—so they look black.) Small grains of salt

or sugar, talcum powder, chalk, the white spots of moths and butterflies, white paper, fog, snow, beaten egg white, and beer foam all look white for the same reason. The white pattern in star rubies and sapphires, and in tiger's-eye quartz are similarly due to scattering from large inclusions. Likewise, the clear albumen becomes white as an egg is cooked because the protein molecules are freed of their surface water and are then able to coagulate into large clumps, which scatter nonselectively. When the watery whey from milk is made into cheese (such as ricotta), the cheese is white because of the same coagulation process.

Yellow lights are often used as fog lights or headlights on cars because yellow is as easily detected by your eyes as white but is scattered less in a fine mist, when the droplets are very small compared to the wavelength of visible light. Unfortunately, more often the droplets are larger, and the yellow is then scattered as well.

Particles may, of course, also produce colors by selective *absorption* and *reflection*. The smog we all know and love takes its brown color because of absorption by particles of nitrous oxide, which has resonances at visible frequencies. Not all particles in the sky produce ugly colors, however. For example, in 1883 the volcano Krakatoa erupted spectacularly, spewing micrometer-size particles into the atmosphere in such abundance that all over the world there were unusually colorful sunrises and sunsets for three years!

Scattering by the air or the particles in it is thus responsible for aerial perspective (Sec. 8.6E), which makes distant dark hills look blue and distant snow-clad peaks look yellow (see Plate 8.4). The purer and more transparent the air, the bluer those dark hills.

FOR SECTION 13.2
TRY IT

Blue skies

A good source of small scattering particles is milk, whose solid particles are much smaller than the wavelength of visible light. You can use these to make a blue "sky" and a red "sunset." In a dark room, shine a light beam, say from a flashlight, through a clear glass of water. Look at the beam from the side so you can see the scattered light. (It helps if there is a black background.) At the same time look at the direct, transmitted light, for example, by reflecting it from a piece of white paper. Now add a little milk, one or two drops at a time, and stir the water. The scattered light will become bluish as the transmitted light becomes yellowish and then reddish. As you add even more milk, the scattered light becomes white because of repeated scattering, and you have made a white "cloud."

You can see the blue of the air directly if you have enough air with a dark background, and something black with which to compare the color of the air. An otherwise dark room viewed from the outside through an open window makes a good dark background. To block all extraneous light, view the window through a long mailing tube. Since you'll want to be as far as possible from the window (30 or 40 m), cover the far end of the tube with some aluminum foil in which you've made a small (several millimeters) hole. To avoid light entering the hole at an angle, wrap a piece of black paper around the far end of the tube, so it extends 15 or 20 cm beyond the foil. Look toward the window when the sun is to one side or overhead. Then the only light entering the tube will be the sunlight that is scattered by the air between you and the window. This should look distinctly bluish compared to the dark surrounding of the aperture in the foil. What happens to the blue as you change your distance from the window?

Since Rayleigh scattering also occurs at night (except that there is less light to scatter), a moonlight photograph taken with a long enough exposure time should show the same colors as one taken by daylight. Try it on a clear night during a full moon. The moonlight is then almost a factor of 10^6 weaker than sunlight. Because of the failure of reciprocity of the film at long exposures (Sec. 4.6), you should make an exposure about 10^7 times as great as you would

need by sunlight. For example, if a good exposure by sunlight was $\frac{1}{1000}$ sec, then, with the moon in the same position as the sun was, you might open your lens an additional three f-stops and make a 20-minute exposure. Most color film won't give accurate color with such a long exposure—Kodak recommends Kodacolor 400 if you want prints, or Kodachrome 25 (daylight) with a CC10M filter if you want slides.

13.3
POLARIZATION DUE TO SCATTERING

The same Rayleigh scattering that gives us the blue sky also polarizes the scattered light. In order to see how this comes about, let's return to some of the basic ideas of polarization, which were first introduced in Section 1.3B.

A. Polarized light

Light, recall, is a *transverse* wave— the electric field is always perpendicular to the direction of propagation of the wave (the direction of the ray). For example, if the wave is traveling in the z-direction (Fig. 13.2a), the electric field may be in the x-direction, in the y-direction, or in any other direction within the x-y plane. If the light wave's electric field is always parallel to the x-axis, we say the light is **linearly polarized** in the x-direction (Figs. 13.2b and c). Similarly, a wave whose electric field is always parallel to the y-axis is linearly polarized in the y-direction.

Suppose the wave is linearly polarized in some other direction, say at an angle of 45° between the x- and y-axes. We may nevertheless think of this wave as consisting of two *in-phase* waves (Fig. 13.2d), one linearly polarized in the x-direction (the x-**component**) and one in the y-direction (the y-component). There are a number of gadgets that illustrate this concept. One, called "Etch-A-Sketch®," allows you to draw a picture by turning two dials.

One dial causes a horizontal line to be drawn on a screen, the other makes a vertical line. If you simultaneously turn both dials in the same direction ("in phase"), you draw a 45° line. Another toy is a maze with two knobs that control its tilt in two perpendicular directions. Simultaneous control of both knobs allows you to roll a ball in any direction through the maze. Analogously, any direction of the electric field in the x-y plane can be thought of as consisting of two components that lie in two mutually perpendicular directions. Further, these directions can be any pair that we choose, not necessarily the x- and y-directions (Fig. 13.2e). While this seems like only a way to think about things now, we'll see that Nature thinks about them in just this way (Sec. 13.6).

What about **unpolarized light?** Such light is polarized in all different directions (perpendicular to the direction of propagation); the direction of the electric field *varies rapidly and randomly*—it has no preferred direction. We can think of this as a wave with two components that have a rapid and random variation of their relative phase— the two components are incoherent. It is like the result of turning the two dials of the Etch-A-Sketch® back and forth with no particular relation between them—you then get a line that wiggles around in all directions, going no place special. Figure 13.2f shows how we'll indicate such unpolarized light.

With this information, let's see why Rayleigh scattering produces polarized light.

B. Polarization due to Rayleigh scattering

A simple model can illustrate how scattering produces polarization. Figure 13.3 shows two long jump ropes that are joined together at their centers by a ring and then stretched out tightly at right angles, so as to form a (horizontal) cross. The ring is like a scatterer in this sense: If one end of one rope is wig-

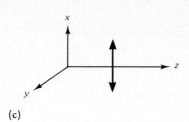

(a)

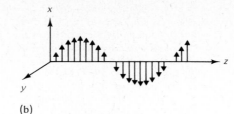

(b)

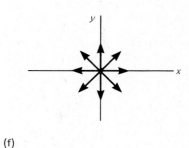

(c)

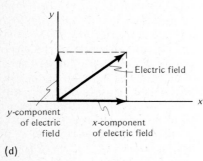

(d)

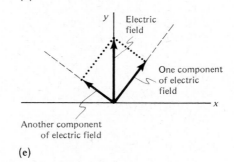

(e)

(f)

FIGURE 13.2

(a) A light wave traveling in the z-direction can have its electric field only in the x-y plane (or in any parallel plane). (b) The electric field of a light wave that is linearly polarized in the x-direction (also commonly, but less appropriately, called **plane polarized**). (c) A shorthand notation for the same wave as in (b). (d) Any electric field in the x-y plane can be thought of as a combination of a field in the x-direction with one in the y-direction. (e) An electric field in one direction within a plane can be thought of as a combination of fields in *any two* mutually perpendicular directions within that plane. (f) Unpolarized light (traveling in the z-direction).

gled up and down (linearly polarized in the x-direction), a wave propagates down that string, wiggling the ring up and down (Fig. 13.3a). This causes a (scattered) wave oscillating up and down to develop on the cross rope. However, if the first rope is wiggled sideways (linearly polarized in the y-direction), it cannot set up a wave in the cross rope because the ring *slides along* that rope and doesn't move it—there is no motion transverse to the cross rope (Fig. 13.3b). Thus, if we wiggle the first rope in both x- and y-directions, making an *unpolarized* wave, the only wave resulting in the cross rope would be *linearly polarized* in the vertical, x-direction.

The same idea applies to scatter-

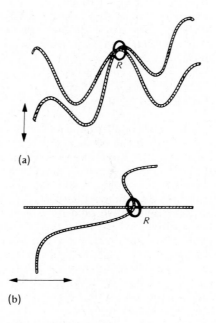

(a)

(b)

FIGURE 13.3

Two crossed ropes are joined at their centers by a very light ring that can slide freely along the ropes. (a) Wiggling one rope up and down (at the arrow) produces a wave. This wave, in turn, creates an up-and-down wave in the cross rope. (b) The first rope is wiggled sideways and simply slides over the cross rope, producing no wave in it.

ing of electromagnetic waves. Imagine that an unpolarized wave traveling in the z-direction strikes a small scatterer at O (Fig. 13.4).

Since the electric field of the unpolarized wave points in all directions *in the* x-y *plane,* the charges in the scatterer will oscillate in all those directions, but *not* in the z-direction. As usual, we may think of any such oscillations in the x-y plane as consisting of just an x- and a y-component. Only the x-component of oscillation radiates in the y-direction—the y-component cannot (because the scattered wave is transverse) and there is no z-component (because the incident wave is transverse). Hence, any light *radiated* in the y-direction (to E_1 or E_2) is *linearly polarized* in the x-direction. Similarly, the light reaching observers at E_3 and E_4 is linearly polarized in the y-direction. An observer at E_5, however, sees the incident wave as well as light scattered in the forward direction, both of which are unpolarized.

For the blue, Rayleigh scattered sky light this means that the blue light is linearly polarized for light coming from points in the sky 90° away from the sun (observers E_1 to E_4). However, the light coming from directions near the sun (observer E_5) or opposite the sun is unpolarized. For regions in between, the light is **partially polarized**—a mixture of polarized and unpolarized light. If there are large particles in the air (as in smog), the forces *within* them may cause the charges

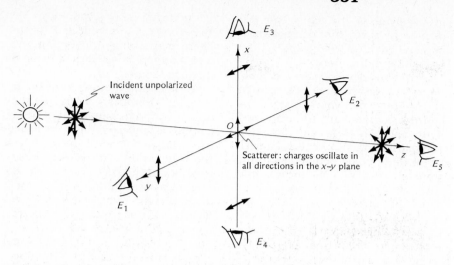

FIGURE 13.4

An unpolarized wave traveling in the z-direction strikes a scatterer at O. The wave that scatters along the y-axis is linearly polarized in the x-direction, while that which scatters along the x-axis is linearly polarized in the y-direction. That which scatters in the z-direction is unpolarized.

to oscillate in other directions than that of the electric field of the incident wave, so the scattered light is polarized less, if at all. Repeated scattering, as in clouds, causes the light to come out polarized in all directions—that is, unpolarized—so light from clouds is not polarized. (The TRY IT tells you how to verify some of these statements.)

It is possible to determine the location of the sun by measuring the polarization of the light from the blue sky. Many insects and arachnids seem to use this as a navigational device. For example, the wiggle dance of bees, by which they communicate the direction of food, depends on the direction of the light's polarization. The sensitivity of the bee's eye to polarization is different at different wavelengths, reaching a maximum at 355 nm. This suggests that a pattern of polarized light appears as a colored pattern to the bee. (Why?) These and other insects can locate the sun even when it is behind a cloud by detecting the polarization of the light from a patch of blue sky.

While *our* eyes are not sufficiently sensitive to the polarization of light for this navigational trick, we can still use it if we have the help of polarizing devices, which we'll describe below. With one such device (cordierite, a dichroic crystal—Sec. 13.5), which they called a "sun stone," Vikings are said to have used the polarization of the sky's light to navigate. Even today, the polarization of the sky in the twilight is of use to airplane navigators who fly over the poles.

A polarizing device that transmits light of one polarization, but not light of a perpendicular polarization, is called a ***polarizing filter***. Because of the polarization of the light from the blue sky, a properly oriented polarizing filter in front of your camera can block out that light, increasing the contrast between the sky and the white clouds.

Light can also undergo Rayleigh scattering from small scatterers *under water*. The resulting polarized light may be, as for airborne insects, a useful clue to underwater denizens for navigation. It may also provide a stable reference to help the animal stay in one place (station keeping). It may help to improve the animal's visual contrast (as it helped the camera "see" the clouds). For these and possibly other reasons, numerous underwater animals show a sensitivity to polarized light: crustaceans, cephalopods, fishes, and amphibi-

ans. Octopuses, for example, have even been trained to respond to 90° changes in the direction of polarization—so if you can't get a polarizing filter for the TRY IT, you can use a trained octopus.

TRY IT

Polarization of the sky

You will need a polarizing filter such as used in polarizing sun glasses, viewers for 3-D movies, or polarizing camera filters. Alternatively, Figure 13.6 shows how you can make one. You can detect the presence of polarized light by noting if the light becomes alternately dark and bright as you rotate the filter around your line of sight.

First use the filter to convince yourself that sunlight is unpolarized. In order to avoid looking directly at the sun, use a pinhole to project its image. Cover the pinhole with the filter and rotate the filter while you look at the image. Does the image of the sun become darker and lighter? Next look through the filter at some blue sky, 90° away from the sun. Again, rotate the filter. Is the light from that part of the sky polarized, as it's supposed to be? Try other parts of the sky. The more the polarization, the greater the difference between dark and light as you rotate the filter. Where is the polarization greatest and where least? How would you locate the sun by using the polarization of sky light? Also look at clouds and smog.

Check these ideas with the blue "sky" you made for the TRY IT for Section 13.2. From the side and from above, look through your polarizing filter at the blue scattered light. Also look end on into the transmitted beam. Additionally, place the filter between the flashlight and the milky water. Look from the side and notice what happens to the scattered light as you rotate the polarizer. Also notice what happens to the transmitted red "sunset." Explain what you see, using Figure 13.4. Repeat these experiments as you add more milk. In particular, when you've added more milk and you look through the filter at the beam from the side, you should notice that there is polarized light coming from the water close to the flashlight, but not from the water farther from the flashlight. Why does this happen?

13.4

POLARIZATION DUE TO REFLECTION

Polarized light may be made in other ways beside scattering. Probably the second most common source of polarized light also relies on the transverse nature of light—polarization by *reflection.*

When light in air strikes a smooth glass surface at an angle of incidence θ_i (Fig. 13.5), it wiggles the charges at the surface of the glass. There is a direction θ_r in which the radiation emitted from all these charges is in phase. This is the reflected beam, at $\theta_r = \theta_i$. In any other direction (in the air) the radiation from the different charges interferes destructively. Similarly, there is a direction θ_t of constructive interference between the incident radiation and that from the glass atoms. This is the transmitted beam, given by Snell's law. Again destructive interference eliminates rays in any other direction in the glass. We've drawn the figure for a special case we want to examine, where the *transmitted and reflected rays* just happen to be at *right angles to each other.* If the incident light is unpolarized, we can consider it as consisting of two components: one linearly polarized

in the plane of the figure (the *x-y* plane), as drawn, and one polarized perpendicular to the plane of the figure (the *z*-direction).

Let's consider the first component. The direction of polarization of the incident beam must be, as shown, perpendicular to the incident ray. Similarly, the direction of polarization of the transmitted beam must be perpendicular to the transmitted ray. This means that the electric field in the glass, and thus the direction in which the *charges oscillate* there, is *perpendicular to the transmitted ray.* But it is the radiation from these oscillating charges that produces the *reflected ray.* Because light is transverse, these charges cannot radiate along their direction of oscillation. Hence there *cannot* be a reflected ray perpendicular to the transmitted ray. Thus, the intensity of the reflected ray is zero for this special angle of incidence, called **Brewster's angle,** for which the reflected beam is perpendicular to the transmitted beam. (This is named after Sir David Brewster, who also invented the kaleidoscope.) Since angles of refraction depend on the two media involved, so does Brewster's angle. (Appendix L gives a mathematical expression.) Brewster's angle is typically near 56°, the value for light in air incident on glass.

Now consider the other component of the incident light—linearly polarized in the *z*-direction (i.e., perpendicular to the plane of the figure). Here the charges in the glass oscillate in the *z*-direction and

are perfectly free to radiate in the direction of the reflected beam. There is nothing unusual in this case, so there *is* a reflected beam of this polarization.

Thus, if an *unpolarized* beam arrives at Brewster's angle of incidence, only one component of polarization is reflected. The reflected beam is then linearly polarized in the *z*-direction. At nearby angles this is almost true—that is, of the light polarized in the *x-y* plane, very little is reflected. Hence, the reflected light is partially polarized, consisting of a large component of one polarization and a small component of the other. Since you often look at objects at angles near Brewster's angle, much of the reflected light that you see is polarized. (The TRY IT invites you to check this.)

The situation is different for light reflected from *metals.* Visible light is not transmitted into metals because there are so many electrons free to move parallel to the surface, which cancel any internal electric fields (Sec. 2.3B). Electrons moving freely parallel to the surface can radiate in all directions away from the surface and can hence create reflected beams of *both* polarizations—the reflected light is not linearly polarized.

Because much of the specularly reflected light from nonmetallic surfaces is at least partially polarized, polarizing filters are often used as sunglasses. If they are oriented so as to remove the component of light that is polarized horizontally, such sunglasses eliminate *specular* reflections (glare) from roadways, lakes, and other horizontal surfaces. On the other hand, light reflected *diffusely* will have been reflected or scattered several times and will thus be unpolarized. Only half of such light is blocked by your sunglasses (the horizontal component), so you still see the road itself but with the glare reduced. For the same reason, airport control towers and the bridges of ocean liners often have sheets of polarizing filters over their windows.

The surface reflections shown in

FIGURE 13.5

Incident, transmitted, and reflected ray directions for the case when the angle of incidence is equal to Brewster's angle.

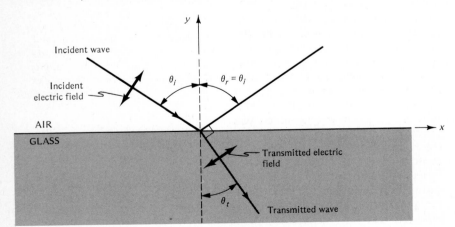

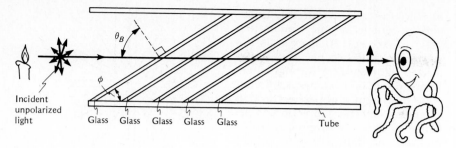

Plate 9.7a were eliminated in Plate 9.7b by means of a polarizing filter in front of the camera lens, oriented so as to block them.

PONDER

What was the orientation of the polarizing filter?

Several other devices take advantage of the polarization at Brewster's angle. At normal incidence, each surface of a piece of glass reflects 4% of the incident intensity, so 92% is transmitted. But suppose you want to transmit light through 100 pieces of glass. Then 0.92 of the incident beam is transmitted by the first piece, 0.92 of *that* by the second, etc. After 100 pieces, only

$$(0.92)^{100} = 2.4 \times 10^{-4}$$

of the original intensity has been transmitted—that is, hardly anything. You can do much better if you slant the pieces of glass so that the light is incident on them at Brewster's angle. At this angle, about 15% of one polarization component is reflected at each of the 200 surfaces. However, *none* of the other polarization component is reflected. Since *all* of that polarization is transmitted, it can comfortably pass through all 100 pieces of glass. Hence, if unpolarized light is incident on this arrangement, half of the intensity is transmitted.

This idea is often used in gas lasers (Sec. 15.4). In such a laser, the light is reflected back and forth between two mirrors about 100 times. For precise adjustment, the mirrors are situated outside glass windows that contain the gas. To avoid

FIGURE 13.6

A polarizing filter made of Brewster windows. Each piece of glass is at Brewster's angle, θ_B, to the incident light direction. The light that reaches the observer consists predominantly of the component of polarization in the plane of the figure. The inside of the tube should be black, so the reflected light is absorbed. The angle ϕ is equal to $90° - \theta_B$. For a glass in air, $\theta_B = 56.3°$, so $\phi = 33.7°$.

losses by reflection, these windows are positioned so the beam strikes them at Brewster's angle. They are then called **Brewster windows.** After 100 passages back and forth through these windows, light of one polarization is completely lost, but the other polarization is not diminished by reflections—the laser light produced this way is therefore polarized.

The idea of the Brewster window can be used to make a polarizing filter of the type needed in many of the TRY IT's. About five pieces of glass (microscope slides work well) are positioned one behind the other, each at Brewster's angle to the incident beam (Fig. 13.6). While not as effective as 100 pieces of glass, this device considerably diminishes the component of light polarized perpendicular to the plane of the figure, but still passes the component polarized in that plane—it therefore acts as a polarizing filter. Like all other polarizing filters, you can use this to detect polarized light. Unlike other filters, however, you can use this to tell the *direction of polarization* of the light, since the easily visible slant of the glass tells you which component the Brewster window passes.

TRY IT

FOR SECTION 13.4
Polarization of reflected light

You will need a polarizing filter, as in the TRY IT for Section 13.3B. Look through this filter at light reflected from a shiny surface (not metal) such as a polished desk or floor, a shiny table top, a piece of smooth white paper, or a piece of glass with some black paper behind it. Position yourself and a light so you see a good specular reflection of the light. Rotate the filter to see if the light is polarized, as you did before. Compare the amount of polarization that you see when the light strikes the surface at near normal incidence, at intermediate angles, and at glancing incidence. At which angle do you expect to see the greatest polarization? Do you?

Repeat the experiment with a diffusely reflecting surface such as a piece of cloth or a piece of white paper with a matte finish. Why isn't the light polarized here? Also try a shiny metal surface, such as a polished cooking pan, a stainless steel knife blade, or the chrome surfaces of a car.

Light reflected from the road has both a diffuse part (by which you see the road) and a specular part (the glare). On a sunny day, look at a road through your polarizing filter, and notice what happens to the glare as you rotate the filter.

Light in the primary rainbow has been reflected once (Sec. 2.6B), so you can expect it to be polarized also. If you think of how a mirror located at some point on the rainbow would have to be oriented if it were to reflect sunlight to your eyes (as the droplets do), you should be able to convince yourself that the light from the rainbow must be polarized along the direction of the arc (rather than radially). That is, light from the (horizontal) top of the bow should be polarized horizontally, whereas light from the vertical arms of the bow (near the pot of gold) should be polarized vertically. Check this with a polarizing filter the next time you see a rainbow (or make a rainbow with your garden hose). What do you expect happens in the secondary rainbow? Does it?

13.5

POLARIZATION DUE TO ABSORPTION

One of the most commonly used visible-light linear polarizers *absorbs* one component of polarization, while transmitting the perpendicular components. **Polaroid** consists of a parallel array of long-chain molecules whose electrons cannot freely move *across* the narrow molecules. When incident light is polarized so its electric field pulls across the molecules, it can't make the electrons move. Hence they don't radiate, and the incident wave continues unaffected—that component of polarization is transmitted. However, when the electric field of the light drives electrons *along* the long molecules, the electrons do move and *absorb* the light's energy. Thus only light of one polarization (across the molecules) is transmitted.

This type of polarizing filter was invented by Edwin H. Land in 1928 when he was a 19-year-old undergraduate. (His interest had been stimulated by reading about some polarizing crystals that had been discovered in the 1850s when a physician's pupil, for some peculiar reason, put drops of iodine in the urine of a dog who had previously been fed quinine.) Sheets of Polaroid material, usually mounted between thin sheets of glass or plastic, form the polarizing filters you often find in sunglasses. Whereas they absorb most of the visible light of the appropriate polarization, they are not completely effective at the shorter wavelengths. For this reason very bright horizontally polarized light (e.g., bright sunlight polarized by reflection from water) may look deep blue or violet through your polarizing sunglasses, instead of black as it would for an ideal polarizing filter.

Various naturally occurring crystals absorb one component of polarization more than they do the perpendicular component. The semiprecious stone tourmaline is one example. However, tourmaline's absorption is more selective than Po-

laroid's—not all wavelengths are absorbed equally, so the transmitted light is colored. Such crystals are called **dichroic** because even unpolarized light passing through them in one direction becomes a different color than light that passes through them in another direction.[*] Because of the color produced, these dichroic crystals are not usually used as polarizing filters. The word dichroic has come to mean any material that produces polarized light by absorption, so Polaroid is considered dichroic.

The back of your eye contains a dichroic material: the yellow pigment (macula lutea) that absorbs light between 430 and 490 nm (blue) and covers your fovea (the place corresponding to the center of your field of view—see Sec. 5.2B). The dichroism in this case consists of stronger absorption when the direction of polarization of light is perpendicular to the pigment fibers than when it is parallel. The pigment in the macula is arranged radially, like the spokes of a wheel, as shown schematically in Figure 13.7a. Suppose that white light incident on your eye is polarized vertically (i.e., parallel to fibers *6* and *12* in Fig. 13.7a). As a consequence of the dichroism, the vertical fibers *(6* and *12)* absorb the blue part of this light least, and fibers perpendicular to this direction (fibers *3* and *9)* absorb it most strongly. Therefore, you see a horizontal dark yellow line (Fig. 13.7b) known as **Haidinger's brush.** As shown in the figure, the yellow brush is often accompanied by adjacent blue regions, presumably due to simultaneous color contrast (Sec. 10.6A). If, instead, the direction of polarization were horizontal, the yellow brush would be vertical—it always lies perpendicular to the direction of polarization. The TRY IT tells you how to look for Haidinger's brush.

---

[*]Greek *dis*, twice, plus *chros*, color. Crystals exhibiting three colors, for light passing through them in three different directions, are called trichroic (Greek *treis*, three). Generally, all such crystals are called pleochroic (Greek *pleion*, more).

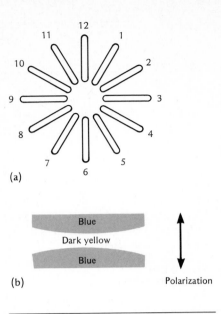

(a)

(b)   Polarization

FIGURE 13.7

(a) The dichroic pigment in the macula lutea is arranged radially. **(b)** Haidinger's brush, when the light is polarized vertically (arrow).

TRY IT

FOR SECTION 13.5
Haidinger's brush

The best way to look for Haidinger's brush is to use a linear polarizing filter and a light blue, nonshiny background, such as a large piece of blue construction paper. Brightly illuminate the paper, and look through the filter at a point near the center of the paper. Because the image will fade as your eye adapts, it helps to rotate the filter occasionally about your line of sight. If you don't see the yellow-brown brush at first, don't be dismayed; it took Helmholtz twelve years to see it after he first learned of the phenomenon from Haidinger. You should be able to do much better—perhaps a few minutes. Once you see the brush, notice how it rotates as you slowly rotate the filter. This provides another technique for determining the direction of polarization of the filter, because the brush is always perpendicular to it.

Another good background is the blue sky. Use your polarizing filter to see the brush there. Because the light from the blue sky is already polarized (Sec. 13.3B), some people are able to see Haidinger's brush in the sky without the aid of a filter. In what direction should you look? Try it!

13.6
POLARIZING
AND ANALYZING

A polarizing filter can be used in two different ways. It can be used as a **polarizer**—because it transmits only one component of polarization, incident unpolarized light striking a polarizing filter results in polarized transmitted light. Alternatively, it can be used as an **analyzer**—you can detect the *presence* of polarized light with it, as in the TRY IT for Section 13.3B, and even the *direction* of polarization once the filter is calibrated. (You can calibrate a polarizing filter—i.e., find the direction of polarization that it passes—by looking at the light reflected near Brewster's angle from a shiny floor, since that light must be horizontally polarized.)

Figure 13.8a shows a set-up using two polarizing filters, one as a polarizer and one as an analyzer. Let's start with the incident light

unpolarized, and the polarizer oriented to pass only *vertically* polarized light. Suppose the analyzer is oriented to pass only *horizontally*

FIGURE 13.8

(a) Incident light strikes the first polarizing filter (the polarizer). The light that passes through the polarizer (the intermediate light) strikes the second polarizing filter (the analyzer). The amount of light that passes through the analyzer (the transmitted light) depends on the relative angle of the polarizer and analyzer. **(b)** Crossed polarizer and analyzer: incident unpolarized light striking a vertically oriented polarizer produces vertically polarized light, which is not passed by a horizontally oriented analyzer. **(c)** Parallel polarizer and analyzer: the same as in **(b)**, but now the vertically polarized light is passed by the vertically oriented analyzer. **(d)** Now the analyzer, oriented at an intermediate angle, passes only that component of the intermediate light polarized at that same angle. A weaker transmitted light results. Note that **(b)** to **(d)** represent head-on views of the light beam and of the filters.

---

polarized light (the polarizer and analyzer are **crossed**—Fig. 13.8b). Since only vertically polarized light strikes the analyzer, no light passes through it—there is no transmitted beam reaching the observer. If instead the analyzer is oriented *vertically* (the polarizer and analyzer are **parallel**—Fig. 13.8c), it passes all the light striking it—the observer then sees the full intermediate intensity (assuming ideal polarizing filters).

Thus you can use an analyzer to tell if the polarizer is doing its job. Makers of polarizing sunglasses often give you a little analyzer so you can convince yourself that the glasses do polarize (without breaking the pair in half).

PONDER

What transmitted light results if the polarizer is oriented horizontally in Figures 13.8b and c?

Suppose now that the analyzer is oriented at some other angle to the polarizer (Fig. 13.8d). Now we must think of the vertically polarized intermediate light as consisting of two components (as in Fig. 13.2e), one *along* the angle of the analyzer, and one oriented *perpendicular* to it. Only the former component (that parallel to the analyzer's orientation) passes through the analyzer. There *is* a transmitted beam in this case, but it has less intensity than in the case where the analyzer and polarizer are parallel. (Malus's law, the mathematical relation between the intensity of the transmitted light and the angle between the analyzer and polarizer, is given in Appendix M.)

Thus, by adjusting the angle of orientation between the polarizer and analyzer, you can control the intensity of the transmitted light. This is often a convenient way of controlling light intensity because it doesn't modify the size or shape of the beam (as a mechanical diaphragm would), and, at least for ideal polarizing filters, it doesn't affect the color temperature (as dimming an incandescent bulb would).

Since the analyzer tests the polar-

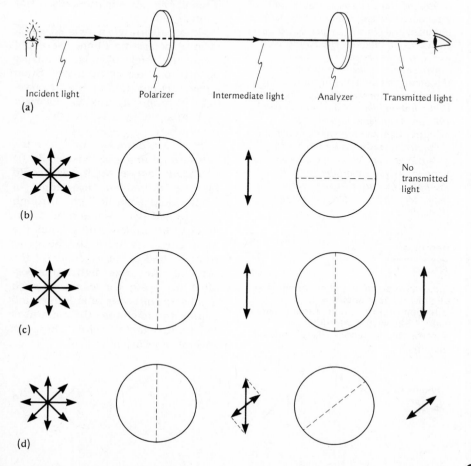

Incident light   Polarizer   Intermediate light   Analyzer   Transmitted light
(a)

(b)   No transmitted light

(c)

(d)

ization of the intermediate light, we can use this set-up to examine the effect of various types of obstacles on polarized light. For example, in Section 13.3B we mentioned that polarized light that suffers repeated scatterings, as in clouds, becomes unpolarized—it is **depolarized.** The TRY IT shows how you can verify that statement using this set-up and depolarizers such as waxed paper or ground glass.

A remarkable phenomenon occurs when you introduce a *third* polarizing filter, inserted between the polarizer and the analyzer. Initially, with only two filters present, no light passed through the crossed polarizer and analyzer; when you looked through them they appeared dark. But now you insert the third polarizing filter between them, oriented at an angle of 45°, an orientation midway between those of the polarizer and the analyzer. When you look through this *three*-filter combination, you see light. This is quite remarkable when you think about it; the filter you insert does nothing but absorb some of the light, yet some light is now transmitted after you inserted an absorber!

We can understand this magic by simply repeating our previous analyses. At each polarizing filter we must think of the light as consisting of two components: one along the polarizing filter's orientation and one perpendicular to it. Thus you start with unpolarized light incident on the vertically oriented polarizer. As in Figure 13.8, only the vertical component passes. When only the crossed analyzer is present, the vertical component incident on it is blocked, so no light is transmitted (Fig. 13.8b). When the third polarizing filter is inserted at a 45° angle, however, some of the vertically polarized light incident on it is transmitted by it *as light polarized at 45°* (as in Fig. 13.8d). *This* light, now incident on the horizontally oriented analyzer, can be thought of as consisting of *both* a vertical and a horizontal component, as in Figure 13.2d. Since there is now a horizontal component present, *it* can pass through

the analyzer—some of the light is therefore transmitted through the entire system. (If the filters are ideal, the transmitted intensity is one-fourth of the original intensity.)

This is a dramatic demonstration that not only is it *useful* to think of any electric field in terms of components, it is *necessary*—it is the way Nature works.

TRY IT

FOR SECTION 13.6
Depolarization by multiple scattering

You will need two polarizing filters: a polarizer and an analyzer, as in Figure 13.8. Place one behind the other with their polarization directions crossed, so no light passes through them. A good source of multiple scattering is a piece of waxed paper or ground glass (which are translucent rather than transparent because they repeatedly scatter the light). If the light is depolarized by this multiple scattering, the analyzer will pass some of it because one component of the then unpolarized light will be parallel to the analyzer's orientation. Hold a piece of waxed paper between the crossed polarizer and analyzer and look through this arrangement at a light source. The waxed paper should intercept only a part of the intermediate beam, so you can still see the polarizer through the analyzer. Can you see any light passing through the waxed paper when the polarizer and analyzer are crossed? What happens as you rotate the analyzer? Does the polarized light that strikes the waxed paper remain polarized after passing through it?

FIGURE 13.9

Successive polarizing filters oriented at slightly different angles rotate the direction of polarization. The intensity of light is unchanged after filter a if successive orientations change only slightly.

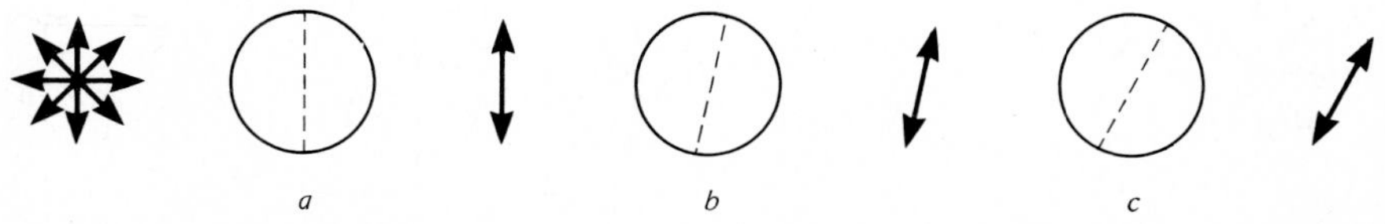

*13.7
CONTROLLING POLARIZATION WITH ELECTRIC FIELDS

We can carry the "magic" of the last section further and arrange that *all* of the intensity be transmitted.

Suppose the polarizer *(a)* is oriented vertically (Fig. 13.9). Light passing through it will then be polarized vertically. If the next polarizing filter *(b)* is oriented at a *slight* angle from the vertical, the light that passes through it will be polarized in this tipped direction, but the reduction of its intensity is negligible. If now we add another polarizing filter *(c)* that is tipped still further from the vertical, at a small angle from the previous one, it will pass light polarized in this new direction. If we continue this process, with enough filters we can rotate the angle of polarization to any direction we choose; and if the angle between successive polarizing filters is small enough, the intensity is not diminished.

Thus, with such a twisted stack of ideal polarizing filters, we could rotate the angle of polarization by 90° without loss of intensity. Such a device could then be used to pass light through a crossed polarizer and analyzer when inserted between them.

If we could make such a stack whose total angle of twist could be *changed* fast enough, we could have a *shutter*—a zero angle of twist would pass no light through the crossed polarizer and analyzer, while a 90° angle would pass all the light emerging from the polarizer (that is, half of the intensity of the original unpolarized light). It is possible to create the effect of such stacks out of layers of suitable *molecules* and to change the amount of twist with *electric fields*. There are several ways of doing this.

a   b   c

A. Liquid crystal displays

Liquid crystals are liquid because their molecules move about within the liquid; they are crystals because their molecules orient themselves in an array. In one type of liquid crystal (called **nematic**), all the molecules in a given layer are oriented parallel to each other and to the layer (Fig. 13.10a). The direction of orientation depends on the environment at the surface. For example, a piece of glass whose surface has been rubbed with a cloth or paper in a given direction will have microscopic grooves (a few atoms deep) in that direction. Nematic molecules adjacent to that surface of the glass will then orient themselves along these grooves. If the liquid crystal lies between *two* glass plates whose surfaces have been rubbed in a given direction, then all the molecules will lie oriented in that direction, as in the figure. If now one of the glass plates is rotated, the molecules closest to that plate will be rotated along with it, producing a twist in the layers of the molecules (Fig. 13.10b). If the thickness of this layer of liquid crystal is large compared with the wavelength of visible light, this **twisted nematic cell** is similar to a twisted stack of polarizing filters in the following sense—if light is polarized in the plane of the figure and incident on the top of the cell, it emerges from the bottom polarized perpendicular

---

FIGURE 13.10

(a) The elongated molecules of a nematic liquid crystal between two suitably prepared glass plates. **(b)** A twisted nematic cell: the same as **(a),** but now the bottom plate has been rotated 90°, so the molecules of the liquid crystal closest to the bottom are now oriented perpendicular to the plane of the figure. **(c)** The same cell as in **(b)**, but with an electric field between the plates.

(a)   (b)   (c)

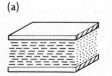
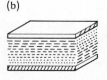
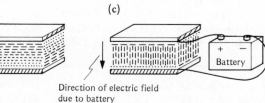

Direction of electric field due to battery

Battery

(a)

Polarizer

Analyzer

(b)

Polarizer

Battery

Analyzer

No light transmitted

---

to the plane of the figure. However, if a battery of a few volts is connected to transparent conductors on the two plates, the molecules tend to line up parallel to the electric field—that is, perpendicular to the plates (Fig. 13.10c). Light incident on the top of the cell now will *not* have its polarization altered. If the battery is then disconnected, the molecules return to the alignment of Figure 13.10b.

This cell thus provides a convenient switch for light. The cell is placed, suitably oriented, between a crossed polarizer and analyzer. When there is no electric field, light passes through the system (Fig. 13.11a). When the electric field is turned on, however, no light passes (Fig. 13.11b).

This arrangement is commonly used in display devices, such as the display in digital watches **(liquid crystal displays, LCD's)**. In that case, the same set-up shown in the figure is used with a mirror after the analyzer. Thus, with no applied

---

FIGURE 13.11

(a) A twisted nematic cell allows light to pass through a crossed polarizer and analyzer. **(b)** With an electric field applied between the plates of the cell, no light emerges from the analyzer.

electric field, incident light is polarized vertically, say, by the polarizer, is rotated 90° by the cell, passes through the analyzer, is reflected by the mirror back through the analyzer (still polarized horizontally), is rotated 90° by the cell to become vertically polarized again, and passes back out through the polarizer. If you look at this display from the front, you see this reflected light, and the display looks bright. When the electric field is turned on, however, no light reaches the mirror and hence none is reflected—the display looks dark. If the electric field is applied only in certain regions that form the segments of a number, you see a dark number against a bright display. Figure 13.12 shows one way of controlling the location of the electric field to form numerals.

B. Pockels and Kerr cells

Certain solids, whose crystal structures are sufficiently asymmetric,

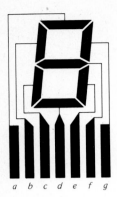

a   b   c   d   e   f   g

FIGURE 13.12

An arrangement of transparent conductors used to display numerals. The seven terminals, *a* through *g*, are located above the liquid crystal and can be separately connected to the battery. For example, to display the numeral 2, terminals *a, g, f, c,* and *d* are connected together to one side of the battery, and a conductor below the liquid crystal is connected to the other side of the battery. Because no electric current actually flows across the cell, this arrangement makes very little drain on the battery, so it is better suited to watches than other devices that require a light source.

can also be used to control light's polarization. A transparent cell made of such a crystal, along with a means of applying a large field (voltages of 10^3 volts across a centimeter-size cell are required), is called a ***Pockels cell.*** The electric field (which for some cases is parallel to the direction of light propagation and for others is perpendicular) can cause the cell to change the polarization direction by 90° and thus control transmission of light through a crossed polarizer and analyzer. Such cells are used for very fast shutters that allow light to pass for as brief a time as a nanosecond!

With even larger fields, the molecules of some solids or liquids (possibly even gases) that normally are not sufficiently asymmetric can be made so and aligned. We then have a ***Kerr cell,*** which, like a Pockels cell, responds extremely rapidly.

13.8
BIREFRINGENCE

Many substances can affect the polarization of incident light without absorbing or reflecting one of the components. They do this because their structures are sufficiently asymmetric that the *speed* of light in these materials is different for one polarization component than it is for the perpendicular component. That is, there may be a direction in which it is hardest to make electrons oscillate or one in which it is easiest to make them oscillate—just as you may find it easier to set your bed springs vibrating up and down than side to side. Such substances are called ***birefringent**** or ***double refracting.*** (Pockels and Kerr cells affect the polarization because they become birefringent when the electric field is turned on. The aligned molecules of a nematic liquid crystal are birefringent even in the absence of an electric field.) Because (transverse) light has both a *propagation* and a *polarization* direction, different phenomena result for different values of each of these directions.

Let's consider only the simplest case, where light polarized in one direction (call it the *z*-direction) travels with one speed, while light polarized in the *x*- or *y*-direction travels with a *different* speed—the same for any direction in the *x-y* plane. The *z*-direction is then called the ***optic axis.*** The phenomena we see now depend on the direction in which we send the light. Suppose first that the light propagates in the *z*-direction. It can then be polarized only in the *x*- and *y*- directions. But the speed of light is the same for polarizations in these two directions, so nothing special happens.

But now suppose that the light propagates in, say the *y*-direction. It can then be polarized in the *x*- or *z*-direction, and each of *these* polarizations travels with a different speed. The formula in Section 1.3A

*Latin *bi*, double, plus *refringere*, break, (as in refract).

tells us that the faster component has a longer wavelength (since the frequency of both components is the same). Hence (while inside the material) light polarized in the *x*-direction has a *different wavelength* than light polarized in the *z*-direction. After a certain distance, then, the two polarizations will have traveled a different *number* of wavelengths. For example, when passing through a slab of such material, light polarized in the *x*-direction might travel 5 of *its* (longer) wavelengths, while light polarized in the *z*-direction travels $5\frac{1}{2}$ of *its* (shorter) wavelengths (Fig. 13.13a). If the light entering the slab is polarized at an angle between the *x*- and *z*-directions, its *x*- and *z*-components are *in phase*—when the *x*-component is positive, so is the *z*-component (Fig. 13.13b). Because of the extra half wavelength traveled by the *z*-component, when the light leaves the slab these components regain the equal wavelengths they entered with, but now they are *out of phase*—when the *x*-component is positive, the *z*-component is negative (Fig. 13.13c). The first TRY IT shows you how to use this effect on the phase to find, make, and test some common birefringent materials.

Such a slab of birefringent material that retards the phase of one component with respect to the other is called a ***retardation plate.*** The particular example of Figure 13.13 is called a ***half-wave plate*** because it introduces a difference in phase of half a wave. Such a plate changes the direction of polarization, flipping it about one of the two axes as shown in the figure. Whether a given plate is a half-wave plate depends on the wavelength of the light—generally, a half-wave plate for green light is not a half-wave plate for blue or red light.

Notice that *unpolarized light* is *not affected* by this half-wave plate or by any thickness of birefringent material as long as the light travels in this direction (perpendicular to the optic axis). Since each component only has its phase changed, if the incident light is polarized in all directions (i.e., unpolarized), the

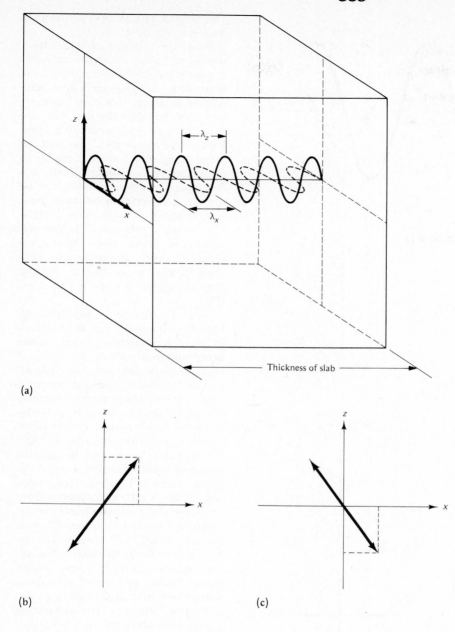

Thickness of slab

(a)

(b)

(c)

FIGURE 13.13

(a) A wave traveling in the *y*-direction through a slab of birefringent material for which light has different speeds when polarized in the *x*- and *z*-directions. Here, light polarized in the *x*-direction has the faster speed, and hence the longer wavelength, λ_x. Light polarized in the *z*-direction has the slower speed, and hence the smaller wavelength, λ_z. In this example, the thickness of the slab happens to be equal to $5\lambda_x$, as well as to $(5\frac{1}{2})\lambda_z$. The two polarization components, which entered in phase with each other, therefore emerge half a wavelength out of phase. (b) Light entering the slab has the *x*- and *z*-components in phase—that

is, when the electric field of the wave has a positive *x*-component it also has a positive *z*-component. (c) Light emerging from the slab then has these two components out of phase—when the electric field has a positive *x*-component it has a negative *z*-component.

emergent light will also be polarized in all directions.

Now suppose we slice the slab of Figure 13.13 in half, so the difference in phase introduced between

the two polarization components is a *quarter* of a wave. (In this example, the *x*-component travels $2\frac{1}{2}$ of its wavelengths, the *z*-component $2\frac{3}{4}$ of its wavelengths, a difference of $\frac{1}{4}$ of a wave.) Such a plate is called a **quarter-wave plate.** To see what this does to the polarizations, consider again an incident wave in which the *x*- and *z*-components are in phase, as in Figure 13.13b. When this light emerges from the quarter-wave plate, these components will be one quarter of a wave out of phase (Fig. 13.14). If we draw the electric fields at different times all on one diagram (Fig. 13.14f), we can see that the field at any point in the wave moves around an ellipse. Such light is said to be **elliptically polarized.** Had the *x*- and *z*-components been of equal size (i.e., had the incident light been linearly polarized along an angle of 45° from the *x*- and *z*-axes), the wave would have been **circularly polarized** (Fig. 13.14g). In that case, if we looked at the wave at any point along the *y*-axis after it emerged from the quarter-wave plate, we would see the electric field at that point simply changing its direction but not changing its size. At neighboring points on the *y*-axis, we'd see the same thing except that the electric field would be rotating a little behind or ahead of the field at the first point. It is as if a long, helical, coiled spring were being rotated about its axis. The helical coils represent the electric field. As they are rotated, the coils seem to move along the spring (like the colors moving up a barber's pole); but seen end on, each coil just rotates around in a circle.

Thus, a linear-polarizing filter followed by a quarter-wave plate oriented at 45° to it becomes a circular-polarizing filter—incident unpolarized light emerges as circularly polarized light.

PONDER

For this to work, the light must first pass through the linear polarizing filter, then through the quarter-wave plate. Why won't it work if the order is reversed?

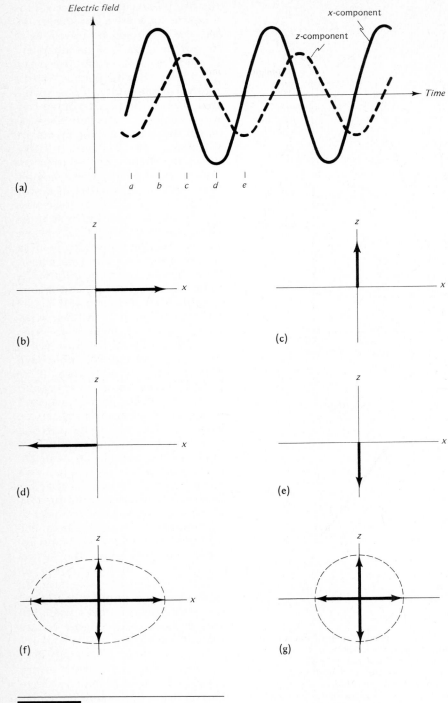

FIGURE 13.14

(a) The x- and z-components of the electric field are a quarter of a wave out of phase. The direction of the electric field at the times marked *b, c, d,* and *e* are shown in **(b), (c), (d),** and **(e),** respectively. **(b)** When the x-component is positive, the z-component is zero. **(c)** Later, when the x-component is zero, the z-component is positive. **(d)** A little later, the x-component is negative and the z-component is zero. **(e)** Still later, the x-component is zero and the z-component is negative. **(f)** The electric fields at all four times are drawn on one graph to show that they travel around on an ellipse. **(g)** If the x- and z-components have equal amplitudes, the electric field moves around on a circle.

The second TRY IT tells you how to demonstrate this.

Since the amount of retardation produced by birefringent materials depends on the wavelength of the light, these materials will often look colored between a crossed polarizer and analyzer. (See the first TRY IT.) Many biological materials (e.g., proteins, nucleic acids), as well as other substances, are not visible in an ordinary microscope, but, because they are birefringent, can be seen in a ***polarizing microscope***— a microscope with built-in polarizing filters, above and below the specimen. When these filters are crossed, the birefringent substances appear as bright, colored objects against a dark background (Plate 13.1).

Some substances, while not originally birefringent, can be made so by the phenomenon called ***stress birefringence;*** when a piece of such a material, say transparent plastic, is under stress it becomes birefringent—the greater the stress, the more the birefringence. If the stress is nonuniform, the birefringence is also nonuniform, and different parts of the plastic exhibit different colors when it is seen between a crossed polarizer and analyzer (Plate 13.2). Transparent plastic models of machine parts can be examined in this way, and the pattern of colors will then map the stress pattern in the parts. You often see colors due to stress polarization when you wear polarizing sunglasses and look through your car's rear window. The polarizing sunglasses serve as analyzers, oriented crossed to the incident polarized light that is reflected from the road (the polarizer). The window was cooled unevenly during its manufacture, so it has a stress pattern in it that becomes visible as a colored array when viewed this way. (Polarized light from the blue sky may also serve as a source for viewing this pattern.) The third TRY IT suggests other ways to see these stress-birefringence colors.

First TRY IT

Birefringent materials

For this you will need two polarizing filters, some cellophane (often used to wrap hard candies), and some household plastic wrap. Place the cellophane behind a polarizing filter and look through the combination. If the intensity doesn't change as you rotate the cellophane, you know it is not a linear polarizing filter. Now place the cellophane between a crossed polarizer and analyzer and again rotate it. You should find certain directions of the cellophane that do not let the light through and certain directions that let most of the light through. This is what you expect from a birefringent material. In the example of Figure 13.13, if the polarizer is oriented along either the x- or the z-direction, only that polarization enters the slab, and its direction is unchanged by the slab. (If only one of the two components is present, there is no relative phase to retard.) No light can then pass through the analyzer. For any other orientation of the polarizer, both an x- and a z-component are present, so the relative phase retardation changes the direction of polarization, allowing some light through the analyzer.

Repeat the experiment with the plastic wrap. If the plastic has not been stretched, you should see very little effect. But if you gently stretch the plastic in one direction (as much as you can without tearing it), you introduce a substantial asymmetry—you make an optic axis along the direction of stretch. Now repeat the experiment and notice that the greatest amount of light passes through when the direction of stretch is at 45° to the crossed polarizer and analyzer directions.

The amount of phase retardation depends on both the wavelength of the light and the direction of the optic axis. Fold over several layers of the stretched plastic wrap (or of transparent sticky tape) so that there are different thicknesses of plastic in different places, and view this between the crossed polarizer and analyzer. You'll need from two to ten layers. Another substance that works very well is a small chip of ice. In each of these cases, the polarizations of different wavelengths of the light are affected differently, so some colors pass more than others—you should see different colors in different parts of the plastic, tape, or ice (the ice colors can be particularly impressive). Holding the polarizer and analyzer fixed, rotate the material between them. Notice how the colors change. (Why?) By cutting out shapes consisting of different numbers of layers of stretched plastic wrap or plastic tape, you can make a picture whose colors change as you rotate it between a crossed polarizer and analyzer. This technique is sometimes used in light shows.

Perhaps surprisingly, you can get similar effects simply by placing a crumpled piece of cellophane or stretched plastic wrap on a smooth surfaced table near a window. Blue sky light that passes through the cellophane and is then reflected from the table will also exhibit different colors in different parts of the reflected image of the cellophane.

PONDER

What serves as the polarizer here? The analyzer? If you have a south window, what times of the day are best for this effect?

Frost on the window, when you view its reflection from the puddle of water on the sill as the frost starts to melt, will appear as colorful frost flowers for the same reason.

Second TRY IT

A circular-polarizing filter

You can make a circular-polarizing filter from a linear-polarizing filter and a piece of plastic wrap that has been sufficiently stretched so it serves as a quarter-wave plate. Hold the plastic wrap behind the filter and immediately in front of a shiny metal surface. Gently stretch the plastic at a 45° angle to the filter. As you stretch the plastic, the metal appears colored. What happened is this—the plastic became a quarter-wave plate, so the light passing through the linear polarizing filter was converted to circularly polarized light. When it was reflected from the metal, the direction of the circular polarization was reversed. (That is, a right-handed helix or screw becomes left-handed when reflected. Circularly polarized light can be either left- or right-handed, depending on the way the electric field moves around the circle in Fig. 13.14g.) This reversed circularly polarized light is converted back into linearly polarized light by the quarter-wave plate. However, it is now polarized perpendicular to the linear polarizing filter's orientation, so it does not pass through. Thus you don't see this reflected light from the metal. For the particular wavelength for which the plastic is a quarter-wave plate, no light is returned from the metal. The effect is most obvious when this wavelength is in the green. Then green light is removed from the reflected light, and the metal looks dark purple.

Try reversing the order of the plastic and filter to verify the claim made in the PONDER.

If you use aluminum foil as the shiny metal, you can fold a vee-shaped crease in it so light will be reflected twice before returning. What happens when you view this crease through your circular-polarizing filter? Why?

Third TRY IT

How to enjoy stress

A solid piece of plastic in which there are holes makes a good candidate for stress birefringence. Clear plastic protractors, right angles, and other plastic drafting devices are often quite suitable. View the plastic between a crossed polarizer and analyzer, looking especially near any sharp angles, where the stress is greatest. If you find a good set of colors, notice how they change as the plastic is rotated while the filters are held still. Try stressing the plastic further by squeezing or twisting it.

Clear, unflavored gelatin also works well and is particularly easy to stress. Notice how the colors change as you wiggle the gelatin, sending waves through it.

13.9
OPTICAL ACTIVITY

Two perpendicular *linear* polarization components travel through birefringent materials, we saw, with different speeds. Similarly, the two different directions of rotation of *circularly* polarized light travel with different speeds in **optically active** materials. Certain crystals, such as quartz, are optically active because their molecules are arranged into a twisted crystal, which allows the electrons to *rotate* more easily one way than the other. Such crystals lose their optical activity when melted or fused. Other materials, such as sugar and turpentine, are optically active because the molecules themselves have a twist in them. These materials are optically active even when in a liquid solution. (In a liquid the molecules are oriented every which way, canceling out any birefringence. But this does *not* cancel out the optical activity. A right-handed screw does not become left-handed when it points the other way—no matter what direction it is pointing, you still have to turn it clockwise to advance it.)

When *linearly* polarized light passes through an optically active material, its direction of polarization is *rotated*. The angle through which the polarization is rotated depends on the *thickness* of the material and the *wavelength* of the light in that material, just as the amount of retardation produced by a birefringent material depends on these quantities. This means that a given thickness of optically active material will rotate the polarization of different colors by different amounts. If you put some optically active material between a polarizer and an analyzer, then, depending on the orientation of the analyzer (compared to that of the polarizer), different colors will be removed from the transmitted light. For example, suppose the thickness and optical activity of the material

are just such that the polarization of blue light is rotated by 90°, while that of red light is hardly rotated at all (Fig. 13.15). If the analyzer is oriented parallel to the polarizer, it passes all of the red light and some of the green, but it blocks the blue light—the transmitted light is yellowish. As the analyzer is rotated, however, the colors change. When the analyzer is oriented perpendicular to the polarizer, it passes all of the blue and some of the green, but blocks the red—now the transmitted light is greenish blue. The TRY IT tells you how to see this lovely, colorful phenomenon. One application of it is a variable color filter for photography. An optically active material between two linear polarizing filters, one of which can be rotated, makes a filter whose color changes as the polarizing filter is rotated.

Optical activity is also used by stereochemists, who want to determine the three-dimensional structure of molecules. Many common and important molecules are optically active, and the direction and amount of twist in them can be studied this way. For example, most sugars are right-handed, while all optically active amino acids found in living objects are left-handed.

FIGURE 13.15

The amount of rotation of the polarization direction depends on the thickness of the optically active material and on the wavelength of the light. Here incident white light is linearly polarized vertically. The emerging light is also linearly polarized, but with a different direction of polarization for each color.

TRY IT

FOR SECTION 13.9
Color around the kitchen

To see the colors produced by optical activity, you need only two linear-polarizing filters, some optically active material, and a light source. Colorless corn syrup contains right-handed sugar molecules and is ideal for this purpose. You can use it right in the bottle if the bottle is clear.

Tape one polarizing filter to a flashlight to make a beam of polarized light. It helps to block any extraneous light from the flashlight with opaque tape or aluminum foil. In a dark room, shine the (polarized) light through the corn syrup bottle and onto a piece of white paper, which will serve as a screen. Now use your other polarizing filter as an analyzer, between the bottle and the screen. Notice the color of the light on the screen as you rotate the analyzer.

Try the experiment with the light passing through different thickness of corn syrup, say by tipping the bottle, Also try using turpentine.

*13.10
MORE BIREFRINGENCE—PROPAGATION AT ODD ANGLES

In Section 13.9 we considered the special cases where light propagated along or perpendicular to the optic axis (as in Fig. 13.13 where the optic axis is in the z-direction, the wave traveling in the y-direction). There is, of course, no reason why we cannot attempt to send light in some other direction through the crystal—at some odd

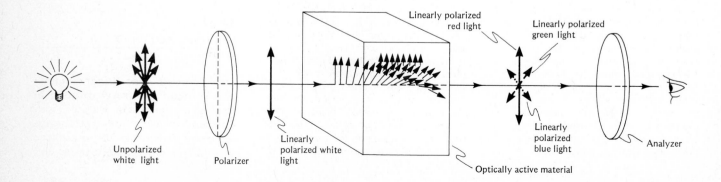

Unpolarized white light   Polarizer   Linearly polarized white light   Linearly polarized red light   Linearly polarized green light   Linearly polarized blue light   Optically active material   Analyzer

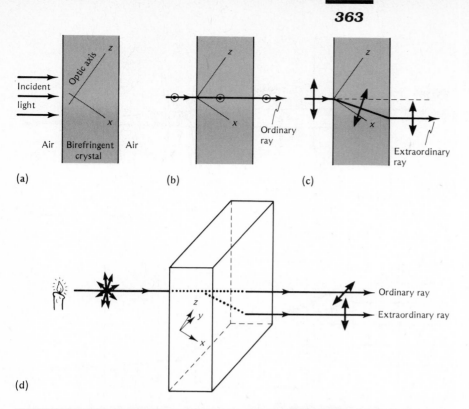

(a) (b) (c)

(d)

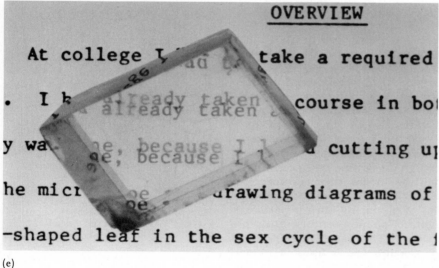

(e)

angle to the optic axis. Let's consider some of the extraordinary consequences of doing this.

Suppose we cut a slab of birefringent material so that its optic axis lies at an odd angle to the surface of the slab. In Figure 13.16a, we have chosen the z-axis to lie along the (skewed) optic axis direction and the y-axis to be perpendicular to the plane of the figure. The speed is the same for light polarized in the x- or y-directions, but different for light polarized in the z-direction. (That's what we mean when we say the z-axis is the optic axis.) Suppose, now, that unpolarized incident light arrives perpendicular to the surface and thus travels neither parallel nor perpendicular to the optic axis. As usual, we must consider this light as consisting of two perpendicular components and examine each component separately.

Let's first consider the component whose polarization is in the y-direction (Fig. 13.16b). Since only one speed of light is involved, nothing special happens to this component—it simply propagates through the crystal in the direction it was traveling (perpendicular to the surface) in a perfectly ordinary way. We call this ray the **ordinary ray.** (Since one doesn't ordinarily give the name "ordinary" to ordinary occurrences, you might expect that there is something extraordinary coming. Stick around!)

The other component of polarization of the incident light lies in the x-z plane—perpendicular to the incident ray, and thus at an odd angle to both the x- and z-directions (Fig. 13.16c). That is, the electric field here is partly in the x-direction and partly in the z-direction. But these two directions correspond to *different* speeds of light—it is easier to oscillate the electrons in one direction than the other. Suppose that, in this particular birefringent crystal, it is easier to oscillate electrons in the z-direction. The amplitude of the electron oscillations in the z-direction will then be greater, compared to those in the x-direction, than would have otherwise been expected. Hence, there will be propor-

---

FIGURE 13.16

(a) A birefringent crystal cut with its surface at an odd angle to its optic axis (here the z-direction). The y-direction is perpendicular to the plane of the figure. In this example, the incident light arrives perpendicular to the crystal. **(b)** The incident light here is linearly polarized in the y-direction (perpendicular to the plane of the figure—we indicate this by the symbol ⊙) and proceeds through the crystal in the direction it was originally traveling. **(c)** The incident light here is linearly polarized in the x-z plane (the plane of the figure). It is bent on entering and leaving the crystal, and thus emerges displaced. **(d)** Unpolarized light breaks up into two components. One, polarized in the y-direction, proceeds through the crystal undeflected, as in **(b)**—the ordinary ray. The other, polarized in the x-z plane, is deflected at the surfaces of the crystal, and emerges displaced as in **(c)**—the extraordinary ray. **(e)** Photograph of the view through such a birefringent crystal.

tionally more radiation from the electron oscillations in the *z*-direction. This means that inside the crystal the electric field is oriented more toward the *z*-direction than it was in the incident beam (see figure)—the electric field inside the crystal is *not* parallel to the electric field of the incident beam. But the direction of *propagation* of the light must be perpendicular to the electric field. Hence the direction of propagation in the crystal is *not* parallel to that of the incident beam—inside the crystal the light travels at an angle to the incident beam. This extraordinary occurrence—that light incident perpendicular to the surface should decide to bend on entering the crystal—is in violation of Snell's law, and earns this ray its name: the **extraordinary ray.** At the second surface the process is reversed (the figure is the same if the light is traveling from the right), so the extraordinary ray emerges parallel to the incident ray, but *displaced.*

The situation for unpolarized incident light is summarized in Figure 13.16d. The polarization components emerge separated from one another. If you look at a point source through such a crystal, you see *two* apparent sources (Fig. 13.16e)!

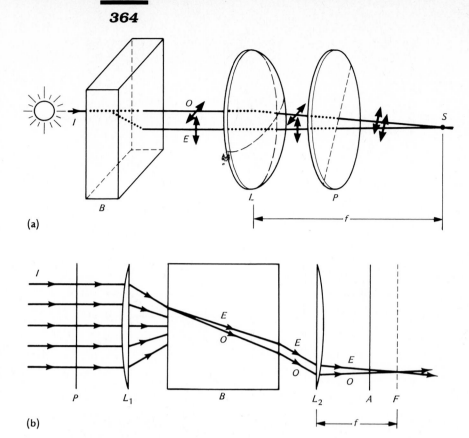

(a)

(b)

Though we have come upon this process toward the end of our discussion, it in fact provided the earliest way of making a polarizing filter. A piece of calcite (a birefringent material also known as Iceland spar or calcium carbonate) was cleverly cut so the extraordinary ray passed though the crystal, but the ordinary ray was totally internally reflected (Sec. 2.5A). This was the standard way of producing linearly polarized light, for about a century, until the advent of Polaroid. (See Table 13.1 for calcite's two indexes of refraction, corresponding to its two different speeds of light.)

FIGURE 13.17

(a) A birefringent crystal *(B)* cut at an angle to its optic axis separates an unpolarized incident beam *(I)* into ordinary *(O)* and extraordinary *(E)* beams. A converging lens *(L)* focuses these beams to a point *(S)* in its focal plane. A polarizing filter *(P)* passes the same component of polarization in each beam, allowing the beams to interfere with each other at *S.* **(b)** Incident unpolarized light is linearly polarized by the polarizer *P.* The lens L_1 makes this light converge within the birefringent crystal *B.* Within the crystal, each ray (only one shown) is broken into an ordinary *(O)* and extraordinary *(E)* beam. After emerging, these beams are brought together at the focal plane *F* of the second converging lens L_2. The analyzer *A* passes only parallel polarization components of the two beams, allowing the beams to interfere.

The emerging ordinary and extraordinary beams of Figure 13.16d can be made to interfere with each other by passing them through a linear polarizing filter oriented somewhere between the directions of polarization of the two beams, and reuniting the two beams (Fig.

TABLE 13.1 *Indexes of refraction for various birefringent crystals*

Material	Index of refraction at 589 nm	
	Electric field parallel to optic axis	*Electric field perpendicular to optic axis*
Calcite	1.658	1.486
Quartz	1.544	1.553
Sapphire, Ruby	1.77	1.76
Ice	1.309	1.313

13.17a). (Only parallel components can interfere with each other—a vertical component cannot cancel a horizontal one.) The amount of interference depends on the phase retardation of one beam with respect to the other, and this in turn depends on the thickness of the material and the wavelength of the light.

It is possible to get a range of both thicknesses and colors, and thus to produce a spectacular interference display, by using the set-up of Figure 13.17b. Here incident unpolarized light is linearly polarized and then made to converge within a birefringent crystal by means of a lens. The birefringent crystal separates the incident rays into ordinary and extraordinary beams, which are brought together by a second lens and passed through an analyzer. The path length through the crystal, and hence the retardation, depends on the angle at which the light strikes the crystal. Thus, different angles result in different amounts of interference—you see rings of interference fringes in the focal plane of the second lens. If white light is used, these rings are quite colorful (Plate 13.3)—the color of each ring depends on which wavelengths suffer destructive interference for light entering the crystal at that particular angle. At certain points, the incident beam is polarized in either the ordinary or extraordinary polarization direction. For these points, there is only one beam in the crystal, and it emerges with the same polarization as it entered. Hence if the analyzer is crossed with the polarizer, this beam will be blocked—for these points, therefore, no light reaches the focal plane. This accounts for the dark crosses on the plate.

The particular pattern seen depends on the nature and orientation of the crystal. Had the crystal in the plate been cut in a different direction, a different pattern would result. Some crystals have *three* different indexes of refraction and can produce patterns that resemble a figure eight. These patterns then reveal information about the nature, structure, and orientation of the crystal, and therefore are a particularly colorful way of identifying mineral specimens.

SUMMARY

Scattering—radiation of light in all directions when it interacts with small, isolated objects—is responsible for many of the colors we see in nature. For instance, **Rayleigh scattering** occurs if the objects are much smaller than the wavelength of the incident light and have their important resonances at frequencies greater than that of the light. Rayleigh scattering is most effective for shorter wavelengths and is responsible for the blue of the sky. In general, the color of the scattered light depends on the size of the scatterer and the frequencies of its resonances.

Unpolarized light, which contains an electric field that varies rapidly and randomly in direction, becomes **linearly polarized** when it suffers Rayleigh scattering to the side, because there is no **component** of the electric field in the direction of propagation of light. (In other directions, the scattered light is **partially polarized.**) Polarized light can be both made and detected with a **polarizing filter.** In some filters one component of polarization is reflected more than the perpendicular component. This occurs when light is reflected from glass, say, at **Brewster's angle.** (In a **Brewster window,** light is transmitted through several layers of glass, all at the same special angle.) Alternatively, one polarization component can be absorbed more than another, as in **Polaroid** and in natural **dichroic** materials. (The dichroic material of your macula lutea is responsible for **Haidinger's brush.**)

After light emerges from a **polarizer,** the amount that is subsequently transmitted by an **analyzer** depends on the relative orientation of the two filters—it ranges from zero when they are **crossed** to 100% when they are **parallel** (for ideal filters), if nothing lies between them. Anything placed between them that **depolarizes** the light will allow some light to pass through a crossed polarizer and analyzer, as will a third polarizing filter, if its orientation is at some angle intermediate to that of the polarizer and analyzer. A twisted stack of polarizing filters rotates the direction of polarization of light. This direction can also be changed by an electric field, which can therefore control the amount of light passing through a crossed polarizer and analyzer, as in **liquid crystal** displays (e.g., in a **twisted nematic cell**), in **Pockels cells,** and in **Kerr cells.**

Birefringent (or **double-refracting**) materials have different indexes of refraction for different polarization components, one for polarization along and one for polarization perpendicular to the **optic axis** of the material. Different thicknesses of birefringent materials form different types of **retardation plates,** depending on the amount of relative change in the phase that two perpendicular polarization components undergo. A phase difference of half a wavelength results from a **half-wave plate,** and a quarter of a wavelength results from a **quarter-wave plate.** Depending on their relative orientation, a quarter-wave plate following a linear polarizing filter may produce **elliptically** or **circularly polarized** light.

A crossed polarizer and analyzer are used in a **polarizing microscope** to observe birefringence. They can also reveal **stress birefringence** and identify **optically active** materials, which rotate the plane of polarization.

Light traveling through a birefringent crystal in a direction neither parallel nor perpendicular to the optic axis is broken into two rays of perpendicular polarization components: an **ordinary ray** and an **extraordinary ray.** Colorful interference fringes result when converging light passes through such a crystal placed between a polarizer and an analyzer.

PROBLEMS

P1 Pictures taken by astronauts on the moon always show the moon sky to be black. Why is this so?

P2 Once in a blue moon there are particles in the air of size such that they scatter red light predominantly. Explain how this affects the color of the moon.

P3 In the figure, the (setting) sun looks red to the traveler and the camel, but the (overhead) sun looks white (or yellow) to the sailors. Why this difference in color?

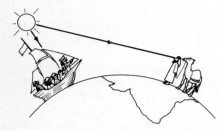

P4 Suppose, for the sake of argument, that instead of scattering blue light, the air *absorbed* yellow light from a large, white sphere surrounding the atmosphere. (a) What color would the sky be? (b) How would the color of the sun change toward sunset?

P5 An astronaut in earth orbit found that "the most spectacular sunrises and sunsets we saw were regions of the atmosphere where the pollution was at its worst." Explain his observation.

P6 Suppose a horizontal beam of light has an electric field that oscillates only in a vertical direction. The light is: (a) projected, (b) perpendicular, (c) polarized, (d) paralyzed, (e) plagiarized. (Choose one.)

P7 The figure is supposed to illustrate how a polarizing filter works. Explain. (Say what represents the light beam and the filter and what the French bread loaves that the man is carrying mean. Neglect the polar bear. Be brief.)

P8 (a) Why do polarizing sunglasses usually reduce the glare of reflected light from a level surface better than ordinary sunglasses? (b) What about glare off the side of a vertical surface, such as a marble column?

P9 Why do fishermen wear polarizing sunglasses?

P10 Suppose you want to look through a window to see what the people inside are up to. Even though you are wearing polarizing sunglasses, you may still see glare from the pane. (a) Why? (b) What can you do to cut this glare?

P11 (a) Redraw Figure 13.5 for the case where the light is traveling from glass to air. (b) Why is Brewster's angle for this case equal to the transmitted angle for the case of Figure 13.5?

P12 (a) Light *reflected* perpendicularly from the surface of a piece of Polaroid is not polarized. Why? (b) Light reflected at an angle from a piece of Polaroid is polarized. Why? (c) If the piece of Polaroid in (b) is horizontal, does the direction of polarization change when you rotate the Polaroid, keeping it horizontal?

P13 Which of the following, when inserted between a crossed polarizer and analyzer, allows light to emerge from the analyzer? (a) A linear polarizing filter with the same orientation as the polarizer. (b) A linear polarizing filter oriented at an angle intermediate to the polarizer and analyzer orientations. (c) A piece of waxed paper. (d) A clear red filter. (e) Rayleigh scatterers (the light being scattered forward).

P14 Suppose you have a twisted stack of ideal polarizing filters, as described in Section 13.7. If the total twist is 90°, you can use this so that 100% of the light that emerges from a polarizer will be transmitted through a crossed analyzer. (a) Draw a diagram similar to Figure 13.11 showing how the stack should be oriented with respect to the polarizer. Show the direction of polarization of the light between the polarizer and the stack, and between the stack and the analyzer. (b) In what direction should you orient the initial polarizing filter of the stack so that *no* light is transmitted between a crossed polarizer and analyzer? Why is no light then transmitted?

P15 A half-wave plate is inserted between a crossed polarizer and analyzer, oriented at an angle of 45° to the orientations of the polarizer and analyzer. Does the half-wave plate allow light to emerge from the analyzer? Explain.

P16 Repeat P15 for a quarter-wave plate.

P17 In Figure 13.15, the direction of polarization of the green light is rotated by 45° in the clockwise direction. If the orientation of the analyzer is rotated from that of the polarizer by 45° in the opposite (counterclockwise) direction, what is the color of the transmitted light? (See Sec. 9.4B.)

P18 Linearly polarized monochromatic light is incident vertically on the top of a glass of clear corn syrup. How does the polarization of the light that emerges from the bottom change as more syrup is added to the glass?

P19 (a) With a diagram similar to Figure 13.16d, show why two images can be seen through an appropriately cut birefringent crystal. (b) How would the apparent separation of these images differ if the crystal were thicker?

P20 When a newspaper is viewed through an appropriately cut piece of calcite, two images of the print can be seen. Describe and explain what you see if you view these images through a linear polarizing filter while rotating the orientation of the filter.

HARDER PROBLEMS

PH1 The bee's sensitivity to polarized light is greatest at 355 nm. Suppose the ultraviolet-sensitive cones respond only to vertically polarized light, while the other two cone types respond equally to all polarizations (Fig. 10.7). (a) As the sun moves, the direction of polarization of the northern sky changes from vertical to horizontal. Describe how the brightness and color of the northern sky changes, as seen by the bee. (Describe the color using ultraviolet, blue, green, or the appropriate combination of these.)

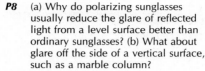

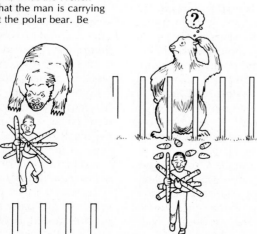

(b) At what time of day is the polarization horizontal?

PH2 A Mach cone analyzer is a simple device for determining whether light is linearly polarized and, if so, its direction of polarization. It consists of a glass cone with the light striking its sides vertically from above its point (see figure). The sides of the cone are slanted so that the light arrives at Brewster's angle to the slanting glass surface. A beam of unpolarized light incident on it from above will be reflected into a circular pattern on a horizontal white screen directly below the cone. If instead the light is linearly polarized, the pattern will no longer form a complete circle. Suppose the downward traveling incident light is linearly polarized in the north-south direction. (a) What will the reflected pattern look like and why? (b) How does the reflected pattern differ if the light is polarized in the east-west direction?

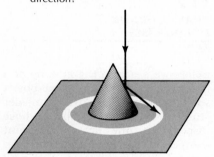

PH3 The reflection of a rainbow is observed in the calm water surface of a lake, at approximately Brewster's angle. Noting that the rainbow's light is polarized (see the TRY IT for Sec. 13.4), describe which parts of the reflection appear brightest and which appear dimmest.

PH4 Consider a digital watch having a display of the type shown in Figure 13.11 but with a mirror on the right, after the analyzer, and the viewer's eye on the left. So its display can also be viewed in the dark, this watch has a (transparent) phosphorescent material (Sec. 15.3B) as a source of light in it. (a) Why should this light source lie between the analyzer and the mirror? (b) If the source of light is a small bulb instead, where can it be located?

PH5 Polarized light is incident on a quarter-wave plate, with its direction of polarization parallel to the optic axis. What is the nature of the polarization of the light when it emerges from the quarter-wave plate?

PH6 Supose the birefringent crystal of Figure 13.16 is cut in two, with the cut parallel to the surfaces shown in the figure (i.e., perpendicular to the ordinary ray). (a) If the two pieces are placed together in their original positions, how many images will be seen by a viewer looking through the crystal? (b) If, instead, one of the pieces is rotated about the ordinary ray, how many images will be seen?

PH7 Linearly polarized white light passes through a substance that changes the direction of polarization of different wavelengths (colors) by different amounts. The figure shows how the polarization varies with color when the beam emerges. (a) What is the color of this light when seen with the naked eye? (b) This light is sent through a polarizing filter that transmits only the vertical component. What is the color of the light transmitted by the filter? (c) The filter of part (b) is rotated 90°. Now what is the color of the transmitted light?

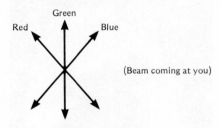

PH8 A long cylinder of quartz is illuminated from one end by a monochromatic source of linearly polarized light. When the quartz cylinder is viewed from one side, scattered light from the beam can be seen only at periodic intervals, with dark regions in between. (a) Explain this phenomenon in terms of the optical activity of the quartz. (b) If the polarization of the incident light is slowly rotated, the dark regions move along the cylinder. Explain this.

PH9 You have one set-up with a twisted stack of polarizers and one with an optically active material. Each material is adjusted to rotate the plane of polarization by 90°. In each case they lie between a crossed polarizer and analyzer in such a way as to pass a particular monochromatic light through the system. You don't know which set-up is which and would like to find out by examining the transmitted light. Explain why each of the following procedures will or will not work: (a) rotating the analyzer only; (b) rotating the polarizer only; (c) rotating both together; (d) changing the wavelength of the light.

MATHEMATICAL PROBLEMS

PM1 What is Brewster's angle for light traveling from air to diamond ($n_t = 2.4$)?

PM2 Consider a plate of glass with parallel faces that is surrounded by air. Show that if light is incident on the glass at the Brewster's angle for the air-glass surface, then the light traveling through the glass will be incident on the second surface at the Brewster's angle for the glass-air surface.

PM3 A third polarizing filter is inserted at 45° between a crossed polarizer and analyzer. Use Equation M.3 to prove that one-fourth of the intensity of the light emerging from the polarizer is transmitted through the analyzer (assuming ideal filters).

PM4 A beam of light shines through a polarizer and analyzer. The analyzer is then rotated once per second. How many pulses of light emerge from the analyzer in a minute?

PM5 A beam of light shines through a crossed polarizer and analyzer. Another polarizing filter is inserted between the polarizer and analyzer and rotated once per second. How many pulses of light emerge from the analyzer in a minute?

PM6 In the set-up of P18, suppose the incident polarization is in the north-south direction. When the height of the corn syrup is 10 cm, the emerging light is polarized in the east-west direction. Give three heights of corn syrup that will result in the emerging beam being polarized in the north-south direction.

Holography

14.1
INTRODUCTION

During the last few decades a new means for storing and displaying images has been developed that is richer and more versatile than ordinary photography. In addition to its many other remarkable achievements, it is the solution (at least in principle) to the problem of the mischievous artist of Chapter 8, who attempted to delude you by replacing your window view of the outside world with a picture. While her picture could convey the ambiguous depth cues, its lack of parallax would ultimately give her away. Why can't the picture record the parallax information that is evidently carried by the light from the actual scene? The picture records only the intensity of the light, but light carries more information; as a wave, it carries both *intensity* and *phase* information. It is the combination of these two that conveys the three-dimensionality of the scene. **Holography*** is a technique for recording both the intensity *and* phase of the light and hence *all* the information that the light carries, including parallax information. Because the **hologram** records the intensity and phase of the light that reaches it, we should be able to use it to **reconstruct** that light by Huygens' construction—the light will then continue on, as if it were still coming from the original object, with all its three-dimensional information.

The three-dimensionality of the

*Greek *holos*, whole, complete.

resultant image seldom fails to surprise and mystify people who have not seen holograms previously. A New York jeweler exploited this in 1972 by displaying a hologram in the store window. A life-like, three-dimensional image of a diamond necklace (Fig. 14.1) hovered over the sidewalk. The hologram caused quite a sensation until an elderly lady cracked the window with her

FIGURE 14.1

Photograph of the hologram that appeared in Cartier's window. A photograph cannot display the eerie, three-dimensional quality of actual holographic images.

umbrella, convinced that the image was the work of the devil.

Let's see how the devil holograms are made and how they work.

14.2
TRANSMISSION HOLOGRAMS

Before we describe the making of a hologram of a complicated object such as a necklace, let's consider simpler holograms—those whose objects are one or a few point sources. These illuminate the relevant physical principles, and we'll find that we've already discussed two simple holograms: the diffraction grating (Sec. 12.3A) and the Lippmann plate (Sec. 12.3D).

A. Transmission holograms of individual point sources

A *monochromatic* light wave leaving an object and passing through a window, say, can be completely characterized by its intensity and phase at all points of the window. Photographic film, however, responds only to the intensity of light. If information about phase is to be recorded on film, we must employ some sort of *interference*. For this reason, holography requires *two coherent light beams*.

How can interference be used to record the phase of a light wave? Consider an idealized case with two plane waves (from two distant coherent point sources) striking a piece of photographic film (Fig. 14.2a). Two beams, the **object**

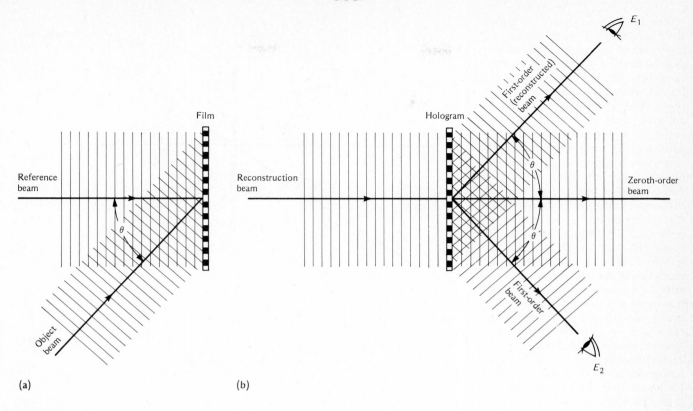

(a) (b)

FIGURE 14.2

(a) Production of a hologram of a distant point source, whose plane wave (the object beam) arrives at angle θ. The reference beam (also a plane wave) arrives perpendicular to the film. The lines represent wavefronts (say crests) of the object and reference beams. The high-resolution photographic film is exposed in regions of constructive interference, where crests of the two beams coincide on the film (black). The film is left unexposed in the regions of destructive interference (white). **(b)** After development, the film acts like a diffraction grating. When illuminated by the reconstruction beam *alone*, the hologram emits a replica of the object beam (reconstructed beam) at the same angle, θ. In addition, the zeroth-order beam and the other first-order beam are emitted by the hologram. Eye E_1 sees an image of a distant point source identical to the original source of the object beam, whereas eye E_2 sees the other first-order beam.

beam and the **reference beam,** arrive with some angle θ between them. There are then regions of constructive interference alternating with regions of destructive interference on the film—fringes are produced in the plane of the film. If its grain is sufficiently fine, the film will record the individual fringes.

Where there is constructive interference the exposure is greatest, so when the film is developed those regions will have black silver grains. At the points of destructive interference there was no exposure, so the developed film is transparent there. Thus the resultant negative consists of many parallel, closely spaced opaque lines (running perpendicular to the plane of the figure) separated by transparent lines. It can therefore act as a diffraction grating. (In fact, many transmission diffraction gratings are made this way.)

This negative is a **transmission hologram** of the distant point source that produced the object beam. If you look at it, however, you won't see a three-dimensional image of that source. You won't even see the interference fringes because they're too fine. In order to regain the information contained in this hologram, you must use it as a diffraction grating. You must illuminate it by the **reconstruction beam,** a beam identical to the reference beam used during exposure. (Note that this is the *only* beam you now use for illumination.) You view this transmission hologram from the side opposite to the source of illumination (Fig. 14.2b), where the various orders of interference fringes are produced. The zeroth-order beam can be considered the reconstruction beam continuing on its merry way through the grating. (If you put your eye in it, you see the reconstruction light source shining in your eye.) The grating also produces two first-order beams. One, we'll see, leaves the hologram at the angle θ, as if it were a continuation of the original object beam. It is therefore a plane wave of the same wavelength and traveling in the same direction as that object beam. It acts in all respects like the continuation of the object beam beyond the film plane. For this reason, we say that the hologram *reconstructs* the object beam. In addition to this **reconstructed beam,** the other first-order beam leaves the hologram at an angle θ on the opposite side of the normal to the hologram.

How do we know that the recon-structed beam indeed leaves the hologram at the *same* angle as that at which the object beam arrived? Consider the close-up view of the film at an instant when a crest from the reference beam strikes it (Fig. 14.3a). There is a fringe at every point of the film where there is constructive interference, that is, wherever a crest from the object beam strikes the film at that in-stant. During reconstruction, the first-order diffraction beam leaves the hologram at an angle that de-pends on the fringe spacing. Be-cause the wavelength λ and fringe spacing (grating constant) d are unchanged for the cases of expo-sure and reconstruction, the angle θ is *also* unchanged. If the object beam had arrived at a large angle, then the fringes would be more closely spaced than before and the reconstructed beam would leave the hologram at the same large angle as its associated object beam arrived (Fig. 14.3b). (Recall Sec. 12.2E. Also see Appendix N for a mathe-matical proof.)

Thus the *phase* information of the object beam is recorded in the *spacing* of the fringes in the holo-gram. The *intensity* information of the object beam is recorded in the *contrast* of the fringes (the differ-ence between the intensities of the brightest and darkest fringes): the more intense the object beam, the greater the contrast.

We have thus, by means of the hologram, reconstructed a light beam. The original beam may have been recorded on film in New York, and the developed hologram used to reconstruct that beam in Vienna, months later. ("Between the lines of photographs I've seen the past," sings Janis Ian.) While this light beam may not be very interesting, it illuminates our point.

Now, let's consider the produc-tion of a transmission hologram of something slightly more interest-ing: a *nearby* point source, S (Fig. 14.4). As before, we use a plane ref-erence beam striking the film per-pendicularly, but now the object beam consists of a wave diverging

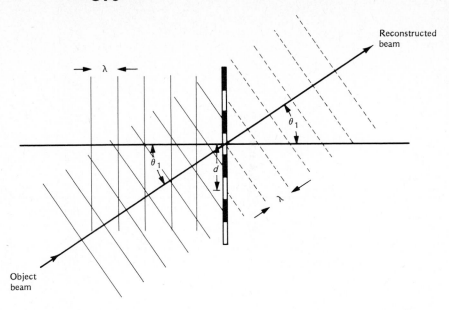

(a)

(b)

FIGURE 14.3

(a) Exposure and reconstruction of a hologram of a point source emitting wavelength λ, placed far from the film at angle θ_1. The reconstructed beam is shown with dashed wavefronts because it and the object beam are *not* present at the same time. **(b)** As in **(a)** except the source is at a larger angle, θ_2. The grating constant here is therefore smaller than in **(a)**, so the reconstructed beam leaves the hologram at the larger angle θ_2.

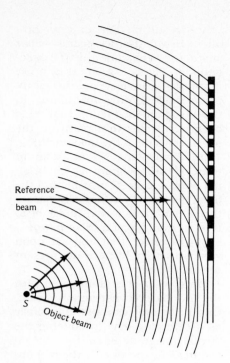

FIGURE 14.4

Production of a hologram of a point source S that is near the film. The object beam here is a wave diverging from the source.

from S. Again, there are interference fringes formed on the film—only here, they are *not* evenly spaced. This is because at different parts of the film the object and reference beams arrive with different angles between them—at the top of the film in Figure 14.4, the beams are traveling almost perpendicular to each other, hence the fringe spacing is narrow there. Closer to the bottom of the film, the beams are traveling in the same direction, hence the fringe spacing is greater there.

After the exposure, the film is developed to produce unequally spaced opaque and transparent regions—a nonuniform diffraction grating. This hologram, like those before, is illuminated by the reconstruction beam alone (Fig. 14.5). Each small patch of this hologram acts as a diffraction grating. Of course, all parts of the hologram pass the zeroth-order beam straight through, but more important, the hologram reconstructs the diverging object beam; at the top of the film, the first-order beam leaves at a large angle, whereas at the bottom of the film, the beam leaves at a smaller angle (Fig. 14.5a). The emitted wavefront is continuous across the face of the hologram, and hence a diverging wave leaves the front of the hologram, just as if the point source S were present and there were no hologram (Fig.

FIGURE 14.5

Reconstruction using the hologram produced in Figure 14.4. **(a)** The reconstructed beam is shown leaving a small portion of the top of the hologram (small grating constant) and a small portion of the bottom of the hologram (large grating constant). **(b)** Because the fringe spacing varies gradually along the hologram, the actual reconstructed beam wavefronts are smooth—they form a replica of the original diverging object beam. Eye E_1 can see the resulting virtual image at S, whereas eye E_2 cannot.

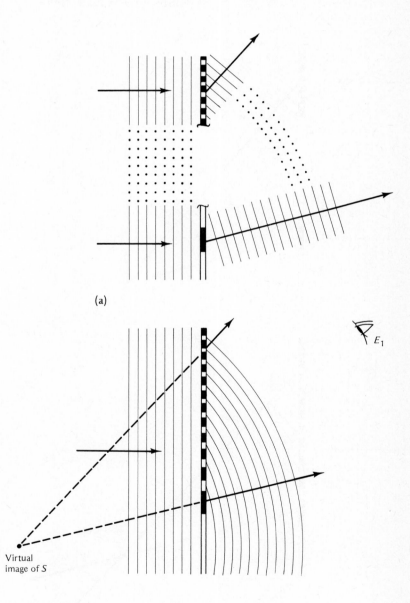

(a)

(b)

Virtual
image of S

E_1

E_2

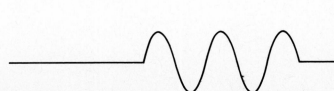

FIGURE 14.6

Reconstruction using the hologram produced in Figure 14.4. **(a)** The same as in Figure 14.5a, but now showing the other first-order beam, which leaves the top of the hologram at a large angle *downward* and leaves the bottom of the hologram at a smaller angle. **(b)** Because the fringe spacing varies gradually along the hologram, the wavefronts are smooth and converge at a point S' in front of the hologram. Eye E_2 can see the real image at S', whereas eye E_1 cannot.

14.5b). When you view this hologram from the far side, you see a point source of light at the original position of S—a virtual image of the original point source. If you move up and down, you get different views of the point source (parallax) because you see it through different parts of the hologram. Moreover, your two eyes must converge to see the point of light. In short, it is as if you were looking at the original point source of light, even though no point source is really there.

But a diffraction grating produces *two* first-order beams—what about the other one? In the hologram of Figure 14.2b, this other first-order beam leaves at the appropriate angle θ, only *downward*. In the hologram of Figure 14.4, this beam *converges* to a point on the observer's side of the hologram (Fig. 14.6). This point, S', is as far in front of the hologram as the original point source, S, was behind the film. Because light actually passes through S', it is a *real* image of S. It is an example of a ***false image***— an extra image that appears upon reconstruction, but was not present during exposure.

Thus, when it is illuminated with the normal reconstruction beam, this hologram acts simultaneously as a diverging lens (producing the virtual image at S) *and* as a converging lens (producing the real image at S').

Another way to produce an image from this hologram is to illuminate it with the ***conjugate beam***—a beam for reconstruction that is opposite to the normal reconstruction beam (Fig. 14.7). The conjugate beam strikes the hologram from the side opposite that of the reference beam and produces a *real* image at the location of the original object.

B. Transmission holograms of extended objects

Real objects of any interest consist of more than one point. Suppose an object beam comes from *two* separate point sources. The wave from each source will interfere with the reference beam. The film simultaneously records the two resulting sets of interference fringes, and the final hologram reconstructs *both* sources (and produces two false images). Likewise, a hologram of three, four, or more point sources properly reconstructs each source in its proper position.

We can think of any illuminated three-dimensional object, such as a tea cup, as being made up of myriad such point sources. By Huy-

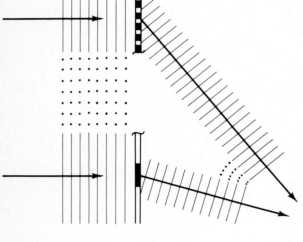

(a)

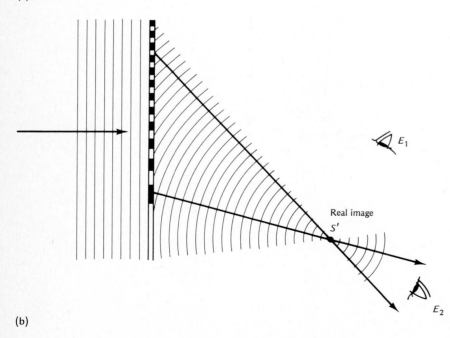

(b)

FIGURE 14.7

Formation of a real image from the hologram produced in Figure 14.4, using the conjugate beam for reconstruction. Because the reference beam consisted of a plane wave striking the film perpendicularly, the conjugate beam here is also a plane wave striking the film perpendicularly, but from the opposite side. (The conjugate beam also produces a virtual image, not shown.)

Conjugate beam

S

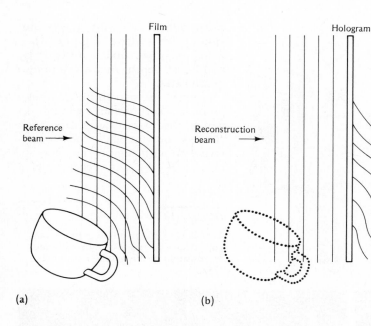

Film

Hologram

Reference beam

Reconstruction beam

(a)

(b)

FIGURE 14.8

Production of a transmission hologram of an extended source, a teacup.
(a) During exposure, the reference beam interferes on the film with the very complex wavefront resulting from the many individual point sources on the cup. **(b)** During reconstruction, the reconstruction beam passes through the hologram and interferes to form a replica of the original object beam.

gens' principle, we may consider the complex wavefront leaving the cup as the sum of many individual Huygens wavelets, one from each point on the cup. Suppose a hologram is made of the cup (Fig. 14.8). We illuminate the cup with a coherent beam, and each point of the cup then emits a diverging wave. Each of these Huygens wavelets interferes with the reference beam on the film. Upon reconstruction, each source point is reconstructed, and thus so is the entire tea cup (Fig. 14.8b). The entire complex wavefront from the tea cup that struck the film is reproduced by the interference that results when the reconstruction beam strikes the hologram. Because the wave leaving the cup is

quite complicated, the interference fringes on the film and the resulting pattern of opaque and clear regions of the hologram are also quite complicated (Fig. 14.9). Nonetheless, when the hologram is illuminated properly, order arises from the apparent chaos of the interference pattern; when you then view the hologram, you see a realistic image of a cup (Fig. 14.10).

We can now see some fundamental differences between holography and photography. A photographic image lies in the plane of the film, so you must view it from at least 25 cm away. A holographic image, however, may be some distance *behind* the plane of the hologram. If that distance is more than 25 cm, you can put your eye right up to the hologram and see the image perfectly well, just as you can put your eye right up to a window and still see the objects behind it.

This window analogy reveals another difference between holography and photography. If you cover one part of a window, you can still see all the objects on the far side, though from a more restricted viewpoint. Likewise, you can cover up part of the hologram and look through the rest of it to see *all* parts of the scene. In fact, you can break the hologram into small pieces and *each* piece will reconstruct the entire scene behind, when properly illuminated. (Naturally, the view through a small piece is more restricted. Further, the fewer the fringes contained in the piece, the more the image will be degraded.) Of course, this is quite unlike what happens in photography; if you cut a photograph into pieces, each piece portrays only its part of the image.

If a photograph of a cup does not reveal the bottom of the inside of the cup, looking at the photograph from another angle won't enable you to see the cup's bottom. With a large hologram of the cup, however, all you have to do is change your position and peer into the image of the cup to see the bottom, just as you would if you were looking at the original cup through a window. Be-

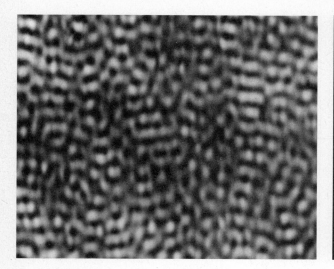

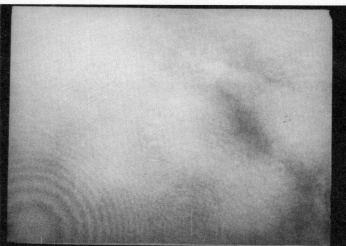

FIGURE 14.9

(a) Microscopic view of the fringes in a transmission hologram. (b) Photograph of transmission hologram illuminated by diffuse white light. The fringes that carry the holographic information are not visible. The large interference pattern that *is* visible is due to dust particles or other slight imperfections in the set-up during exposure. These do not seriously affect the image quality during reconstruction. Note that under this illumination there is no reconstructed image.

"Oh, that's not Dr. Zakheim. That's a hologram."

cause different viewing angles provide different views (Fig. 14.11), a hologram offers binocular disparity, and hence depth, in a way a photograph can't.

In short, we can think of a hologram as a sort of magic window, in which the object beam has been "frozen" or "locked." Upon reconstruction with the proper reference beam (the "key"), the reconstructed

FIGURE 14.11

Two views through a hologram of a tea-cup. Notice the parallax by comparing the position of the handle in the two views.

FIGURE 14.10

FIGURE 14.12

(a) The false image produced by a simple transmission hologram is pseudoscopic—points most distant from the film *(A, C)* produce real images *closest* to the viewer *(A', C')*. **(b)** The false image of an extended object is "inside out." The handle of the teacup is farthest from the viewer but is nonetheless visible. Because light from the back of the cup, *A*, never struck the film (having been blocked by the front of the cup), the corresponding points are not produced in the real image (dashed lines).

beam is "unlocked," released, and produces the image of the subject.

But there is an important difference between a window and our hologram. In addition to the virtual image, our hologram produces a (real) false image of every point of the cup. In this example, the false image of the cup floats in *front* of the hologram at a distance equal to that at which the original source

FIGURE 14.13

(a) Real image projected onto a white screen by illuminating the transmission hologram with the reconstruction beam. **(b)** As in **(a),** except the reconstruction beam strikes only a small piece of the hologram. The image is sharper here because only one view of the three-dimensional object is projected onto the screen.

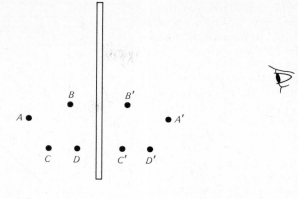

(a)

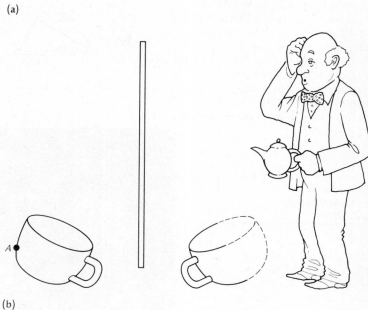

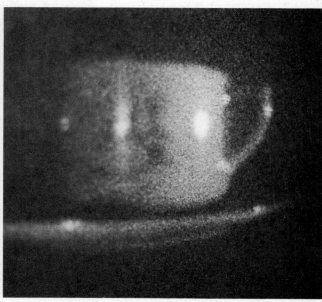

(b)

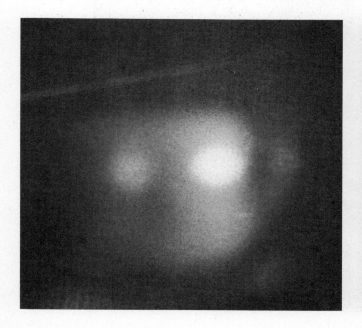

was behind the film. This real image is somewhat unusual. The parts of the cup that are most distant from the film have images farthest in front of the hologram (Fig. 14.12a). Hence the perspective of this false image is reversed—the image is *pseudoscopic* (Sec. 8.5A). In the space in front of our hologram, we get a curious, inside-out cup.

This real image can be projected onto a screen. You need only place a white screen at any position in the image, and the cup will be seen (Fig. 14.13a). As with the virtual image, the real image can be produced from a small part of the hologram (Fig. 14.13b). It is like the formation of an image by means of a lens; even if most of the lens is covered (by stopping down) the entire image is produced.

14.3
PRODUCTION OF TRANSMISSION HOLOGRAMS

So far we have not said *how* to produce the coherent object and reference beams necessary to make a hologram. Indeed, although the principles of holography were put forth by Dennis Gabor in 1947 (earning him the 1971 Nobel Prize),

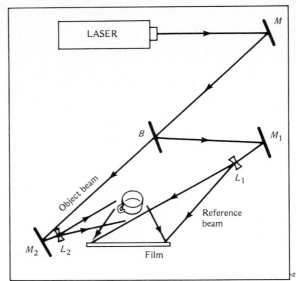

(a)

FIGURE 14.14

A possible set-up for exposing a transmission hologram. **(a)** Coherent, monochromatic light from the laser is reflected by a mirror, M, and passes to a beam splitter B. Part of the beam reflects off the beam splitter and travels to another mirror, M_1, then through a diverging lens, L_1, and finally to the film. The lens spreads the beam so that the entire film will be exposed. This beam is the reference beam. The other part of the original beam passes directly through the beam splitter to a mirror, M_2, and another diverging lens, L_2 that spreads the beam to illuminate the entire object—here, our friend the teacup. Finally, this light is reflected by the cup, and strikes the film—it is the object beam. The film is exposed in an otherwise dark room. **(b)** Photograph of set-up. (The optical components rest on half-ton isolation table.)

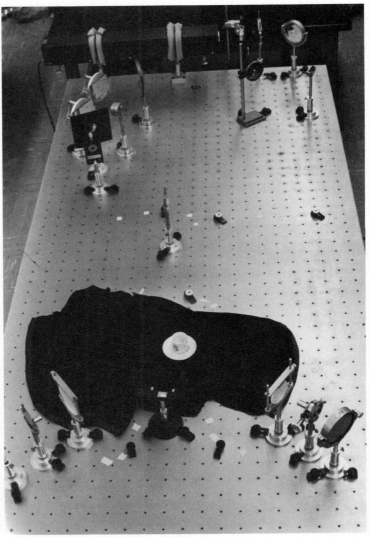

(b)

this problem delayed progress and thwarted interest in holography for 15 years. Gabor used filtered light from a mercury lamp for making his holograms, which were of low quality (at least by today's standards) because such light is dim and not very coherent. Interest in holography only revived with the 1960 invention of the laser (Sec. 15.4)—a powerful source of monochromatic light that maintains its coherence for a relatively long time.

The technique for producing holograms we'll discuss in this section was first proposed by Emmett Leith and Juris Upatnieks in 1960 and is the method most commonly used. In it, the object *and* reference beams strike the film at angles, so that the reconstructed image will not lie in the reconstruction beam. As before, the reconstruction beam must arrive at nearly the same angle as the reference beam so that the object beam will be reconstructed faithfully. The zeroth-order transmitted beam will then leave the hologram at an angle, thereby missing the viewer's eyes when she looks at the reconstructed image. Further, in this technique, it is possible to get the real image out of the way or even to eliminate it.

Figure 14.14 shows one set-up for making a transmission hologram. The reference and object beams are derived from the same monochromatic light source, so they are coherent and interefere on the film. (The reference beam need not be a plane wave. However, since the reconstruction beam must be nearly the same as the reference beam, relatively simple beams are generally used so that reconstruction is not difficult to set up.)

Reconstruction of the light from the tea cup requires the proper reconstruction beam (Fig. 14.15). When this beam alone strikes the hologram, the interference of the light passing through the hologram forms a replica of the complicated wave of the original object beam, creating a three-dimensional virtual image of the cup. Because the reconstruction beam strikes the hologram at a large angle, its continuation (the zeroth-order

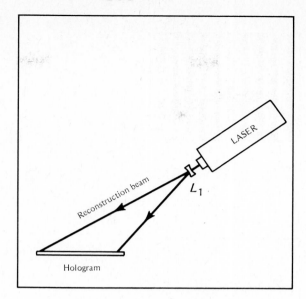

FIGURE 14.15

Reconstruction of the hologram that was made using the set-up of Figure 14.14. Note that the reconstruction beam comes at the same angle as did the reference beam. (The original set-up of Fig. 14.14 could have been used simply by blocking the object beam.)

beam) passes out of the hologram at a large angle and hence does not obscure the image due to the first-order beam. Moreover, since the *other* first order beam is deviated yet further, the false image is then well out of the way.

If you move your eye while viewing the virtual image, you see the effects of parallax. The shadows cast by the handle, say, when the object beam illuminated the cup are fully captured too. Such lighting is one aspect that holographic artists can use; just as a portrait photographer carefully adjusts the number, direction, and intensity of the lights falling on his subject, so too a holographer can insert other beam splitters and further split the laser beam to illuminate the object from several directions simultaneously. Moreover, she can adjust the relative intensities of these beams by means of filters. She can

even get "soft" illumination by placing a translucent diffusing screen in the object beam before the object.

There are several technical aspects of our set-up that are important. All the optical components must rest on a heavy **isolation table**—a heavy block of metal, concrete, or box of sand supported by inflated inner tubes or a similar device that isolates the set-up from most outside vibrations. If there were any shaking of the optical components relative to each other during the exposure by so much as half the wavelength of the light, the regions of constructive and destructive interference would smear over one another, preventing the formation of crisp fringes in the final hologram and ruining any image.

Similarly the subject itself (the cup) must not move during the exposure. Motion of the subject, like motion of the optical components, would lead to smeared fringes in the final hologram. Cups sit still, unlike cats, gerbils, and people; it is impossible for anything living to hold sufficiently rigid throughout an average holographic exposure period (perhaps ten seconds or longer) such that a good, crisp fringe pattern results on the film. Hence, holographic portraiture must use *extremely brief* (nanosecond) bursts of intense laser light. Even the blood flowing through the subject's head can cause his skin to

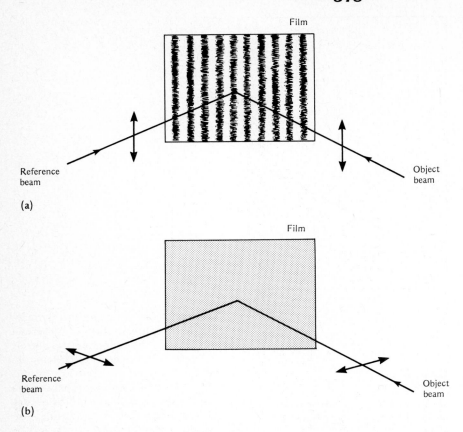

FIGURE 14.16

(a) Both the reference and object beams are polarized vertically, leading to good interference fringes. **(b)** Both the reference and object beams are polarized in the (horizontal) plane of the isolation table, leading to reduced interference. For example, if the two beams cross at a right angle, as shown, *no* interference fringes result.

move more than half a wavelength of the laser light, thereby washing out the interference pattern due to those points of his head. Portraits made with slightly too long exposures reveal ghoulish images of people with black veins running over their faces.

Lasers used for much of holography emit polarized light (see Sec. 13.4), and we must take this into account. Two beams interfere well only if they have the same polarization—a horizontal electric field cannot cancel a vertical one. If the reference and object beams both have vertical polarization (Fig. 14.16a), they arrive at the film with parallel

polarization, and they will produce good interference fringes there. If, on the other hand, the reference and object beams have horizontal polarization, when they arrive at the film they won't have parallel polarizations because they arrive from different angles (Fig. 14.16b). The interference on the film will then generally be poor. Thus, the laser should be oriented to emit light polarized in the vertical direction—the one direction that is perpendicular to *both* the object and reference beams.

PONDER

Does the direction of polarization of the laser light used for reconstruction make any difference?

Because it must record the pattern of tiny interference fringes, holography film must have extremely fine grain; it must be able to record as many as about 2000 fringes per millimeter. Such resolution necessarily comes at the expense of film

sensitivity (Sec. 4.7G)—holography film is among the slowest film made. One common holography film has ASA 1!

A good hologram produces a reconstructed beam with the same variations in intensity as in the object beam. Because of the complicated shape of the H & D curve (Sec. 4.7F), this can only be done if the fringes are of *low* contrast. This requires that the object beam be *weaker* than the reference beam (which has the additional advantage that we can then ignore the dim fringes produced when light from one part of the object acts as an additional "reference beam" for another part of the object). The adjustment of the relative intensities of the reference and object beams can be done with filters, but is more often done using a variable beam splitter. (A variable beam splitter consists of a piece of glass full-silvered on one part, the right, say, gradually changing to half-silvered in the middle and unsilvered on the left. By positioning the beam splitter properly, then, any desired ratio of beam intensities can be achieved.)

Another problem concerns the coherence of the beams. Even in a laser beam, the waves at different points along the beam are coherent with each other only if the points are not too far apart. The distance along the beam over which waves are still in step, or coherent, is called the **coherence length.** The object and reference beams must be coherent with each other when they reach the film in order to have interference there. Therefore when making a hologram, the path lengths of the object and reference beams must differ by less than the coherence length and preferably by as little as possible. (Typically, the coherence length of a laser is roughly the length of its tube, perhaps $\frac{1}{3}$ m.)

PONDER

Are the lengths of the object and reference beams in Figure 14.14 equal? (Use a ruler to determine this.)

The path lengths can generally be equal only if the subjects all lie at about the same distance from the film. If the subjects lie at widely different distances, only some of them will be properly recorded. Hence holograms have **depth of field**—a range of object distances that will be recorded. The greater the coherence length of the laser, the greater the depth of field. Of course, the *wave* origin of depth of field in holography is quite different from the *ray* origin in photography (Fig. 4.30).

In the holograms we've discussed so far, the fringes lead to variations in the *transparency* of the film, that is, the intensity (or equivalently *amplitude*) of the transmitted wave. Such holograms are therefore called **amplitude holograms.** These are somewhat inefficient because much of the reconstruction light is absorbed by the dark silver grains. To increase the brightness of the reconstructed images, holograms can be chemically **bleach processed.** This process removes the dark silver grains, replacing them by a transparent material of a different index of refraction or leaving a different emulsion thickness. Such **phase holograms** produce variations in the *phase* (rather than the amplitude) of the transmitted wave, but still produce the interference pattern. (Notice, in Fig. 14.3, that if the opaque regions emitted light out of phase with that emitted by the transparent regions, that light would add *constructively* to the reconstructed beam.)

14.4
APPLICATIONS OF TRANSMISSION HOLOGRAPHY

Because of the unusual properties and recording procedures of holography, it can be used to solve many problems of image recording and storage.

Multichannel holography is a way of storing *several* three-dimensional images, each of which can be made separately available with the proper reconstruction beam. A multichannel hologram is exposed several times, each time with a *different* reference beam. One exposure may be of a cup using a reference beam at one angle; another may be of a necklace using a reference

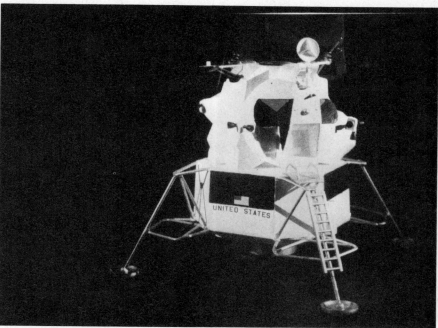

(a)

(b)

FIGURE 14.17

(a) A two-channel transmission hologram illuminated with the proper reconstruction beam for one of the channels. (b) The same hologram illuminated with the proper reconstruction beam for the other channel.

beam at a different angle; another of a shell with yet a different reference beam; and so forth. The film records all the fringe patterns produced during the various exposures. After development, the different three-dimensional images are individually reconstructed by using the appropriate reconstruction beam. To see a different image you simply change to a different reconstruction beam (Fig. 14.17).

Another application concerns secret codes. By looking at the fringes on a hologram (Fig. 14.9) you cannot tell what the subject in the hologram is; only the *proper reconstruction beam* will reveal it. If the reconstruction beam is complex (for instance, if it passed through a specific, precisely aligned frosted piece of glass) then it is extremely unlikely that any spy would be able to reproduce this beam exactly and use it as his reconstruction beam. Hence secret information stored with a complex reference beam is virtually impervious to decoding by spies, while those who have the right "key"—the proper reconstruction beam—can "unlock" and release the image directly.

We've noted that motion of the subject *during* exposure of the hologram smears regions of constructive and destructive interference into one another. Such washed-out interference patterns do not reconstruct the part of the object that moved. Suppose, instead, *two brief* exposures are made, and that some part of the subject moves *between* the exposures so that the path of the object beam to the viewer changes by half a wavelength. The fringes on the hologram produced by that part are then shifted between exposures—those produced in the second exposure just filling in the spaces between those from the first exposure. Thus the fringes recorded in these two exposures cancel each other, and the part that produced them is not visible during reconstruction. (Similarly for parts where the change was $3\lambda/2$, $5\lambda/2$, . . .) However, parts where the change was *a full wavelength* (or 2λ, 3λ, . . .) will have bright fringes that coincide in both expo-

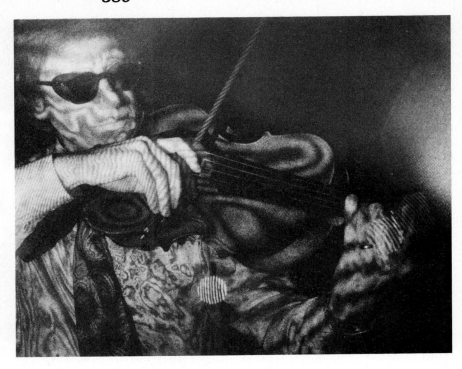

FIGURE 14.18

Photograph of a two-exposure hologram of a man playing a viola. The motion that occurred between exposures is shown by the visible interference fringes, which connect points whose motion shifted the object beam by the same amount.

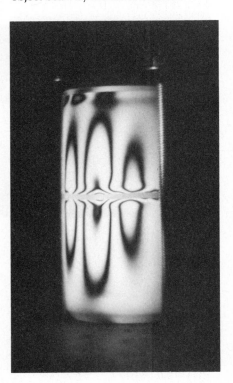

sures, and these parts will therefore appear during reconstruction. Figure 14.18 shows a photograph of such a hologram. These holograms can be used to study, say, the acoustical properties of a viola.

This powerful technique—called **holographic interferometry** because the reconstructed beams from the successive exposures interfere with each other—has also been used by engineers to study the stress patterns of structures. A model of the structure, such as the arch of a bridge, is used as the subject and, as before, two exposures are made on the same piece of film. (These need not be brief since the subject is stationary). For the second exposure, however, stress is applied to the structure, for example by means of a clamp. The stress makes the structure warp and bend slightly, producing dark bands in the final hologram (Fig. 14.19).

A variation of this procedure involves only *one* holographic expo-

FIGURE 14.19

A rubber band stresses an aluminum can. The stress is revealed by holographic interferometry.

FIGURE 14.20

Another form of holographic interferometry. A can and its holographic image are superposed, and the light from them interferes. The resultant fringe pattern is shown for different amounts of stress as air is pumped into the can.

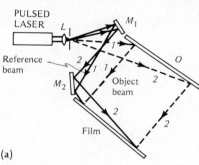

(a)

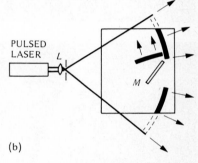

(b)

FIGURE 14.21

(a) Light-in-flight technique for recording the actual propagation of a wave of light. All rays leave the laser simultaneously and are spread by lens *L*. Part of the beam strikes mirror M_1, then mirror M_2, and finally the film. This is the reference beam. Ray *1* of the reference beam arrives at the left side of the film before ray *2* arrives at the right side. The object beam passes from the lens *L* to a vertical white card (the diffusely reflecting object card *O*), and then reflects to the film. Ray *1* of the object beam strikes the left side of the film at the same instant ray *1* of the reference beam does, and so is recorded there. A split-nanosecond later, ray *2* of the reference beam reaches the right side of the film at the same instant ray *2* of the object beam does, which is thus recorded in a different place than ray *1*. **(b)** Side view of the object card, with a small mirror, *M*, that breaks up the object-beam wave. The (broken) wave and its direction of motion are shown.

sure. After the exposure, the developed hologram is placed in the original set-up and illuminated by the reconstruction beam, thus producing the reconstructed beam. The actual object, which was left in its original position, is also illuminated by the laser light, as it was during exposure. Consequently, the reconstructed beam and the light from the actual object are coherent and interfere. They can be adjusted to add constructively so you see a bright image of the object. As the actual object is then stressed, the wave from the object is shifted slightly, producing dark bands that move as more and more stress is applied (Fig. 14.20).

A remarkable procedure—**light-in-flight recording**—uses light itself to act as a "shutter," and probably represents the ultimate in high-speed image recording. Figure 14.21a shows the basic set-up. The laser emits an extremely short pulse of coherent light, which is broken up and recombined in such a way

that different object rays are recorded at different times as the reference pulse sweeps across the film. The developed film is illuminated solely by an ordinary reconstruction beam. When you view the hologram through its *left* portion, you see the wavefront that struck the left side of the card. When you move and look through the right part of the hologram, you see the wavefront that struck the right side of the card. If an object such as a small mirror (Fig. 14.21b) had been placed next to the object card to disturb the wavefront, the disturbed wavefront would have been recorded. Such recordings are shown in Figure 14.22. The wavefronts of Figure 12.44b were recorded in a similar way, only using a lens within the object beam, so part of the wavefront of the object beam was focused before it struck the white card.

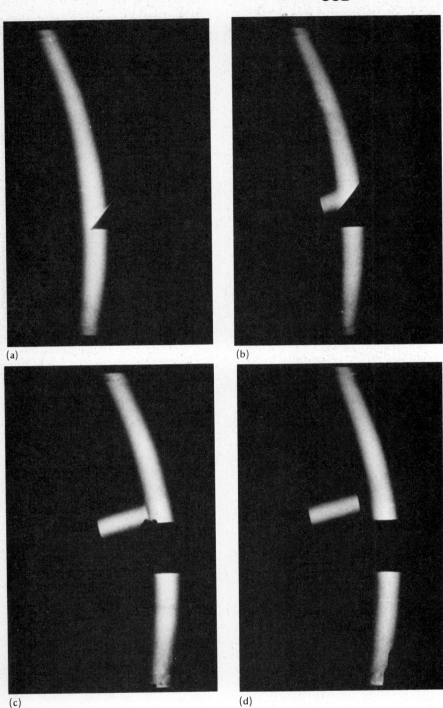

(a)

(b)

(c)

(d)

FIGURE 14.22

Four views through the hologram produced using the set-up of Figure 14.21. **(a)** The diverging object pulse strikes the white card, **(b)** the object beam begins to reflect off the angled mirror, **(c)** later, **(d)** still later, the part of the wave that reflected from the mirror has traveled vertically.

14.5
WHITE-LIGHT HOLOGRAMS

In the holograms we've considered, the reconstruction beam had to be monochromatic and at the same wavelength as the light used for exposing the film. However, in Section 12.3D we saw that a Lippmann plate—multiple thin layers made photographically—*selects* one wavelength from incident broad-band white light. This suggests that it may be possible to construct a hologram that uses a *white-light* reconstruction beam. In fact, this is the case.

A. Reflection hologram of a distant point source

Figure 14.23a shows the formation of a Lippmann plate; the standing wave produced by two plane waves incident from opposite sides is recorded in a piece of *thick*-emulsion film. (The emulsion might be 15 μm thick and must not have an anti-halation backing.) When this film is then developed, silver grains will lie in layers throughout the emulsion; if it is subsequently bleach processed, layers of varying index of refraction will be formed. If we regard one of the two beams as the object beam (from a distant point source) and the other as the reference beam, the developed film can be thought of as a hologram. Because the information of the interfering waves is recorded throughout the volume of the emulsion and not simply on its surface, such a hologram is called a **volume hologram.**

Suppose this hologram is illuminated by a reconstruction beam of the same wavelength as was used for exposure (Fig. 14.23b). The waves that *reflect* from the successive layers interfere constructively to form a replica of the original object beam. We thus have a **reflection hologram**—it must be viewed from the side from which the reference beam comes.

What happens if we use broad-band white light (arriving from the

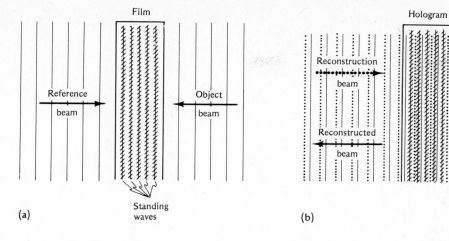

FIGURE 14.23

(a) Exposure of thick-emulsion film to form a reflection hologram of a distant point source that is on the axis. The hatched lines represent the regions of standing waves where the developed silver will be. **(b)** Reconstruction of object beam. Here the hatched regions represent the developed silver.

same direction as the reference beam did) as a reconstruction beam for this hologram? In general, the waves reflected from many successive layers interfere *destructively*, except for the waves of the one wavelength used for exposure. Hence (as we saw in Sec. 12.3D), the stack of silver layers reflects only the desired wavelength and no others. Thus by illuminating the hologram with *white light*, we can reconstruct the object beam—we don't need a laser for viewing. We have made a **white-light reflection hologram** (also called a **Lippmann hologram**). (Unless special chemicals are used, film shrinkage during development and bleaching will cause the wavelength reconstructed to be shorter than that used for exposure.)

Usually, white-light reflection holograms are exposed with the reference beam arriving at an angle, rather than perpendicularly, so the viewer's head doesn't block the reconstruction beam. Of course, the object beams from an extended object strike the film at a variety of angles. Let's check that proper reconstruction occurs in this more complicated case, when the reference and object beams each strike the film at an arbitrary angle.

What points of the emulsion in Figure 14.24a will be exposed when these two beams interfere? The solid lines show the crests of the reference and object beams at one instant. There is constructive interference—the waves remain in phase—at each point where these crests coincide (marked with dots). The silver halide at these points is exposed.

But the dots represent only some points that are exposed. To determine which other points in the emulsion are exposed, you must find all those points that receive crests from both waves simultaneously. To do this, place the edge of a piece of paper along a wavefront of the reference beam and the edge of another piece of paper along a wavefront of the object beam in Figure 14.24a. The edges cross at a dot—hence the emulsion is being exposed there. Now "propagate" each wavefront; slowly advance each piece of paper to the adjacent wavefront in the forward direction (lower right for the reference beam, lower left for the object beam). Notice that the places where your (edge) wavefronts intersect lie along one of the slanted, dashed lines shown. The silver halide along these dashed lines is exposed (Fig. 14.24b).

When the developed film is illuminated at the proper angle (θ_{ref}) with a reconstruction beam of

white light, this light reflects from the layers to form the reconstructed beam—a replica of the object beam (Fig. 14.24c). The law of reflection is obeyed for light reflecting from each of the layers, and this guarantees that the reconstructed beam travels in the proper direction. A single wavelength is selected as a result of interference between the reflections from the *many* stacked layers of silver. If there is no shrinkage, this is the same wavelength as that used for exposure.

When a complicated object beam is used, such as that reflected by our tea cup, the film records a complicated sequence of layers throughout its volume. When illuminated by the reconstruction beam, the entire complicated object is reconstructed.

B. Production of white-light reflection holograms

Figure 14.25a shows one possible set-up for making white-light reflection holograms. Note that the reference beam and object beam strike the thick-emulsion film from opposite sides. During reconstruction, white light (from an ordinary spotlight, say) strikes the hologram at the same angle as the reference beam (Fig. 14.25b).

A different set-up for producing white-light reflection holograms is shown in Figure 14.26. Here no beam splitter is necessary; the beam passes *through* the emulsion and is reflected back by the object— the object beam. This beam interferes with part of the original beam that did not yet pass through the film—the reference beam. The two beams thus strike the film from opposite sides. (Since the lengths of the reference and object beams can never be equal, a laser with adequate coherence length must be used.)

C. White-light transmission holograms

Unlike the reflection volume holograms we've been discussing, the

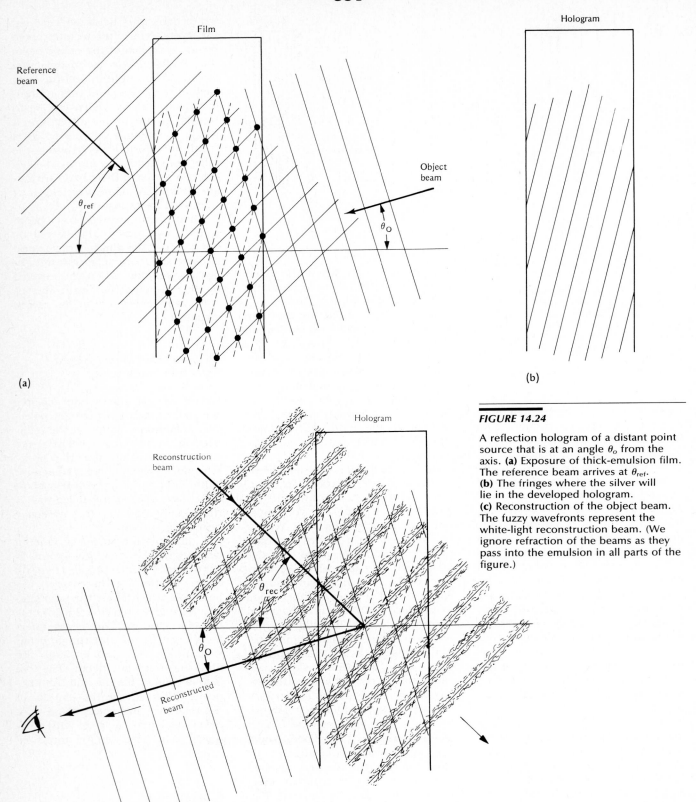

(a)

(b)

(c)

FIGURE 14.24

A reflection hologram of a distant point source that is at an angle θ_o from the axis. **(a)** Exposure of thick-emulsion film. The reference beam arrives at θ_{ref}. **(b)** The fringes where the silver will lie in the developed hologram. **(c)** Reconstruction of the object beam. The fuzzy wavefronts represent the white-light reconstruction beam. (We ignore refraction of the beams as they pass into the emulsion in all parts of the figure.)

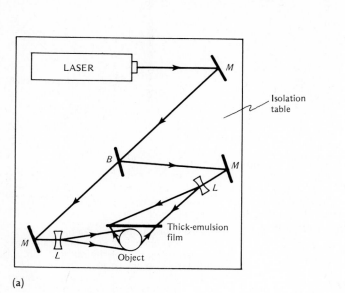

(a)

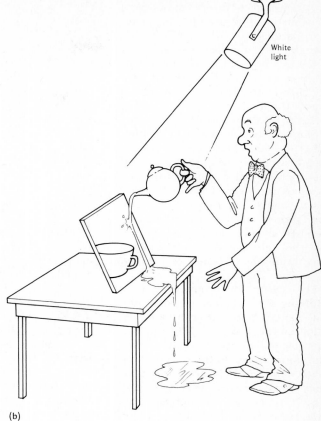

(b)

FIGURE 14.25

(a) One set-up for production of a white-light reflection hologram using three mirrors, *M*, two diverging lenses, *L*, and a beam splitter, *B*. All the optical components lie on an isolation table, and a laser is used as the light source. As usual, the intensity of the reference beam must be greater than that of the object beam on the film and is controlled by the beam splitter. (b) The hologram mounted and illuminated with ordinary white light. (c) Photograph of white-light reflection hologram. The actual holographic image is green.

(c)

FIGURE 14.26

Side view of a simple, single-beam set-up for production of white-light reflection holograms.

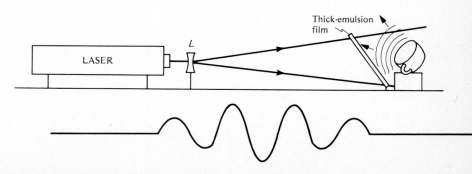

transmission holograms of Sec. 14.3 required monochromatic light for viewing. Is it possible to make a *transmission* hologram that is *white-light viewable?* Yes—all it requires is that a *thick*-emulsion film be exposed with reference and object beams arriving from the *same* side of the film.

Figure 14.27 illustrates the basic principle. Just as with the white-light reflection hologram of Figure 14.24, layers of silver are exposed in the thick emulsion. Light of the reconstruction beam reflects off successive layers and interferes to form the proper reconstructed beam. Here the layers of exposed silver are just the right separation to provide maximum *transmission* of the desired wavelength.

The set-up for production of a white-light transmission hologram is the same as for the laser-viewable transmission hologram (Fig. 14.14), the sole difference being that the film must have a thick emulsion (*with* an antihalation backing). Most holograms of this type are bleach processed to increase the brightness of the final image. Plate 14.1 shows such a hologram illuminated with broad-band white light. The color blurring, or "chromatic aberration," occurs because each wavelength interferes to form its own image at a slightly different

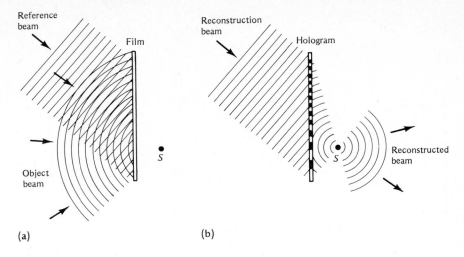

(a)   (b)

FIGURE 14.27

(a) Exposure of thick-emulsion film to form a white-light transmission hologram. (b) When white light is used for reconstruction (here represented by fuzzy wavefronts), only the wavelength used for exposure leads to proper reconstruction of the image.

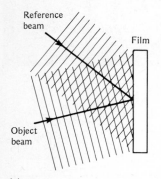

(a)   (b)

position. (The blurring is less in white-light *reflection* holograms.)

*14.6
OTHER METHODS OF DISPLAY HOLOGRAPHY

Many recent advances in display holography have an appeal that is aesthetic as well as scientific. Indeed, few fields show as much cross-fertilization of the sciences and the visual arts as does holography.

A. Image-plane holograms

An **image-plane hologram** is a hologram of a mixed breed; part of its image is real, while part is virtual—an image of an extended object straddles the plane of the hologram. Both parts have normal perspective (no false image). The hologram produces a *diverging* wave to recon-

FIGURE 14.28

(a) Exposure of film using a plane reference beam and an object beam that is converging toward the point *S*.
(b) During reconstruction, a beam identical to the converging object beam is reconstructed and passes through *S*.

struct the *virtual* image *behind* the plane of the hologram, as in Figure 14.5. It also produces a *converging* wave so as to reconstruct the *real* image out in *front* in much the same way. A hologram doesn't care what kind of wave it records—it can freeze and reconstruct a converging wave just as well as it can a diverging wave (Fig. 14.28).

Thus, to make an image-plane hologram, we must somehow get a subject that produces both diverging *and* converging waves. Real objects won't work—they emit only diverging waves. However, when we use a lens to project an image we *do* produce both types of wave (e.g., see Fig. 3.27). Hence we can use a lens to form a *real* image of an extended object, then "slice" this image with our film (Fig. 14.29a). On the one hand, those points of the image that lie in *front* of the film serve as sources of diverging waves and are recorded on the film as in Figure 14.4. On the other hand, light from the lens converging toward points of the image *behind* the film is first intercepted by the film and recorded as in Figure 14.28a. The reconstructed image is then part real and part virtual,

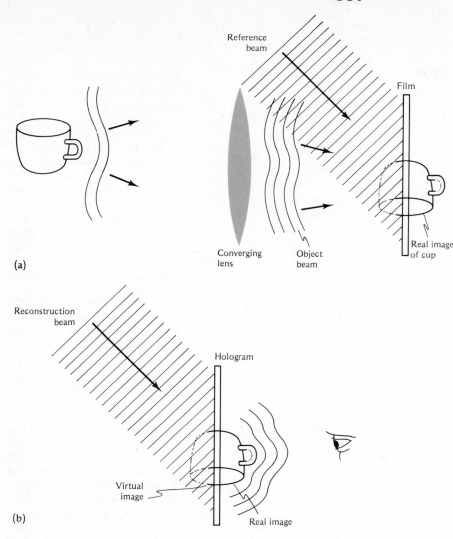

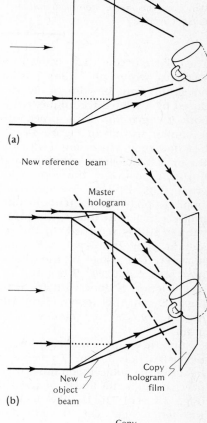

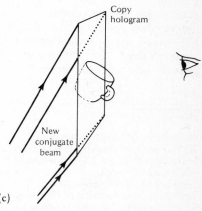

FIGURE 14.29

(a) Exposure of a focused-image hologram. The object beam here leaves the converging lens and would produce a complete, three-dimensional, real image of the cup if the film were not there. **(b)** During reconstruction, the image of the cup is produced. The image is real at those parts of it to the right of the hologram, virtual at those parts to the left. Here we have both real and virtual parts because the film "sliced" the image during exposure. (Of course, no part of the cup can be seen in this image that wasn't visible on the original cup when viewed from the lens.)

straddling the film (Fig. 14.29b). This technique for producing an image-plane hologram is called **focused-image holography.**

Instead of a lens, another hologram—the **master hologram**—can be used to produce the real image. We can either use the false image (Fig. 14.6) or illuminate the hologram with a conjugate beam (Fig. 14.7). In either case, the resulting real image is pseudoscopic (Fig. 14.30a). We then "slice" this image with the film that will become the **copy hologram** (Fig. 14.30b). The light from the master hologram thus becomes the new object beam and interferes on the copy hologram with a reference beam. After development, the copy hologram must be illuminated with its *conjugate* beam in order that the resultant final image have normal perspective (Fig. 14.30c).

FIGURE 14.30

Production of a white-light transmission copy hologram. **(a)** First, a real (pseudoscopic) image is produced using a transmission master hologram. **(b)** The film for the copy hologram is placed at the location of this image and exposed by the light forming the image, as well as by a new reference beam. **(c)** The developed copy hologram is turned around and illuminated with the beam conjugate to the reference beam in **(b)**. The resulting image then has normal perspective.

Why does this image then have normal perspective?

The color blurring that occurs in a standard white-light transmission hologram can be drastically reduced by a related technique, **rainbow holography.** This procedure is similar to that of Figure 14.30, but now only a horizontal strip of the master hologram is illuminated during exposure of the copy hologram (Fig. 14.31a). Because any small region of the master hologram projects the entire image, the entire subject still is recorded. In the resultant copy (rainbow) hologram, however, the image of the slit is now also produced. Since the reconstruction is done with the conjugate beam, this image of the slit lies in *front* of the rainbow hologram.

Suppose we use a *monochromatic* reconstruction beam. To see the image of the subject, you must then peer through the image of the slit (Fig. 14.31b). It is like looking through a mail slot—if you move your head up or down, you no longer view through the slot and the image disappears. (Since light went *through* the slit during exposure of the hologram, when

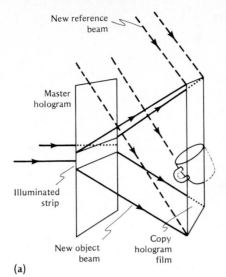

(a)

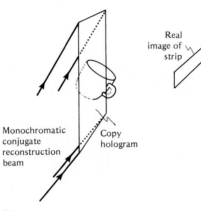

(b)

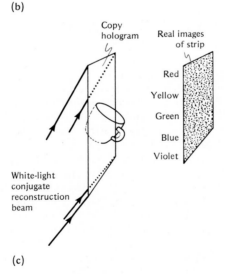

(c)

FIGURE 14.31

Production of a rainbow hologram.
(a) The master hologram is masked to a horizontal strip and illuminated to form a real image of the object. The film for the copy hologram is placed (in this example) at the location of this image and exposed by the light forming the image, as well as by a new reference beam. **(b)** If the developed hologram is illuminated with a monochromatic beam conjugate to the reference beam in **(a)**, a normal-perspective image of the original object is formed. In addition, a real image of the horizontal strip is formed in front of the hologram. **(c)** If the conjugate reconstruction beam consists of *white light,* each wavelength produces its own image of the strip in front of the hologram. These images of the strip are smeared together vertically in the same way that the colors of the spectrum blend together smoothly (Plate 2.1a).

the hologram is viewed using the conjugate beam, the light must come back *out* the image of the slit.) When your two eyes *do* look through the image of the slit, you experience binocular disparity due to *horizontal* parallax, hence you see the object in depth.

With a *white-light* reconstruction beam, however, each wavelength in the white light produces its own image of the slit at a different angle—as you move your head up and down, you see the same image first in red, then in green, and so forth (Fig. 14.31c). This color effect gives rainbow holography its name. Thus the slit allows you to see only one wavelength at a time during reconstruction—there is less blurring of the image. (For good wavelength separation, the reference beam must come from above or below the object beam, not from the sides.)

B. True-color holograms

The principles for making true-color holograms are similar to those for color film (see Sec. 11.2)—expose black and white film with light from three lasers (red, green, and blue) in order to make three color records, and somehow keep these records separate when they are illuminated by the reconstruction beams, so the red record is "played back" by only the red light, and similarly for the blue and green records. In holography, however, since each part of a hologram is exposed to light from the *entire* object, the three sets of fringes that constitute the three records are intermingled everywhere on the hologram.

One way to keep the records fairly separate is to use thick-emulsion Lipmann holograms (Fig. 14.32). Because of the wavelength selectivity of these holograms, when white light is used as the reconstruction beam the particular red, green, and blue wavelengths used for exposure reconstruct their appropriate images (Plate 14.2). Of course, because most of the reconstruction beam is thrown away, such holograms are very inefficient.

Another technique for keeping the records separate is similar to that used in additive color film (see Sec. 11.3). A mask consisting of narrow, alternating red, green, and blue filters lies directly in front of the film (Fig. 14.33). The red reference and object beams can interfere only at points of the film that lie behind the red filters, and similarly for the green and the blue beams. Upon reconstruction, the red reconstruction beam passes through only the red filters, reaching only those parts of the film that contain the red record. Since each part of a hologram reconstructs the entire image, the red record reconstructs the entire object in red, and similarly for the green and blue records. The three colored images lie at the same place, and hence you see one three-dimensional image in full color.

C. Integral holograms

Integral holograms (sometimes known as ***cylindrical holographic stereograms***) wed the principle of

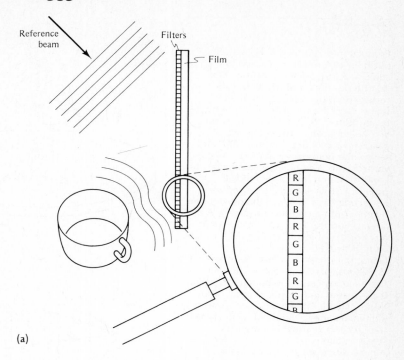

(a)

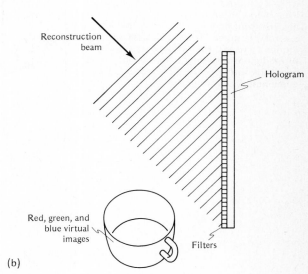

(b)

FIGURE 14.33

(a) Production of a true-color hologram using many vertical filter strips aligned side-by-side to keep the red, green, and blue beams (from three lasers) separate on the film. **(b)** Reconstruction of the color hologram requires that the reconstruction beam pass through the thin filters as in **(a)**.

FIGURE 14.32

Set-up for production of true-color reflection holograms using three lasers, mirrors *M*, beam splitters *B*, and lenses *L*, and thick-emulsion film.

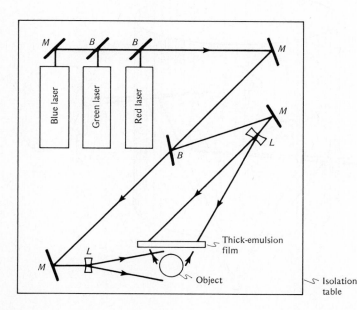

FIGURE 14.34

Production of an integral holographic portrait. **(a)** A movie is made of the subject, *S*, on a slowly rotating table. **(b)** The developed movie film is then recorded holographically, frame by frame. Light from the laser is split by the beam splitter, *B*, to form the reference and object beams. The reference beam passes to a cylindrical mirror, *CM*, and strikes the holographic film in a thin vertical strip. The object beam passes to a mirror *M* and then into a movie projector. This laser light is used for projecting each frame of the movie, one by one, onto a translucent screen. A large cylindrical lens, *CL*, ensures that most of the object beam strikes the same vertical strip illuminated by the reference beam, but does not affect the image (Sec. 6.6A). **(c)** After each vertical strip has been exposed, the holographic film is processed and mounted in a cylindrical display stand. The reconstruction beam illuminates the integral hologram, which then reveals a virtual image of the subject floating within the cylinder. Almost every integral hologram is a white-light hologram and is illuminated from below by an unfrosted incandescent light bulb (which provides a narrow source). The two viewers see different images. Each viewer sees a three-dimensional image because his eyes look through strip holograms containing different images.

(a)

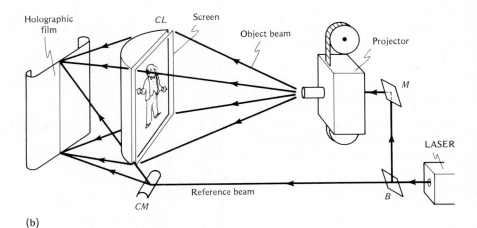

(b)

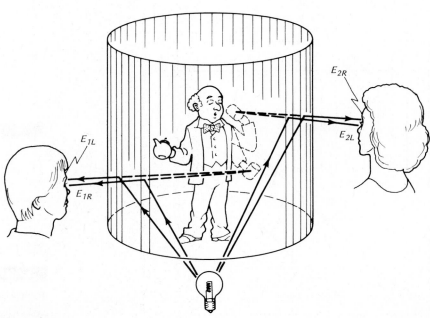

(c)

the stereoscope to holography. The initial recording stage is photographic, and the final is holographic. The depth within the image that you experience when viewing such a hologram arises from binocular disparity; each of your eyes actually looks through a different, thin, component hologram that shows a different flat image. Because there are many such images, it is possible to record some limited motion of the subject with this type of hologram.

A movie film is first made of the subject (Fig. 14.34a). Since this is ordinary photography involving no interference, any subject can be used. Integral holograms have been made of such subjects as people, street scenes, imaginary three-dimensional "objects" stored in a computer memory, and even the image from an electron microscope. For portraiture, the subject is rotated $\frac{1}{3}°$ between exposures of successive movie frames. Hence for an integral hologram that forms a complete cylinder around the subject (360°), 1080 frames are exposed. Next, each frame of the developed film is used as the (flat) *object* to make a narrow strip hologram (Fig. 14.34b)—usually a rainbow hologram, so it can be reconstructed with white light. In this way, the 1080 frames are recorded in thin vertical strips side by side, each strip a hologram of an individual frame of the movie film. Hence no strip hologram by itself contains depth information of the original subject but, because it's a hologram, the entire subject is visible through it. The integral hologram is curved into a cylinder so that the images from each strip lie at the center. When you view such a hologram your eyes look through different individual strip holograms (Fig. 14.34c). Since these strip holograms recorded different images (different frames that were exposed from different angles), binocular disparity results—you see one image in depth. As you walk around the cylinder you see that image from different sides.

(a)

(b)

PONDER

Is there vertical parallax in an integral hologram?

These holograms can portray some motion of the subject. For this, the original subject moves slightly between successive frames (in addition to being rotated). As you walk around the integral hologram, you then see an image of the subject taken from different angles and at different times, so the three-dimensional image seems to move (Fig. 14.35). Such motion cannot be too great, however, otherwise the two images in the hologram seen at

FIGURE 14.35

(a) View of an integral hologram of baseball players as seen from E_{1L} of Figure 14.34c. **(b)** The same hologram as viewed from E_{2L} of Figure 14.34c. Notice that the subject has moved between the views.

one time by your two eyes will differ too much for you to fuse them.

There are yet other types of holograms, as well as combinations of the types we've described. Holography is still in its infancy, but it already has been a fertile field for artists and scientists to explore and extend its expressive and technical possibilities.

SUMMARY

Whereas a painting or a photograph records only the intensity of its subject's light, a **hologram** records both the *intensity* and *phase* of such light. This is achieved by using interference between a **reference beam** and an **object beam** reflected by the subject. So that they are coherent, these beams come from the same, monochromatic source, usually a laser. The film is exposed to the resulting interference fringes and developed to form the hologram. Later, light from the **reconstruction beam** (which must be similar to the reference beam) passes through the hologram and its interference forms three beams: the zeroth-order beam, a first-order beam—the **reconstructed beam** (a replica of the original object beam)—and another first-order beam (which yields a **false image**).

Even a laser beam has a limited **coherence length,** which places a restriction on the range of distances that can be recorded holographically—the **depth of field.** To record good fringes, the intensity of the reference beam is somewhat greater than that of the object beam. Most holograms are exposed on motionless **isolation tables.** Motion of the optical components during the exposure leads to washout of the microscopic fringe pattern on the film, ruining the reconstructed images. **Holographic interferometry,** in which two exposures (or one exposure of sufficient duration) are used to reveal motion, exploits such washout. Other applications of holography include storage of many images, as in **multichannel holography,** or recording certain extremely fast events, as in **light-in-flight recording.**

White-light holograms are made using film that has a *thick* emulsion **(volume holograms).** If the reference and object beams strike the film from *opposite* sides of the film, we get a **white-light reflection hologram** (or **Lippmann hologram**); if the two beams strike from the *same* side, we get a **white-light transmission hologram,** which (unlike a reflection hologram) displays some color blurring. **Image-plane holograms** are made when a real image produced by a lens is recorded on film **(focused-image holography).** Alternatively, the real image from another hologram—the **master hologram**—can be used to produce a **copy hologram.** If the reconstruction beam illuminates only a horizontal strip of the master hologram when the copy hologram is exposed, the resulting **rainbow hologram** allows you to view only one color from a given viewing location.

True-color holograms employ laser beams of three colors to expose three records. These records must be kept separate during exposure and reconstruction—for example, by using thick emulsions or by placing a three-color filter directly in front of the film. **Integral holograms** (or **cylindrical holographic stereograms**) employ movie photography as the initial recording step. Each movie frame is then recorded holographically onto thin, vertical, strip holograms aligned side by side. This collection of strip holograms, curved to form a vertical cylinder, reveals depth through binocular disparity. Because the subject can move somewhat during the photographic recording, integral holograms can be made of virtually any subject that can be recorded on movie film.

PROBLEMS

P1 A laser-viewable transmission hologram is exposed with the reference and object beams striking the film nearly perpendicularly. A second hologram is exposed with the object beam arriving at a large angle while the reference beam still arrives nearly perpendicularly. Which of these two holograms has its fringes spaced more closely?

P2 In making a certain transmission hologram, the angle of incidence of the object beam is 40°, and the angle of incidence of the reference beam is 0° (straight on). (a) If the angle of incidence of the object beam is now decreased to 20° (toward the reference beam), is the spacing of the bright fringes on the film larger or smaller than in the original case? (b) Suppose instead that the angles of incidence are 40° and 0° as before, but longer wavelength laser light is now used. Is the spacing of the bright fringes on the film larger or smaller than in the original case?

P3 In developing his film, a holographer mistakenly uses reversal development instead of ordinary negative development. What does he see when he views this hologram with his usual reconstruction beam? (Hint: consider Babinet's principle.)

P4 Which of the following depth perception cues is most important for enabling you to see a holographic image in three dimensions? (a) Variation in color, (b) variation in sharpness, (c) binocular disparity, or (d) previous knowledge.

P5 In making a laser-viewable transmission hologram, from where to where does the reference beam go, and what is its purpose?

P6 A standard transmission hologram must be made on a virtually motionless table because: (a) there can be only constructive interference on a motionless table, (b) there can be only destructive interference on a motionless table, (c) lasers can operate only on motionless tables, (d) motion of the set-up would smear out the interference pattern on the film, (e) the table makers' union demands it. (Choose one.)

P7 (a) Describe and explain the differences between a hologram and a photograph. (b) Describe and explain the similarities between a hologram and a window.

P8 You wish to make a standard transmission hologram of a room full of people. List *several* problems you have, different from those encountered when making a hologram of a chess piece, and indicate how (ideally) you might overcome these problems.

P9 If a standard transmission hologram is viewed in ordinary red light rather than in laser light: (a) no image is seen, (b) the image is visible but appears fuzzier than with the laser illumination, (c) the image loses its three-dimensional appearance, (d) lasers are not, in general, used for viewing holograms. (Choose one.)

P10 (a) Draw diagrams similar to those of Figure 14.14a showing how to make a two-channel hologram. (b) How would you view each recorded image separately?

P11 Suppose you are making a standard transmission hologram as in Figure 14.14a, but you want to eliminate some shadows on the object. You can do this by illuminating the object with *two* coherent beams coming from different directions. Sketch a modification of the set-up of Figure 14.14a to accomplish this. You'll have to insert extra optical elements to bring some of the laser light to the object from another direction. Label the parts carefully.

P12 (a) What results if an ordinary (thin-emulsion) transmission hologram is viewed with the proper reconstruction beam, except that the beam contains broad-band white light? (b) Suppose it is viewed with the proper reconstruction beam, except that the beam contains white light made of two complementary wavelengths, say blue (480 nm) and yellow (580 nm). What is seen then?

P13 Even though the individual frames in the movie film may be sharp, a final integral hologram made from this film may be confusing if the subject moved too quickly. Consider what each of your eyes sees in this integral hologram to explain the effect.

HARDER PROBLEMS

PH1 One way to make a photographic silhouette is to have the subject stand between the camera and an extended light source, such as an illuminated screen, as in the figure. One holographic artist specializes in the holographic equivalent of this. (a) Following Figure 14.14a, sketch a possible set-up for making a laser-viewable transmission "silhouette" hologram of a scene consisting of a vase and a beer can. Be sure to check the path lengths of the two beams. (b) Is the "silhouette" in the final image three dimensional? Explain what we mean by a three-dimensional silhouette.

PH2 The hologram produced using light-in-flight recording (Figs. 14.21 and 22) did *not* actually employ brief pulses of laser light. Instead, a longer exposure was used and the laser had a very *short coherence length*. Explain how using a laser having very short coherence length reveals the progression of a pulse along the object just as the pulsed laser does.

PH3 (a) Following the approach of Figure 14.24a, draw a large diagram showing the fringes within the thick emulsion of Figure 14.27b. (b) Suppose that the emulsion in (a) is developed to form a volume hologram. Following the approach of Figure 14.24c, show that when the reconstruction wave strikes your hologram, the proper reconstructed beam results. Do this by drawing the ray and the wavefronts of the reconstruc*tion* beam and of the reconstruc*ted* beam (reflected from the developed silver layers).

PH4 You wish to produce a white-light hologram of a toy soldier using the set-up of Figure 14.25a. The hologram is to be hung on a wall and illuminated with an overhead spotlight, as in Figure 14.25b. The final holographic image should be upright. What must be the orientation of the toy soldier during exposure in order that the displayed image be upright?

PH5 Suppose a master hologram is illuminated with its conjugate beam. Suppose further that when the resulting real image (along with a plane-wave reference beam) is used to make a copy hologram, it lies between the master and copy hologram. When this copy hologram is then viewed, using its (plane-wave) reconstruction beam: (a) its real image is pseudoscopic, (b) its real image has normal perspective, (c) its virtual image has normal perspective, (d) no image is

visible because a hologram can only be made using an actual object, not an image. (Choose one.)

PH6 In the production of an integral hologram (see Fig. 14.34b) the image of the movie film frame is projected onto the cylindrical lens. The reference beam strikes the hologram film from *below*. In some set-ups, however, the reference beam strikes the holographic film from *above*. For such a set-up, should the movie film be erect or inverted in the projector if the viewing system shown in Figure 14.34c is to be used unchanged?

MATHEMATICAL PROBLEMS

PM1 A certain transmission hologram is constructed with a plane-wave reference beam arriving perpendicular to the film, and a plane-wave object beam arriving at an angle of 39.27° from the reference beam. Light of 633 nm is used. (a) What is the spacing of the resultant dark fringes on the film? (b) The resolution of film is usually characterized by the number of lines per millimeter it can record. What must be the minimum resolution of the film to record the hologram of (a)? (c) If the film is capable of resolving only 500 lines per millimeter, what range of angles of the object beam (with the same perpendicular reference beam) can be holographically recorded?

PM2 (a) About how many layers of exposed silver will result if the 633-nm red light from a helium-neon laser is used to make a volume hologram as in Figure 14.23, and the emulsion is 15 μm thick? (Assume that the wavelength is the same in the emulsion as in the air.) (b) Repeat part (a) making the more realistic assumption that the index of refraction of the emulsion is about 1.3.

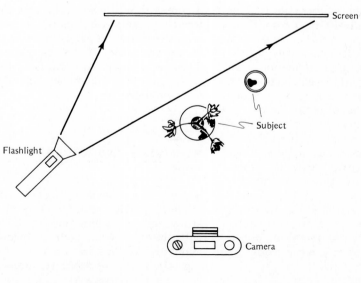

Screen

Flashlight

Subject

Camera

Light in Modern Physics

15.1
INTRODUCTION

Most of the basic properties of light that we've discussed so far were well understood by the turn of this century; it seemed that only details needed to be filled in and technological applications worked out. But in the first few decades of this century, a whole series of discoveries forced physicists to recognize a new reality; the old description of nature *(classical physics)* is only an idealization—rather like geometrical optics is only an idealized, special case of wave optics. Although many of the new principles that developed from this realization now date back over three score and ten years, they are commonly known as *modern physics.*

Light played a paramount role in the discoveries of modern physics. One reason was simply that our curiosity was reaching ever farther beyond the earth and the solar system, and light is the only reliable and available messenger from these distant places. Another reason is that light is the most easily observable *elementary* phenomenon; that is, it does not reveal layers upon layers of structure as it is examined in more and more detail. By way of contrast, sound waves consist of motion of a medium—usually air. The air, however, consists of molecules, which are made of atoms, which in turn have constituents such as electrons, and so forth— each layer of structure opens a new field of physics. Not so with light; there need be no medium, nor is light made of any other constituents.

A sound wave in air doesn't make sense at wavelengths that are smaller than the distance between air molecules. Light, being elementary, encounters no such limitations. Thus some unexpected features of wave behavior itself that become increasingly important at short wavelengths were first encountered with light waves, as we'll see.

Although there is no medium for light in vacuum, it nevertheless travels at a single, constant speed, as experiments from the turn of the century showed. In contrast, everything else that was known had a speed that depended on the speed of its source, its detector, or the medium in which it traveled. The notions of space and time, as then understood, did not allow for anything else. Ultimately it was space, time, and the laws of mechanics that had to be modified and made consistent with the properties of light.

Because light needs no medium, it comes to us through the vast interstellar space and thus provides a sensitive probe of space and time on a grand scale. This leads us to a discussion of galaxies and the universe—so that this book can end with a bang.

15.2
PARTICLES AND WAVES

Some of the basic discoveries of modern physics resulted from the study of the interaction of light with matter. We've talked before about the various ways of interacting, such as reflection, refraction, diffraction, scattering, etc.; here we are interested in *absorption* and *emission.* We already know that absorption can occur when light shakes the charges in matter, transferring energy to them by increasing their motion—heating them. However, sometimes we'd rather convert the light's energy into something else, say an electric current to run a camera's exposure meter, or a nerve impulse from eye to brain. One way such a conversion may happen occurs when light knocks an electron loose from matter.

A. The photoelectric effect

Electrons, as we know, are found in all matter and are particularly available in metals (conductors), where they can travel around freely. However, it is not easy for them to *escape* from the metal. If an electron with its negative charge leaves the (originally uncharged) metal, the metal becomes charged positively and hence tends to pull the electron back. So we can think of the metal as a zone of free travel surrounded by a barrier for electrons. To liberate an electron we must give it enough energy to carry it over this barrier. One way to do this is to heat the metal, literally boiling the electrons off, as is done in the electron gun in your TV. Another method is to hit the electron with light— the *photoelectric effect* (Fig. 15.1).

When we shine light on the metal, we expect that the light wave will rock the electrons back and forth, increasing their energy until some can make it over the top of the barrier. (These liberated electrons are

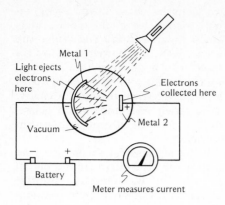

FIGURE 15.1

An evacuated glass tube containing two pieces of metal, which are connected to a battery and a meter that measures the electric current. In the dark no current flows in the circuit because the barrier of metal *1* prevents electrons from leaving it to get to metal *2*. However, when light strikes metal *1*, it knocks electrons out of the metal. These electrons flow to metal *2*, completing the circuit. A current then flows, as shown by the meter. The more light falls on metal *1*, the more photoelectrons are produced, so the meter reading is proportional to the light's intensity.

called **photoelectrons.**) We shall then have converted light into an electric current. Generally this is indeed what happens, but the details don't make sense in terms of classical physics.

One puzzle is the role of the light's wavelength. Long-wavelength (low-frequency) light does not release any photoelectrons at all, no matter how intense the light. Photoelectrons are given off only for light frequencies higher than a certain cut-off frequency, which depends on the metal and on how great its electron barrier is. (For most metals the cut-off frequency is in the UV.) At those higher frequencies, the energy of each photoelectron is independent of the light's intensity (energy). Greater light intensity only gives more (not more energetic) photoelectrons, and even at low intensities some photoelectrons are emitted immediately after the light is turned on. On the basis of classical wave theory we would expect that the light's intensity should influence the photoelectron

energy, and for sufficiently low intensity we would expect a considerable delay before any electron has collected enough energy from the light to get across the barrier.

By way of analogy, imagine your car is stuck in a snowbank, and you try to get it out by rocking it back and forth. If your car behaved as electrons do, you'd find you could always get it going by pushing at a high enough frequency. If you pushed rapidly but not very hard, sometimes you'd find that the car would jump the snowbank right away, after you barely touched it. But if you rocked the car too slowly, it would never move, no matter how hard you pushed. Clearly, then, the photoelectric effect doesn't behave the way classical physics would lead us to expect.

The part of modern physics that explains this odd behavior is called **quantum theory,**[*] and the person who first explained the photoelectric effect was called Albert Einstein. He had a remarkable knack for taking some apparently paradoxical result and making it the basis of a new, successful theory. In the case of the photoelectric effect the relevant principle of quantum theory is:

Every monochromatic electromagnetic wave can transfer energy only in discrete units (quanta). The size of the energy quantum is proportional to the wave's frequency.

So, different electrons don't each get the same amount of energy as they dance in the wave's field—some get it all (a whole quantum) and others get nothing. If the quantum is too small (frequency too low), even the lucky electrons that get a quantum are not carried across the barrier; but if the quantum exceeds the barrier energy (frequency sufficiently high), they will become photoelectrons (Fig. 15.2).

---

[*]Latin *quantum* (plural *quanta*), literally "how much."

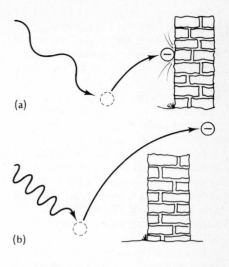

FIGURE 15.2

(a) An electron struck by a low-frequency photon gets insufficient energy to surmount the barrier. **(b)** An electron struck by a high-frequency photon receives enough energy to leave the metal.

In the high-frequency case, the more intense the light, the more quanta it can transfer, hence the more photoelectrons and the larger the resulting electric current. Of course, to compute this current we don't need to know which particular electrons are "lucky." In fact, quantum theory discourages attempts to make pictures in ordinary language (i.e., in terms of classical physics) of the intermediate steps in any physical process. But we like to have some mental picture anyway, and here is how it is usually described. The energy in a light beam is not only transferred in quanta, it is already present in the beam as discrete quanta, called **photons,** each carrying a quantum of energy. The more intense the beam (of a given frequency), the more photons it has. Common light beams, of the kind we've been discussing throughout this book, carry so many photons that we don't generally notice the quantum nature of the light. It required special circumstances, such as the photoelectric effect, before that became apparent.

Photons act in some ways like particles—they are discrete concentrations of energy, you can never absorb a fraction of a photon, you can count how many you have, and so on. However, unlike ordinary particles they have no definite position, but are spread out—just as a wave is spread out. Suppose now that a photon interacts with a material photon detector, for example a photographic film or a TV camera tube. The photon then concentrates and delivers all its energy to an electron at *one* place. Those places where the wave has high intensity are more likely to receive the energy, but wherever the photon goes, *all of it* goes.

Situations where light acts as photons include wave effects, such as the interference pattern of Young's fringes. For the whole interference pattern to be visible many photons must contribute to it, with most of the photons landing on the bright places and none at the dark places (Fig. 15.3). However, the same interference pattern results if the light is so faint that photons travel through an interferometer *one at a time* (and their effects are then added, say by exposing a photographic film for a long time). Each photon may therefore be said to interfere with itself, just as a wave does.

The amount of energy absorbed by any photon detector equals the entire energy of the photon. By the quantum principle this energy, in turn, is proportional to the photon's frequency. Therefore, high-frequency photons carry more energy than low-frequency ones. That's why, for instance, blue light would expose all photographic emulsions (Sec. 4.7C) and why the yellow filter is needed in color film (Sec. 11.2). In Table 1.1, showing the electromagnetic spectrum, you find the high frequencies near the top. Thus gamma rays and x-rays carry a relatively large amount of energy per quantum. On an everyday scale this amount still seems tiny—it would take 10^7 gamma ray photons to equal the energy of a pin dropped from a height of 1 cm. Yet on the scale of energies that are typically exchanged within cells of our body, the gamma ray energy is very

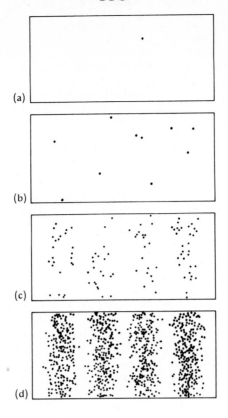

FIGURE 15.3

Simulated picture of interference fringes exposed with **(a)** one photon, **(b)** 10 photons, **(c)** 100 photons, and **(d)** 1000 photons.

high. Therefore, gamma rays and x-rays (and even UV light, to some extent) can do damage to living cells. The high quantum energy also means that it is relatively easy to observe *single* photons at these high frequencies. Nearer the bottom of the table are radiations with very much smaller photon energies. A microwave photon, for example, carries a billion times less energy than an x-ray photon.

Our understanding of light as photons helps us, at last, clear up the mystery of Section 1.4B: why the black-body spectrum of Figure 1.19 has the shape it does. That figure shows the intensity of light at each wavelength coming from a cavity whose walls have a given temperature. The curve has a peak and falls off for both large and small wavelengths. The fall-off at large

wavelengths is a wave phenomenon, which we could have understood much earlier: The larger the wavelength, the harder it is to fit a wave into a given volume. Hence, we expect the longer wavelengths to be relatively rare inside the cavity. This much was understood classically, before the advent of quantum theory.

But the classical theory gave no hint of why the curve should fall off at shorter wavelengths. Shorter wavelengths mean higher frequencies, and the quantum theory now tells us that the higher the frequency of light, the more energy it takes to make a photon. Since the walls of the cavity are at a definite temperature, they only have a certain amount of thermal energy to emit as light—they cannot radiate very energetic, short-wavelength photons. That's why the spectrum falls off at *short* wavelengths. Of course, the hotter the walls, the more thermal energy is available, and therefore the more short-wavelength photons can be radiated. Thus as we heat the walls, the peak of the spectrum moves towards the short-wavelength end, as the figure shows.

*B. Applications of the photoelectric effect

An electric current is produced by the set-up of Figure 15.1 whenever light strikes the metal surface. Such a device is called a **phototube,** and it is useful for all sorts of gadgets, for example to make a simple electric eye (see Sec. 7.10).

Phototubes (or their solid-state equivalents, photocells) are also used to decode the sound track of movie films. The sound oscillations are recorded as alternate bright and dark bands on the edge of the film (see Fig. 4.16). After a given picture on the film has been jerked through the projection beam, its jerky motion is smoothed out by a flywheel, and it then moves past a light source that focuses a slit of light on the sound track. The transmitted light intensity varies in accordance with the sound oscillations. A pho-

FIGURE 15.4

Mechanical analogue of one type of photoconductor. A semiconductor is represented by a mountain landscape with many lakes and an aqueduct. Electrons are represented by water. The lakes are separated by mountains, so the water cannot flow from one lake to the others. The aqueduct connects distant regions of the landscape, but it carries no water. A photon corresponds to the man with the bucket who raises water from one of the lakes into the aqueduct. Now a current can flow across the landscape.

totube converts these light variations into variations of electrical current, which are fed into an amplifier and loudspeaker to reproduce the sound.

Instead of letting light knock electrons all the way out of a metal, you can use it to break the bonds that restrain electrons within a nonconductor. These electrons can then move freely through the material, which thus behaves as a conductor when light strikes it. The result is called a **photoconductor.** Some semiconductors, which conduct electricity only poorly in the dark, are very good photoconductors. (Fig. 15.4 shows a mechanical analogue of one type of photoconductor.) Such solid-state devices can serve the same purpose as a phototube, but in less space. They are used in cameras as light meters, in detectors of infrared radiation, in light-sensitive switches, in TV cam-

era tubes, in photocopying machines, and in many other applications.

A TV camera tube (Fig. 15.5) can be thought of as a large number of photoconductor elements, one for each spot of the picture; positive charge placed on the front of a photoconductor is conducted to the back of those points that are exposed to light. The back of the photoconductor then contains a kind of latent image consisting of positive charges. This positive charge is "read" by means of an electron beam that sweeps over the photoconductors and is "connected" to each picture element in turn, thus scanning the picture.

A Xerox photocopier (Fig. 15.6) also forms a latent image of positive charge on a photoconductor. To develop this picture, the photoconductor is dusted with a negatively charged black powder called a **toner.** The toner is attracted to the regions that were not exposed. It is then transferred to paper, to which it is heat fused, yielding a positive copy.

In photoconducting semiconductors the electrons remain in the material, but light enables them to move freely. By combining two types of such semiconductors, it is possible to get all these freely moving electrons to travel in one direction and thus to create a kind of battery that converts light energy to electrical energy. Figure 15.7a shows a schematic diagram of a **photovoltaic cell.** They are familiar from exposure meters that work without batteries and from **solar**

FIGURE 15.5

The front of a TV camera tube carries a target consisting of a transparent conducting film that is charged positively and a thin photoconductive layer deposited on it. The layer becomes conductive where light hits it, so at those points some positive charge moves across the photoconductive layer, forming a latent image on its back side. An electron beam scans the photoconductor and deposits negative electrons only at those positively charged places. The circuit is completed as shown, so a current—the video signal—flows when the electron beam hits the places that were exposed to light, but not otherwise.

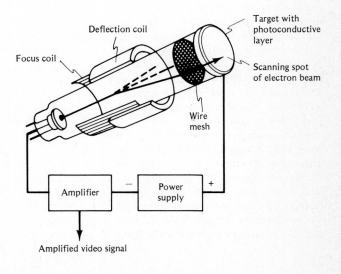

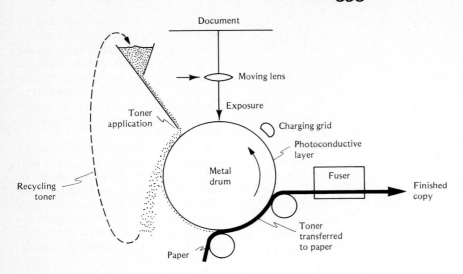

FIGURE 15.6

The photoconductive layer on the metal drum of a photocopier is given a positive charge (by corona discharge—Sec. 11.6) and is then exposed, strip by strip, to an image of the document. The exposing light makes the coating conduct and lose its charge to the metal drum. A latent (photographically negative) image of positive charge remains on the coating only where there is no light during exposure. Negatively charged black toner is used to develop a (photographically positive) image.

cells. Here photons are absorbed near the junction of two types of semiconductors. As in photoconductivity, the photons make some electrons conducting, allowing them to move freely. But the junction is constructed so that these conducting electrons are pulled into one of the semiconductors by charges within the junction. The resultant flow of the electrons is the useful output of the solar cell. (Fig. 15.7b gives a mechanical analogue of this process.) Thus the photoelectric effect can not only turn on a current from some other source of electricity, but it can be used as the generator of the current itself.

What voltage can we get out of a solar cell? A visible photon from the sun (at wavelength 550 nm) has as much energy as an electron that has been accelerated by a 2-volt battery. Hence 2 volts is the most we can expect to get from a solar cell illuminated by *this* wavelength.

FIGURE 15.7

(a) Diagram of a photovoltaic ("solar") cell, consisting of two different types of semiconductors in thin layers and in close contact. One semiconductor is transparent so light can reach the junction between the two semiconductors. When light strikes the cell, part of its energy is converted to an electric current that drives the motor. If the motor were replaced by a meter that measures the current, this device could be used as a photoelectric exposure meter. **(b)** Mechanical analogue of a photovoltaic cell. The situation is similar to Figure 15.4, but rather than being horizontal, the aqueduct has a slope at the junction. When the man who represents the photon raises the water into the aqueduct near the junction, the water flows in one direction only, so part of the energy he puts in can be used to drive the water wheel.

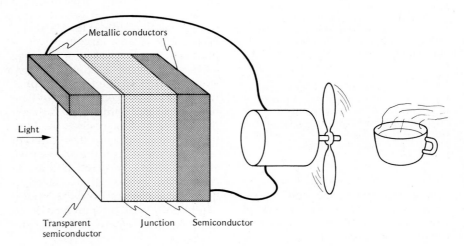

(a)

(b)

In practice you want longer-wavelength light also to contribute to the cell's power, so you choose substances that require less energy to become photoconducting. Silicon, for example, can be made photoconducting by photons of half as much energy (1100 nm wavelength), so the maximum possible voltage you can get from such a solar cell is about 1 volt. In silicon, then, photons with wavelength less than 1100 nm carry more energy than needed, and this extra energy is wasted as heat. Such losses make it difficult to convert a large fraction of the sun's energy incident on a photovoltaic cell into electrical power.

C. Particle waves

Quantum theory not only accounts for the photoelectric effect, but also explains other phenomena, some of which you have already observed. In the TRY IT for Section 12.3A, you used a diffraction grating to look at various light sources—such as sodium highway lights or fluorescent tubes—in which there were bright lines. Plate 15.1 shows similar lights when photographed through a grating spectroscope. Clearly some colors are more favored than others in these spectra, but there is nothing about *light* itself that prefers one color or frequency over another. The reason that certain frequencies stand out in the spectra of these sources is that *matter* does prefer certain frequencies; the lines you see are the radiation from individual sodium or mercury atoms. Each type of atom emits only sharply defined frequencies, called the atom's **line spectrum.** For example, whenever you see that particular type of yellow light (Plate 15.1b) you can be sure that there is some sodium present in the light source.

Earlier we explained these characteristic *frequencies* as resonances of the atom. We have since learned, from the photoelectric effect, that each frequency of light corresponds to photons of one particular energy. Hence, now we can say that when

an atom resonates (in response to receiving some energy), it emits a photon of well-defined *energy*. Why does the atom emit only certain energies? Presumably, because it only *has* certain energies. This suggests that an atom, unlike a pendulum, cannot store any arbitrary amount of energy. We say that an atom has only a discrete set of **energy levels**—that is, possible energies. Whenever it emits a photon, the atom changes from one energy level to another, and the photon carries away the amount of energy lost by the atom. Emission of the photon takes place rather rapidly, so an atom spends most of its time sitting in one of its energy states, doing nothing. This is difficult to understand if we think of the atom's electrons as classical particles whirling about a central nucleus; classically the electrons would radiate all the time, with a frequency equal to their rotation frequency.

It becomes more reasonable if we pretend, without justification for the moment, that *electrons* are a kind of *wave*. If electrons are to be waves, the electron-wave in an atom cannot be a traveling wave, since the electron is supposed to stay in the vicinity of the nucleus. Instead, it must be a standing wave that is confined to the region near the nucleus. But standing waves in confined regions have *discrete* frequencies, as you know if you are familiar with a musical instrument. In such instruments, standing waves are confined (Fig. 12.34). The vibrating string of a violin or a piano is confined to a certain length. The vibrating air in a bugle is confined by the dimensions of the bugle's tube. The violin string plays a definite note (a given set of frequencies)—to change the note, you change the length of the string with your finger. The bugle can produce a discrete *set* of notes, if you blow hard enough, but only those. An electron-wave confined in an atom similarly has a discrete set of possible frequencies. The energies associated by quantum theory with these discrete frequencies are the *atom's energy levels*.

But are electrons (and other forms of matter) really a type of wave? At first sight this seems a very odd idea, since we tend to think of electrons as particles—like tiny billiard balls. But remember that in all of geometrical optics, as well as in the photoelectric effect, light also acts like particles. In other words, a wave can hide its true nature to the casual observer and only show it when it is examined in a carefully designed *interference* experiment (the ultimate test of waviness). Sure enough, by means of delicate experiments it is indeed possible to observe **electron interference.** One version uses wavefront splitting by means of a thin, negatively charged wire (Fig. 15.8a). Since the wire repels the (negative) electrons, it splits the wavefront of an electron beam into two parts, pushing one part to each side. A positively charged wire can then be used in a similar way to recombine the beams. The fringes recorded on a photographic plate (Fig. 15.8b) are analogous to Young's fringes; they not only show conclusively the reality of electron waves, but also give us an idea of the size of their wavelength.

For a more accurate measurement of electron wavelengths we can use electron diffraction gratings, which fortunately are easier to come by. As in the case of photons, the wavelength of electrons depends on the electrons' energy, and for typical electron beams this wavelength is rather small, comparable to the wavelength of x-rays. Therefore, just as for x-rays, crystals can be used as gratings for electrons. When an electron beam is sent through a crystal, a diffraction pattern is seen that is very similar to an x-ray diffraction pattern (Fig. 15.9).

In fact, you can't tell whether the patterns of Figures 15.8b or 15.9a were made by light or by electrons or neutrons or other particles of matter. That is, on a microscopic scale, single particles and photons behave similarly, both exhibiting particle *as well as* wave behavior.

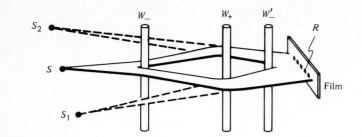

(a)

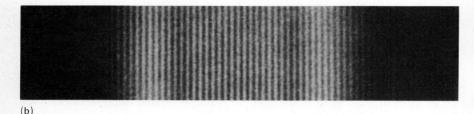

(b)

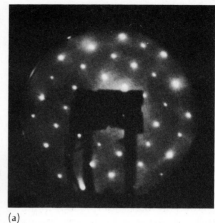

(a)

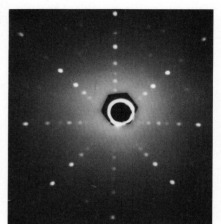

(b)

(c)

FIGURE 15.8

An electron interferometer. **(a)** Schematic diagram of apparatus, showing the negatively charged wire W_-, the positively charted wire W_+, and two paths taken by electrons from source S. Young's fringes are observed in the region R, where the paths cross, corresponding to two virtual coherent sources at S_1 and S_2. (In the actual experiment, another negatively charged wire W'_- is used to make the beams more nearly parallel and thus give wider fringes.) **(b)** Photograph of the fringes—compare with Young's fringes, Figure 12.9.

Why is it then that macroscopically you never have any trouble distinguishing a billiard ball from an electromagnetic wave? To experience a macroscopic ball, or a macroscopic wave, *many* electrons or photons have to contribute. When you get a whole crowd of matter particles together, they *do* behave differently from a crowd of photons (which never roll themselves up into a ball). For example, the effects of the electrical charge carried by matter show up only when you get at least two particles together. The details of how to make a billiard ball out of particle-waves are not on the light side of nature studied here. However, we shall see how a crowd of photons behaves when we discuss lasers.

15.3
ATOMIC SPECTRA

We've seen that atoms have line spectra because atoms have discrete energy levels. An atom's energy is related to the distance between its electrons and its nuleus—the higher the energy level, the farther the electrons are from the nucleus, on the average. When an atom has one of these discrete energies it is in a **stationary state,** that is, its charge is not moving back and forth, and it therefore emits no electromagnetic wave. If it is disturbed, for example by a collision with another atom or by an electromagnetic field, it may change from one stationary state to another. In making the transition to a *higher* energy level, an atom must *absorb* the necessary energy from whatever is causing the disturbance. If the transition is to a *lower* energy level, it *emits* the energy difference as a photon. If we know the energies of the stationary states (the energy levels of the atom), then we know the energies of all the photons that may be emitted (or absorbed) by the atom, since *the energy of these photons must equal the energy difference between some pair of stationary states.*

But why does changing from one stationary state to another

FIGURE 15.9

(a) Pattern produced by chlorine atoms on a crystal of silver when diffracting electrons, **(b)** x-ray diffraction pattern produced by a crystal of table salt (NaCl), which has a similar structure, and **(c)** neutron diffraction pattern produced by a crystal of table salt.

cause an electron to emit light? After all, light is emitted by *oscillating* charges. We can think of the process in this way. During the transition, the electron waves of both energy levels are present and interfere. Since their frequencies are unequal, the interference pattern shifts in time. Since this pattern is due to waves of *charged* electrons, the charge also shifts around. This oscillating charge, as usual, changes the electromagnetic field—most frequently by absorbing or emitting just one photon.

Thus, an atom can absorb light (for instance, when the atom starts in its lowest energy level—the **ground state**) and emit light (when the atom starts in a higher energy level—an **excited state**). Suppose the atom is initially in a stationary state, and a photon comes along. The photon disturbs the atom a little, so that the charge starts shifting around. If the photon wave pushes the charge into stronger oscillations, the atom gains energy. Since this energy must come from the photon, **absorption** takes place—finally there is no photon, and the atom is in a higher energy level.

The emission process requires that the atom *start* in an excited state and can occur in two ways. A photon that is already present (say because we shine light on the atom) can disturb the atom, causing it to emit an additional photon. The energy of the additional photon must then come from the atom, which therefore must end up in a lower energy level. This process is called **stimulated emission.** On the other hand, even if we don't shine light on it, the atom will eventually fall to a lower energy level anyway, just as a pencil balanced on its tip will eventually fall over. When the atom falls to a lower energy state, it emits a photon that carries off the energy—a process called **spontaneous emission.** Typically an undisturbed atom spends a **lifetime** of only 10^{-8} seconds in an excited state before it returns to the ground state by spontaneous emission. The only state in which an atom can re-

main for a long time is its ground state.

Absorption and emission merely spell out in more detail the effects we already discussed that happen when an electromagnetic wave shakes a charge. For example, scattering occurs when absorption and emission go on at the same time.

A. Emission spectra

Most emission of light happens by spontaneous transitions in atoms. An atom receives some energy, for example by colliding with another atom in a gas or by being hit by an electron in an electrical discharge (Sec. 1.4C). The collision deforms the atom and places it in an excited energy state (car owners can sympathize). After the collision is over, the atom starts falling back to the ground state; the excited state's wave becomes weaker and the ground state's wave grows. During the transition time the electron charge oscillates and emits a light wave. The light wave's **coherence length** (the distance along the beam over which the light wave remains in step) is about the distance it travels during the excited state's lifetime, that is, about a meter. Generally the atom is disturbed, say by another collision, long before it has finished emitting the wave, and the coherence length is then correspondingly much shorter.

If an atom emits a photon, the frequency of that photon must be just right so that its quantum of energy corresponds to the difference in energy between the atom's initial and final level. Often when an atom is in a highly excited state, there are several possible routes by which it can reach the ground state, and thus there is more than one possible line in the atom's spectrum of emitted light. Figure 15.10 gives an example for the case of mercury. The lifetime is not the same along these different routes, so the atom usually picks the fastest route down to the ground state. Transitions that take a long time are seldom seen in the spectrum if

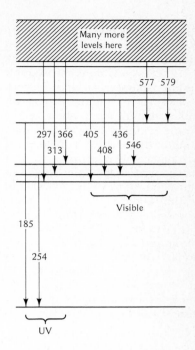

FIGURE 15.10

Energy-level diagram for mercury. The transitions indicated by arrows have lifetimes of about 10^{-8} sec and correspond to the spectral lines of Plate 15.1c. The wavelength of these lines is given on the diagram in nanometers. Transitions with longer lifetimes give lines too faint to be seen in the plate.

there are competing faster transitions. So we do not expect to see spectral lines corresponding to *all* the possible transitions between energy levels. However, many of the transitions *are* seen in the spectrum, so an energy level diagram is an efficient summary of a complicated spectrum with many lines **(emission lines)** by means of a smaller number of energy levels.

Since the energy levels are determined by the atom's structure, the spectrum of each type of atom is different from that of any other. Atoms can therefore be identified by their spectrum. We don't need a very large amount of a material to be able to excite its atoms and cause them to emit their characteristic spectrum. (The atoms can be excited by electrical discharge, say.) Thus the constituents of even small samples can be analyzed by means of **spectroscopy**—examining the spectrum of the emitted light (e.g., with the instrument of Fig. 12.22)—as is done for example in crime laboratories. Spectroscopy can also tell us the atomic constituents of distant stars, about which we have no other information than the light they send us. Before the development of spectroscopy, this was thought to be impossible—as August Comte wrote in 1835 about celestial bodies: "We understand the possibility of determining their shapes, their distances, their sizes and motion. whereas never, by any means, will we be able to study their chemical composition." In fact, the element helium* was first detected in the sun by its spectrum, before it was found on earth.

B. Absorption and luminescence

An atom can absorb a photon, provided the photon's frequency is just right—its quantum of energy must be just enough to lead the atom from its initial state (usually the ground state) to one of its excited states. Thus, like emission, absorption also occurs at discrete frequencies. To observe this we let white light pass through a region filled with the atoms of interest and then analyze the light by spectroscopy. Since the atoms absorb light of only these discrete frequencies, the transmitted light looks like the usual continuous white-light spectrum, but with *dark* **absorption lines** at these discrete frequencies (Plate 15.1d). (Usually the atom soon returns to the ground state, e.g., by emitting a photon similar to the one it absorbed, but in a different direction—these "scattered" photons do not reach the spectroscope.)

The sun's spectrum provides examples of both emission and absorption spectra. The overall, blackbody shape of the spectrum (Fig. 1.19) is a broad-band *emission*

*Greek *helios*, sun.

spectrum from a hot layer of gases on the sun's surface. So many types of atoms contribute to this light that few of the individual emission lines stick out of this continuous distribution. But farther out from the sun there are cooler, less concentrated gases that *absorb* discrete frequencies from the solar spectrum. Thus, if examined closely, the solar spectrum shows a large number of dark lines, which tell us most of what we know about the elements present in the sun (see the first TRY IT).

The whole process whereby an atom absorbs some energy and then returns to its ground state by emission of one or several photons is called **luminescence.** Of particular interest are those cases when the emitted light has a *different* frequency than the exciting light. This can happen if the atom drops down to the ground state by way of an intermediate excited state. Since the energy originally gained is then emitted in several parts, each emitted photon has less energy and hence a *lower* frequency than the photon that was absorbed. If the luminescence process happens rapidly, that is, in a time that is comparable to lifetimes of typical excited states (10^{-8} sec), it is called **fluorescence.** We have already mentioned its use in making visible light out of UV (Sec. 1.4C). Similarly, fluorescent screens can make the presence of x-rays visible.

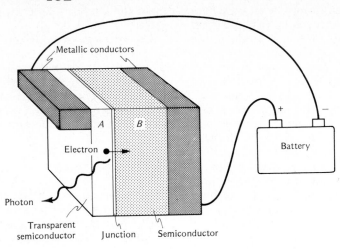

FIGURE 15.11

Schematic diagram of a light-emitting diode. Here electrons are pushed by the battery through a barrier of charges at the junction of two semiconductors. Once past this barrier, these electrons drop down to the ground state in the second semiconductor, emitting light in the process.

PONDER

Why are there no fluorescent substances that can make infrared radiation visible?

Of course, the atom's original excitation need not be due to *electromagnetic* waves. For instance, the atoms on the fluorescent screens of TV tubes and video display terminals are excited by *electron* beams. Some materials can also be excited and made to luminesce by an electric current **(electroluminescence)**. This effect is used in **light-emitting diodes (LED's).** An LED is made by putting two types of semiconductor in contact (Fig. 15.11). One type *(A)* has conducting electrons on its side of the junction, whereas the other type *(B)* has empty states for them to fall into on the other side of the junction. Charges at the junction prevent electrons from flowing from *A* to *B*. However, when a battery is connected to this so-called diode, electrons from *A* can be driven past this barrier. When they arrive at *B* they drop down to the ground state and emit light. This will be visible in suitable materials, in which the drop in energy is large enough. As we saw for solar cells, the energy corresponding to visible light can be supplied by a battery of only about 2 volts. LED's can display numerals and other symbols by a technique similar to that used with liquid crystals (Fig. 13.12). Their power requirement is low, though higher than that of liquid crystals. They find great use in optical communications systems (Sec. 2.5B).

It can happen that one of the transitions involved in luminescence has a long lifetime. For example, the absorption process may have transferred enough energy to remove an electron entirely from an atom and entrap it in another atom. It may then take considerable time for the electron to get back to the level in the original atom from which it can drop to the ground state by emission of light. Such a process is called **phosphorescence,** and the afterglow it produces may last for hours or longer. Some toys and watch dials glow in the dark after being exposed to light (see the second TRY IT). (Usually some unstable substance that emits electrons is also put on the dial to give additional excitation, so it will glow even during long periods in the dark.)

First TRY IT

FOR SECTION 15.3B
Solar absorption spectrum

You can see some of the sun's absorption lines with the grating spectroscope of Figure 12.24. Set the two razor blades as close together as possible (less than $\frac{1}{2}$ mm) to make a very fine slit. If the slit is very narrow, you can point it directly at the sun to examine the solar spectrum. Don't look at the slit, but rather at the spectrum to one side. You'll see at least a few dark absorption lines in the first-order spectrum (the finer the slit, the sharper the lines), for example, the "yellow" sodium lines (here seen as an absence of this particular yellow light). To identify them, compare with Plate 15.1b.

Second TRY IT

FOR SECTION 15.3B
Phosphorescence

You'll need an object that glows in the dark (for example, the luminous dial of an alarm clock) and colored filters, particularly red. "Charge" your object with white light, then turn off the light and observe how the glow decays. Next "charge" your object with light of various colors by covering it with a filter before reexposing. Explain why some colors can evoke phosphorescence and some cannot.

Your TV screen is somewhat phosphorescent, decaying rather quickly, but just slowly enough to make successive scans of the picture blend into each other. You can excite the phosphorescence with your camera's electronic flash unit. Have the TV turned off in a completely dark room. Hold some object, maybe your hand, on the screen. Then close your eyes, and fire off the flash toward the screen at close range. Immediately open your eyes and remove your hand from the screen to see a fleeting shadow of your hand on the TV screen. Try replacing your hand with filters of various colors. Why do some filters seem to throw good shadows and others hardly any?

15.4
LASERS

Now that we understand how atoms emit light, we can consider the special arrangement that allows atoms in a laser to emit coherent light.

When many atoms emit spontaneously, there is no particular relationship between one atom and the next, so the light from them is *incoherent*. In order to get coherent light, we must make the atoms emit light in a correlated manner. This occurs in a radio antenna, in which all the radiating electrons are made to move up and down *together*. Therefore, radio waves are very monochromatic and very coherent.

Emitting atoms can be correlated if they radiate by *stimulated* emission, the key principle of the **la-ser.**[*] In a laser, many atoms are initially in their excited state. If only one atom emits a photon, it does so spontaneously, in a random direction. If two atoms emit, the second photon is more likely to be similar to the first, because of the effects of stimulated emission. In emission from three atoms, the third photon is even more likely to be identical to the first pair, and so on. When *many* atoms emit while stimulated by each other's photons, the result is one *coherent wave*, traveling in one direction, containing about as many photons as there were excited atoms in the beginning. This can be a large number, so the light can be quite intense as well as coherent.

To be as strong as possible, the coherent wave ideally should encounter all the atoms in the laser. But if it encounters atoms that are *not* excited, it can excite those atoms and thereby *lose* photons to them. Therefore, the laser must contain more excited atoms than atoms in the ground state—more radiating atoms than absorbing atoms. We call such a preponderance of excited atoms a **population inversion**—a population of atoms in the two states that is reversed (inverted) from what it would normally be.

To make a laser we need many atoms in a population inversion. (In fact, such lasers have been found to occur naturally in the atmosphere of Mars. This has been used to argue against laser patents, since natural phenomena cannot be patented—or maybe the Martians should get the patent!) If we are to make a population inversion, we must have some way of exciting atoms from the ground state to an excited state where they will remain until we're ready to use them. The trouble is, if an excited state is one that doesn't decay rapidly, it is also one that is hard to get to from the ground state. Alternatively, if it's easy for us to excite the atom to a state, that

---

[*]Light amplification by stimulated emission of radiation.

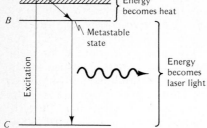

FIGURE 15.12

Energy-level diagram of the emitting atoms used in a three-level laser (for example, the chromium atoms in a ruby-rod laser).

state won't remain long enough for us to use it. Figure 15.12 shows how this problem is solved in a **three-level laser.** The excitation of the atoms is similar to that of fluorescence. By electrical discharge (or other means) many atoms are placed in a broad upper energy level, A. This level, which is easily accessible, decays rapidly—but to another excited state, B, rather than to the ground state. The transition from A to B is made *not* by emitting a photon (which could reverse this transition in other, excited atoms) but by dissipating the energy as heat. The excited state B, reached after the first transition, must be a **metastable state***—one that has a relatively long lifetime, and hence many atoms can accumulate in it. The result of this **pumping** process via the upper level is the desired population inversion—more atoms are in the metastable state B than in the ground state C.

The atoms can now produce coherent radiation via stimulated emission, once the process gets started by one or a few spontaneous emissions and provided all the atoms are exposed to each other's photons. For practical lasers we need a way to keep the emitted photons around long enough to interact with most

*Greek *meta*, after, plus Latin *stabilis*, stable.

of the atoms. To achieve this, the photons are reflected back and forth many times between two parallel mirrors. One of the mirrors is not quite 100% reflecting, say 98%, so that some of the light leaks out. If the excited state is being pumped continuously, the radiation inside is continuously being replenished. As long as the stimulating wave stays around, all the light that leaks out is coherent and in phase. Thus, a standing wave is set up between the mirrors, and during each pass back and forth through the gas, 2% of the coherent radiation energy present between the mirrors leaks out and an equal amount is replenished by the pumping. The beam that leaks out is the laser's useful output. We can think of it as photons that have gone back and forth between the mirrors 50 times, remaining coherent with the standing wave inside. The coherence length of these photons is therefore about 100 times the length of the laser. (In many common lasers, however, several of the standing waves that fit between the mirrors are set up. These differ by half a wavelength over the length of the laser, so they are no longer in step with each other after traveling that distance—the coherence length of such a laser is thus about the length of the laser tube.)

Because a laser's output beam has passed many times along the tube before emerging, it is not only coherent but also very directional—the laser beam reflected off the moon (Fig. 2.41b), for example, leaves the telescope with a diameter

of about 2 m and spreads so little that its spot is only a few kilometers across by the time it gets to the moon, 380,000 km away. (Even this small amount of spread is caused mainly by fluctuations in the atmospheric refraction.)

Figure 15.13 shows the parts of a slightly more complicated laser—the common **helium-neon laser.** This laser is continuously pumped by an electrical discharge. The energy is first taken up by the helium and then transferred to the neon when the atoms collide, thereby placing neon in its metastable state. Helium-neon lasers emit red coherent light of wavelength 633 nm, generally with a power of only a few milliwatts. This is enough to make a visible spot at a distance of many meters in broad sunlight (which you cannot do even with strong automobile headlights)!

For a device that is younger than the transistor, the laser enjoys a remarkable gamut of applications. Since a laser beam is very close to

FIGURE 15.13

(a) Schematic diagram of a laser that uses a mixture of helium (He) and neon (Ne) gases in a glass tube. The reflecting mirrors are located outside of the tube. Since the light's intensity increases only by a small amount on each pass through the tube, transmission and reflection losses must be kept to a minimum. The ends of the tube therefore carry Brewster windows (Sec. 13.4), and the mirrors are coated to increase reflection (Sec. 12.3C). (b) Energy-level diagram of the levels in helium and neon that take part in the pumping and stimulated emission of this laser.

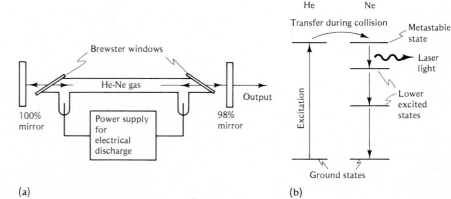

an ideal plane wave, the entire laser power can be focused to an almost ideal point. This concentration of power into a tiny region creates high temperatures, and such precisely controlled heating is the principle behind many laser uses—vaporization of small samples or selected layers of larger samples (e.g., for spectroscopy), microirradiation of cells (in genetic studies, for example), welding objects (as delicate as detached retinas and as massive as automobiles and pipelines), fusing of filling material in dentistry, and even tattoo removal. By heating and melting a material very rapidly, lasers (often under computer control) can cut rapidly and precisely, as for example in the garment industry or in awkward places (as in the movie "Goldfinger"). Laser cutting has special advantages in surgery (e.g., tonsillectomy) because the heat can cauterize small capillaries and thus reduce bleeding. Focused laser beams of low power, on the other hand, can interact very gently but precisely, as they do in the "optical stylus" of readout heads in videodisk systems. An unfocused laser beam spreads very little, making a bright, straight line over a long distance, which is useful in surveying. Finally, the laser light's high coherence allows extremely precise interferometric measurements, as well as communication by means of light waves in fiber optics. Of course, the laser finally made holography practical, and science fiction movies would still be in the dark ages without lasers.

15.5
NEW LIGHT ON MECHANICS

Simultaneously with the development of the quantum theory at the beginning of this century, another revolution was brewing. By then, the electromagnetic wave nature of light had been firmly established, and people began asking questions about the way their understanding of light fit into their previous understanding of mechanics.* They found that theories of that time gave contradictions. For example, any measurement of the speed of light also involves mechanics, because speed is *defined* by means of mechanics—to measure the speed of, say, a runner we need to measure the *time* it takes her to run some *distance,* that is, two quantities based on mechanics. However, various ways of measuring her speed give different results. If she were running on a treadmill, she wouldn't get anywhere, so her speed would be zero. Why is it, then, that she is working just about as hard on the treadmill as when she runs on the road? Well, the floor of the treadmill is continually moving backwards, so *relative* to that floor she is running just as fast. Her speed, then, depends on the context in which it is measured. We call this context a **frame.** She is *not* moving relative to the room's frame, but she *is* moving with respect to the floor-of-treadmill frame.

Now light waves also have a speed, so it is natural to ask, in what frame does light have its constant speed, *c*? What would happen if you shined a light from a moving car—would it move faster in the direction the car is moving, slower in the opposite direction? Or, could you catch up with a light beam by running very fast? What would happen if you carried a mirror along as you were racing the light beam—would the light be reflected?

A. Special relativity

When he was a boy, Albert Einstein asked questions like those above. As understood at that time, Nature herself seemed to be confused on these questions. For example, if the speed of light on earth were affected by the earth's motion, then that speed would be different for light beams traveling in different directions, and the amount of difference would indicate how fast the earth was moving. But no experiment comparing the speed of light in different directions has been able to reveal anything about the motion of the earth in its orbit—light appears to have the same speed in all frames. In fact, no one understood properly how light behaves on moving bodies, so poor Albert had to figure out the answer to his question himself. Characteristically, he accepted the paradox that light has a unique speed but not a unique frame—it doesn't matter in what frame its speed is measured. He made this one of the principles of his **special theory of relativity:**

Light moves the same way, with the same speed, in *every* frame.

So even if you chase after a light beam, you'll still find it moving at precisely *c*, no matter how fast you run.

The theory built on this principle has all sorts of consequences. However, it says nothing new about light itself, because the theory of light said light didn't need a medium to travel in—it traveled in vacuum with speed *c*, no matter how fast its source or detectors moved. Rather, therefore, all other theories (such as mechanics) had to be reconciled with this principle about the behavior of light. But we are primarily mechanical creatures, so we believe that we really understand and have intuition about mechanics and that we know what space and time are. So when Einstein first challenged the long-ingrained beliefs about mechanics, there was considerable opposition; but today Einstein's special relativity is so firmly established that is equations are written on tee shirts, cartoons, and stained glass windows (Fig. 15.14).

Since we are not primarily interested in mechanics here, we shall give only one example of a consequence from the principle of relativ-

---
*Greek *mechane,* machine. Mechanics deals with motion of material bodies under the influence of forces.

FIGURE 15.14

ity. Classical mechanics allows all speeds, no matter how large. Accordingly, nothing (except the expense of large rockets) prevents us from making a whole laboratory (i.e., a frame) move faster than light. Suppose Prof. Blitzschnell has built a laboratory that is zipping past you, going to the right, faster than the speed of light. At the instant the good professor is even with you, you shoot off your flash camera, hoping to get his picture (Fig. 15.15). But the light of your flash never catches up with him, since he is moving faster than that light. Your flash bulb, then, sends off light to the right and to the left, but both of these flashes of light re-main to the *left* of the superlumi-nary professor.

In *his* lab, Prof. Blitzschnell's assistants observe the flashes.

PONDER

The assistants must be spread out, must note when the flashes coincide with their position, and later compare notes to figure out how the flashes moved. Why can't Blitzschell just look *at the flashes?*

The professor finds two flashes on his left that started out together and then spread apart. Therefore, they must both move to the left at *different* speeds. (Note that no mechanical measurements of speed are needed for this conclusion—only the distinction between forward and backward and between coincidence and separation.) But

this is impossible according to the principle of relativity; light has only one possible speed, so two flashes that move in the *same* direction and start out together must stay together. The only way to avoid this contradiction is to conclude that:

No object can move faster than light.

This conclusion may seem strange, because most of us have no experience with really rapidly moving bodies. To builders of particle accelerators or other investigators of high-energy particles, however, this is a well-known fact; no matter how powerful the accelerator, the speed of the particles it produces is always less than the speed of light (even if ever so slightly).

B. Doppler shift

Another consequence of the principle of relativity is that if you add energy to a given flash of *light* (in vacuum), you won't increase its speed. By contrast, if you add energy to (i.e., accelerate) a slowly moving *particle*, you *do* increase its speed. How does light manage to take up energy without going any faster? Giving a beam of light additional energy can be accomplished in various ways. You could, for example, have the source moving toward you, or equivalently you could be moving toward the source. If light were particles, the particles would then surely come at you with greater speed.

But according to quantum theory, light *is* particles—photons. Having recalled this, it is easy to see what must change when source and observer are approaching each other. Since the photon energy is proportional to the light's frequency, the *frequency* must increase. Thus, if someone throws a baseball at you while running toward you, the baseball comes *faster;* but if he shines a light at you while running toward you, the light becomes (slightly) *bluer*. Similarly, when source and observer are receding from each

FIGURE 15.15

(a) Diagram (constructed from eyewitness reports, *not* a snapshot taken with light) of an observer shooting off a flash as Prof. Blitzschnell is passing by toward the right in his superluminary laboratory.
(b) A little later, the light *L* from the flash has spread out from the camera, but Blitzschnell has moved even farther. His assistants observe the flashes.

other, the light's energy decreases, hence so does the frequency, and the light becomes redder. This frequency shift is familiar from sound. A train whistle sounds higher while the train is approaching and lower after it has passed you (Fig. 15.16). This change in frequency (for both sound and light) is called the **Doppler shift,** after the nineteenth-century physicist Christian Doppler who realized that this phenomenon applies to *all* kinds of waves. (It was first tested by using just such a train, filled with trumpeters.) Police use the Doppler shift to catch you speeding; when electromagnetic radiation (here, radar) is reflected by a moving car, its frequency changes by an amount that depends on the speed of the car. The police officer has a device that measures this frequency change, so she can quickly determine your car's speed.

*C. Matters of gravity

Special relativity tells us that the speed of light in vacuum is a universal constant—nothing can affect it. But there is a universal force that affects everything—gravitation. What happens when light meets gravity? The force of gravity has been understood since the time Newton realized it is universal—the same force that makes an apple drop keeps the moon going around the earth. But again it was Einstein who emphasized that gravity must affect not only material bodies like apples and moons, but *every* physical phenomenon, including light. This was one of the basic principles of his **general theory of relativity,** which is today's accepted description of gravitation. For many years this theory was based on just three experimental tests. One is the explanation of a small peculiarity in the motion of the planet Mercury; the other two involve the behavior of light.

If gravity affects light, as Einstein argued, then the light should not trace out an exactly straight line when it passes a strong center of attraction (a very massive object), but instead it should be deflected. The deflection is very tiny, only 0.0005° for light passing the edge of the sun. To observe such a glancing beam of light (from a star), it is best

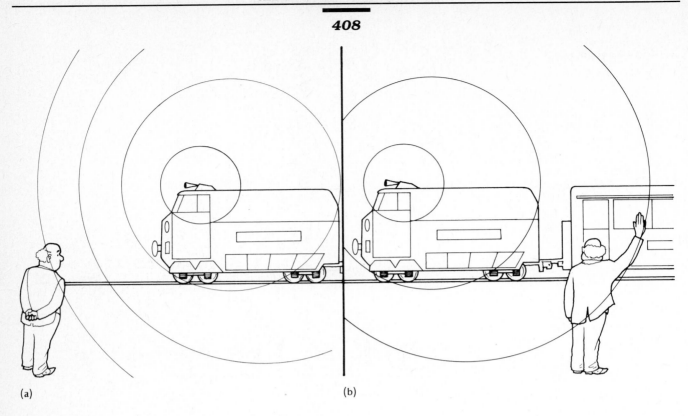

(a) (b)

FIGURE 15.16

A train travels toward the left, past a listener. **(a)** Wavefronts from the train's whistle before the engine has reached the listener. The wavefronts arrive more rapidly (higher frequency) than if the train were not moving, because successive wavefronts are emitted from closer points. **(b)** After the engine has passed the listener. Now successive wavefronts are emitted from successively more distant points, so the time between their arrival is greater than it would be if the train were at rest (i.e., they arrive with lower frequency).

to eliminate the sun's own light by observing during a solar eclipse. When Einstein first presented his ideas on light deflection, an expedition to the location of the next eclipse was prevented by war. This was lucky for Einstein, because his prediction at that time was a factor of two too small. A few years later, in 1919, he had improved his theory so it gave the right value, the war was over, and the verification by an English expedition of the work of a Jewish scientist working in Germany caught the world's attention as a symbol of science joining all nations in peaceful effort— Einstein became world famous.

Thus, according to general relativity, light is bent (refracted) in regions where gravity acts. Even near the sun, however, the effective index of refraction corresponding to this bending is only 1.000004, so the gravitational bending due to the entire sun is much less than the already small bending caused by the air on the earth's surface (Sec. 2.5C). Nonetheless, like the latter, the gravitational light bending can occasionally have remarkable consequences that are analogous to mirages, but on an intergalactic scale. That is, if a source of radiation and a massive galaxy, say, are properly aligned with respect to the earth, we may see several images of the source (Fig. 15.17), much like in an ordinary mirage (Fig. 2.61). A candidate for such a gravitational mirage (also called a **gravitational lens**) was found in the early 1980s.

When a photon passes a massive object, gravity usually pulls the photon forward during the approach as strongly as it pulls it backward after the encounter. Therefore, the whole process changes only the photon's direction, not its energy. However, when a photon falls *downward* in a gravitational field it picks up energy—its frequency therefore increases. This is called a **blue shift.** Conversely, photons climbing *out* of a gravitational field have their frequency decreased, because they give up energy as they climb—they are said to be **red shifted.**

FIGURE 15.17

A galaxy's gravitational field bends light rays from a star-like light source so that several images can be seen by an observer on earth (rays for only two of the images are shown).

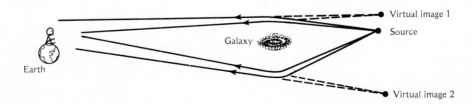

PONDER

Why these names?

On the earth this effect is quite small. Spectral lines from the sun and stars, however, come to us by means of light that had to climb out of a strong gravitational field (and then fell down the earth's weaker gravitational field). They are therefore red shifted—in the case of the sun, by 0.0002% of their frequency. This is how the gravitational red shift was first observed, and this was the other test of Einstein's general theory of relativity.

To get a larger gravitational red shift, we need an object on which gravity is stronger. In fact, it is theoretically possible to have objects so massive and compact that the frequency of any light traveling away from them would be shifted all the way to zero. But of course there is no such thing as zero-frequency light. Hence, no light could escape from the "surface" (called the **event horizon**) of such a body—the body would look completely black. (You could see no "events" beyond that horizon.) Due to its strong gravitational attraction, it would tend to swallow up matter. Such hypothetical bodies are called **black holes** (Fig. 15.18), because they let nothing escape and swallow everything; once any matter has crossed the event horizon, it cannot ever escape—it would have to outrun light.

. . . let me get to some theoretical considerations. First there is the question of black holes. . . . I believe it would be possible to send a ship there with reasonable assurance that it would arrive without serious damage. Return, however, is another question.

Frederick Pohl,
Beyond the Blue Event Horizon

There are objects in the sky that may be black holes. Even if we don't have a firm candidate, black holes have inspired many literary and movie efforts.

*D. More shifty ideas

Most stars are moving at high speeds—they seem fixed in the sky only because they are so far away. When the spectrum of a star is Doppler shifted compared to the earth-bound spectrum of the elements present in the star, we can conclude that the star is moving with respect to us. (The star's gravitational field would have a similar effect, but it is much smaller—unless the star is collapsing into a black hole.) Thus the light from a star not only tells us where a star is and what it is made of, but also how fast it is moving.

This Doppler shift is an important clue for understanding the universe as a whole. It was once thought that, throughout a suffi-ciently big chunk of space, motion of the stars averaged out—the universe was at rest. It was therefore a great surprise when, around 1930, Edwin Hubble found frequency shifts even in the spectra of whole galaxies. He correctly interpreted this as a common motion of these agglomerations of 10^{10} or more stars. Another surprise was the direction of this motion; nearly all galaxies are moving *away* from us—and thus nearly all the galaxies' spectra are shifted toward the *red*. Furthermore, the fainter (and hence farther) the galaxy, the greater this red shift. The shift is by no means small (Fig. 15.19); the fastest (and most distant) observed galaxies have their frequencies reduced by about

FIGURE 15.18

Artist's conception of a black hole (detail).

FIGURE 15.19

Emission spectrum of galaxies at various distances from the earth. The two dark lines seen are absorption lines of calcium—calcium in cool gases pervading the galaxies has absorbed this light. These lines are shifted from the UV (393 and 397 nm) into the visible, by the amounts shown by the arrows (ranging up to about 80 nm). Seen below is a terrestrial iron emission spectrum, used for comparison because it has many lines.

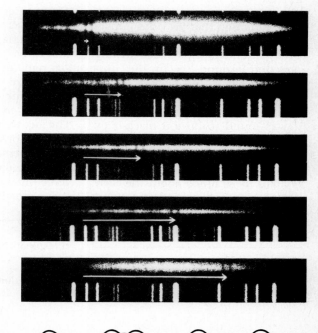

DISTANCE IN LIGHT-YEARS	RED-SHIFTS
78,000,000	
1,000,000,000	
1,400,000,000	
2,500,000,000	
3,960,000,000	

50%—they appear to be receding from us at 60% the speed of light!

This motion of all the galaxies away from us (and presumably away from each other) has led to the comparison of the present large-scale structure of the universe to a yeast bread dough with raisins, each raisin representing a galaxy. As the dough rises, *all* the raisins move away from each other—we in our galaxy raisin (which, however, we call the Milky* Way) are not uniquely repulsive to everyone else.

When we look at distant objects we see the universe at an earlier epoch, because it has taken over a billion years for light to get to us from the most distant of these objects (Fig. 1.5). The red shift of the light from these distant objects tells us that the universe was expanding when that light was emitted, and the light from closer objects tells us it has been expanding ever since. The huge Doppler shift of the light from the most distant galaxies means that the universe must have been about half its present size when the most distant galaxies emitted the light we now see from them.

Although we cannot see any galaxies from even earlier times, there is good reason to believe that the expansion has been going on ever since the very beginning of the universe—the universe apparently started with a bang, rather than a whimper. Again, it is electromagnetic radiation that provides important evidence for this scenario (Fig. 15.20). Namely, as we look in any direction around us in the universe, we see a faint glow of microwave radiation. When this radiation background was measured in 1965, it was found to have a black-body spectrum at the rather cool temperature of 2.7°K (corresponding to microwave frequencies). This discovery tells us so much about the early universe that it earned the Nobel Prize for Arno Penzias and Robert Wilson.

That black-body radiation pervades the universe appears to be puzzling, because there is no black body to give it its temperature. The matter in the universe today is so rarified and spread out that it hardly interacts with radiation at all, and we see light that has traveled undisturbed for over 10^9 years—the universe is essentially transparent to electromagnetic radiation. However, at earlier epochs the universe was denser. Because the universe is expanding, all the light we see is red shifted, including the background 2.7°K radiation. When it was emitted, then, this light had a higher frequency (more energy) and hence the universe must have been hotter. The most reasonable explanation of the background radiation observed today is therefore that it is a remnant of a time when the universe was dense and hot enough so it was *not* transparent to light. Before that time, the universe was so hot that electrons and atomic nuclei were not combined into elements; they were moving around as free, charged particles with which the light could interact strongly. As the universe expanded there was a rather definite time (called ***recombination time***—about ten billion years ago) when it got "cool" enough (about 3000°K) that electrons and nuclei could combine* into neutral atoms (mainly hydrogen). Because neutral atoms don't interact much with light, the universe became transparent at that time, but the radiation from this early, primordial fireball of hot gas remained. When we observe its microwave background today, we are looking from the inside at this fireball, enormously red shifted. The fact that the radiation was once in equilibrium with the matter in the universe, rather than isolated as it is today, means that the universe must have been at least a thousand times hotter and smaller

than it is today. This implies an explosion at the beginning to drive the matter in the universe apart—it must have started with a hot ***big bang.***

As we look farther and farther into the universe, examining fainter and fainter objects, we find more and more distant galaxies with spectra shifted more and more toward low frequencies. We see them as they were at the time they emitted their light, so we can deduce from them the history of the universe, for example, how its expansion has proceeded. But at the enormous frequency shift corresponding to the cosmic background radiation we can see no more detail, because everything appears equally bright, at the same 2.7°K temperature. The glow of the same fireball that gives us evidence of a hot, compressed early universe prevents us from looking back to even earlier stages of the universe—it seems to be the ultimate limit of what we can explore with light about the universe's origin.

So the story told by electromagnetic waves reaches from tiny, intricate detail on the scale of atoms to the large-scale structure of our cosmos. This radiation quite literally surrounds us and fills the universe. Light, our most important way to appreciate and communicate with the outside world, also limits that world—it limits the speed with which news can reach us, the distances over which we can communicate, and the earliest time from which we can receive it. Indeed, seeing the universe is in large measure seeing the light.

SUMMARY

One of the tenets of ***classical physics*** was that waves and particles are separate and distinct phenomena. But light behaves according to *both* of these models, as shown by the ***photoelectric effect.*** This effect is useful in ***photoconductive*** de-

---

*English, meaning *galaktodos* (in Greek, as in galaxy).

---

*Presumably for the first time—there is no evidence that they combined before. Nonetheless, it is usually called *recombination.*

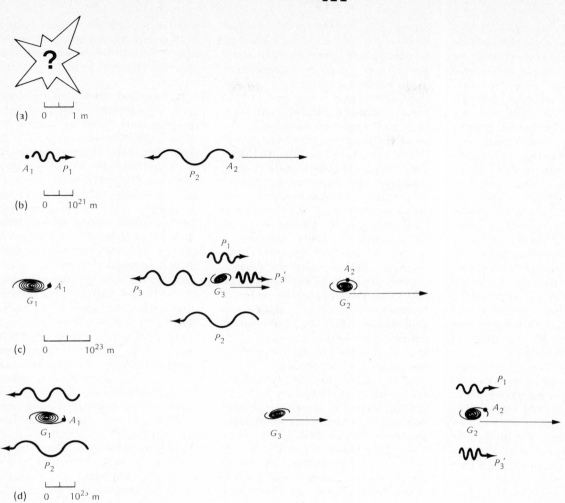

FIGURE 15.20

Schematic representation of some important events in the history of the universe, as reconstructed by an observer in our galaxy. **(a)** The big bang, when the universe was very hot and dense, cannot be directly seen by light, because the universe was not transparent to light for the first 10^5 years of its history. This era is largely *universum incognitum*. **(b)** At recombination time (10^5 years after the bang). As the universe expanded, its temperature dropped until oppositely charged electrons and nuclei combined into neutral atoms that no longer influenced radiation very much. We show an atom A_1 that will be part of our galaxy, and an atom A_2 at some distance, moving rapidly away from A_1. Each atom has just interacted for the last time with one of the photons flying around the universe. These photons then have the black-body spectrum associated with the temperature of the universe at that time. (They are predominately visible photons.) Since A_2 is rapidly moving away from our observer, its photon (P_2) has a frequency that will appear very much decreased by the Doppler shift when it reaches us. **(c)** After the matter has formed into galaxies (about 5 billion years after the bang). A_2 is now part of a galaxy G_2; its photon still has not got to us, but passes a galaxy G_3 along the way that has not been moving away from us as fast as G_2 and is therefore closer to us. A star in G_3 emits visible photons P_3 and P_3'. Since G_3 is not moving as fast as A_2 was, its photon P_3 is not as much red shifted as P_2. **(d)** At present (20 billion years after the bang). Both P_2 and P_3 reach us. The former comes from farther away and is most red shifted—it is part of the background radiation. The red shift of P_3 corresponds to the distance of G_3 (i.e., to the speed G_3 was moving away from us when it emitted P_3). An observer on the planet Zorg in galaxy G_2 would make similar observations. For example, since she is moving rapidly away from us, she receives our photon P_1 with a large red shift—it is part of the radiation background that she sees.

vices such as TV camera tubes and photocopiers and in **photovoltaic cells** or solar cells. When light ejects **photoelectrons** from a metal, the amount of energy given to each electron depends on the light's frequency, *not* on its intensity. According to the **quantum theory** this happens because electromagnetic waves carry energy in discrete units, called **quanta,** of size proportional to the wave's frequency. The smallest unit of light carries one energy quantum and is called a **photon.** To observe light's optical properties, we need a beam of many photons, because each photon can contribute at most one spot of light to an image, with photons being most likely to fall where classical optics assigns a high intensity. The nature of light as both waves and photons accounts for the shape of the **black-body spectrum.**

That electrons also have both wave and particle properties is shown by **electron interference** experiments. The electron's wave properties also explain why atoms have discrete **energy levels,** which cause an atom's **line spectrum;** when an atom changes from one energy level to another it emits or absorbs a definite amount of light energy. This

photon therefore has a definite frequency, that of one of the **emission** or **absorption lines** that are characteristic of the atom. If an atom is in an **excited state** (i.e., in an energy level higher than that of the lowest, **ground state**), emission can occur **spontaneously,** or it can be **stimulated** by light already present before the emission. The **lifetime** of an excited state before spontaneous emission is usually very short. In **fluorescence,** an atom is excited by a photon and then decays to its ground state via an intermediate state, emitting a photon of less energy (lower frequency) than the exciting energy. The exciting energy may instead be due to an electric current **(electroluminescence),** as in a **light-emitting diode (LED).** In **phosphorescence** the intermediate state is long-lived, giving an afterglow when the excitation has ceased.

Since light due to stimulated emission is coherent with the stimulating wave, intense coherent light beams can be produced when many atoms are stimulated by the same wave. In a **laser** the excited atoms are in a **population inversion** (more excited atoms are present than atoms in their ground state),

and their radiation is kept around to stimulate further atoms by bouncing it many times between two mirrors surrounding the excited atoms.

The **special theory of relativity** recognizes another feature of light; its (vacuum) speed is always the same, regardless of the speed of the source. But the motion of a source (observer) changes the frequency of the light **(Doppler shift),** increasing it for approaching sources and decreasing it for receding sources. The **general theory of relativity** shows that light is also affected by gravity; light moving horizontally is bent downward slightly; when it is moving up, its frequency is decreased **(red shift);** when moving down, its frequency is increased **(blue shift). Black holes** are hypothetical objects that would red shift light leaving their surface **(event horizon)** to zero frequency—that is, no light (or matter) can escape them. Light from distant galaxies has its frequency decreased because the galaxies are moving away from us. The shift is largest from the light that is left over from the primordial fireball, which occurred during the **big bang** at the creation of the universe.

PROBLEMS

P1   Does a photon of yellow light have more, less, or the same energy as one of cyan light?

P2   Infrared photons cause a particular piece of metal to emit photoelectrons. (a) Will photoelectrons be emitted when the metal is illuminated by visible light? (b) If not, explain. If they will, what properties of the emitted photoelectrons will be different from the case of infrared illumination?

P3   Phototubes cut off (i.e., do not respond to photons) at sufficiently small wavelengths. This is because the phototube's glass does not transmit UV. They also cut off at sufficiently large wavelengths, where the glass *does* transmit the light. Why?

P4   We have frequently mentioned that a glass surface transmits about 96% of the incident light, and reflects about 4%. How many reflected photons are detected if: (a) 100 photons are incident on the glass? (b) Just one photon is incident? Explain.

P5   You shine light at a mirror that is moving toward you. Due to its momentum the light presses on the mirror, so that the mirror slows down. The energy lost by the mirror is gained by the reflected light. (a) Is the speed of the reflected light larger, the same as, or less than that of the incident light? (b) Is the frequency of the reflected light larger, the same as, or less than that of the incident light?

P6   Is there a relation between the lines seen in the emission and the

absorption spectra of a given type of atom? Explain.

P7   What determines the color of an LED? What happens to the color and intensity of the emitted light and why, when you decrease the current through an LED?

P8   What conditions are necessary for a gas of excited atoms to emit laser radiation?

P9   How does laser light differ from the light from a point source, made parallel by a lens?

P10   What properties of a laser beam are used in (a) holography? (b) Punching the holes in baby-bottle nipples? (c) Monitoring the location of the crater of a dormant volcano by interferometry? (d) Laying straight railroad tracks?

P11 What is the answer to Einstein's boyish question, whether someone traveling at close to the speed of light can see himself in a mirror that he is carrying along?

P12 Comment on any difference between the frequency detection devices needed for the American radar speed control, involving reflection from the front of the car, and the European method, where the radar is reflected from the back of the car.

P13 Light from a massive star in a distant galaxy can be red shifted for *two* different reasons. What are these reasons?

HARDER PROBLEMS

PH1 Prof. Zweifelstein does not believe in quantum theory. To prove his point he takes an incandescent bulb, an exposure meter, and two color filters. When he puts the red filter in front of the meter, the meter reading is pretty high. With the blue filter in front instead, the meter reading is pretty low. Zweifelstein says: according to quantum theory, the blue photons have more energy and should therefore give more current, that is, a *higher* meter reading. Who is wrong and why?

PH2 An ultraviolet and an infrared light source have the same intensity. (This means, for example, that if their energy were converted completely to heat, you'd get the same amount of heat out of them.) (a) Which one has photons of higher energy? (b) Which one sends out photons at a higher rate?

PH3 The barrier that electrons have to overcome in the photoelectric effect gets smaller with increasing temperature. How does increasing temperature affect the long-wavelength cutoff of a phototube?

PH4 If two frequencies are seen in a spectrum, the difference between the two is sometimes also seen as a spectral line. Use an energy level diagram (with three levels) to account for this.

PH5 When an atom in a normal transparent medium (such as glass) resonates due to incident light, it starts out in the ground state and is raised to some excited state. Suppose a ground state atom has resonances only in the UV. (This is the case of normal dispersion as discussed in Chapter 2.) Draw a possible energy level diagram for such an atom that also permits emission of visible light,

and identify which transitions emit which type of light.

PH6 Some stars are double, orbiting about a common center. Suppose we look at the orbit edge on. How do the wavelengths in the spectrum coming to us from one of the stars vary in time?

PH7 A (stationary) light source is reflected in a mirror that is moving toward you. (a) Pretend that the reflected image is an actual light source. How would the frequency of the reflected light change (compared to that of the incident light)? (b) Although pretending gives the right answer, the reflected image cannot really act in all respects like an actual light source. To convince yourself, design an experiment where an image moves faster than *c*.

PH8 Assume there are two stars in the sky bright enough to be seen even when the sun is near them. As the sun moves between these stars, does their apparent separation increase or decrease?

PH9 When the gravitational frequency shift was measured terrestrially, photons were "dropped" (i.e., sent in a downward direction) from a 22.5-m high tower. (a) At the bottom of the tower, was the photons' frequency higher or lower than when they started? (b) Due to the frequency shift, an atom at the bottom could not absorb a photon emitted by an identical atom at the top. Why? (c) Therefore, the bottom atom was *moved* to compensate, by the Doppler shift, for the gravitational red shift. Did the bottom atom have to move with an upward or a downward speed to be able to absorb the photon?

PH10 A daring explorer claims to have discovered a black hole in our solar system and visited it. He tells of his experience: "The closer my ship came to the hole, the more its darkness seemed to engulf me. The sun behind me was only a faint red disk in the darkness, and suddenly ahead of me I could make out the event horizon, like a smooth sphere of steel glowing in the light from my retro-rockets. When I saw that, I knew it was time to pour on the power and get out." Would you buy a used spaceship from this man? Why not?

MATHEMATICAL PROBLEMS

PM1 How many radio-wave photons are needed to equal the energy of a

visible photon? (Use approximate frequency values read off Table 1.1.)

PM2 What is the ratio of the energy of the most energetic *visible* photon to that of the least energetic visible photon?

PM3 A beam of light of wavelength 500 nm that delivers 10^{10} photons per second carries an energy per second of 3×10^{-9} watts. How much energy per second would the beam carry if it delivered the same number of photons, but at 633 nm?

PM4 A square meter on the earth's surface receives about 500 watts when the sun shines on it. Assume that visible sunlight consists of photons of wavelength 500 nm. One such photon arriving every second would carry 3×10^{-19} watts. How many photons does the square meter receive each second when the sun shines?

PM5 Your answer to the previous problem probably sounds like a large number. Let's find out whether these photons crowd each other. (You will need your result from PM4.) (a) Since light goes 3×10^8 m in one second, the photons that hit our square meter in one second came from a volume of 3×10^8 cubic meters. How much volume does that give for each photon? (b) The range of the sun's spectrum (not very monochromatic) means that photons from the sun are only about 0.5 μm long—that is, this is the coherence length of sunlight. The sun's angular size ($\frac{1}{2}°$) means that the photons are about 60 μm across. How much volume does one photon occupy? (c) Compare your answers to (a) and (b). Are the solar photons crowded together or well separated?

PM6 Repeat the calculations of PM4 and 5 for a 1-watt laser of beam diameter (photon width) 1 cm, with coherence length 10 cm and wavelength 500 nm.

PM7 An electromagnetic wave, coming from some opening, must spread as it travels. The minimum amount by which it spreads is the size of the interference pattern due to that opening (the distance between the first minima), and is given by Equation K.9. Unfocused laser beams typically spread by no more than this minimum amount. Compute the size of the spot of light made by a He-Ne laser beam at a distance of 1 km, if the diameter of the beam coming out of the laser is 1 mm and the wavelength is 633 nm.

APPENDIX A
POWERS-OF-TEN NOTATION

Large numbers crop up as soon as you talk quantitatively about light waves; the largest frequency we will mention will be a hundred billion billion times as large as the smallest, and the smallest wavelength will be about one ten-thousandth of a billionth of an inch. We would like to find a way to avoid these rather cumbersome forms of numbers.

An extremely useful and fairly simple shorthand for dealing with numbers that span a large range is the **powers-of-ten notation.** Instead of writing:

one hundred billion billion
= 100,000,000,000,000,000,000
(20 zeros)

we write:

10^{20} (which means, a 1 followed by 20 zeros)

How do we get this? It is just a familiar rule of raising a number to a power. For example, consider:

one thousand = 1000 = 10^3

according to the notation we just introduced. By the usual rule of what an exponent means, 10^3 is the "cube of ten," which gives the same result:

$10^3 = 10 \times 10 \times 10 = 1000$

So 10 raised to any power is 10 multiplied by itself that many times, which is 1 with that many zeros after it.

What about 1 itself—that is, a 1 with no zeros after it? In our notation we must write this as:

$1 = 10^0$

How about small numbers, say 0.1? Instead of multiplying 1 by powers of 10, this requires dividing 1 by powers of 10. The number of 10's we divide by will be indicated by negative powers, for example

one tenth = 0.1 = $\frac{1}{10}$ = 10^{-1}

or:

one ten thousandth of a billionth
= 1/10,000,000,000,000
= 10^{-13} (count the zeros!).

In addition to brevity, this notation has another advantage: multiplication and division of such "powers-of-ten" numbers is easy—you need only add or subtract the exponents. For example, $10^{13} \times 10^4$ is 13 tens times 4 tens all multiplied together, so the product is (13 + 4) tens, 17 tens, or 10^{17}:

$10^3 \times 10^4 = 10^{13+4} = 10^{17}$

Similarly $10^{-2} \times 10^{-1}$ is

$$\frac{1}{10 \times 10} \times \frac{1}{10} = \frac{1}{(10 \times 10) \times 10},$$

or:

$10^{-2} \times 10^{-1} = 10^{-2+(-1)} = 10^{-3}$

Thus multiplication becomes simple addition. Division is just as easy when we note, for example $1/10^{13} = 10^{-13}$, so dividing by 10 to a power is the same as multiplying by 10 to the negative power. Thus,

$$\frac{10^{13}}{10^{13}} = 10^{13} \times 10^{-13} = 10^0$$

= 1 (of course)

APPENDIX B
THE MATHEMATICAL FORM OF SNELL'S LAW

To state the mathematical form of Snell's law, we shall first need the trigonometric function called the *sine* of an angle θ, written as sin θ. To do this, we construct a right triangle with one angle equal to θ (Fig. B.1a). Then sin θ is equal to the length of the side opposite θ, divided by the length of the hypotenuse:

$$\sin \theta = \frac{p}{r} \qquad (B1)$$

For any value of the angle θ, this function can be found in tables or calculators.

Now, consider a beam of light incident on a boundary at angle θ_i, in material of index of refraction n_i, and then transmitted at angle θ_t, in material of index of refraction n_t (Fig. B.1b). We have drawn the wavefront AA', which is perpendicular to the incident beam. Hence (since the surface AB' is perpendicular to its own normal) the angle $\sphericalangle A'AB'$ must equal θ_i. Then:

$$\sin \theta_i = \frac{\overline{A'B'}}{\overline{AB'}}$$

or;

$$\frac{1}{\overline{AB'}} = \frac{1}{\overline{A'B'}} \sin \theta_i \qquad (B2)$$

Drawing the wavefront BB', perpendicular to the transmitted beam,

and going through the same steps, gives us:

$$\frac{1}{\overline{AB'}} = \frac{1}{\overline{AB}} \sin \theta_t \qquad (B3)$$

Comparing Equations B2 and B3 gives us:

$$\frac{1}{\overline{A'B'}} \sin \theta_i = \frac{1}{\overline{AB}} \sin \theta_t \qquad (B4)$$

Suppose that it took light a time T to travel between the two wave fronts, that is, from A' to B' for the part still on the incident side, and from A to B for the part on the transmitted side. If the light travels on the incident side with speed $v_i = c/n_i$, and on the transmitted side with speed $v_t = c/n_t$, then we have:

$$\overline{A'B'} = v_i T = \frac{cT}{n_i}, \quad \text{and}$$

$$\overline{AB} = v_t T = \frac{cT}{n_t}$$

Putting these results in Equation B4 gives us:

$$\frac{n_i}{cT} \sin \theta_i = \frac{n_t}{cT} \sin \theta_t$$

or, *Snell's law:*

$$\boldsymbol{n_i \sin \theta_i = n_t \sin \theta_t} \qquad (B5)$$

FIGURE B.1

(a) Right triangle with one angle equal to θ. (b) Light incident from a fast medium to a slow medium ($n_i < n_t$) at angle θ_i. The refracted beam bends toward the normal; that is, $\theta_t < \theta_i$. Two wavefronts, AA' and BB' are shown.

If we refer back to Figure B.1a, we notice that p can never be larger than r—the hypotenuse is always the largest side. Hence, from Equation B1, sin θ can never be greater than 1. When sin θ_t equals 1, we have the condition of *total internal reflection*. That is, the incident angle is then the *critical angle* θ_c, where (from Eq. B5, setting sin θ_t = 1):

$$\boldsymbol{n_i \sin \theta_c = n_t} \qquad (B6)$$

Any incident angle greater than θ_c will result in a totally internally reflected light beam. There is then no transmitted beam—Equation B5 cannot be satisfied. Since sin θ_c must also be less than 1, we can only satisfy Equation B6 if n_i is greater than n_t. That is, only when going from a slower medium (say glass) toward a faster medium (say air) can light be totally internally reflected.

APPENDIX C
THE FOCAL POINT OF A CONVEX MIRROR

To prove the relation given in Section 3.3A, that the focal length is one half the radius of a spherical convex mirror: $f = OF = \frac{1}{2}\overline{OC}$, consider Figure C.1. An incident ray,

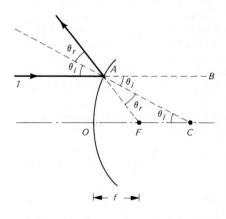

FIGURE C.1

Ray *1*, traveling parallel to the axis, strikes a convex mirror at *A*, and leaves as if it had come from the focal point, *F*.

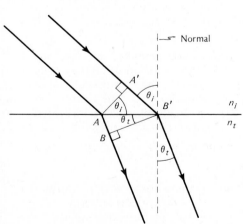

(b)

(a)

ray *1*, traveling parallel to the axis, strikes the mirror at *A*, with angle of incidence θ_i. It is reflected with angle of reflection θ_r, as shown. Since *AB* is a continuation of the incident ray, and *AC* is the normal at *A*, the angle $\sphericalangle CAB$ is equal to θ_i. The angle $\sphericalangle ACF$ must then also be equal to θ_i, because it and $\sphericalangle CAB$ are opposite interior angles between two parallel lines. Similarly, *AF* is a continuation of the reflected ray, so the angle $\sphericalangle CAF$ must equal θ_r, which by the law of reflection is equal to θ_i. Hence the triangle $\triangle CAF$ is an isosceles triangle, having two equal angles ($\sphericalangle CAF$ and $\sphericalangle ACF$), and therefore two equal sides

$$\overline{AF} = \overline{FC}$$

Now, for *paraxial* rays, the point *A* must be close to the point *O*, so $\overline{AF} \simeq \overline{OF}$. Hence, we can write:

$$\overline{OF} = \overline{FC} \text{ (paraxial rays)}$$

But since $\overline{OC} = \overline{OF} + \overline{FC}$, we have, finally:

$$\overline{OF} = \tfrac{1}{2}\,\overline{OC}$$

That is, the focal length of the spherical mirror is half its radius.

FIGURE D.1

Object *PQ* of size s_o emits ray *1* parallel to the axis and ray *3* through the focal point, *F*. The (inverted) image *P'Q'* has size $-s_i$.

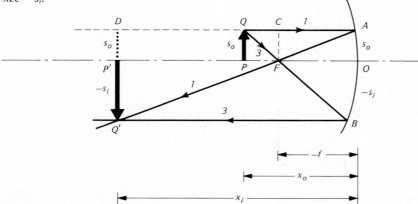

APPENDIX D
THE MIRROR EQUATION

To derive the **mirror equation,** we redraw Figure 3.17, showing only rays *1* and *3*, but extending ray *1* backward (Fig. D.1). We then note pairs of *similar* triangles (treating *AOB* as a straight line): $\triangle ABQ$ and $\triangle CFQ$, also $\triangle CFA$ and $\triangle DQ'A$. From the first pair of similar triangles, we may write the equal ratios:

$$\frac{\overline{CF}}{\overline{AB}} = \frac{\overline{QC}}{\overline{QA}}$$

We note that $\overline{CF} = s_o$, the object size; $\overline{AB} = s_o - s_i$, where $-s_i$ is the image size (the minus sign indicates that the image points *downward*); $\overline{QA} = x_o$, the distance of the object in front of the mirror; and $\overline{QC} = x_o - (-f) = x_o + f$, where f is the focal length of the mirror and is negative for this concave mirror. Hence the above equation reads:

$$\frac{s_o}{s_o - s_i} = \frac{x_o + f}{x_o}$$
$$= 1 + \frac{f}{x_o} \qquad \text{(D1)}$$

The second pair of similar triangles gives:

$$\frac{\overline{CF}}{\overline{DQ'}} = \frac{\overline{CA}}{\overline{DA}}$$

But $\overline{CF} = s_o$, $\overline{DQ'} = s_o - s_i$, $\overline{CA} = -f$, and $\overline{DA} = x_i$, the distance of the image in front of the mirror. So this equation reads:

$$\frac{s_o}{s_o - s_i} = \frac{-f}{x_i} \qquad \text{(D2)}$$

Since the left-hand sides of Equations D1 and D2 are the same, their right-hand sides must be equal:

$$\frac{-f}{x_i} = 1 + \frac{f}{x_o}, \quad \text{or} \quad -\frac{f}{x_o} - \frac{f}{x_i} = 1$$

Dividing both sides of the latter form of the equation by f then gives:

$$-\frac{1}{x_o} - \frac{1}{x_i} = \frac{1}{f} \qquad \text{(D3)}$$

This is the *mirror equation,* derived here for the case of the *concave* mirror, where f is *negative*. It also is valid for *convex* mirrors: you just put in a *positive f*. If you should find x_i to be *negative*, it simply means that the image is that distance *behind* the mirror (a *virtual* image).

Equation D3 locates the image; what about the size of the image? If we write Equation D2 upside down, we get:

$$-\frac{x_i}{f} = \frac{s_o - s_i}{s_o} = 1 - \frac{s_i}{s_o}$$

or:

$$\frac{s_i}{s_o} = 1 + \frac{x_i}{f} \qquad \text{(D4)}$$

Hence, if we know the location of the image x_i, then Equation D4 tells us its size s_i. We can simplify this by using the mirror equation, Equation D3. Multiplying it by x_i gives:

$$-\frac{x_i}{x_o} = 1 + \frac{x_i}{f}$$

Using this in Equation D4 gives:

$$\frac{s_i}{s_o} = -\frac{x_i}{x_o} \qquad \text{(D5)}$$

This equation tells us how big the image is, compared to the object—how much it is *magnified*. The minus sign tells us that if x_i and x_o are both positive (in front of the mirror, as in Fig. D.1), then s_i will have the opposite sign to s_o—it will be *inverted*, as in Figure D.1.

Thus, Equations D3 and D4 tell us the size, position, and orientation of the image.

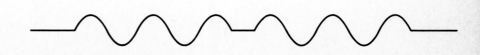

APPENDIX E
THE LENS EQUATION

To derive the **lens equation,** we redraw Figure 3.26, showing only rays *1* and *3*, and use the technique of Appendix D (Fig. E.1). The pairs of similar triangles here are $\triangle ABQ$ and $\triangle CFQ$, as well as $\triangle ABQ'$ and $\triangle AOF'$. The first pair gives:

$$\frac{\overline{CF}}{\overline{AB}} = \frac{\overline{QC}}{\overline{QA}}$$

Here $\overline{CF} = s_o$, the object size; $\overline{AB} = s_o - s_i$, where $-s_i$ is the image size (again, it has a minus sign because the image points *downward*); $\overline{QA} = x_o$, the distance of the object in *front* of the lens; and $\overline{QC} = x_o - f$, where f is the focal length of the lens. Hence the above equation reads:

$$\frac{s_o}{s_o - s_i} = \frac{x_o - f}{x_o} = 1 - \frac{f}{x_o}. \quad \text{(E1)}$$

The second pair of similar triangles gives:

$$\frac{\overline{AO}}{\overline{AB}} = \frac{\overline{OF'}}{\overline{BQ'}}$$

But $\overline{AO} = s_o$, $\overline{AB} = s_o - s_i$, $\overline{OF'} = f$, and $\overline{BQ'} = x_i$, the distance of the image *past* the lens. So this equation reads:

$$\frac{s_o}{s_o - s_i} = \frac{f}{x_i} \quad \text{(E2)}$$

FIGURE E.1

Object *PQ* of size s_o emits ray *1* parallel to the axis and ray *3* through the focal point, *F*. The (inverted) image *P'Q'* has size $-s_i$.

Comparing Equations E1 and E2, we get:

$$\frac{f}{x_i} = 1 - \frac{f}{x_o}, \quad \text{or} \quad \frac{f}{x_o} + \frac{f}{x_i} = 1$$

Dividing both sides of the latter form of the equation by f then gives:

$$\frac{1}{x_o} + \frac{1}{x_i} = \frac{1}{f} \quad \text{(E3)}$$

This is the *lens equation*, derived here for a *converging* lens, where f is *positive*. It also is valid for *diverging* lenses; you just put in a *negative f*. If you should find x_i to be *negative*, it simply means that the image is in *front* of the lens (a *virtual* image).

Equation E3 locates the image; what about the size of the image? As in Appendix D, we write Equation E2 upside down:

$$\frac{x_i}{f} = \frac{s_o - s_i}{s_o} = 1 - \frac{s_i}{s_o}$$

or:

$$\frac{s_i}{s_o} = 1 - \frac{x_i}{f} \quad \text{(E4)}$$

Multiplying Equation E3 by x_i gives:

$$\frac{x_i}{x_o} + 1 = \frac{x_i}{f}, \quad \text{or} \quad 1 - \frac{x_i}{f} = -\frac{x_i}{x_o}$$

which, combined with Equation E4, gives:

$$\frac{s_i}{s_o} = -\frac{x_i}{x_o} \quad \text{(E5)}$$

That is, as in the case of the mirror, the *magnification* is just the negative of the ratio of the image and object distances. Again, positive x_i and x_o (as in Fig. E.1) means there is a negative magnification—the image is *inverted*, as in Figure E.1.

APPENDIX F
TWO THIN LENSES TOUCHING

Consider two thin lenses with focal lengths f_1 and f_2. Start with them separated by a distance t, with an object at the first focal plane of the first lens. Light from that object must then emerge from the first lens in a parallel beam (Fig. F.1a). But a parallel beam incident on the second lens will be focused in that lens' second focal plane, as shown in the figure.

This is true no matter what the separation t is. If we let the separation vanish, so the two lenses touch, we can think of them as one *effective* lens, and the diagram looks like Figure F.1b. But now we know that the object distance is $x_o = f_1$, and the image distance is $x_i = f_2$. The lens equation for this effective lens is (Eq. E3):

$$\frac{1}{x_o} + \frac{1}{x_i} = \frac{1}{f}$$

where f is the effective focal length of the combined lens. With the known values of the object and image distances, this becomes:

$$\frac{1}{f_1} + \frac{1}{f_2} = \frac{1}{f} \quad \text{(F1)}$$

But $(1/f_1) = P_1$ is the *power* of the first lens; $(1/f_2) = P_2$ is the power of the second lens; and $(1/f) = P$ is the power of the combined lens. So Equation F1 becomes:

$$P_1 + P_2 = P \quad \text{(F2)}$$

—the *powers add* for the two touching lenses, as advertised.

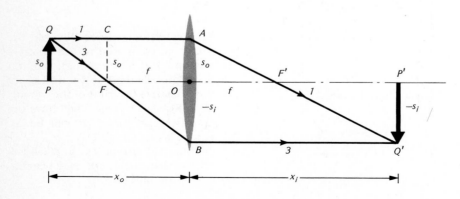

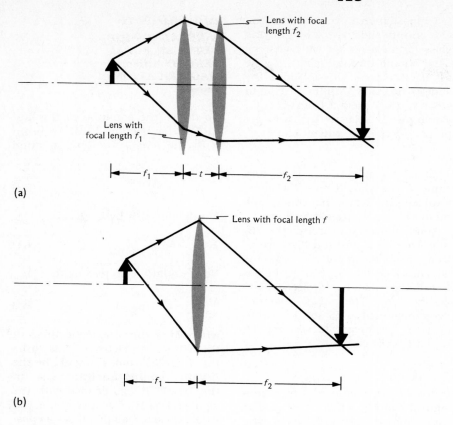

(a)

(b)

FIGURE F.1

(a) Two thin lenses separated by a distance t. (b) Two thin lenses touching can be thought of as one effective lens.

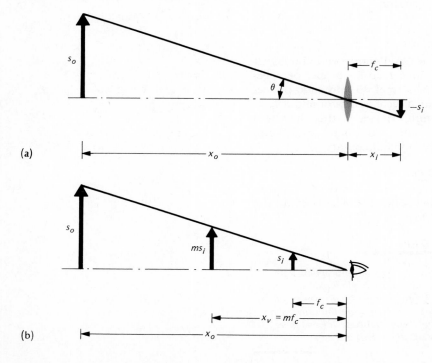

(a)

(b)

APPENDIX G
PHOTOGRAPHIC PERSPECTIVE

When you look at a scene, each object in your field of view produces an image of a particular size on your retina. For the perspective in a *photograph* of the scene to appear correct to you, the retinal image sizes must be the same as when you viewed the *scene* directly. Another way to say this is that the visual angle subtended by each image on the photograph must be the same as that subtended by the corresponding original object. Clearly this condition depends on the focal length of the lens used for photographing the scene, the magnification of any resulting print, and the distance from which you view the photograph.

Figure G.1a shows a camera photographing a distant object in a scene. Recall that the relation between the object size, s_o, and the image size, s_i, depends only upon the object distance, x_o, and the image distance, x_i. As Equation E5 states:

$$\frac{s_i}{s_o} = -\frac{x_i}{x_o} \qquad (E5)$$

For distant objects, the image lies in the focal plane of the converging lens being used, hence $x_i \simeq f_c$. Here f_c is the focal length of the camera lens. Equation E5 is then:

$$s_i = -f_c \frac{s_o}{x_o} \qquad (G1)$$

FIGURE G.1

(a) A camera whose lens has focal length f_c photographs a distant object whose size is s_o. The object subtends an angle θ and produces an (inverted) image of size $-s_i$ on the film. (b) An eye (represented by a lens) views a photographic image of the original object. If there is no magnification ($m = 1$), the eye should view the photo from a distance of f_c for the image to subtend the same angle θ as in (a). If the photo is enlarged by a factor m, then the eye must view it from a distance mf_c. This will guarantee the proper perspective.

Thus f_c sets the *scale* of the sizes of the images; all image sizes are proportional to f_c. Hence your telephoto lens gives you a larger image.

Suppose you now look directly at the developed photographic image. Figure G.1b shows that you must view it from a distance f_c in order that each photographic image subtends the same angle as the corresponding original object did; that is, so that the perspective appears normal. Usually you can't focus from so close, so you must enlarge the image.

Suppose you enlarge your photograph, *increasing its size* by a magnification factor m. As Figure G.1b shows, you must now look at the print from a *larger viewing distance* for the photographic image to subtend the same angle and hence the perspective to appear correct. As the figure illustrates, the proper viewing distance is:

$$x_v = m f_c \qquad (G2)$$

Note that this result is independent of the size and distance of the object in the scene. That is, if you view your print from the proper distance for one of the objects, *all* objects will appear the proper size and hence the perspective will appear correct. If you view the print from a smaller distance, it will have telephoto perspective (see the TRY IT for Sec. 4.3B); if you view it from a larger distance, it will have wide-angle perspective.

Let's consider some applications of Equation G2. Suppose first you take a 35-mm picture and blow it up to 12 × 18 cm (a 5 × 7" print). This means a magnification of about $m = 5$ (since a 35-mm negative is 24 × 36 mm). If you wish to view this print from 25 cm, what focal-length camera lens should you have used so the perspective looks correct? According to Equation G2, correct perspective is achieved if:

$$f_c = \frac{x_v}{m}$$

and $x_v = 25$ cm while $m = 5$, so $f_c = 5$ cm $= 50$ mm. Thus, for these conditions, a 50-mm camera lens gives correct perspective.

Suppose instead you wish to look at a 35-mm slide with a magnifying glass (as in a pocket slide viewer); what should its focal length, f_m, be for correct perspective? Since the slide is not blown up ($m = 1$), and since you will view it from $x_v = f_m$, Equation G2 tells us that you get proper perspective when:

$$f_c = f_m$$

That is, the magnifying glass and camera lens should have the same focal length. (Notice that this is true no matter what size film you use.)

Now suppose you project the 35-mm slide on a screen a distance x_p from the *projector*. If the projector lens has focal length f_p, then the object distance (slide to projector lens) is about f_p. The image distance (lens to screen) is x_p, so the magnification produced (described by Equation E5) is:

$$m = \frac{x_p}{f_p} \qquad (G3)$$

(Here we have ignored the minus sign in Eq. E5, which tells us that the image is inverted—that's why you put the slide in upside down.) If you view the projected image from a distance x_v from the screen, using the value of m in Equation G3 tells us that the perspective will be correct providing:

$$x_v = x_p \frac{f_c}{f_p}$$

This is the distance you should sit from the screen for proper perspective. A normal 35-mm slide projector has a focal length of about $f_p = 100$ mm. Hence if the slide was taken with a normal $f_c = 50$-mm lens, you should sit at:

$$x_v = \frac{x_p}{2}$$

halfway between the screen and the projector.

PONDER

Suppose the original photograph was taken with a wide-angle lens, say $f_c = 25$ mm. Where should you sit (if proper perspective were the only consideration)? Suppose it was taken with a 200-mm telephoto lens?

APPENDIX H
A RELATIONSHIP BETWEEN FOCAL LENGTH AND MAGNIFICATION

To derive the equation of the first TRY IT for Section 4.4B, we begin with the *lens equation*, Equation E3:

$$\frac{1}{x_o} + \frac{1}{x_i} = \frac{1}{f} \qquad (E3)$$

Multipyling this by x_o gives:

$$1 + \frac{x_o}{x_i} = \frac{x_o}{f} \qquad (H1)$$

Now, Equation E5 tells us that:

$$M = \frac{x_i}{x_o} = -\frac{s_i}{s_o} \qquad (H2)$$

where M is the quantity defined in the TRY IT. (Strictly, if M is to be the *magnification*, it should be the negative of this quantity: s_i/s_o. In the TRY IT it was defined with the sign of Eq. H2 because there you would measure all distances as *positive*, despite the fact that the image is inverted.) Using Equation H2 in Equation H1, we get:

$$\frac{x_o}{f_o} = 1 + \frac{1}{M} = \frac{M+1}{M}$$

or, writing this upside down:

$$\frac{f}{x_o} = \frac{M}{1+M}$$

or, finally:

$$f = x_o \frac{M}{1+M}$$

APPENDIX I
LOGARITHMS

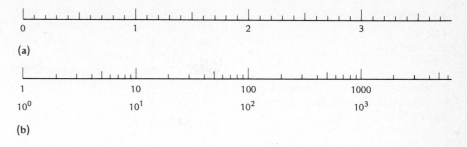

(a)

(b)

Ordinary scales used for graphs, such as that shown in Figure I.1a, have equal values between markings. In the example shown, when you go from one marking to the next, you *add* 1, no matter which marking you started at. Such a scale is called a *linear* scale. However, we have seen cases where such a scale would be very inconvenient. For example, in Table 1.1, if we had let a distance on the paper of 1 mm represent 100 Hz, we would have needed a piece of paper that stretched from here to beyond the star Sirius in order to show the frequency range included in the table.

We managed to include such a tremendous range by using a scale in which each step differed from the previous one by a *multiplicative factor* of 10 (Fig. I.1b). Here the second marking corresponds to a number 10 times as big as the first, the third to a number 10 times as big as the second, and so on. We've indicated the marking numbers in the powers-of-ten notation, and you see that the markings correspond to equal steps in the *power*—from 10^1 to 10^2 to 10^3, etc.

The *logarithm* is a device to display that power. The logarithm of a number y, written as log y, is defined as follows:

$$\log y = x \quad \text{means} \quad 10^x = y$$

Thus, the logarithm of the markings in Figure I.1b would be $\log 10 = 1$, $\log 100 = 2$, . . . , because $10^1 = 10$, $10^2 = 100$, ... The markings, then, correspond to equal steps on a *logarithmic* (or *log*) scale.

Logarithmic scales find use in many places where you are interested in the percentage change of something rather than the actual value. A common example, from the newspaper, is the stock market (Fig. I.2). It doesn't matter whether your stock cost $1 per share or $100 per share; if you bought $5000 worth of shares and their

FIGURE I.1

(a) In a linear scale, successive equal increments correspond to the addition of a constant. (b) In a logarithmic scale, successive equal increments correspond to the multiplication by a constant.

FIGURE I.2

Stock Exchange indexes plotted on a logarithmic scale versus time, which is plotted on a linear scale.

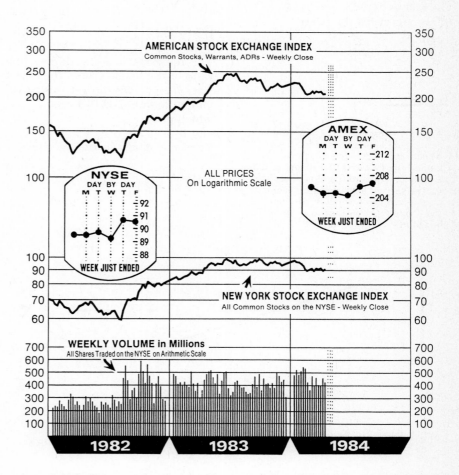

value doubled, your shares would then be worth $10,000 in either case. On a linear scale, however, an increase from $1 to $2 per share would look very small compared to an increase from $100 to $200 per share. On a log scale, the size of the increase would be the same in both cases because each increase was a *factor* of 2. The log scale enables you to compare how well different stocks are doing with respect to each other, even though their prices may be quite different.

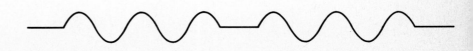

APPENDIX J
TELESCOPE MAGNIFICATION

A telescope (Fig. J.1) converts incoming parallel rays, from some object PQ at a very large distance D, to outgoing parallel rays at a different, larger, angle. An eye looking through the telescope then sees the distant virtual image PQ'' still at a large distance, but subtending a larger angle than the object did without the telescope (that is, $\theta_2 > \theta_1$).

In the figure, we see pairs of similar triangles: ΔQPO_1 and $\Delta Q'P'O_1$, as well as $\Delta Q''PO_2$ and $\Delta Q'P'O_2$. The first pair yields the relation:

$$\frac{\overline{QP}}{\overline{Q'P'}} = - \frac{\overline{PO_1}}{\overline{O_1P'}}$$

FIGURE J.1

A simple telescope consisting of two converging lenses is used to view a distant object PQ. The image $P'Q'$ is behind the first lens (the objective lens), a distance equal to the lens' focal length, f_o. The second lens (the eyepiece) is used as a magnifying glass for viewing the image $P'Q'$, and forms a virtual image PQ'' very far away.

(The minus sign indicates that $\overline{Q'P'}$ points in the opposite direction to that of $\overline{QP}$.) Here $\overline{PO_1} = D$, the large distance to the object; and $\overline{O_1P'} = f_o$, the focal length of the objective. Rewriting this equation then gives:

$$\overline{QP} = - \overline{Q'P'}\frac{D}{f_o} \tag{J1}$$

The second pair of triangles yields:

$$\frac{\overline{Q''P}}{\overline{Q'P'}} = \frac{\overline{PO_2}}{\overline{P'O_2}}$$

where $\overline{P'O_2} = f_e$, the focal length of the eye piece; and $\overline{PO_2} = D + f_o + f_e \simeq D$, if D is large compared to the focal lengths f_o and f_e. Hence, this equation gives:

$$\overline{Q''P} = \overline{Q'P'}\frac{D}{f_e} \tag{J2}$$

The telescope magnification is the ratio of the image size to the object size: $\overline{Q''P}/\overline{QP}$. We can get this by dividing Equation J2 by Equation J1, which gives:

$$\text{Telescope magnification} = - \frac{f_o}{f_e} \tag{J3}$$

APPENDIX K
POSITIONS OF INTERFERENCE AND DIFFRACTION FRINGES

Let's derive the *locations* of the intensity maxima (bright fringes) and minima (dark fringes) in the screen patterns considered in Chapter 12. Maxima occur at points of constructive interference (waves in phase) and minima at points of destructive interference (waves out of phase). We shall assume that all the *sources* are *in phase*. A phase difference then occurs only when the interfering beams travel *different distances* to reach a point on the screen. Recall that the *waves* from two sources are:

> in phase
> if their path difference ϵ is
> $0, \pm\lambda, \pm2\lambda, \pm3\lambda, \ldots$
>
> out of phase
> if their path difference ϵ is
> $\pm\lambda/2, \pm3\lambda/2, \pm5\lambda/2, \ldots$
> $\tag{K1}$

PONDER

If the two sources are out of phase, however, the out-of-phase case and the in-phase case are interchanged in Equation K1. That is, maxima become minima and vice versa. Why?

1. Two Sources. The path difference ϵ can be computed by plane geometry. For example, let's examine the case of *two sources* as shown in Figure K.1. In this figure we can consider a point P a distance h to the right of the center point O on the screen. (A negative h means that P is to the left of O.) We can now find ϵ by similar triangles. The two angles marked ϕ are equal, because they are alternate interior angles for parallel lines (the screen and the line connecting the sources). The angles marked θ are therefore also equal, because they are complementary to the angles ϕ. From Appendix B, where the sine function was defined, we have (using ΔS_1S_2Q):

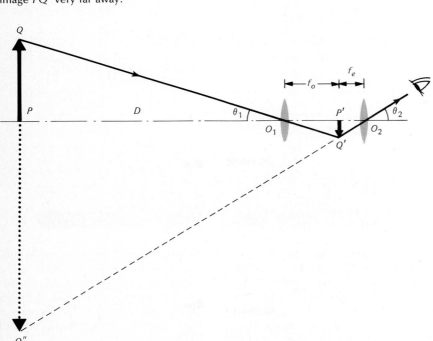

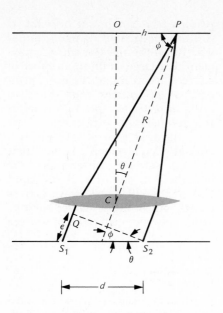

FIGURE K.1

Geometry of the rays from two sources S_1 and S_2 that interfere in the focal plane of a lens of focal length f, centered at C.

$$\frac{\epsilon}{d} = \sin \theta$$

or:

$$\epsilon = d \sin \theta \qquad (K2)$$

We can find $\sin \theta$ by using $\triangle COP$:

$$\sin \theta = \frac{h}{R}$$

In many interference set-ups the angle θ is quite small, so that $R \simeq f$, and hence $\sin \theta \simeq h/f$. Equation K2 then gives:

$$\epsilon = \frac{dh}{f} \text{ (for small } \theta) \qquad (K3)$$

Now we can apply the criterion of Equation K1. There are maxima when:

$$\frac{dh}{f} = 0, \pm \lambda, \pm 2\lambda, \pm 3\lambda, \ldots$$

that is, for points on the screen where:

$$h = 0, \pm \lambda \frac{f}{d}, \pm 2\lambda \frac{f}{d}, \pm 3\lambda \frac{f}{d}, \ldots \quad (K4)$$
(position of intensity maxima)

Similarly, we find the minima when:

$$h = \pm \lambda \frac{f}{2d}, \pm 3\lambda \frac{f}{2d}, \pm 5\lambda \frac{f}{2d} \ldots (K5)$$
(position of intensity minima)

The fringe spacing is the distance between successive maxima. Hence:

fringe spacing $= \lambda \dfrac{f}{d}$ $\qquad$ (K6)

If the focal length f of the lens is very large, the lens is very weak, and we can simply remove it without significant change in the geometry, leaving the screen at the same *large* distance, which we now call D. Equation K6 then says that the fringes spacing is $\lambda D/d$—the formula used in Section 12.2E.

Note, however, that for high order fringes, the angle θ is *not* small, and we *cannot* replace R by f. In fact, since $\sin \theta$ never exceeds 1 (no matter what the angle θ), Equation K2 shows that ϵ can never exceed d. Hence there can only be a finite number of fringes on the screen, and our approximate formula, Equation K6, for the fringe spacing ceases to be valid for high orders. (The fringe spacing increases for higher order fringes, and the total number of fringes is finite.)

2. Thin Film. Equation K1 for two-beam interference can also be used to find the wavelengths that give intensity maxima and minima when light is *reflected by a thin film*. We assume that the first and second reflections are both hard or both soft. (If one is hard and the other soft, we need only interchange "maxima" and "minima" in the following.) If the thickness of the film is t, the extra path length ϵ of the beam reflected from the second surface is $2t$ (at near normal incidence). Hence we get:

maxima at

$$\lambda' = 2t, \frac{2t}{2}, \frac{2t}{3}, \frac{2t}{4}, \frac{2t}{5}, \ldots$$
$$(K7)$$

minima at

$$\lambda' = 4t, \frac{4t}{3}, \frac{4t}{5}, \frac{4t}{7}, \ldots$$

FIGURE K.2

Reflected intensity versus wavelength for **(a)** a film with both reflections in phase, **(b)** a different film with reflections out of phase. The inserts in the graph show the arrangement of the film.

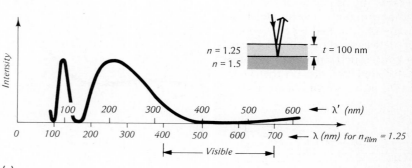

(a)

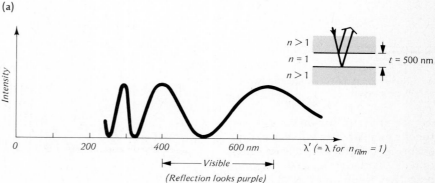

(b)

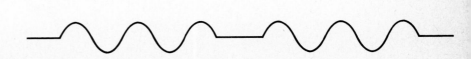

Here λ' is the wavelength *in the* film. To relate λ' to λ, the wavelength in air, we recall that when a light wave enters a medium of index of refraction n, the frequency of the wave is not changed (Sec. 1.3A), but the speed of the wave changes from c to c/n (Sec. 2.5). From the formula of Section 1.3A that relates speed and frequency to wavelength we then find $\lambda' = \lambda/n$.

Thus, for a given thickness, Equation K7 tells us those wavelengths that are most strongly reflected (maxima), and those that are not reflected (minima). At intermediate wavelengths the reflected intensity varies smoothly between these maxima and minima. For example, Figure K.2a shows this distribution for $t = 100$ nm, a typical thickness for an antireflective coating where the first and second reflections are both hard. Shown in Figure K.2b is the distribution for $t = 500$ nm in the case where the first reflection is soft and the second hard.

3. Wedge. If the light is *monochromatic* we can use Equation K7 to find the film *thicknesses t* that give maxima or minima. In a film of variable thickness, then, Equation K7 tells us the positions of the maxima and minima. For example, consider the film formed by the *wedge* of air between two plane sheets of glass that touch on one edge and are separated by a gap on the other edge (Fig. K.3). By similar triangles

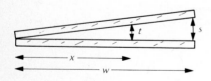

FIGURE K.3

A wedge-shaped air film between two sheets of glass.

we see that the distance x across the air film satisfies $x/t = w/s$. Substituting for t the values from Equation K7 and noting that here one reflection is soft and the other hard, we find:

minima at

$$x = 0, \frac{\lambda w}{2s}, 2\frac{\lambda w}{2s}, 3\frac{\lambda w}{2s}, \ldots \quad \text{(K8)}$$

(Since the wedge is made of air, we have here $n = 1$ and $\lambda' = \lambda$.) The minima appear as dark fringes (parallel to the edge of the wedge), and there are bright fringes in between.

How many fringes are there across the entire width, that is, when $x = w$? Since x increases by $(\lambda w/2s)$ for each fringe, there will be a total number of $w/(\lambda w/2s) = 2s/\lambda$ fringes. For example, if $s = 0.1$ mm and $\lambda = 500$ nm, we find that there are 400 fringes. Changing s by only $\lambda/2 = 250$ nm changes the number of fringes by one, a change that is easily observable. Thus such a small distance can be very accurately measured by this easily visible effect.

4. Single slit. The intensity minima in the Fraunhofer diffraction pattern of a *single slit* of width b can also be treated by using Equation K5. Although the slit is equivalent to a large (infinite) number of sources according to Huygens' principle, these sources must cancel in *pairs* in order to obtain destructive interference. As explained in the text, the first zero of intensity occurs at such a screen position that, for each source in the left half of the slit, there is a source in the right half of the slit whose wave arrives exactly out of phase. Since the distance between such pairs of sources

FIGURE K.4

(a) Intensity versus screen position in the Fraunhofer diffraction pattern of a slit of width b. **(b)** Intensity versus screen position for Young's fringes when the separation of the two slits is ten times the width of the individual slits. (The same slit width is used in both plots.)

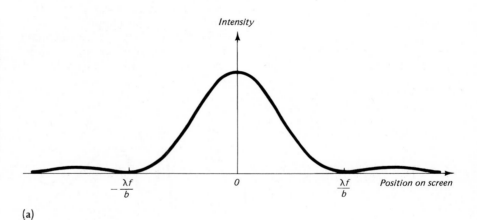

(a)

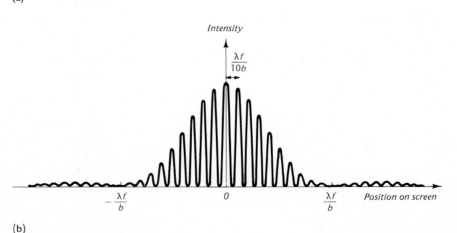

(b)

is $d = b/2$, this value substituted in Equation K5 gives the first minimum (zero) of intensity for the slit:

$$h = \pm\lambda\frac{f}{b} \qquad \text{(K9)}$$

(Again, for a distant screen we can remove the lens and replace f by D. This gives the formula of Sec. 12.5A.) We get further minima whenever we can divide the slit width b into an *even* number of regions such that each Huygens' source in one region has a destructively interfering partner in the next region. This corresponds to pairs of sources at distances $d = b/2$ (as in Eq. K9), or $d = b/4$, or $d = b/6$, . . . which, by Equation K5, have their *first minima* (in the small-angle approximation) at:

$$h = \pm\frac{\lambda f}{\underset{2}{2b}}, \pm\frac{\lambda f}{\underset{4}{2b}}, \pm\frac{\lambda f}{\underset{6}{2b}}, \cdots$$

$$\qquad \text{(K10)}$$

$$= \pm\frac{\lambda f}{b}, \pm\frac{2\lambda f}{b}, \pm\frac{3\lambda f}{b}, \cdots$$

This equation gives the locations of *all* the minima in the Fraunhofer diffraction pattern of a slit of width b. (The higher-order minima for any one of the source distances are already contained in this sequence, so they do not have to be listed separately.)

Thus the dark fringes due to a single slit are regularly spaced about the origin O by $\lambda f/b$, except at the origin itself ($h = 0$). The central fringe is of course bright, because all the sources contribute in phase there. Other bright fringes occur between the dark fringes. Unlike in the Young's fringe pattern, these single-slit side maxima have intensities that *decrease* rapidly with h. For example, the intensity of the first bright fringe near the central fringe (at $h \simeq 3\lambda f/2b$) is less than $\frac{1}{20}$ of that of the central fringe. Figure K.4a shows a plot of the intensity of the single-slit Fraunhofer diffraction pattern. Figure K.4b shows how this diffraction pattern modifies the Young's fringes produced by two slits.

APPENDIX L
BREWSTER'S ANGLE

To find Brewster's angle, we must find the incident angle for the special case when the transmitted beam is perpendicular to the reflected beam. Figure 13.5 tells us that this happens when the reflection angle, θ_r, and the transmission angle, θ_t, satisfy the relation:

$$\theta_r + 90° + \theta_t = 180°, \quad \text{or}$$

$$\theta_t = 90° - \theta_r$$

Since the reflection angle equals the incidence angle ($\theta_r = \theta_i$), this means that we are at Brewster's angle when:

$$\theta_t = 90° - \theta_i \qquad \text{(L1)}$$

But Snell's law (Eq. B5) gives another relation between θ_t and θ_i:

$$n_i \sin \theta_i = n_t \sin \theta_t \qquad \text{(B5)}$$

Therefore, at Brewster's angle both Equations L1 and B5 must be true. Using Equation L1, we can then rewrite Equation B5 as:

$$\frac{\sin\theta_i}{\sin(90° - \theta_i)} = \frac{n_t}{n_i} \qquad \text{(L2)}$$

Recall how the sine of an angle is defined (Eq. B1) and consider Figure L.1. From this figure you can see that:

$$\sin \theta_i = \frac{p}{r}$$

and

$$\sin(90° - \theta_i) = \frac{q}{r}$$

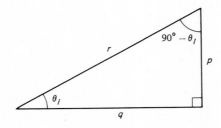

FIGURE L.1

A right triangle with one angle equal to θ_i.

Hence:

$$\frac{\sin \theta_i}{\sin(90° - \theta_i)} = \frac{p}{q} \qquad \text{(L3)}$$

The ratio p/q represents another trigonometric function, the **tangent** of θ_i:

$$\tan \theta_i = \frac{p}{q} \qquad \text{(L4)}$$

Hence, when the angle of incidence is Brewster's angle ($\theta_i = \theta_B$), we see from Equations L4, L3, and L2, that:

$$\tan \theta_B = \frac{p}{q} = \frac{\sin\theta_i}{\sin(90° - \theta_i)} = \frac{n_t}{n_i}$$

or:

$$\tan \theta_B = \frac{n_t}{n_i} \qquad \text{(L5)}$$

As an example, for light traveling *from air* ($n_i = 1.0$) *to water* ($n_t = 1.3$), Equation L5 and a pocket calculator tells us that $\theta_B = 52.4°$. For light traveling *from air to glass* ($n_t = 1.5$), Brewster's angle is $\theta_B = 56.3°$.

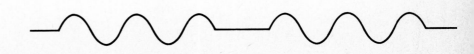

APPENDIX M
MALUS'S LAW

We want to know how the intensity of the transmitted light varies as the angle between the polarizer and the analyzer is changed. The polarizer transmits only light polarized in one direction (arbitrarily drawn vertical in Fig. M.1a). This polarized light is then incident on the analyzer. Suppose the analyzer is oriented at an angle θ with respect to the polarizer. We must then break up the electric field incident at the analyzer (E_{in}) into two components: one that the analyzer will pass (E_{out}), and a perpendicular component that it won't.

From Figure M.1a we see that:

$$\frac{E_{in}}{E_{out}} = \cos \theta \qquad (M1)$$

where the **cosine** of an angle is defined, using Figure L.1, as:

$$\cos \theta_t = \frac{q}{r} \qquad (M2)$$

Since the intensity of light is proportional to the *square* of the electric field, the intensity of the light transmitted by the analyzer is seen to be (upon squaring Eq. M1):

$$\frac{I_{in}}{I_{out}} = \cos^2 \theta \qquad (M3)$$

Equation M3 is called **Malus's law,** and is plotted in Figure M.1b.

FIGURE M.1

(a) Only the component of the incident polarized light that is parallel to the analyzer's orientation will be transmitted by the analyzer. (b) A graph showing the ratio of the intensity of the light transmitted by the analyzer to that of the light incident on the analyzer, for different values of the angle θ between the direction of polarization of the incident light and the orientation of the analyzer.

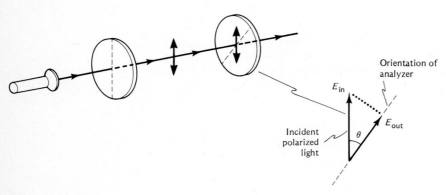

(a)

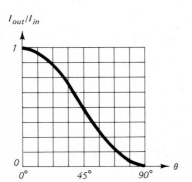

(b)

APPENDIX N
HOLOGRAPHIC RECORDING AND RECONSTRUCTION OF WAVES

Let's check that a transmission hologram such as described in Figure 14.3 indeed forms a plane reconstructed beam at the *same* angle as that of the original object beam. Figure N.1a shows plane reference and object beams striking the film. The reference beam strikes the film perpendicularly and is shown at an instant when one of its crests lies along the thin emulsion. The object beam arrives at angle θ_o, so $\sphericalangle OAB = \theta_o$. The wavelength of the light in both of these beams is λ. The regions of constructive interference on the film (the bright fringes) are spaced a distance d. Using ΔOAB and the definition of the sine (Eq. B1), we can see that d, λ, and θ_o are related in the following way:

$$\sin \theta_o = \frac{\lambda}{d} \qquad (N1)$$

During reconstruction (Fig. N.1b), the hologram acts as a diffraction grating. The wavelength of the reconstruction beam is λ, the same as that used for exposure. The grating constant here is d, the same value as the fringe spacing during exposure. We get constructive interference when the path difference between two different slits is equal to a wavelength. That is, according to Equations K1 and K2, we get a first-order beam when:

$$\epsilon = \pm \lambda = d \sin \theta_{rec} \qquad (N2)$$

or:

$$\sin \theta_{rec} = \frac{\pm \lambda}{d} \qquad (N3)$$

Comparing Equations N1 and N3 gives, for the + sign in Equation N3:

$$\sin \theta_{rec} = \sin \theta_o \qquad (N4)$$

or:

$$\theta_{rec} = \theta_o$$

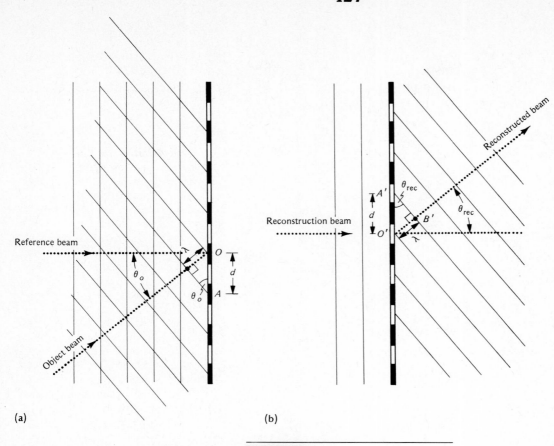

(a)

(b)

Hence the reconstructed beam leaves the hologram at the same angle as that of the object beam.

PONDER

The − sign in Equation N3 gives $\theta_{rec} = -\theta_o$. What does this mean?

FIGURE N.1

(a) Exposure of film by two plane waves. The reference beam arrives perpendicularly to the film while the object beam arrives at angle θ_o.
(b) During reconstruction, the first order (reconstructed) beam leaves the hologram at angle θ_{rec}, where $\theta_{rec} = \theta_o$.

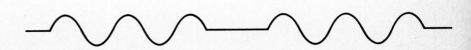

References

The bracketed numbers below indicate the chapter in this book to which the references relate. An excellent general source is *Scientific American*, which has articles of relevance in almost every issue. The Troubador Press (385 Fremont Street, San Francisco, CA 94105) puts out a number of inexpensive "Model Books" with instructions and patterns similar to our TRY IT's.

GENERAL OPTICS

American Association of Physics Teachers. *Polarized Light.* American Institute of Physics, 1963. [13]

Cagnet, Michel; Françon, Maurice; and Thriers, Jean Claude. *Atlas of Optical Phenomena.* Prentice-Hall, 1962. [12]

Cagnet, Michel; Françon, Maurice; and Thriers, Jean Claude. *Atlas of Optical Phenomena, Supplement.* Prentice-Hall, 1962. [12]

Faughn, Jerry S., and Kuhn, Karl F. *Physics for people who think they don't like physics.* Saunders, 1976. [1,2,3,5,6,9,10,12,13,14,15]

Gamow, George. *Mr. Tompkins In Paperback.* Cambridge University Press, 1965. [15]

Gilmore, C. P. *The Unseen Universe: Photographs From The Scanning Electron Microscope.* Schocken, 1974. [12]

Hecht, Eugene, and Zajac, Alfred. *Optics.* Addison-Wesley, 1974. [1,2,3,6,12,13,14,15]

Jenkins, Francis A., and White, Harvey E. *Fundamentals of Optics,* 4th edition. McGraw-Hill, 1976. [1,2,3,4,5,6,12,13,14,15]

Kallard, Thomas. *Exploring Laser Light: Laboratory Exercises And Lecture Demonstrations With Low-Power Helium-Neon Gas Lasers.* Optosonic Press, 1977. [12,14,15]

Mueller, Conrad G.; Rudolf, Mae; and the Editors of *Life. Light And Vision.* Time, Inc., 1966. [1,2,3,4,5,6,7,8,9,10,11,12,13,14,15]

Pirenne, Maurice Henri Léonard. *Optics, Painting And Photography.* Cambridge University Press, 1970. [1,2,3,4,7]

The Science Foundation Course Team. *The Wave Nature of Light.* Open University Press, 1971. [12]

Shurcliff, William A., and Ballard, Stanley S. *Polarized Light.* Van Nostrand, 1964. [13]

Simon, Hilda. *The Splendor Of Iridescence.* Dodd, Mead, 1971. [12]

Texereau, Jean. *How to Make a Telescope.* Doubleday Anchor, 1963. [6]

Todd, Hollis N. *Polarization Of Light: Basic Theory And Experiments.* Bausch and Lomb, 1960–64. [13]

Tolansky, Samuel. *Curiosities Of Light Rays And Light Waves.* American Elsevier Publishing, 1966. [1,2,12,15]

Tolansky, Samuel. *Revolution In Optics.* Penguin, 1968. [6,12,14,15]

ATMOSPHERIC OPTICS

Greenler, Robert. *Rainbows, Halos And Glories.* Cambridge University Press, 1980. [2,12,13]

Middleton, W. E. K. *Vision Through The Atmosphere.* University of Toronto Press, 1968. [13]

Minnaert, M. *The Nature Of Light And Colour In The Open Air.* Dover Publications, 1954. [1,2,3,5,7,12,13]

Introduction by Walker, Jearl. *Light From The Sky* (readings from *Scientific American*).W. H. Freeman, 1980. [2,12,13]

PHOTOGRAPHY

Coe, Brian. *Colour Photography: The first hundred years 1840–1940.* Ash and Grant Ltd., 1978. [11]

Crawford, William. *The Keepers of Light: A History and Working Guide to Early Photographic Processes.* Morgan and Morgan, 1979. [4,11]

Editors of Eastman Kodak Company. *More Joy Of Photography: 100 Techniques For More Creative Photographs.* Addison-Wesley, 1981. [4,11]

Edgerton, Harold E., and Killian, James R., Jr. *Moments Of Vision: The Stroboscopic Revolution in Photography.* MIT Press, 1979. [4,11]

Hunt, Robert W. G. *The Reproduction Of Colour.* Wiley, 1975. [9,10]

Langford, Michael J. *Basic Photography.* Amphoto, 1974. [4,11]

Langford, Michael J. *Advanced Photography.* Amphoto, 1972. [4,11]

Mueller, Conrad G.; Rudolf, Mae; and the Editors of *Life. Light And Vision.* Time, Inc., 1966. [1,2,3,4,5,6,7,8,9,10,11,12,13,14,15]

Neblette, C. B. *Photography, Its Materials And Processes.* Van Nostrand Reinhold, 1962. [4,11]

Pirenne, Maurice Henri Léonard. *Optics, Painting And Photography.* Cambridge University Press, 1970. [1,2,3,4,7]

Scharf, Aaron. *Art And Photography.* Penguin, 1974. [4,11]

Schechter, Harold, and Everitt, David. *Film Tricks: Special Effects In The Movies.* Harlin Quist, 1980. [FO]

Swedlund, Charles. *Photography.* Holt, Rinehart and Winston, 1974. [4,11]

Editors of Time-Life Books. *The Life Library Of Photography—Light And Film.* Time, Inc., 1970. [4,11]

Editors of Time-Life Books. *The Life Library Of Photography—The Camera.* Time, Inc., 1970. [4,11]

Editors of Time-Life Books. *The Life Library Of Photography—The Print.* Time, Inc., 1970 [4,11]

VISION

Arnheim, Rudolf. *Art And Visual Perception: A Psychology of the Creative Eye.* University of California Press, 1974. [7]

Arnheim, Rudolf. *Toward A Psychology Of Visual Art.* University of California Press, 1966. [7,10]

Barrett, Cyril. *Op Art.* Viking Press, 1970. [7,10]

Barrett, Cyril. *An Introduction To Optical Art.* Studio Viste/Dutton, 1971. [7]

Begbie, G. Hugh. *Seeing And The Eye: An Introduction To Vision.* Natural History Press, 1969. [5,7,10]

Bloomer, Carolyn M. *Principles Of Visual Perception.* Van Nostrand Reinhold, 1976. [7,10]

Boynton, Robert M. *Human Color Vision.* Holt, Rinehart and Winston, 1979. [5,10]

Coren, Stanley, and Girgus, Joan Stern. *Seeing Is Deceiving: The*

Psychology Of Visual Illusion. Wiley, 1978. [7]

Cornsweet, Tom N. *Visual Perception.* Academic Press, 1970. [7,9,10]

Descargues, Pierre (translated by Z. Mark Paris). *Perspective.* Harry N. Abrams, 1974. [8]

Edgerton, Harold E., and Killian, James R., Jr. *Moments Of Vision: The Stroboscopic Revolution in Photography.* MIT Press, 1979. [4,11]

Falk, Nat. *How To Make Animated Cartoons.* Foundation Books, 1941. [7]

Frisby, John P. *Seeing: Illusion, Brain And Mind.* Oxford University Press, 1980. [5,7,8]

Gombrich, Ernst Hans. *Art and Illusion,* 2nd edition. Princeton University Press, 1969. [7,10]

Gregg, James R. *Experiments In Visual Science.* Ronald Press, 1966. [5,7,10]

Gregory, Richard L. *Eye And Brain: The Psychology Of Seeing,* 3rd edition. McGraw-Hill, 1978. [5,7,9,10]

Held, Richard (introductions by). *Image, Object, And Illusion* (readings from *Scientific American*). W. H. Freeman, 1974. [7,10]

Held, Richard, and Richards, Whitman (introduction by). *Perception: Mechanisms And Models* (readings from *Scientific American*). W. H. Freeman, 1972. [5,7,10]

Held, Richard, and Richards, Whitman (introductions by). *Recent Progress In Perception* (readings from *Scientific American*). W. H. Freeman, 1976. [7,10]

Hurvich, Leo M. *Color Vision.* Sinauer Associates, Inc., 1981. [10]

Kaufman, Lloyd. *Sight And Mind.* Oxford University Press, 1974. [7,10]

Lighting Business Group. *Light And Color.* General Electric, 1978. [9,10]

Lindsay, Peter H., and Norman, Donald A. *Human Information Processing: An Introduction To Psychology.* Academic Press, 1972. [5,7,10]

Ludel, Jacqueline. *Introduction To Sensory Processes.* W. H. Freeman, 1978. [5,7,9,10]

Mueller, Conrad G.; Rudolf, Mae; and the Editors of *Life. Light And Vision.* Time, Inc., 1966. [1,2,3,4,5,6,7,8,9,10,11,12,13,14,15]

Robinson, J. O. *The Psychology Of Visual Illusion.* Hutchinson, 1972. [7]

Seitz, William Chapin. *The Responsive Eye.* Museum of Modern Art, 1965. [7,10]

DEPTH

Frisby, John P. *Seeing: Illusion, Brain And Mind.* Oxford University Press, 1980. [5,7,8]

Mueller, Conrad G.; Rudolf, Mae; and the Editors of *Life. Light And Vision.* Time, Inc., 1966. [1,2,3,4,5,6,7,8,9,10,11,12,13,14,15]

COLOR

Agoston, G. A. *Color Theory And Its Application In Art And Designs.* Springer-Verlag, 1979. [9,10]

Barrett, Cyril. *Op Art.* Viking Press, 1970. [7,10]

Begbie, G. Hugh. *Seeing And The Eye: An Introduction To Vision.* Natural History Press, 1969. [5,7,10]

Bellman, Willard F. *Stage Lighting, Sound, Costume, and Makeup.* Harper & Row, 1983. [1,3,9]

Billmeyer, F. W., and Saltzman, M. *Principles Of Color Technology,* 2nd edition. Wiley, 1981. [9]

Bloomer, Carolyn M. *Principles Of Visual Perception.* Van Nostrand Reinhold, 1976. [7,10]

Bockus, H. William, Jr. *Advertising Graphics,* 2nd edition. Macmillan, 1974. [9]

Bouma, Pieter Johannes. *Physical Aspects Of Colour: An Introduction To The Scientific Study Of Colour Stimuli And Colour Sensations,* 2nd edition. Macmillan, 1971. [9]

Boynton, Robert M. *Human Color Vision.* Holt, Rinehart and Winston, 1979. [5,10]

Burnham, Robert W.; Hanes, Randall M.; and Bartleson, C. James. *Color: A Guide To Basic Facts And Concepts.* Wiley, 1963. [9]

Clulow, F. W. *Colour Its Principles And Their Applications.* Fountain Press, 1972. [9,10]

Coe, Brian. *Colour Photography: The first hundred years 1840–1940.* Ash and Grant Ltd., 1978. [11]

Cornsweet, Tom N. *Visual Perception.* Academic Press, 1970. [7,9,10]

Eastman Kodak Co. *Color As Seen and Photographed.* 1966. [2,9,10,11]

Evans, Ralph Merrill. *An Introduction To Color.* Wiley, 1948. [9,10]

Evans, Ralph Merrill. *The Perception Of Color.* Wiley, 1974. [9,10]

Gregory, R. L., and Gombrich, E. H., eds. *Illusion In Nature And Art.* Scribner's, 1973. [7,10]

Gregory, Richard L. *The Intelligent Eye.* McGraw-Hill, 1970. [5,7,9,10]

Gregory, Richard L. *Eye And Brain: The Psychology Of Seeing,* 3rd edition. McGraw-Hill, 1978. [5,7,9,10]

Held, Richard (introductions by). *Image, Object, And Illusion* (readings from *Scientific American*). W. H. Freeman, 1974. [7,10]

Held, Richard, and Richards, Whitman (introduction by). *Perception: Mechanisms And Models* (readings from *Scientific American*). W. H. Freeman, 1972. [5,7,10]

Held, Richard, and Richards, Whitman (introductions by). *Recent Progress In Perception* (readings from *Scientific American*). W. H. Freeman, 1976. [7,10]

Hunt, Robert W. G. *The Reproduction Of Colour.* Wiley, 1975. [9,10]

Hurvich, Leo M. *Color Vision.* Sinauer Associates, Inc., 1981. [10]

Itten, Johannes. *The Art Of Color.* Van Nostrand Reinhold, 1973. [9,10]

Lighting Business Group. *Light And Color.* General Electric, 1978. [9,10]

Lindsay, Peter H., and Norman, Donald A. *Human Information Processing: An Introduction To Psychology.* Academic Press, 1972. [5,7,10]

Ludel, Jacqueline. *Introduction To Sensory Processes.* W. H. Freeman, 1978. [5,7,9,10]

MacAdam, D. L. *Color Measurement: Theme And Variations.* Springer-Verlag, 1981. [9,10]

Mueller, Conrad G.; Rudolf, Mae; and the Editors of *Life. Light And Vision.* Time, Inc., 1966. [1,2,3,4,5,6,7,8,9,10,11,12,13,14,15]

Optical Society of America (ed.). *The Science Of Color.* Optical Society of America, 1963. [9,10]

Osborne, Roy. *Light And Pigments: Colour Principles For Artists.* John Murray, 1980. [9,10]

Pease, Paul L., ed. *Color and Color Vision: Selected Reprints.* American Association of Physics Teachers, 1982. [10]

Robinson, J. O. *The Psychology Of Visual Illusion.* Hutchinson, 1972. [7]

Sargent, Walter. *The Enjoyment And Use Of Color.* Dover, 1964. [9]

Wright, W. D. *The Measurement Of Colour,* 4th edition. Van Nostrand Reinhold, 1969. [9]

Wyszecki, G., and Stiles, W. S. *Color Science.* Wiley, 1967. [9,10]

ART

Agoston, G. A. *Color Theory And Its Application In Art And Designs.* Springer-Verlag, 1979. [9,10]

Arnheim, Rudolf. *Art And Visual Perception: A Psychology of the Creative Eye.* University of California Press, 1974. [7]

Arnheim, Rudolf. *Toward A Psychology Of Visual Art.* University of California Press, 1966. [7,10]

Barrett, Cyril. *An Introduction To Optical Art.* Studio Viste/Dutton, 1971. [7]

Bockus, H. William, Jr. *Advertising Graphics,* 2nd edition. Macmillan, 1974. [9]

Coren, Stanley, and Girgus, Joan Stern. *Seeing Is Deceiving: The Psychology Of Visual Illusion.* Wiley, 1978. [7]

Descargues, Pierre (translated by Z. Mark Paris). *Perspective.* Harry N. Abrams, 1974. [8]

Fleming, Stuart J. *Authenticity In Art: The Scientific Determination of Forgery.* Crane, Russack, 1976. [9]

Gombrich, Ernst Hans. *Art and Illusion,* 2nd edition. Princeton University Press, 1969. [7,10]

Gregory, R. L., and Gombrich, E. H., eds. *Illusion In Nature And Art.* Scribner's, 1973. [7,10]

Gregory, Richard L. *The Intelligent Eye.* McGraw-Hill, 1970. [5,7,9,10]

Hunter, Sam, and Jacobus, John. *American Art Of The 20th Century.* Harry N. Abrams, 1973. [7,8,10]

Itten, Johannes. *The Art Of Color.* Van Nostrand Reinhold, 1973. [9,10]

Leeman, Fred. *Hidden Images: Games Of Perception, Anamorphic Art, Illusion.* Harry N. Abrams, 1976. [3,4,8]

Osborne, Roy. *Light And Pigments: Colour Principles For Artists.* John Murray, 1980. [9,10]

Pirenne, Maurice Henri Léonard. *Optics, Painting And Photography.* Cambridge University Press, 1970. [1,2,3,4,7]

Scharf, Aaron. *Art And Photography.* Penguin, 1974. [4,11]

Schechter, Harold, and Everitt, David. *Film Tricks: Special Effects In The Movies.* Harlin Quist, 1980. [FO]

Seitz, William Chapin. *The Responsive Eye.* Museum of Modern Art, 1965. [7,10]

HOLOGRAPHY

Hecht, Eugene, and Zajac, Alfred. *Optics.* Addison-Wesley, 1974. [1,2,3,6,12,13,14,15]

Jenkins, Francis A., and White, Harvey E. *Fundamentals of Optics*, 4th edition. McGraw-Hill, 1976. [1,2,3,4,5,6,12,13,14,15]

Klein, H. Arthur. *Holography.* Lippincott, 1970. [14]

Saxby, Graham. *Holograms: How To Make And Display Them.* Focal Press, 1980. [14]

Unterseher, Fred; Hansen, Jeannene; and Schlesinger, Bob. *Holography Handbook: Making Holograms The Easy Way.* Ross Books, 1982. [14]

MODERN PHYSICS

Gamow, George. *Mr. Tompkins In Paperback.* Cambridge University Press, 1965. [15]

Hecht, Eugene, and Zajac, Alfred. *Optics.* Addison-Wesley, 1974. [1,2,3,6,12,13,14,15]

Mueller, Conrad G.; Rudolf, Mae; and the Editors of *Life. Light And Vision.* Time, Inc., 1966. [1,2,3,4,5,6,7,8,9,10,11,12,13,14,15]

The Science Foundation Course Team. *The Wave Nature Of Light.* Open University Press, 1971. [12]

Tolansky, Samuel. *Curiosities Of Light Rays And Light Waves.* American Elsevier, 1966. [1,2,12,15]

Text

p.1: © Sidney Harris. p.3: Courtesy Culver Pictures. p.4: *(bottom)* G.F. Stork. p.5: *(bottom)* Palomar Observatory photograph. p.18: *(top left)* R. Kerr, *Science,* vol. 202, pp. 1172–1174, December 15, 1978. © 1978 American Association for the Advancement of Science. *(top right)* Equipment courtesy Sumio Uematsu. *(center, left and right)* M.W. Frolich, *Science,* vol. 194, pp. 839–841, November 19, 1976. © 1976 American Association for the Advancement of Science. *(bottom left)* Photograph by L.F. Ehrke and C.M. Slack, from L. Moholy-Nagy, *Vision in Motion,* Paul Theobald and Company, Chicago. *(bottom right)* Courtesy Ervin Kaplan, M.D., Chief, Nuclear Medicine Service, Veterans Administration Hospital, Hines, IL. p.20: Courtesy U.S. Steel Corporation. p.22: *(top)* Courtesy General Electric Co., Lighting Business Group.p.29: *(bottom, right)* Courtesy H. and N. Laughon. p.31: Neg. No. 323048. Courtesy American Museum of Natural History. p.41: *(top right)* Courtesy National Film Archive, London. *(bottom right)* Photograph Helene Adant, © 1985 S.P.A.D.E.M., Paris/V.A.G.A., New York. p.42: *(top)* Courtesy Abby Aldrich Rockefeller Folk Art Center, Williamsburg, VA. p.43: Flip Schulke/Black Star. p.46: *(top)* Robert Greenler, *Rainbows, Halos, and Glories,* Cambridge University Press, Cambridge, MA, 1980. Photograph by Robert Greenler. p.47: *(top)* © Alistair B. Fraser. p.48: *(bottom)* Gene Taylor, University of Maryland. p.49: *(bottom)* © M.C. Escher Heirs. c/o Cordon Art, Baarn, Holland. p.50: *(right, top)* Courtesy NASA. *(right, bottom)* Courtesy University of Texas at Austin, News and Information Service. p.54: *(right)* © Allen Bronstein. p.58: Courtesy AT&T and Bell Laboratories. Photograph by Erich Hartmann. p.59: *(bottom)* © Alistair B. Fraser. pp.60–61: *(all)* © Alistair B. Fraser. p.62: Photograph courtesy David K. Lynch. p.66: *(top left)* Eberhard Froehlich. p.67: *(top and center)* © Alistair B. Fraser. p.72: Courtesy Portland Center for the Visual Arts, OR. Photograph by Charles Rhyne. p.74: *(center, bottom)* © Allen Bronstein. p.75: Courtesy Albright-Knox Art Gallery, Buffalo, NY. Gift of Seymour H. Knox, 1966. p.76: *(bottom)* © Allen Bronstein. p.78: *(left and right)* G. Frederick Stork. p.79: *(bottom)* © M.C. Escher Heirs. c/o Cordon Art, Baarn, Holland. p.80: *(top)* Eberhard Froehlich. *(bottom)* Allen Bronstein. p.81: Reproduced by courtesy of the Trustees, The National Gallery, London. p.85: *(top)* © Allen Bronstein. p.87: *(bottom)* "Broomhilda." Reprinted by permission: Tribune Company Syndicate, Inc. p.89: *(center)* © Alistair B. Fraser. *(right)* Courtesy Huntington Barclay. p.93: *(bottom, left)* Courtesy Robert C. Lautman. *(bottom, right)* Allen Bronstein. p.99: *(bottom)* Allen Bronstein. p.100: *(top and bottom)* Allen Bronstein. p.102: *(top and right)* Allen Bronstein. p.104: Courtesy U.S. Department of Energy. p.107: *(top)* Camera courtesy Tom Beck, Edward L. Bafford Photography Collection, Albin O. Kuhn Library and Gallery, University of Maryland, Baltimore County. p.109 *(bottom, right)* Camera courtesy Tom Beck, Edward L. Bafford Photography Collection, Albin O. Kuhn Library and Gallery, University of Maryland, Baltimore County. p.115: *(all)* G. Frederick Stork. p.117: "B.C." By permission of Johnny Hart and Field Enterprises, Inc. p.119: Courtesy Twentieth Century-Fox. p.121: *(bottom)* Courtesy Vivitar Corporation. p.123: *(all)* G. Frederick Stork. p.124: *(top center)* G. Frederick Stork. *(bottom)* Courtesy Tom Beck, Edward L. Bafford Photography Collection, Albin O. Kuhn Library and Gallery, University of Maryland, Baltimore County. p.125: Courtesy V.A.G.A., New York. p.126: William G. Hyzer, Janesville, WI. p.128: *(left, top and bottom)* Allen Bronstein. p.134: *(center and bottom)* Daguerreotype courtesy Tom Beck, Edward L. Bafford Photography Collection, Albin O. Kuhn Library and Gallery, University of Maryland, Baltimore County. p.135: Courtesy Hans P. Kraus, Jr. p.136: Courtesy Research Division, Kodak Limited, Middlesex, England. p.138: *(top and bottom)* Philip C. Geraci. p.140: *(top and bottom)* G. Frederick Stork. p.146: *(left)* Csaba L. Martonyi. The W.K. Kellogg Eye Center, University of Michigan, Ann Arbor. *(right)* Philip Clark, courtesy Richard L. Gregory. p.147: *(top)* Courtesy Robert Rush Miller and the American Society of Ichthyologists and Herpetologists. Reproduced by permission from *Copeia,* 1979(1), 85, Fig. 2. p.149: Csaba L. Martonyi. The W.K Kellogg Eye Center, University of Michigan, Ann Arbor. p.150: *(left, bottom)* Csaba, L. Martonyi. The W.K. Kellogg Eye Center, University of Michigan, Ann Arbor. p.153: *(top)* Courtesy Amray, Inc., Bedford, MA. p.154: *(top, right)* Deric Bownds and Stan Carlson. *(bottom)* John E. Dowling. p.155: "Snuffy Smith." Courtesy King Features Syndicate. p.163: *(bottom)* Csaba Martonyi, The W.K. Kellogg Eye Center, University of Michigan, Ann Arbor. p.165: British Crown Copyright, Science Museum, London. p.168: *(bottom)* Sashi B. Mohanty. p.173: *(top)* Palomar Observatory photograph. *(bottom)* Michael F. Land. p.174: *(top)* Allen Winkelmann, Aerospace Engineering Department, University of Maryland, College Park. p.181: © 1973 Harper & Row. p.186: *(left, top and bottom)* Edwin H. Land. p.190: *(left, top)* Floyd Ratliff. p.193: Courtesy New Jersey State Museum and Reginald Neal. p.196: © Gordon Gahan/Prism for SCIENCE 83. p.197: Courtesy Fergus Campbell and John Robson. p.198: © John P. Frisby. From *Seeing: Illusion, Brain and Mind,* Oxford University Press, New York, 1980. Courtesy Roxby and Lindsey Press. p.202: © John P. Frisby. From *Seeing: Illusion, Brain and Mind,* Oxford University Press, New York, 1980. Courtesy Roxby and Lindsey Press, p.204: D. Marr, *Phil. Trans. Roy. Soc.,* vol. 275, p.483, Courtesy Lucia Vaina and the Royal Society, London. p.207: Courtesy The Brooklyn Museum, NY. Dick S. Ramsey Fund. p.208: © 1985 AD-AGP, Paris. p.210: *(bottom left)* Courtesy Imperial War Museum, London. p.212: *(bottom)* Courtesy Tom Beck, Edward L. Bafford Photography Collection, Albin O. Kuhn Library and Gallery, University of Maryland, Baltimore County. p.214: *(top)* Philip Clark, courtesy Richard L. Gregory. p.217: *(bottom, left to right)* Property of Sir Charles Abraham Elton, Bart. Photograph by Philip Clark. p.220: *(center, left)* © M.C. Escher Heirs, c/o Cordon Art, Baarn, Holland. p.221: *(bottom, left)* Courtesy National Gallery of Art, Washington, D.C. p.222: *(top, right)* G. Frederick Stork. *(bottom,*

right) © Alistair B. Fraser. p.224: *(top)* Courtesy William Vandivert and *Scientific American. (bottom, left and right)* © DuMont Buchverlag, Koln, 1975, Fred Leemann, Anamorphosen. p.225: *(bottom left)* Courtesy Art Resource, New York. *(bottom right)* Courtesy National Gallery of Art, Washington, D.C. p.226: *(center, middle)* Courtesy Bibliotheque National, Paris. *(center, right)* Private collection. *(bottom)* Private collection U.S.A. p.227: *(top, left)* Pencil drawing, collection Arpiar Saunders. *(top, right)* © 1985 S.P.A.D.E.M., Paris/V.A.G.A., New York. *(center, left)* All rights reserved. The Metropolitan Museum of Art, New York. *(bottom)* Private collection. p.228: *(top, left)* Lillian Heidenberg Gallery, New York. *(top, right)* Courtesy Richard A. Johnson. p.230: Collection, The Museum of Modern Art, New York. Gift of G. David Thompson. p.231: *(top, center and right)* Courtesy Museum of Fine Arts, Boston. *(center, left)* Philip Clark, courtesy Richard L. Gregory. *(center, right)* The Salvador Dali Museum, St. Petersburg, FL. p.232: All rights reserved. The Metropolitan Museum of Art, New York. p.233: © The New York Times Company. Reprinted by permission. Drawing by David Suter. p.238: "B.C." By permission of Johnny Hart and Field Enterprises, Inc. p.258: Courtesy Bertha R. Mole. p.316: Courtesy Josef Schreiner. p.321: *(bottom)* Courtesy Helen Ghiradella. p.322: *(left, top and bottom)* From *Illusion in Nature and Art,* R.L. Gregory and E.H. Gombrich, eds., Charles Scribner's Sons, New York. © 1973 H.E. Hinton. Reproduced with permission of the publisher. *(right)* Courtesy J.V. Sanders, CSIRO, Division of Materials Science, Parkville, Victoria, Australia. p.323: *(left, bottom)* J. Robert Anderson. p.325: *(bottom, center)* Victor E. Scherrer and Edgar A. McLean, Naval Research Laboratory. p.326: *(bottom)* From *Illusion in Nature and Art,* R.L. Gregory and E.H. Gombrich, eds., Charles Scribner's Sons, New York. © 1973 H.E. Hinton. Reproduced with permission of the publisher. p.332: *(right, bottom)* Courtesy Nils Abramson.

p.335: Courtesy M. Francon and Springer-Verlag, Heidelberg. p.337: *(right, top to bottom)* Courtesy M. Francon and Springer-Verlag, Heidelberg. p.338: *(left)* Courtesy J.G. Rosenbaum. p.341: *(left, top and bottom)* J.D. Gaskill, *Linear Systems, Fourier Transforms, and Optics,* John Wiley and Sons, New York. © John Wiley and Sons. p.342: *(all)* J.D. Gaskill, *Linear Systems, Fourier Transforms, and Optics,* John Wiley and Sons, New York. © John Wiley and Sons. p.343: *(all)* Photomicrographs by Dr. Cecil H. Fox, Laboratory of Pathology, National Cancer Institute–National Institutes of Health, Bethesda, MD 20205. p.363: Crystal courtesy Bryn Mawr College Mineral Collection. p.368: Courtesy Robert Schinella. p.374: *(center)* © Sidney Harris. p.379: *(top and bottom)* Courtesy Tung H. Jeong. p.380: *(top)* Courtesy Keith Hodgkinson of the Open University (UK) and John Cookson. *(bottom)* Courtesy Tony Hsu, Newport Corporation. p.381: *(top, left to right)* Courtesy Tony Hsu, Newport Corporation. p.382: *(all)* Courtesy Nils Abramson. p.391: *(top and bottom)* Courtesy Multiplex Co., San Francisco. p.400: *(left, bottom)* G. Möllenstedt and H. Dücker, Zeitschrift für Physik, vol. 145, p.385, 1956. *(right, top)* Courtesy Ellen Williams. Photograph by David E. Taylor, University of Maryland. *(right, center)* J. Robert Anderson. *(right, bottom)* Photograph by C.G. Shull and E.O. Wollan. p.406: © Sidney Harris. p.409: *(right)* © Charles E. Long. Reprinted by permission of Harper & Row, Publishers, Inc. p.415: "Shoe." © Tribune Company Syndicate, Inc. p.421: Courtesy Ricky D. Wood.

Color Plates

Plate 2.1: Allen Bronstein. Plate 2.2: Allen Bronstein. Plate 2.3: © Alistair B. Fraser. Plate 3.1: Allen Bronstein. Plate 5.1: © 1981 Allen Bronstein. Plate 5.2: Ellsworth Kelly. *Green, Blue, Red.* 1964. Oil on canvas 73 × 100 in. Collection

We list here the Greek letters and special symbols used within the text, giving the page of their first occurrence.

ϵ (epsilon), used here for a small path difference, **317**
Δ (delta), used here for a difference in pointing angle, **338**
$\Delta\lambda$ (delta lambda), used here for a just noticeable difference in wavelength, **272**
λ (lambda), used here for a wavelength, **13**
γ (gamma), used here for the slope on an H & D curve, **140**
μ (mu), used here for the metric prefix "micro-", **16**
ν (nu), used here for a frequency, **13**
θ (theta), used here for an angle, **40**
    θ_c, critical angle, **55**
    θ_i, angle of incidence, **40**
    θ_r, angle of reflection, **40**
    θ_t, angle of transmission, **53**

$\equiv$, looks the same as, **242**
$<$, is less than, **63**
$>$, is greater than, **422**
$\simeq$, is approximately equal to, **7**
$\angle A'AB'$, denotes the angle between the lines $A'A$ and AB', **416**
$\triangle CAF$, denotes the triangle with points C, A, and F at its vertexes, **417**
$\ll$, is much less than, **311**
$[X]$, $[Y]$, $[Z]$, imaginary primaries used for the C.I.E. chromaticity diagram, **246**
$\overline{OF}$, the distance from point O to point F, **76**

436